Methods for Testing and Evaluating Survey Questionnaires

WILEY SERIES IN SURVEY METHODOLOGY

Established in Part by WALTER A. SHEWHART AND SAMUEL S. WILKS

The ***Wiley Series in Survey Methodology*** covers topics of current research and practical interests in survey methodology and sampling. While the emphasis is on application, theoretical discussion is encouraged when it supports a broader understanding of the subject matter.

The authors are leading academics and researchers in survey methodology and sampling. The readership includes professionals in, and students of, the fields of applied statistics, biostatistics, public policy, and government and corporate enterprises.

BIEMER, GROVES, LYBERG, MATHIOWETZ, and SUDMAN · Measurement Errors in Surveys
BIEMER and LYBERG · Introduction to Survey Quality
COCHRAN · Sampling Techniques, *Third Edition*
COUPER, BAKER, BETHLEHEM, CLARK, MARTIN, NICHOLLS, and O'REILLY (editors) · Computer Assisted Survey Information Collection
COX, BINDER, CHINNAPPA, CHRISTIANSON, COLLEDGE, and KOTT (editors) · Business Survey Methods
*DEMING · Sample Design in Business Research
DILLMAN · Mail and Internet Surveys: The Tailored Design Method
GROVES and COUPER · Nonresponse in Household Interview Surveys
GROVES · Survey Errors and Survey Costs
GROVES, DILLMAN, ELTINGE, and LITTLE · Survey Nonresponse
GROVES, BIEMER, LYBERG, MASSEY, NICHOLLS, and WAKSBERG · Telephone Survey Methodology
GROVES, FOWLER, COUPER, LEPKOWSKI, SINGER, and TOURANGEAU · Survey Methodology
*HANSEN, HURWITZ, and MADOW · Sample Survey Methods and Theory, Volume I: Methods and Applications
*HANSEN, HURWITZ, and MADOW · Sample Survey Methods and Theory, Volume II: Theory
HARKNESS, VAN DE VIJVER, and MOHLER · Cross-Cultural Survey Methods
HEERINGA and KALTON · Leslie Kish Selected Papers
KISH · Statistical Design for Research
*KISH · Survey Sampling
KORN and GRAUBARD · Analysis of Health Surveys
LESSLER and KALSBEEK · Nonsampling Error in Surveys
LEVY and LEMESHOW · Sampling of Populations: Methods and Applications, *Third Edition*
LYBERG, BIEMER, COLLINS, de LEEUW, DIPPO, SCHWARZ, TREWIN (editors) · Survey Measurement and Process Quality
MAYNARD, HOUTKOOP-STEENSTRA, SCHAEFFER, VAN DER ZOUWEN · Standardization and Tacit Knowledge: Interaction and Practice in the Survey Interview
PRESSER, ROTHGEB, COUPER, LESSLER, MARTIN, MARTIN, and SINGER (editors) · Methods for Testing and Evaluating Survey Questionnaires
RAO · Small Area Estimation
SIRKEN, HERRMANN, SCHECHTER, SCHWARZ, TANUR, and TOURANGEAU (editors) · Cognition and Survey Research
VALLIANT, DORFMAN, and ROYALL · Finite Population Sampling and Inference: A Prediction Approach

*Now available in a lower priced paperback edition in the Wiley Classics Library.

Methods for Testing and Evaluating Survey Questionnaires

Edited by

STANLEY PRESSER
University of Maryland, College Park, MD

JENNIFER M. ROTHGEB
U.S. Bureau of the Census, Washington, DC

MICK P. COUPER
University of Michigan, Ann Arbor, MI

JUDITH T. LESSLER
Research Triangle Institute, Research Triangle Park, NC

ELIZABETH MARTIN
U.S. Bureau of the Census, Washington, DC

JEAN MARTIN
Office for National Statistics, London, UK

ELEANOR SINGER
University of Michigan, Ann Arbor, MI

A JOHN WILEY & SONS, INC., PUBLICATION

Published by John Wiley & Sons, Inc., Hoboken, New Jersey.
Published simultaneously in Canada.

Library of Congress Cataloging-in-Publication Data:

Methods for testing and evaluating survey questionnaires / Stanley Presser ... [et al.].
p. cm.—(Wiley series in survey methodology)
Includes bibliographical references and index.
ISBN 0-471-45841-4 (pbk. : alk. paper)
1. Social surveys—Methodology. 2. Questionnaires—Methodology. 3. Social sciences—Research—Methodology. I. Presser, Stanley II. Series.

HM538.M48 2004
300′.72′3—dc22

2003063992

Printed in the United States of America.

10 9 8 7 6 5 4 3 2 1

To the memory of
Charles Cannell and Seymour Sudman,
two pretesting pioneers whose contributions shaped the field
of survey research

Contents

PART V MODE OF ADMINISTRATION

PART VI SPECIAL POPULATIONS

PART VII MULTIMETHOD APPLICATIONS

Contributors

Reginald P. Baker, Market Strategies, Inc., Livonia, MI

Paul Beatty, National Center for Health Statistics, Hyattsville, MD

Paul Biemer, Research Triangle Institute, Research Triangle Park, NC

Johnny Blair, Abt Associates, Washington, DC

Natacha Borgers, Utrecht University, The Netherlands

Anna Chan, U.S. Bureau of the Census, Washington, DC

Frederick G. Conrad, University of Michigan, Ann Arbor, MI

Mick P. Couper, University of Michigan, Ann Arbor, MI

Scott Crawford, Market Strategies, Inc., Livonia, MI

Edith de Leeuw, Utrecht University, The Netherlands

Theresa J. DeMaio, U.S. Bureau of the Census, Washington, DC

Wil Dijkstra, Free University, Amsterdam, The Netherlands

Don A. Dillman, Washington State University, Pullman, WA

Pat Doyle, U.S. Bureau of the Census, Washington, DC

Stasja Draisma, Free University, Amsterdam, The Netherlands

Jennifer Dykema, University of Wisconsin, Madison, WI

Barbara Forsyth, Westat, Rockville, MD

Floyd Jackson Fowler, Jr., University of Massachusetts, Boston, MA

Irmtraud Gallhofer, University of Amsterdam, The Netherlands

Julia Klein Griffiths, U.S. Bureau of the Census, Washington, DC

Sue Ellen Hansen, University of Michigan, Ann Arbor, MI

Janet Harkness, Zentrum fur Umfragen Methoden und Analysen, Mannheim, Germany

John P. Hoehn, Michigan State University, East Lansing, MI

Lilli Japec, Statistics Sweden, Stockholm, Sweden

Michael D. Kaplowitz, Michigan State University, East Lansing, MI

Ashley Landreth, U.S. Bureau of the Census, Washington, DC

Judith T. Lessler, Research Triangle Institute, Research Triangle Park, NC

Frank Lupi, Michigan State University, East Lansing, MI

Lars Lyberg, Statistics Sweden, Stockholm, Sweden

Elizabeth Martin, U.S. Bureau of the Census, Washington, DC

Jean Martin, Office for National Statistics, London, United Kingdom

Louise C. Mâsse, National Cancer Institute, Bethesda, MD

Danna L. Moore, Washington State University, Pullman, WA

Jeffrey Moore, U.S. Bureau of the Census, Washington, DC

Joanne Pascale, U.S. Bureau of the Census, Washington, DC

Beth-Ellen Pennell, University of Michigan, Ann Arbor, MI

Stanley Presser, University of Maryland, College Park, MD

Cleo D. Redline, National Science Foundation, Arlington, VA

Bryce B. Reeve, National Cancer Institute, Bethesda, MD

Jennifer M. Rothgeb, U.S. Bureau of the Census, Washington, DC

Willem E. Saris, University of Amsterdam, The Netherlands

Nora Cate Schaeffer, University of Wisconsin, Madison, WI

Alisú Schoua-Glusberg, Research Support Services, Evanston, IL

Eleanor Singer, University of Michigan, Ann Arbor, MI

Johannes H. Smit, Free University, Amsterdam, The Netherlands

Tom W. Smith, National Opinion Research Center, Chicago, IL

Astrid Smits, Statistics Netherlands, Heerlen, The Netherlands

Janice Swinehart, Market Strategies, Inc., Livonia, MI

John Tarnai, Washington State University, Pullman, WA

Roger Tourangeau, University of Michigan, Ann Arbor, MI

Diane K. Willimack, U.S. Bureau of the Census, Washington, DC

William van der Veld, University of Amsterdam, The Netherlands

Johannes van der Zouwen, Free University, Amsterdam, The Netherlands

Patricia Whitridge, Statistics Canada, Ottawa, Ontario, Canada

Gordon B. Willis, National Cancer Institute, Bethesda, MD

Preface

During the past 20 years, methods for testing and evaluating survey questionnaires have changed dramatically. New methods have been developed and are being applied and refined, and old methods have been adapted from other uses. Some of these changes were due to the application of theory and methods from cognitive science and others to an increasing appreciation of the benefits offered by more rigorous testing. Research has begun to evaluate the strengths and weaknesses of the various testing and evaluation methods and to examine the reliability and validity of the methods' results. Although these developments have been the focus of many conference sessions and the subject of several book chapters, until the 2002 International Conference on Questionnaire Development, Evaluation and Testing Methods, and the publication of this monograph, there was no conference or book dedicated exclusively to question testing and evaluation.

Jennifer Rothgeb initially proposed the conference at the spring 1999 Questionnaire Evaluation Standards International Work Group meeting in London. The Work Group members responded enthusiastically and encouraged the submission of a formal proposal to the organizations that had sponsored prior international conferences on survey methodology. One member, Seymour Sudman, provided invaluable help in turning the idea into a reality and agreed to join the organizing committee, on which he served until his death in May 2000. Shortly after the London meeting, Rothgeb enlisted Stanley Presser for the organizing committee, as they flew home from that year's annual meetings of the American Association for Public Opinion Research (and, later, persuaded him to chair the monograph committee). The members of the final organizing committee, chaired by Rothgeb, were Mick P. Couper, Judith T. Lessler, Elizabeth Martin, Jean Martin, Stanley Presser, Eleanor Singer, and Gordon B. Willis.

The conference was sponsored by four organizations: the American Statistical Association (Survey Research Methods Section), the American Association for Public Opinion Research, the Council of American Survey Research Organizations, and the International Association of Survey Statisticians. These organizations provided funds to support the development of both the conference and the monograph. Additional financial support was provided by:

Abt Associates
Arbitron Company
Australian Bureau of Statistics
Iowa State University
Mathematica Policy Research
National Opinion Research Center
National Science Foundation
Nielsen Media Research
Office for National Statistics (United Kingdom)
Research Triangle Institute
Schulman, Ronca & Bucuvalas, Inc.
Statistics Sweden
University of Michigan
U.S. Bureau of Justice Statistics
U.S. Bureau of Labor Statistics
U.S. Bureau of the Census
U.S. Bureau of Transportation Statistics
U.S. Energy Information Administration
U.S. National Agricultural Statistics Service
U.S. National Center for Health Statistics
Washington State University
Westat, Inc.

Without the support of these organizations, neither the conference nor the monograph would have been possible.

In 2000, the monograph committee, composed of the editors of this volume, issued a call for abstracts. Fifty-three were received. Authors of 23 of the abstracts were asked to provide detailed chapter outlines that met specified goals. After receiving feedback on the outlines, authors were then asked to submit first drafts. Second drafts, taking into account the editors' comments on the initial drafts, were due shortly before the conference in November 2002. Final revisions were discussed with authors at the conference, and additional editorial work took place after the conference.

A contributed papers subcommittee, chaired by Gordon Willis and including Luigi Fabbris, Eleanor Gerber, Karen Goldenberg, Jaki McCarthy, and Johannes van der Zouwen, issued a call for submissions in 2001. One hundred five were received and 66 chosen. Two of the contributed papers later became monograph chapters.

The International Conference on Questionnaire Development, Evaluation and Testing Methods—dedicated to the memory of Seymour Sudman—was held in Charleston, South Carolina, November 14–17, 2002. There were 338 attendees, with more than one-fifth from outside the United States, representing 23 countries on six continents. The Survey Research Methods Section of the American Statistical Association funded 12 conference fellows from South Africa, Kenya, the Philippines, Slovenia, Italy, and Korea, and a National Science Foundation grant funded 10 conference fellows, most of whom were U.S. graduate students.

Over half of the conference participants attended at least one of the four short courses that were offered: Methods for Questionnaire Appraisal and Expert Review by Barbara Forsyth and Gordon Willis; Cognitive Interviewing by Eleanor Gerber; Question Testing for Establishment Surveys by Kristin Stettler and Fran Featherston; and Behavior Coding: Tool for Questionnaire Evaluation by Nancy Mathiowetz. Norman Bradburn gave the keynote address, "The Future of Questionnaire Research," which was organized around three themes: the importance of exploiting technological advances, the increasing challenges posed by multicultural, multilanguage populations, and the relevance of recent research in sociolinguistics. The main conference program included 32 sessions with 76 papers and 15 poster presentations.

Conference planning and on-site activities were assisted by Linda Minor, of the American Statistical Association (ASA), and Carol McDaniel, Shelley Moody, and Safiya Hamid, of the U.S. Bureau of the Census. Adam Kelley and Pamela Ricks, of the Joint Program in Survey Methodology, developed and maintained the conference Web site, and Robert Groves, Brenda Cox, Daniel Kasprzyk and Lars Lyberg, successive chairs of the Survey Research Methods Section of the ASA, helped to promote the conference. We thank all these people for their support.

The goal of this monograph is a state-of-the-field review of question evaluation and testing methods. The publication marks a waypoint rather than an ending. Although the chapters show great strides have been made in the development of methods for improving survey instruments, much more work needs to be done. Our aim is for the volume to serve both as a record of the many accomplishments in this area, and as a pointer to the many challenges that remain.

We hope the book will be valuable to students training to become the next generation of survey professionals, to survey researchers seeking guidance on current best practices in questionnaire evaluation and testing, and to survey methodologists designing research to advance the field and render the current chapters out of date.

After an overview in Chapter 1 of both the field and of the chapters that follow, the volume is divided into seven parts

I. Cognitive Interviews: Chapters 2 to 5
II. Supplements to Conventional Pretests: Chapters 6 to 8
III. Experiments: Chapters 9 to 11
IV. Statistical Modeling: Chapters 12 to 14
V. Mode of Administration: Chapters 15 to 18
VI. Special Populations: Chapters 19 to 22
VII. Multimethod Applications: Chapters 23 to 25

Each of the coeditors served as a primary editor for several chapters: Rothgeb for 3 to 5; Singer for 6 to 8; Couper for 9, 10, 15, 16, and 18; Lessler for 11, 12, 14, and 17; E. Martin for 2, 13, and 19 to 22; and J. Martin for 23 to 25. In addition, each coeditor served as a secondary editor for several other chapters. We are grateful to the chapter authors for their patience during the lengthy process of review and revision, and for the diligence with which they pursued the task.

We are also indebted to Rupa Jethwa, of the Joint Program in Survey Methodology (JPSM), for indefatigable assistance in creating a final manuscript from materials provided by dozens of different authors, and to Robin Gentry, also of JPSM, for expert help in checking references and preparing the index.

Finally, for supporting our work during the more than four years it took to produce the conference and book, we thank our employing organizations: the University of Maryland, U.S. Bureau of the Census, University of Michigan, Research Triangle Institute, and U.K. Office for National Statistics.

August 2003

Stanley Presser
Jennifer M. Rothgeb
Mick P. Couper
Judith T. Lessler
Elizabeth Martin
Jean Martin
Eleanor Singer

CHAPTER 1

Methods for Testing and Evaluating Survey Questions

Stanley Presser
University of Maryland

Mick P. Couper
University of Michigan

Judith T. Lessler
Research Triangle Institute

Elizabeth Martin
U.S. Bureau of the Census

Jean Martin
Office for National Statistics, United Kingdom

Jennifer M. Rothgeb
U. S. Bureau of the Census

Eleanor Singer
University of Michigan

1.1 INTRODUCTION

An examination of survey pretesting reveals a paradox. On the one hand, pretesting is the only way to evaluate in advance whether a questionnaire causes problems for interviewers or respondents. Consequently, both elementary textbooks

Methods for Testing and Evaluating Survey Questionnaires, Edited by Stanley Presser, Jennifer M. Rothgeb, Mick P. Couper, Judith T. Lessler, Elizabeth Martin, Jean Martin, and Eleanor Singer
ISBN 0-471-45841-4

and experienced researchers declare pretesting indispensable. On the other hand, most textbooks offer minimal, if any, guidance about pretesting methods, and published survey reports usually provide no information about whether questionnaires were pretested and, if so, how, and with what results. Moreover, until recently, there was relatively little methodological research on pretesting. Thus, pretesting's universally acknowledged importance has been honored more in the breach than in the practice, and not a great deal is known about many aspects of pretesting, including the extent to which pretests serve their intended purpose and lead to improved questionnaires.

Pretesting dates either to the founding of the modern sample survey in the mid-1930s or to shortly thereafter. The earliest references in scholarly journals are from 1940, by which time pretests apparently were well established. In that year, Katz reported: "The American Institute of Public Opinion [i.e., Gallup] and *Fortune* [i.e., Roper] pretest their questions to avoid phrasings which will be unintelligible to the public and to avoid issues unknown to the man on the street" (1940, p. 279).

Although the absence of documentation means we cannot be certain, our impression is that for much of survey research's history, there has been one conventional form of pretest. Conventional pretesting is essentially a dress rehearsal in which interviewers receive training like that for the main survey and administer a questionnaire as they would during a survey proper. After each interviewer completes a handful of interviews, response distributions (generally univariate, occasionally bivariate or multivariate) may be tallied, and there is a debriefing in which the interviewers relate their experiences with the questionnaire and offer their views about the questionnaire's problems.

Survey researchers have shown remarkable confidence in this approach. According to one leading expert: "It usually takes no more than 12–25 cases to reveal the major difficulties and weaknesses in a pretest questionnaire" (Sheatsley, 1983, p. 226), a judgment similar to that of another prominent methodologist, who maintained that "20–50 cases is usually sufficient to discover the major flaws in a questionnaire" (Sudman, 1983, p. 181).

This faith in conventional pretesting was probably based on the common experience that a small number of conventional interviews often reveals numerous problems, such as questions that contain unwarranted suppositions, awkward wordings, or missing response categories. But there is no scientific evidence justifying the confidence that this type of pretesting identifies the major problems in a questionnaire.

Conventional pretests are based on the assumption that questionnaire problems will be signaled either by the answers that the questions elicit (e.g., don't knows or refusals), which will show up in response tallies, or by some other visible consequence of asking the questions (e.g., hesitation or discomfort in responding), which interviewers can describe during debriefing. However, as Cannell and Kahn (1953, p. 353) noted: "There are no exact tests for these characteristics." They go on to say that "the help of experienced interviewers is most useful at this point in obtaining subjective evaluations of the questionnaire." Similarly, Moser and Kalton (1971, p. 50) judged that "almost the most useful evidence of all on the adequacy of a questionnaire is the individual fieldworker's [i.e., interviewer's] report on how the interviews went, what difficulties were encountered, what alterations should be made, and so forth." This emphasis on interviewer perceptions is nicely illustrated in Sudman and Bradburn's (1982, p. 49) advice for detecting unexpected word meanings: "A careful pilot test conducted by *sensitive* interviewers is the most direct way of discovering these problem words" (emphasis added).

Yet even if interviewers were trained extensively in recognizing problems with questions (as compared with receiving no special training at all, which is typical), conventional pretesting would still be ill suited to uncovering many questionnaire problems. This is because certain kinds of problems will not be apparent from observing respondent behavior, and the respondents themselves may be unaware of the problems. For instance, respondents can misunderstand a closed question's intent without providing any indication of having done so. And because conventional pretests are almost always "undeclared" to the respondent, as opposed to "participating" (in which respondents are informed of the pretest's purpose; see Converse and Presser, 1986), respondents are usually not asked directly about their interpretations or other problems the questions may cause. As a result, undeclared conventional pretesting seems better designed to identify problems the questionnaire poses for interviewers, who know the purpose of the testing, than for respondents, who do not.

Furthermore, when conventional pretest interviewers do describe respondent problems, there are no rules for assessing their descriptions or for determining which problems that are identified ought to be addressed. Researchers typically rely on intuition and experience in judging the seriousness of problems and deciding how to revise questions that are thought to have flaws.

In recent decades, a growing awareness of conventional pretesting's drawbacks has led to two interrelated changes. First, there has been a subtle shift in the goals of testing, from an exclusive focus on identifying and fixing overt problems experienced by interviewers and respondents to a broader concern for improving data quality so that measurements meet a survey's objectives. Second, new testing methods have been developed or adapted from other uses. These include cognitive interviews (the subject of Part I of this volume), behavior coding, response latency, vignette analysis, and formal respondent debriefings (all of which are treated in Part II), experiments (covered in Part III), and statistical modeling (Part IV). In addition, new modes of administration pose special challenges for pretesting (the focus of Part V), as do surveys of special populations, such as children, establishments, and those requiring questionnaires in more than one language (all of which are dealt with in Part VI). Finally, the development of new pretesting methods raises issues of how they might best be used in combination, as well as whether they in fact lead to improvements in survey measurement (the topics of Part VII).

1.2 COGNITIVE INTERVIEWS

Ordinary interviews focus on producing codable responses to the questions. Cognitive interviews, by contrast, focus on providing a view of the processes elicited by the questions. Concurrent or retrospective *think-alouds* and/or probes are used to produce reports of the thoughts that respondents have either as they answer the survey questions or immediately after. The objective is to reveal the thought processes involved in interpreting a question and arriving at an answer. These thoughts are then analyzed to diagnose problems with the question.

Although he is not commonly associated with cognitive interviewing, William Belson (1981) pioneered a version of this approach. In the mid-1960s, Belson designed "intensive" interviews to explore seven questions that respondents had been asked the preceding day during a regular interview administered by a separate interviewer. Respondents were first reminded of the exact question and the answer they had given to it. The interviewer then inquired: "When you were asked that question yesterday, exactly what did you think the question meant?" After nondirectively probing to clarify what the question meant to the respondent, interviewers asked, "Now tell me exactly how you worked out your answer from that question. Think it out for me just as you did yesterday—only this time say it aloud for me." Then, after nondirectively probing to illuminate how the answer was worked out, interviewers posed scripted probes about various aspects of the question. These probes differed across the seven questions and were devised to test hypotheses about problems particular to each of the questions. Finally, after listening to the focal question once more, respondents were requested to say how they would now answer it. If their answer differed from the one they had given the preceding day, they were asked to explain why. Six interviewers, who received two weeks of training, conducted 265 audiotaped, intensive interviews with a cross-section sample of residents of London, England. Four analysts listened to the tapes and coded the incidence of various problems.

These intensive interviews differed in a critical way from today's cognitive interview, which integrates the original and follow-up interviews in a single administration with one interviewer. Belson assumed that respondents could accurately reconstruct their thoughts from an interview conducted the previous day, which is inconsistent with what we now know about the validity of self-reported cognitive processes (see Chapter 2). However, in many respects, Belson moved considerably beyond earlier work, such as Cantril and Fried (1944), which used just one or two scripted probes to assess respondent interpretations of survey questions. Thus, it is ironic that his approach had little impact on pretesting practices, an outcome possibly due to its being so labor intensive.

The pivotal development leading to a role for cognitive interviews in pretesting did not come until two decades later with the Cognitive Aspects of Survey Methodology (CASM) conference (Jabine et al., 1984). Particularly influential was Loftus's (1984) postconference analysis of how respondents answered survey questions about past events, in which she drew on the think-aloud technique used by Herbert Simon and his colleagues to study problem solving (Ericsson and Simon, 1980). Subsequently, a grant from Murray Aborn's program at the National Science Foundation to Monroe Sirken supported both research on the technique's utility for understanding responses to survey questions (Lessler et al., 1989) and the creation at the National Center for Health Statistics (NCHS) in 1985 of the first "cognitive laboratory," where the technique could routinely be drawn on to pretest questionnaires (e.g., Royston and Bercini, 1987).

Similar laboratories were soon established by other U.S. statistical agencies and survey organizations.[1] The labs' principal, but not exclusive activity involved cognitive interviewing to pretest questionnaires. Facilitated by special exemptions from Office of Management and Budget survey clearance requirements, pretesting for U.S. government surveys increased dramatically through the 1990s (Martin et al., 1999). At the same time, the labs took tentative steps toward standardizing and codifying their practices in training manuals (e.g., Willis, 1994) or protocols for pretesting (e.g., DeMaio et al., 1993).

Although there is now general agreement about the value of cognitive interviewing, no consensus has emerged about best practices, such as whether (or when) to use think-alouds versus probes, whether to employ concurrent or retrospective reporting, and how to analyze and evaluate results. In part, this is due to the paucity of methodological research examining these issues, but it is also due to lack of attention to the theoretical foundation for applying cognitive interviews to survey pretesting.

In Chapter 2, Gordon Willis addresses this theoretical issue, and in the process contributes to the resolution of key methodological issues. Willis reviews the theoretical underpinnings of Ericsson and Simon's original application of think-aloud interviews to problem-solving tasks and considers the theoretical justifications for applying cognitive interviewing to survey tasks. Ericsson and Simon concluded that verbal reports can be veridical if they involve information a person has available in short-term memory, and the verbalization itself does not fundamentally alter thought processes (e.g., does not involve further explanation). Willis concludes that some survey tasks (for instance, nontrivial forms of information retrieval) may be well suited to elucidation in a think-aloud interview. However, he cautions that the *general* use of verbal report methods to target cognitive processes involved in answering survey questions is difficult to justify, especially for tasks (such as term comprehension) that do not satisfy the conditions for valid verbal reports. He also notes that the social interaction involved in interviewer administered cognitive interviews may violate a key assumption posited by Ericsson and Simon for use of the method.

Willis not only helps us see that cognitive interviews may be better suited for studying certain types of survey tasks than others, but also sheds light on the different ways of conducting the interviews: for instance, using think-alouds versus

[1]Laboratory research to evaluate self-administered questionnaires was already under way at the Census Bureau before the 1980 census (Rothwell, 1983, 1985). Although inspired by marketing research rather than cognitive psychology, this work foreshadowed cognitive interviewing. For example, observers asked respondents to talk aloud as they filled out questionnaires. See also Hunt et al. (1982).

probes. Indeed, with Willis as a guide we can see more clearly that concurrent think-alouds may fail to reveal how respondents interpret (or misinterpret) word meanings, and that targeted verbal probes should be more effective for this purpose. More generally, Willis's emphasis on the theoretical foundation of testing procedures is a much-needed corrective in a field that often slights such concerns.

Chapter 3, by Paul Beatty, bears out Willis's concern about the reactivity of aspects of cognitive interviewing. Beatty describes NCHS cognitive interviews which showed that respondents had considerable difficulty answering a series of health assessment items that had produced no apparent problems in a continuing survey. Many researchers might see this as evidence of the power of cognitive interviews to detect problems that are invisible in surveys. Instead, Beatty investigated whether features of the cognitive interviews might have created the problems, problems that the respondents would not otherwise have had.

Transcriptions from the taped cognitive interviews were analyzed for evidence that respondent difficulty was related to the interviewer's behavior, in particular the types of probes posed. The results generally indicated that respondents who received reorienting probes had little difficulty choosing an answer, whereas those who received elaborating probes had considerable difficulty. During a further round of cognitive interviews in which elaborating probes were restricted to the post-questionnaire debriefing, respondents had minimal difficulty choosing an answer. This is a dramatic finding, although Beatty cautions that it does not mean that the questions were entirely unproblematic, as some respondents expressed reservations about their answers during the debriefing.

Elaborating and reorienting probes accounted for only a small fraction of the interviewers' contribution to these cognitive interviews, and in the second part of his chapter, Beatty examines the distribution of all the interviewers' utterances aside from reading the questions. He distinguishes between cognitive probes (those traditionally associated with cognitive interviews, such as "What were you thinking ...?" "How did you come up with that ...?" "What does [term] mean to you?"); confirmatory probes (repeating something the respondent said in a request for confirmation); expansive probes (requests for elaboration, such as "Tell me more about that"); functional remarks (repetition or clarification of the question, which included all reorienting probes); and feedback (e.g., "Thanks; that's what I want to know" or "I know what you mean"). Surprisingly, cognitive probes, the heart of the method, accounted for less than 10% of interviewer utterances. In fact, there were fewer cognitive probes than utterances in any of the other categories.

Taken together, Beatty's findings suggest that cognitive interview results are importantly shaped by the interviewers' contributions, which may not be well focused in ways that support the inquiry. He concludes that cognitive interviews would be improved by training interviewers to recognize distinctions among probes and the situations in which each ought to be employed.

In Chapter 4, Frederick Conrad and Johnny Blair argue that (1) the raw material produced by cognitive interviews consists of verbal reports; (2) the different techniques used to conduct cognitive interviews may affect the quality of these verbal reports; (3) verbal report quality should be assessed in terms of problem detection and problem repair, as they are the central goals of cognitive interviewing; and (4) the most valuable assessment data come from experiments in which the independent variable varies the interview techniques and the dependent variables are problem detection and repair.

In line with these recommendations, they carried out an experimental comparison of two different cognitive interviewing approaches. One was uncontrolled, using the unstandardized practices of four experienced cognitive interviewers; the other, more controlled, used four less-experienced interviewers, who were trained to probe only when there were explicit indications that the respondent was experiencing a problem. The authors found that the conventional cognitive interviews identified many more problems than the conditional probe interviews.

As with Beatty's study, however, more problems did not mean higher-quality results. Conrad and Blair assessed the reliability of problem identification in two ways: by interrater agreement among a set of trained coders who reviewed transcriptions of the taped interviews, and by agreement between coders and interviewers. Overall, agreement was quite low, consistent with the finding of some other researchers about the reliability of cognitive interview data (Presser and Blair, 1994). But reliability was higher for the conditional probe interviews than for the conventional ones. (This may be due partly to the conditional probe interviewers having received some training in what should be considered "a problem," compared to the conventional interviewers, who were provided no definition of what constituted a "problem.") Furthermore, as expected, conditional interviewers probed much less than conventional interviewers, but more of their probes were in cases associated with the identification of a problem. Thus, Conrad and Blair, like Willis and Beatty, suggest that we rethink what interviewers do in cognitive interviews.

Chapter 5, by Theresa DeMaio and Ashley Landreth, describes an experiment in which three different organizations were commissioned to have two interviewers each conduct five cognitive interviews of the same questionnaire using whatever methods were typical for the organization, and then deliver a report identifying problems in the questionnaire and a revised questionnaire addressing the problems (as well as audiotapes for all the interviews). In addition, expert reviews of the original questionnaire were obtained from three people who were not involved in the cognitive interviews. Finally, another set of cognitive interviews was conducted by a fourth organization to test both the original and three revised questionnaires.

The three organizations reported considerable diversity on many aspects of the interviews, including location (respondent's home versus research lab), interviewer characteristics (field interviewer versus research staff), question strategy (think-aloud versus probes), and data source (review of audiotapes versus interviewer notes and recollections). This heterogeneity is consistent with the findings of Blair and Presser (1993) but is even more striking given the many intervening years in which some uniformity of practice might have emerged. It does, however, mean that differences in the results of these cognitive interviews across organization cannot be attributed unambiguously to any one factor.

There was variation across the organizations in both the number of questions identified as having problems and the total number of problems identified. Moreover, there was only modest overlap in the particular problems diagnosed (i.e., the organizations tended to report unique problems). Similarly, the cognitive interviews and the expert reviews overlapped much more in identifying which questions had problems than in identifying what the problems were. The organization that identified the fewest problems (both overall and in terms of number of questions) also showed the lowest agreement with the expert panel. This organization was the only one that did not review the audiotapes, and DeMaio and Landreth suggest that relying solely on interviewer notes and memory leads to error.[2] However, the findings from the tests of the revised questionnaires did not identify one organization as consistently better or worse than the others.

All four of these chapters argue that the methods used to conduct cognitive interviews shape the data they produce. This is a fundamental principle of survey methodology, yet it may be easier to ignore in the context of cognitive interviews than in the broader context of survey research. The challenge of improving the quality of verbal reports from cognitive interviews will not be easily met, but it is akin to the challenge of improving data more generally, and these chapters bring us closer to meeting it.

1.3 SUPPLEMENTS TO CONVENTIONAL PRETESTS

Unlike cognitive interviews, which are completely distinct from conventional pretests, other testing methods that have been developed may be implemented as add-ons to conventional pretests (or as additions to a survey proper). These include behavior coding, response latency, formal respondent debriefings, and vignettes.

Behavior coding was developed in the 1960s by Charles Cannell and his colleagues at the University of Michigan Survey Research Center and can be used to evaluate both interviewers and questions. Its early applications were almost entirely focused on interviewers, so it had no immediate impact on pretesting practices. In the late 1970s and early 1980s, a few European researchers adopted behavior coding to study questions, but it was not applied to pretesting in the United States until the late 1980s (Oksenberg et al.'s 1991 article describes it as one of two "new strategies for pretesting questions").

Behavior coding involves monitoring interviews or reviewing taped interviews (or transcripts) for a subset of the interviewer's and respondent's verbal behavior in the question asking and answering interaction. Questions marked by high frequencies of certain behaviors (e.g., the interviewer did not read the question verbatim or the respondent requested clarification) are seen as needing repair.

[2]Bolton and Bronkhorst (1996) describe a computerized approach to evaluating cognitive interview results, which should reduce error even further.

Behavior coding may be extended in various ways. In Chapter 6, Johannes van der Zouwen and Johannes Smit describe an extension that draws on the sequence of interviewer and respondent behaviors, not just the frequency of the individual behaviors. Based on the sequence of a question's behavior codes, an interaction is coded as either paradigmatic (the interviewer read the question correctly, the respondent chose one of the alternatives offered, and the interviewer coded the answer correctly), problematic (the sequence was nonparadigmatic but the problem was solved, e.g., the respondent asked for clarification and then chose one of the alternatives offered), or inadequate (the sequence was nonparadigmatic and the problem was not solved). Questions with a high proportion of nonparadigmatic sequences are identified as needing revision.

Van der Zouwen and Smit analyzed a series of items from a survey of the elderly to illustrate this approach as well as to compare the findings it produced to those from basic behavior coding and from four *ex-ante methods*, that is, methods not entailing data collection: a review by five methodology experts; reviews by the authors guided by two different questionnaire appraisal systems; and the quality predictor developed by Saris and his colleagues (Chapter 14), which we describe in Section 1.5. The two methods based on behavior codes produced very similar results, as did three of the four ex ante methods—but the two sets of methods identified very different problems. As van der Zouwen and Smit observe, the ex-ante methods point out what *could* go wrong with the questionnaire, whereas the behavior codes and sequence analyses reveal what actually *did* go wrong.

Another testing method based on observing behavior involves the measurement of response latency, the time it takes a respondent to answer a question. Since most questions are answered rapidly, latency measurement requires the kind of precision (to fractions of a second) that is almost impossible without computers. Thus, it was not until after the widespread diffusion of computer-assisted survey administration in the 1990s that the measurement of response latency was introduced as a testing tool (Bassili and Scott, 1996).

In Chapter 7, Stasja Draisma and Wil Dijkstra use response latency to evaluate the accuracy of respondents' answers, and therefore, indirectly to evaluate the questions themselves. As they operationalize it, *latency* refers to the delay between the end of an interviewer's reading of a question and the beginning of the respondent's answer. The authors reason that longer delays signal respondent uncertainty, and they test this idea by comparing the latency of accurate and inaccurate answers (with accuracy determined by information from another source). In addition, they compare the performance of response latency to that of several other indicators of uncertainty.

In a multivariate analysis, both longer response latencies and the respondents' expressions of greater uncertainty about their answers were associated with inaccurate responses. Other work (Chapters 8 and 23), which we discuss below, reports no relationship (or even, an inverse relationship) between respondents' confidence or certainty and the accuracy of their answers. Thus, future research needs to develop a more precise specification of the conditions in which different measures of respondent uncertainty are useful in predicting response error.

Despite the fact that the interpretation of response latency is less straightforward than that of other measures of question problems (lengthy times may indicate careful processing, as opposed to difficulty), the method shows sufficient promise to encourage its further use. This is especially so, as the ease of collecting latency information means that it could be routinely included in computer-assisted surveys at very low cost. The resulting collection of data across many different surveys would facilitate improved understanding of the meaning and consequences of response latency and of how it might best be combined with other testing methods, such as behavior coding, to enhance the diagnosis of questionnaire problems.

Chapter 8, by Elizabeth Martin, is about vignettes and respondent debriefing. Unlike behavior coding and response latency, which are "undeclared" testing methods, respondent debriefings are a "participating" method, which informs the respondent about the purpose of the inquiry. Such debriefings have long been recommended as a supplement to conventional pretest interviews (Kornhauser, 1951, p. 430), although they most commonly have been conducted as unstructured inquiries improvised by interviewers. Martin shows how implementing them in a standardized manner can reveal both the meanings of questions and the reactions that respondents have to the questions. In addition, she demonstrates how debriefings can be used to measure the extent to which questions lead to missed or misreported information.

Vignette analysis, the other method Martin discusses, may be incorporated in either undeclared or participating pretests. Vignettes—hypothetical scenarios that respondents evaluate—may be used to (1) explore how people think about concepts, (2) test whether respondents' interpretations of concepts are consistent with those that are intended, (3) analyze the dimensionality of concepts, and (4) diagnose other question wording problems. Martin provides examples of each of these applications and offers evidence of the validity of vignette analysis by drawing on evaluations of questionnaire changes made on the basis of the method.

The three chapters in this part suggest that testing methods differ in the types of problems they are suited to identify, their potential for diagnosing the nature of a problem and thereby for fashioning appropriate revisions, the reliability of their results, and the resources needed to conduct them. It appears, for instance, that formal respondent debriefings and vignette analysis are more apt than behavior coding and response latency to identify certain types of comprehension problems. Yet we do not have good estimates of many of the ways in which the methods differ. The implication is not only that we need research explicitly designed to make such comparisons, but also that multiple testing methods are probably required in many cases to ensure that respondents understand the concepts underlying questions and are able and willing to answer them accurately.

1.4 EXPERIMENTS

Both supplemental methods to conventional pretests and cognitive interviews identify questionnaire problems and lead to revisions designed to address the problems. To determine whether the revisions are improvements, however, there is no substitute for experimental comparisons of the original and revised items. Such experiments are of two kinds. First, the original and revised items can be compared using the testing method(s) that identified the problem(s). Thus, if cognitive interviews showed that respondents had difficulty with an item, the item and its revision can be tested in another round of cognitive interviews to confirm that the revision shows fewer such problems than the original. The interpretation of results from this type of experiment is usually straightforward, although there is no assurance that observed differences will have any effect on survey estimates.

Second, original and revised items can be tested to examine what, if any, difference they make for a survey's estimates. The interpretation from this kind of experiment is sometimes less straightforward, but such split-sample experiments have a long history in pretesting. Indeed, they were the subject of one of the earliest articles devoted to pretesting (Sletto, 1950), although the experiments that it described dealt with the impact on cooperation to mail surveys of administrative matters such as questionnaire length, nature of the cover letter's appeal, use of follow-up postcards, and questionnaire layout. None of the examples concerned question wording.

In Chapter 9, Floyd Fowler describes three ways to evaluate the results of experiments that compare question wordings: differences in response distributions, validation against a standard, and usability, as measured, for instance, by behavior coding. He provides six case studies that illustrate how cognitive interviews and experiments are complementary. For each, he outlines the problems that the cognitive interviews detected and the nature of the remedy proposed. He then presents a comparison of the original and revised questions from split-sample experiments that were behavior coded. As he argues, this type of experimental evidence is essential in estimating whether different question wordings affect survey results, and if so, by how much.

All of Fowler's examples compare single items that vary in only one way. Experiments can also be employed to test versions of entire questionnaires that vary in multiple, complex ways. This type of experiment is described in Chapter 10, by Jeffrey Moore, Joanne Pascale, Pat Doyle, Anna Chan, and Julia Klein Griffiths with data from SIPP, the Survey of Income and Program Participation, a large U.S. Bureau of the Census survey that has been conducted on a continuing basis for nearly 20 years. The authors revised the SIPP questionnaire to meet three major objectives: minimize response burden and thereby decrease both unit and item nonresponse, reduce seam bias reporting errors, and introduce questions about new topics. Then, to assess the effects of the revisions before switching to the new questionnaire, an experiment was conducted in which respondents were randomly assigned to either the new or old version.

Both item nonresponse and seam bias were lower with the new questionnaire, and with one exception, the overall estimates of income and assets (key measures in the survey) did not differ between versions. On the other hand, unit nonresponse reductions were not obtained (in fact, in initial waves, nonresponse was higher for the revised version) and the new questionnaire took longer to administer. Moore et al. note that these latter results may have been caused by two complicating features of the experimental design. First, experienced SIPP interviewers were used for both the old and new instruments. The interviewers' greater comfort level with the old questionnaire (some reported being able to "administer it in their sleep") may have contributed to their administering it more quickly than the new questionnaire and persuading more respondents to cooperate with it. Second, the addition of new content to the revised instrument may have more than offset the changes that were introduced to shorten the interview.

In Chapter 11, Roger Tourangeau argues that the practical consideration that leads many experimental designs to compare packages of variables, as in the SIPP case, hampers the science of questionnaire design. Because it experimented with a package of variables, the SIPP research could estimate the overall effect of the redesign, which is vital to the SIPP sponsors, but not estimate the effects of individual changes, which is vital to an understanding of the effects of questionnaire features (and therefore to sponsors of other surveys making design changes). As Tourangeau outlines, relative to designs comparing packages of variables, factorial designs allow inference not only about the effects of particular variables, but about the effects of interactions between variables as well. In addition, he debunks common misunderstandings about factorial designs: for instance, that they must have equal-sized cells and that their statistical power depends on the cell size.

Other issues that Tourangeau considers are complete randomization versus randomized block designs (e.g., should one assign the same interviewers to all the conditions, or different interviewers to different versions of the questionnaire?), conducting experiments in a laboratory setting as opposed to the field, and statistical power, each of which affects importantly the inferences drawn from experiments. Particularly notable is his argument in favor of more laboratory experiments, but his discussion of all these matters will help researchers make more informed choices in designing experiments to test questionnaires.

1.5 STATISTICAL MODELING

Questionnaire design and statistical modeling are usually thought of as being worlds apart. Researchers who specialize in questionnaires tend to have rudimentary statistical understanding, and those who specialize in statistical modeling generally have little appreciation for question wording. This is unfortunate, as the two should work in tandem for survey research to progress. Moreover, the two-worlds problem is not inevitable. In the early days of survey research, Paul Lazarsfeld, Samuel Stouffer, and their colleagues made fundamental contributions to both questionnaire design and statistical analysis. Thus, it is fitting that the first of our three chapters on statistical modeling to evaluate questionnaires draws on a technique, latent class analysis, rooted in Lazarsfeld's work. In Chapter 12, Paul Biemer shows how estimates of the error associated with questions may be made when the questions have been asked of the same respondents two or more times.

Latent class analysis (LCA) models the relationship between an unobservable latent variable and its indicator. Biemer treats the case of nominal variables where the state observed is a function of the true state and of false positive and false negative rates. He presents several applications from major surveys to illustrate how LCA allows one to test assumptions about error structure. Each of his examples produces results that are informative about the nature of the errors in respondent answers to the questions. Yet Biemer is careful to note that LCA depends heavily on an assumed model, and there is usually no direct way to evaluate the model assumptions. He concludes that rather than relying on a single statistical method for evaluating questions, multiple methods ought to be employed.

Whereas Biemer's chapter focuses on individual survey questions, in Chapter 13, Bryce Reeve and Louise Mâsse focus on scales or indexes constructed from multiple items. Reeve and Mâsse note that the principles of classical test theory usually yield scales with many items, without providing much information about the performance of the separate questions. They describe Item Response Theory (IRT) models that assess how well different items discriminate among respondents who have the same value on a trait. The authors' empirical example comes from the SF-36 Mental Health, Social Functioning, Vitality and Emotional subscales, widely used health indicators composed of 14 questions.

The power of IRT to identify the discriminating properties of specific items allows researchers to design shorter scales that do a better job of measuring constructs. Even greater efficiency can be achieved by using IRT methods to develop computer adaptive tests (CATs). With a CAT, a respondent is presented a question near the middle of the scale range, and an estimate of the person's total score is constructed based on his or her response. Another item is then selected based on that estimate, and the process is repeated. At each step, the precision of the estimated total score is computed, and when the desired precision is reached, no more items are presented. CAT also offers the opportunity for making finer distinctions at the ends of the range, which would be particularly valuable for the SF-36 scale, since as Reeve and Mâsse note, it does not do well in distinguishing people at the upper end of the range.

In Chapter 14, Willem Saris, William van der Veld, and Irmtraud Gallhofer draw on statistical modeling in a very different fashion. In the early 1980s, Frank Andrews applied the multitrait, multimethod (MTMM) measurement approach (Campbell and Fiske, 1959) to estimate the reliability and validity of a sample of questionnaire items, and suggested that the results could be used to characterize the reliability and validity of question types. Following his suggestion, Saris et al. created a database of MTMM studies that provides estimates of reliability and validity for 1067 questionnaire items. They then developed a coding system to characterize the items according to the nature of their content, complexity, type of response scale, position in the questionnaire, data collection mode, sample type, and the like. Next, they fit two large regression models in which these characteristics were the independent variables and the dependent variables were the MTMM reliability or validity estimates. The resulting model coefficients estimate the effect on the reliability or validity of the question characteristics.

Saris et al. show how new items can be coded and the prediction equation used to estimate their quality. They created a program for the coding, some of which is entirely automated and some of which is computer guided. Once the codes are assigned, the program calculates the quality estimates.

The authors recognize that more MTMM data are needed to improve the models. In addition, the predictions of the models need to be tested in validation studies. However, the approach is a promising one for evaluating questions.

1.6 MODE OF ADMINISTRATION

The introduction of computer technology has changed many aspects of administering questionnaires. On the one hand, the variety of new methods—beginning with computer-assisted telephone interviewing (CATI), but soon expanding to computer-assisted personal interviewing (CAPI) and computer-assisted self-interviewing (CASI)—has expanded our ability to measure a range of phenomena more efficiently and with improved data quality (Couper et al., 1998). On the other hand, the continuing technical innovations—including audio-CASI, interactive voice response, and the Internet—present many challenges for questionnaire design.

The proliferation of data collection modes has at least three implications for the development, evaluation, and testing of survey instruments. One implication is the growing recognition that answers to survey questions may be affected by the mode in which the questions are asked. Thus, testing methods must take the delivery mode into consideration. A related implication is that survey instruments consist of much more than words. For instance, an instrument's layout and design, logical structure and architecture, and the technical aspects of the hardware and software used to deliver it all need to be tested and their possible effects on measurement error explored. A third implication is that survey instruments are increasingly complex and demand ever-expanding resources for testing. The older methods, which relied on visual inspection to test flow and routing, are no longer sufficient. Newer methods must be found to facilitate the testing of instrument logic, quite aside from the wording of individual questions. In summary, the task of testing questionnaires has greatly expanded.

Although Chapter 15, by Don Dillman and Cleo Redline, deals with traditional paper-based methods, it demonstrates that a focus on question wording is insufficient even in that technologically simple mode. The authors discuss how cognitive interviews may be adapted to explore the various aspects of visual language in self-administered questionnaires. They then describe three projects with self-administered instruments that mounted split-sample experiments based on the insights from cognitive interviews. In each case, the experimental results generally confirmed the conclusions drawn from cognitive interviews. (One of the advantages of self-administered approaches such as mail and the Web is that the per unit cost of data collection is much lower than that of interviewer-administered methods, permitting more extensive use of experiments.)

In Chapter 16, John Tarnai and Danna Moore focus primarily on testing computerized instruments for programming errors. With the growing complexity of survey instruments and the expanding range of design features available, this has become an increasingly costly and time-consuming part of the development process, often with no guarantee of complete success. Tarnai and Moore argue that much of this testing can be done effectively and efficiently only by computers, but that existing software is not up to the task—conclusions similar to that of a recent Committee on National Statistics workshop on survey automation (Cork et al., 2003).

Chapter 17, by Sue Ellen Hansen and Mick Couper, concerns usability testing to evaluate computerized survey instruments. In line with Dillman and Redline's chapter, it shows that the visual presentation of information—in this case, to the interviewer—as well as the design of auxiliary functions used by the interviewer in computer-assisted interviewing, are critical to creating effective instruments. As a result, Hansen and Couper maintain that testing for usability is as important as testing for programming errors. With computerized questionnaires, interviewers must manage two interactions, one with the computer and another with the respondent, and the goal of good design must therefore be to help interviewers manage both interactions to optimize data quality. Using four separate examples of usability testing to achieve this end, Hansen and Couper demonstrate its value for testing computerized instruments.

Chapter 18, by Reginald Baker, Scott Crawford, and Janice Swinehart, covers the development and testing of Web questionnaires. They review the various levels of testing necessary for Web surveys, some of which are unique to that mode (e.g., aspects of the respondent's computing environment such as monitor display properties, the presence of browser plug-ins, and features of the hosting platform that define the survey organization's server). In outlining a testing approach, the authors emphasize standards or generic guidelines and design principles that apply across questionnaires. In addition to testing methods used in other modes, Baker and his colleagues discuss evaluations based on process data that are easily collected during Web administration (e.g., response latencies, backups, entry errors, and breakoffs). As with Chapter 16, this chapter underscores the importance of automated testing tools, and consistent with the other two chapters in this part, it emphasizes that testing Web questionnaires must focus on their visual aspects.

1.7 SPECIAL POPULATIONS

Surveys of children, establishments, and populations that require questionnaires in multiple languages pose special design problems. Thus, pretesting is still more vital in these cases than for surveys of adults interviewed with questionnaires in a single language. Remarkably, however, pretesting has been neglected even further for such surveys than for ordinary ones.

Establishments and children might seem to have little in common, but the chapters that deal with them follow a similar logic. They begin by analyzing how the capabilities and characteristics of these special respondents affect the response process, move on to consider what these differences imply for choices about testing methods, and then consider steps to improve testing.

In Chapter 19, Diane Willimack, Lars Lyberg, Jean Martin, Lilli Japec, and Patricia Whitridge draw on their experiences at four national statistical agencies as well as on an informal survey of other survey organizations to describe distinctive characteristics of establishment surveys that have made questionnaire pretesting uncommon. Establishment surveys tend to be mandatory, rely on records, and target populations of a few very large organizations, which are included with certainty, and many smaller ones, which are surveyed less often. These features seem to have militated against adding to the already high respondent burden by conducting pretests. In addition, because establishment surveys are disproportionately designed to measure change over time, questionnaire changes are rare. Finally, establishment surveys tend to rely on postcollection editing to correct data.

Willimack et al. describe various ways to improve the design and testing of establishment questionnaires. In addition to greater use of conventional methods, they recommend strategies such as focus groups, site visits, record-keeping studies, and consultation with subject area specialists and other stakeholders. They also suggest making better use of ongoing quality evaluations and reinterviews, as well as more routine documentation of respondents' feedback, to provide diagnoses of questionnaire problems. Finally, they recommend that tests be embedded in existing surveys so that proposed improvements can be evaluated without increasing burden.

In Chapter 20, Edith de Leeuw, Natacha Borgers, and Astrid Smits consider pretesting questionnaires for children and adolescents. They begin by reviewing studies of children's cognitive development for guidance about the types of questions and cognitive tasks that can be asked of children of various ages. The evidence suggests that 7 is about the earliest age at which children can be interviewed with structured questionnaires, although the ability to handle certain types of questions (e.g., hypothetical ones) is not acquired until later. The authors discuss how various pretesting methods, including focus groups, cognitive interviews, observation, and debriefing, can be adapted to accommodate children of different ages. Finally, they provide examples of pretests that use these methods with children and offer advice about other issues (such as informed consent) that must be specially addressed for children.

Questionnaire translation has always been basic to cross-national surveys, and recently it has become increasingly important for national surveys as well. Some countries (e.g., Canada, Switzerland, and Belgium) must, by law, administer surveys in multiple languages. Other nations are translating questionnaires as a result of increasing numbers of immigrants. In the United States, for instance, the population 18 and older who speak a language other than English at home increased from 13.8% in 1990 to 17.8% in 2000 (U.S. Bureau of the Census, 2003). Moreover, by 2000, 4.4% of U.S. adults lived in "linguistically isolated" households, those in which all the adults spoke a language other than English and none spoke English "very well."

Despite its importance, Tom Smith reports in Chapter 21 that "... no aspect of cross-national survey research has been less subjected to systematic, empirical investigation than translation." His chapter is one of two that address the development and testing of questionnaires to be administered in more than one language. The author describes sources of nonequivalence in translated questions and discusses the problems involved in translating response scales or categories so that they are equivalent. In addition, he points out that item comparability may be impaired because response effects vary from one country to another. Smith reviews different approaches to translation, which he argues must be an integral part of item development and testing rather than an isolated activity relegated to the end of the design process.

The chapter outlines several strategies to address problems arising from noncomparability across languages. One approach involves asking multiple questions about a concept (e.g., well being) with different terms in each (e.g., satisfaction versus happiness), so that translation problems with a single term do not result in measurement error for all the items. Another approach is to use questions that are equivalent across the cultures and languages as well as those that are culture specific. A third strategy is to conduct special studies to calibrate scale terms.

As Janet Harkness, Beth-Ellen Pennell, and Alisú Schoua-Glusberg note in Chapter 22, translation is generally treated as a minor aspect of the questionnaire development process, and pretesting procedures that are often employed for monolingual survey instruments are not typically used for translated questionnaires. These authors provide illustrations of the sources and possible consequences of translation problems that arise from difficulties of matching meaning across questions and answer scales, and from differences between languages, such as whether or not words carry gender. Too-close (word-by-word) translations can result in respondents being asked a different question than intended, or being asked a more cumbersome or stilted question.

Based on their experience, Harkness and her colleagues offer guidance on procedures and protocols for translation and assessment. They envision a more rigorous process of "translatology" than the ad hoc practices common to most projects. They emphasize the need for appraisals of the translated text (and hence do not believe back-translation is adequate), and they argue that the quality of translations, as well as the performance of the translated questions as survey questions, must be assessed. Finally, they recommend team approaches that bring different types of expertise to bear on the translation, and suggest ways to organize the effort of translation, assessment, and documentation (the last of which is particularly important for interpreting results after a survey is completed).

1.8 MULTIMETHOD APPLICATIONS

The importance of multiple methods is a recurring theme throughout this volume. Two of the three chapters in the final section provide case studies of multiple methods, and the third provides an experimental assessment of a combination of methods. Although none of the chapters permits a controlled comparison of the effectiveness of individual methods, each shows how a multimethod approach can be employed to address survey problems. The case studies, which describe and evaluate the methods used at different stages in the design and evaluation process, should help inform future testing programs, as there are few published descriptions of how such programs are conducted and with what results. The experimental comparison is of importance in evaluating pretesting more generally.

In Chapter 23, Nora Cate Schaeffer and Jennifer Dykema describe a testing program designed to ensure that respondents correctly understood a concept central to a survey on child support: joint legal custody. Legal custody is easily confused with physical custody, and an earlier survey had shown that respondents underreported it. The authors' aim was to reduce underreporting of joint custody by improving the question that asked about it. They first convened focus groups to explore the domain, in particular the language used by parents in describing custody arrangements. They then conducted cognitive interviews to evaluate both the question from the earlier survey and the revised versions developed after the focus groups. Finally, they carried out a split-sample field experiment that varied the context of the legal custody question, to test the hypothesis that asking it after questions about physical custody would improve accuracy. The field interviews, which included questions about how sure respondents were of their answers, were also behavior coded. Moreover, the authors had access to official custody records, which allowed them to validate respondents' answers.

Overall, the testing program was successful in increasing accurate reporting among those who actually had joint legal custody. In other words, the initially observed problem of false negatives was greatly reduced. There was, however, an unanticipated reduction in accuracy among those who did not have joint legal custody. That is, false positives, which had not initially been a serious problem, increased substantially. Thus the study serves as a reminder of the danger of focusing on a single dimension of a measurement problem. Fixing one problem only to cause another may be more common than we suppose. It also reminds us of the usefulness of testing the revised questionnaire before implementing it in a production survey.

Chapter 24, by Michael Kaplowitz, Frank Lupi, and John Hoehn, describes multiple methods for developing and evaluating a stated-choice questionnaire to value wetlands. Measuring people's choices about public policy issues poses a dual challenge. The questionnaire must clearly define the issue (in this instance, the nature and role of wetlands) so that a cross section of adults will understand it, and it must specify a judgment about the issue (in this instance, whether restoring wetlands with specified characteristics offsets the loss of other wetlands with different characteristics) that respondents will be able to make meaningfully.

To accomplish these ends, the authors first conducted a series of focus groups, separately with target respondents and a panel of subject matter experts, to explore the subject matter and inform the design of two alternative questionnaires. They then convened additional focus groups to evaluate the instruments, but because the "no-show" rate was lower than expected, they also conducted cognitive interviews with the additional people not needed for the groups. The results from these cognitive interviews turned out to be much more valuable for assessing the questionnaires than the results from the focus groups. Finally, they conducted further cognitive interviews in an iterative process involving revisions to the questionnaire. Kaplowitz et al. analyze the limitations of focus groups for evaluating survey instruments in terms of conversational norms and group dynamics. Thus, the chapter illustrates a method that did not work well with a useful description of what went wrong, and why.

The volume ends with Chapter 25 by Barbara Forsyth, Jennifer Rothgeb, and Gordon Willis. These authors assessed whether pretesting predicts data collection problems and improves survey outcomes. A combination of three methods—informal expert review, appraisal coding, and cognitive interviews—was used to identify potential problems in a pretest of a questionnaire consisting of 83 items. The 12 questions diagnosed most consistently by the three methods as having problems were then revised to address the problems. Finally, a split-sample field experiment was conducted to compare the original and revised items. The split-sample interviews were behavior coded and the interviewers were asked to evaluate the questionnaires after completing the interviews.

The versions of the original questions identified in the pretest as particularly likely to pose problems for interviewers were more likely to show behavior-coded interviewing problems in the field and to be identified by interviewers as having posed problems for them. Similarly, the questions identified by the pretest as posing problems for respondents resulted in more respondent problems, according to both the behavior coding and the interviewer ratings. Item nonresponse was also higher for questions identified by the pretest as presenting either recall or sensitivity problems than for questions not identified as having those problems. These results demonstrate that the combination of pretesting methods was a good predictor of the problems the items would produce in the field.

However, the revised questions generally did not appear to outperform the original versions. The item revisions had no effect on the frequency of behavior-coded interviewer and respondent problems. And while interviewers did rate the revisions as posing fewer respondent problems, they rated them as posing more interviewer problems. The authors suggest various explanations for this outcome, including their selection of only questions diagnosed as most clearly problematic, which often involved multiple problems that required complex revisions to address. In addition, the revised questions were not subjected to another round of testing using the three methods that originally identified the problems to confirm that the revisions were appropriate. Nonetheless, the results are chastening, as they suggest that we have much better tools for diagnosing questionnaire problems than for fixing them.

1.9 AGENDA FOR THE FUTURE

The methods discussed here do not exhaust the possibilities for testing and evaluating questions. For instance, formal appraisal schemes that are applied by coders (Lessler and Forsyth, 1996) are treated only in passing in this volume, and those involving artificial intelligence (Graesser et al., 2000) are not treated at all. In addition, there is only a little on focus groups (Bischoping and Dykema, 1999) and nothing on ethnographic interviews (Gerber, 1999), both of which are most commonly used at an early development stage before there is an instrument to be tested. Nonetheless, the volume provides an up-to-date assessment of the major evaluation methods currently in use and demonstrates the progress that has been attained in making questionnaire testing more rigorous. At the same time, the volume points to the need for extensive additional work.

Different pretesting methods, and different ways of carrying out the same method, influence the numbers and types of problems identified. Consistency among methods is often low, and the reasons for this need more investigation. One possibility is that in their present form, some of the methods are unreliable. But two other possibilities are also worth exploring. First, lack of consistency may occur because the methods are suited for identifying different problem types. For example, comprehension problems that occur with no disruption in the question asking and answering process are unlikely to be picked up by behavior coding. Thus, we should probably expect only partial overlap in the problems identified by different methods. Second, inconsistencies may reflect a lack of consensus among researchers, cognitive interviewers, or coders about what is regarded as a problem. For example, is it a problem if a question is awkward to ask but obtains accurate responses, or is it a problem only if the question obtains erroneous answers? The types and severity of problems that a questionnaire pretest (or methodological evaluation) aims to identify are not always clear, and this lack of specification may contribute to the inconsistencies that have been observed.

In exploring such inconsistencies, the cross-organization interlaboratory approach used in DeMaio and Landreth's chapter (see also Martin et al., 1999) holds promise not only of leading to greater standardization, and therefore to higher reliability, but to enhancing our understanding of which methods are appropriate in different circumstances and for different purposes.

It is also clear that problem *identification* does not necessarily point to problem *solution* in any obvious or direct way. For instance, the authors of Chapters 23 and 25 used pretesting to identify problems that were then addressed by revisions, only to find in subsequent field studies that the revisions either did not result in improvements, or created new problems. The fact that we are better able to identify problems than solutions underscores the desirability of additional testing after questionnaires have been revised.

Many of the chapters contain specific suggestions for future research, but here we offer four general recommendations for advancing questionnaire testing and evaluation. These involve:

- The connection between problem identification and measurement error
- The impact of testing methods on survey costs
- The role of basic research and theory in guiding the repair of question flaws
- The development of a database to facilitate cumulative knowledge

First, we need studies that examine the connection between problem diagnosis and measurement error. A major objective of testing is to reduce measurement error, yet we know little about the degree to which error is predicted by the various problem indicators at the heart of the different testing methods. Chapters 7 and 23 are unusual in making use of external validation in this way. Several of the other chapters take an indirect approach, by examining the link between problem diagnosis and specific response patterns (e.g., missing data, or seam bias), on the assumption that higher or lower levels are more accurate. But inferences based on indirect approaches must be more tentative than those based on direct validation (e.g., record check studies). With appropriately designed validation studies, we might be better able to choose among techniques for implementing particular methods, evaluate the usefulness of various methods for diagnosing different kinds of problems, and understand how much pretesting is "enough." We acknowledge, however, that validation data are rarely available and are themselves subject to error. Thus, another challenge for future research is to develop further indicators of measurement error that can be used to assess testing methods.

Second, we need information about the impact of various testing methods on survey costs. The cost of testing may be somewhat offset, completely offset, or even more than offset (and therefore reduce the total survey budget), depending on whether the testing results lead to the identification (and correction) of problems that affect those survey features (e.g., interview length, interviewer training, and postsurvey data processing) that have implications for cost. Although we know something about the direct costs of various testing methods, we know almost nothing about how the methods differ in their impact on overall costs. Thus, a key issue for future research is to estimate how various testing methods perform in identifying the types of problems that increase survey costs.

Third, since improved methods for diagnosing problems are mainly useful to the extent that we can repair the problems, we need more guidance in making repairs. As a result, advances in pretesting depend partly on advances in the science of asking questions (Schaeffer and Presser, 2003). Such a science involves basic research into the question-and-answer process that is theoretically motivated (Sudman et al., 1996; Tourangeau et al., 2000; Krosnick and Fabrigar, forthcoming). But this is a two-way street. On the one hand, pretesting should be guided by theoretically motivated research into the question-and-answer process. On the other hand, basic research and theories of the question-and-answer process should be shaped by both the results of pretesting and developments in the testing methods themselves [e.g., the question taxonomies, or classification typologies, used in questionnaire appraisal systems (Lessler and Forsyth, 1996) and the type of statistical modeling described by Saris et al.]. In particular, pretesting's focus on aspects of the response tasks that can make it difficult for respondents to answer accurately ought to inform theories of the connection between response error and the question-and-answer process.

Finally, we need improved ways to accumulate knowledge across pretests. This will require greater attention to documenting what is learned from pretests of individual questionnaires. One of the working groups at the Second Advanced Seminar on the Cognitive Aspects of Survey Methodology (Sirken et al., 1999, p. 56) suggested that survey organizations archive, in a central repository, the cognitive interviews they conduct, including the items tested, the methods used, and the findings produced. As that group suggested, this would "facilitate systematic research into issues such as: What characteristics of questions are identified by cognitive interviewing as engendering particular problems? What testing features are associated with discovering different problem types? What sorts of solutions are adopted in response to various classes of problems?" We believe that this recommendation should apply to *all* methods of pretesting. Establishing a pretesting archive on the Web would not only facilitate research on questionnaire evaluation, it would also serve as an invaluable resource for researchers developing questionnaires for new surveys.[3]

ACKNOWLEDGMENTS

The authors appreciate the comments of Roger Tourangeau. A revised version of this chapter is to be published in the Research Synthesis Section of the Spring 2004 issue of *Public Opinion Quarterly*.

[3]Many Census Bureau pretest reports are available at *http://www.census.gov/srd/www/byyear.html*, and many other pretest reports may be found in the Proceedings of the American Statistical Association Survey Research Methods Section and the American Association for Public Opinion Research available at *http://www.amstat.org/sections/srms/proceedings*. But neither site is easily searchable, and the reports often contain incomplete information about the procedures used.

CHAPTER 2

Cognitive Interviewing Revisited: A Useful Technique, in Theory?

Gordon B. Willis
National Cancer Institute

2.1 INTRODUCTION

In contrast to the more empirically oriented chapters in this volume, in this chapter I adopt a more theoretic viewpoint relating to the development, testing, and evaluation of survey questions, stemming from the perspective commonly termed CASM (cognitive aspects of survey methodology). From this perspective, the respondent's cognitive processes drive the survey response, and an understanding of cognition is central to designing questions and to understanding and reducing sources of response error. A variety of cognitive theorizing and modeling has been applied to the general challenge of questionnaire design (see Sudman et al., 1996, and Tourangeau et al., 2000, for reviews). Optimally, an understanding of cognitive processes will help us to develop design rules that govern choice of response categories, question ordering, and so on.

I address here a facet of CASM that is related to such design decisions, but from a more empirical viewpoint that involves questionnaire *pretesting*, specifically through the practice of *cognitive interviewing*. This activity is not carried out primarily for the purpose of developing general principles of questionnaire design, but rather, to evaluate targeted survey questions, with the goal of modifying these questions when indicated. That is, we test survey questions by conducting what is variably termed the cognitive, intensive, extended, think-aloud, or laboratory

Methods for Testing and Evaluating Survey Questionnaires, Edited by Stanley Presser, Jennifer M. Rothgeb, Mick P. Couper, Judith T. Lessler, Elizabeth Martin, Jean Martin, and Eleanor Singer
ISBN 0-471-45841-4

pretest interview,[1] and focus on the cognitive processes involved in answering them. Following Tourangeau (1984), the processes studied are generally listed as *question comprehension, information retrieval, judgment and estimation*, and *response*.

2.1.1 Definition of Cognitive Interviewing

The cognitive interview can be conceptualized as a modification and expansion of the usual survey interviewing process (Willis, 1994, 1999). The interview is conducted by a specially trained cognitive interviewer rather than by a survey field interviewer, and this interviewer administers questions to a cognitive laboratory "subject" in place of the usual survey respondent. Further, the cognitive interview diverges from the field interview through its application of two varieties of *verbal report methods*:

1. *Think-aloud.* The subject is induced to verbalize his or her thinking as he or she answers the tested questions (Davis and DeMaio, 1993; Bickart and Felcher, 1996; Bolton and Bronkhorst, 1996). For example, Loftus (1984, p. 62) provides the following example of think-aloud obtained through testing of the question *In the last 12 months have you been to a dentist?*:

   > "Let's see ... I had my teeth cleaned six months ago, and so ... and then I had them checked three months ago, and I had a tooth ... yeah I had a toothache about March ... yeah. So yeah, I have."

2. *Verbal probing.* After the subject provides an answer to the tested survey question, the interviewer asks additional probe questions to further elucidate the subject's thinking (Belson, 1981; Willis, 1994, 1999). In testing a 12-month dental visit question, the interviewer might follow up the subject's affirmative answer by asking:

   > "Can you tell me more about the last time you went to a dentist?"
   >
   > "When was this?"
   >
   > "Was this for a regular checkup, for a problem, or for some other reason?"
   >
   > "How sure are you that your last visit was within the past 12 months?"

Despite the overt focus on understanding what people are thinking as they answer survey questions, it is important to note that a key objective of cognitive interviewing is not simply to understand the strategies or general approaches that subjects use to answer the questions, but *to detect potential sources of response*

[1]Various labels have been applied by different authors, but these terms appear to be for the most part synonymous. A major source of potential confusion is the fact that the term *cognitive interview* is also used to refer to a very different procedure used in the justice and legal fields to enhance the retrieval of memories by crime victims or eyewitnesses (Fisher and Geiselman, 1992).

error associated with the targeted questions (Conrad and Blair, 1996). Consider another example: "In general, would you say your health is excellent, very good, good, fair, or poor?" From the perspective of Tourangeau's (1984) model, the cognitive interviewer assesses whether the subjects are able to (1) comprehend key terms (e.g., "health in general"; "excellent") in the way intended by the designer; (2) retrieve relevant health-oriented information; (3) make decisions or judgments concerning the reporting of the retrieved information; and (4) respond by comparing an internal representation of health status to the response categories offered (e.g., a respondent chooses "good" because that is the best match to his or her self-assessment of health status). If the investigators determine that problems related to any of these cognitive operations exist, they enact modifications in an attempt to address the deficiencies. Pursuing the example above, cognitive testing results have sometimes indicated that subjects regard physical and mental/emotional health states to be disparate concepts and have difficulty combining these into one judgment (see Willis et al., in press). Therefore, the general health question has for some purposes been decomposed into two subparts: one about physical health, the other concerning mental functioning (e.g., Centers for Disease Control and Prevention, 1998).

2.1.2 Variation in Use of Cognitive Interviewing

Cognitive interviewing in questionnaire pretesting is used by a wide variety of researchers and survey organizations (DeMaio and Rothgeb, 1996; Esposito and Rothgeb, 1997; Friedenreich et al., 1997; Jobe and Mingay, 1989; Thompson et al., 2002; U.S. Bureau of the Census, 1998; Willis et al., 1991). However, core definitions and terminology vary markedly: Some authors (e.g., Conrad and Blair, 1996) describe cognitive interviewing as synonymous with the think-aloud method; some (e.g., Willis, 1994, 1999) consider this activity to include both think-aloud and the use of targeted probing by the interviewer; whereas still others (Bolton and Bronkhorst, 1996) define cognitive interviewing in terms of verbal probing, and the "verbal protocol" method as involving solely think-aloud. There is also considerable variability in practice under the general rubric of cognitive interviewing (Beatty, undated; Forsyth, 1990, Forsyth and Lessler, 1991; Willis et al., 1999a), and there are no generally accepted shared standards for carrying out cognitive interviews (Tourangeau et al., 2000). Cognitive interviewing is practiced in divergent ways with respect to the nature of verbal probing, the coding of the collected data, and other features that may be instrumental in producing varied results (Cosenza, 2002; Forsyth and Lessler, 1991; Tourangeau et al., 2000, Willis et al., 1999b; Chapter 5, this volume).

The variety of approaches raises questions about the conduct of cognitive interviews: How important is it to obtain a codable answer (e.g. "good") versus a response that consists of the subject's own words?; Should we direct probes toward features of the tested *question*, or request elaboration of the *answer* given to that question (Beatty et al., 1996)?; On what basis should the decision to administer a probe question be made (Conrad and Blair, 2001)?

2.1.3 Determining the Best Procedure

The various approaches to cognitive interviewing may differ in effectiveness, and at the extreme, it is possible that none of them are of help in detecting and correcting question flaws. Although several empirical evaluations of these methods have been conducted (Conrad and Blair, 2001; Cosenza, 2002; Chapter 5, this volume), attempts at validation have to date been inconclusive (Willis et al., 1999a), and researchers' views concerning the types of empirical evidence that constitute validation vary widely. In particular, criterion validation data for survey questions (e.g., "true scores" that can be used as outcome quality measures) are generally rare or nonexistent (Willis et al., 1999b).

In the absence of compelling empirical data, it may be helpful to address questions concerning procedural effectiveness through explicit consideration of the role of theory in supporting and guiding interviewing practice (Conrad and Blair, 1996; Lessler and Forsyth, 1996). Attention to theory leads to two related questions:

1. First, is cognitive interviewing a theoretically driven exercise? Theory is often a useful guide to scientific practice, yet Conrad and Blair (1996) have suggested that "cognitive interviews are not especially grounded in theory." Thus, before considering whether theory will help us by providing procedural guidance, it behooves us to consider the extent to which such theory in fact exists.
2. To the extent that theories do underlie the practice of cognitive testing, have these theories been confirmed empirically? If a theory is testable and found to be supported through appropriate tests, the use of specific procedures deriving from that theory gains credence. On the other hand, if the theory appears undeveloped or misapplied, the procedures it spawns are also suspect.

In this chapter I endeavor to determine whether a review of cognitive theory can assist us in restricting the field of arguably useful procedural variants of cognitive interviewing. From this perspective, I examine two theories having an explicit cognitive emphasis. The first of these, by Ericsson and Simon (1980, 1984, 1993), focuses on memory processes and their implications for the use of verbal report methods. The second, *task analysis theory*, strives to develop the Tourangeau (1984) model into an applicable theory that focuses on staging of the survey response process.

2.2 REVISITING ERICSSON AND SIMON: THE VALIDITY OF VERBAL REPORTS

The cognitive perspective on question pretesting has been closely associated with a seminal *Psychological Review* paper by Ericsson and Simon (1980), followed by an elaborated book and later revision (1984, 1993). Ericsson and Simon reviewed the use of verbal report methods within experimental psychology experiments, and spawned considerable interest among survey methodologists in the use of *think-aloud* and *protocol analysis*. Although these terms are sometimes used interchangeably, *think-aloud* pertains to the procedure described above in which experimental subjects are asked to articulate their thoughts as they engage in a cognitive task, whereas *protocol analysis* consists of the subsequent analysis of a recording of the think-aloud stream (Bickart and Felcher, 1996; Bolton and Bronkhorst, 1996). Based on the Ericsson–Simon reviews, survey researchers have made a consistent case for the general application of verbal report methods to cognitive interviewing, and specifically, for selection of the think-aloud procedure (Tourangeau et al., 2000).

2.2.1 Ericsson–Simon Memory Theory and Verbal Reporting

Ericsson and Simon were interested primarily in assessing the usefulness of verbal reports of thinking, given a long tradition of ingrained skepticism by behaviorists concerning the ability of people to articulate meaningfully their motivations, level of awareness, and cognitive processes in general (e.g., Lashley, 1923). After reviewing an extensive literature, they concluded that verbal reports can be veridical if they involve information that the subject *has available in short-term memory at the time the report is verbalized*. To ground their contentions in theory, Ericsson and Simon presented a cognitive model (following Newell and Simon, 1972) emphasizing short- and long-term memory (STM, LTM). Of greatest significance was the issue of *where* information is stored: Information contained in STM during the course of problem solving or other cognitive activity could purportedly be reported without undue distortion. However, reporting requirements involving additional retrieval from LTM, at least in some cases, impose additional cognitive burdens that may produce biased modes of behavior.

Ericsson and Simon asserted that veridical reporting was possible not only because pertinent information was available in STM at the time it was reported, but also because the act of verbalizing this information imposed little additional processing demand. Further, verbalization does not in itself produce *reactivity*; that is, it does not fundamentally alter the "course and structure of the cognitive processes" (Ericsson and Simon, 1980, p. 235). Reactivity is observed when the act of verbal reporting produces results at variance from those of a silent control group; hence Ericsson and Simon emphasized the impact of task instructions that are found to minimize such effects. Specifically, Ericsson–Simon originally proposed that two types of self-reports, labeled level 1 and level 2 verbalizations, require a fairly direct readout of STM contents and are likely to satisfy the requirements of nonreactive reporting. The distinction between levels 1 and 2 concerns the need to recode nonverbal information into verbal form for the latter but not the former (verbalization of solution of a word-based puzzle would involve level 1, and that of a visual–spatial task, level 2).

However, a third type, *level 3 verbalization*, which involves further explanation, defense, or interpretation, is more likely to produce reactivity. Level 3 verbalization is especially relevant to self-reports involving survey questions, as many probe questions used by cognitive interviewers (e.g, "How did you come up with that answer?") may demand level 3 verbalization rather than simply the direct output of the cognitive processes that reach consciousness during the course of question answering.

2.2.2 Support for the Ericsson–Simon Theory

Ericsson and Simon cited numerous research studies demonstrating that when compared to a silent control condition, subjects providing verbal reports did not differ in measures such as task accuracy in solving complex puzzles, or in the solution steps they selected, although there was some evidence that thinking-aloud resulted in an increase in task completion time (Ericsson and Simon, 1980, 1984, 1993). However, the general debate concerning the efficacy of think-aloud methods has not been resolved (Crutcher, 1994; Ericsson and Simon, 1993; Nisbett and Wilson, 1977; Payne, 1994; Wilson, 1994). In brief, theorists have changed their focus from the general question of *whether* verbal reports are veridical to that concerning *when* they are (Austin and Delaney, 1998; Smith and Miller, 1978), and even Ericsson and Simon have modified their viewpoint as additional research results have accumulated. Key issues involve specification of: (1) the types of cognitive *tasks* that are amenable to verbal report methods, and (2) the nature of the specific verbal report *procedures* that are most effective, especially in terms of instructions and probe administration.

Table 2.1 summarizes major task parameters that have been concluded by Ericsson and Simon and other investigators to be instrumental in determining the efficacy of verbal reports; Table 2.2 similarly summarizes key procedural

Table 2.1 Task Variables Cited as Enhancing the Validity of Ericsson–Simon Verbal Protocols

1. The task involves verbal (as opposed to nonverbal, or spatial) information (Ericsson and Simon, 1993; Wilson, 1994).
2. The task involves processes that enter into consciousness, as opposed to those characterized by automatic, nonconscious processing of stimuli (Ericsson and Simon, 1993; Wilson, 1994).
3. The task is novel, interesting, and engaging, as opposed to boring, familiar, and redundant and therefore giving rise to automatic processing (Smith and Miller, 1978).
4. The task involves higher-level verbal processes that take more than a few seconds to perform, but not more than about 10 seconds (Ericsson and Simon, 1993; Payne, 1994).
5. The task to be performed emphasizes problem solving (Fiske and Ruscher, 1989; Wilson, 1994).
6. The task involves rules and criteria that people use in decision making (Berl et al., 1976; Slovic and Lichtenstein, 1971).

Table 2.2 Procedural Variables Cited as Enhancing the Validity of Ericsson–Simon Verbal Reporting Techniques

1. *Task timing*. As little time as possible passes between a cognitive process and its reporting, so that information is available in STM (Crutcher, 1994; Kuusela and Paul, 2000; Smith and Miller, 1978).
2. *Instructions*. Subjects should be asked to give descriptions of their thinking as opposed to interpretations or reasons (Austin and Delaney, 1998; Crutcher, 1994; Ericsson and Simon, 1993; Nisbett and Wilson, 1977; Wilson, 1994).
3. *Training*. Subjects should be introduced to think-aloud procedures but not overtrained, to minimize reactivity effects (Ericsson and Simon, 1993).
4. *Establishing social context*. The procedure should *not* be conducted as a socially oriented exercise that explicitly involves the investigator as an actor (Austin and Delaney, 1998; Ericsson and Simon, 1993).

variables. Again, such results have tended to temper more sweeping generalizations about the efficacy of self-report; for example, several researchers, including Ericsson and Simon (1993), have suggested that materials that are inherently verbal in nature are more amenable to verbal reporting than are nonverbal tasks, which reflects a modification of Ericsson and Simon's (1980) view that level 1 (verbal) and level 2 (nonverbal) vocalizations are equally nonreactive. In aggregate, these conclusions reinforce the general notion that verbal reports are useful for tasks that (1) are at least somewhat interesting to the subject, (2) involve verbal processing of information that enters conscious awareness, and (3) emphasize problem solving and decision making. Further, the verbal reporting environment should be arranged in such a way that (1) reports are given during the course of processing or very soon afterward, (2) social interaction between experimenter and subject is minimized in order to focus the subject's attention mainly toward the task presented, and (3) the instructions limit the degree to which subjects are induced to speculate on antecedent stimuli that may have influenced or directed their thinking processes.

2.2.3 Application of Verbal Report Methods to Survey Question Pretesting

The survey researcher is not interested primarily in the general usefulness of verbal report methods within psychology experiments, but rather, their applicability to the cognitive testing of survey questions. As such, cognitive interviewers should adopt verbal report methods only to the degree that any necessary preconditions are satisfied within this task domain. We must therefore consider whether cognitive testing either satisfies the conditions presented in Tables 2.1 and 2.2, or if not, can be determined to provide veridical self-reports when these conditions are violated. Somewhat surprisingly, these issues have rarely been addressed. To the extent that Ericsson–Simon advocated the general use of verbal report methods, cognitive interviewers have largely assumed that use of these methods to evaluate survey questionnaires represents a natural application, without considering the types of issues within Tables 2.1 and 2.2. However, in several ways, survey questions may pose a more complex cognitive domain than is generally recognized.

Range and Type of Cognitive Processes Considered The case for application of verbal report methods to the CASM area can be traced to Loftus (1984), who applied think-aloud to evaluate survey questions involving recall of past events (e.g., "During the past 12 months, about how many times did you see or talk to a medical doctor?"), and found that subjects appeared to prefer past-to-present retrieval order of such events. However, the research question Loftus posed was much more constrained than that implied by the wholesale adoption of verbal reporting methods; specifically, she advocated their use to study how survey respondents *rely on strategies in order to retrieve information* from memory in response to questions about past events, and made no reference to the use of verbal reports to study other cognitive processes. This limited application can be contrasted with a more general application of verbal reports to target all of the cognitive processes involved in answering survey questions: comprehension, decision/judgment, and response. However, this more general application involves a further conceptual step that has not generally been set forth explicitly and justified in theoretical terms. In particular, the use of verbal reports to study the process of comprehension—or understanding the question or respondent instructions—has not been specifically discussed. Yet comprehension is seen as fundamental to the task of answering survey questions (e.g., interpretation of the term *abdomen* or *full-time employment*). In fact, the pretesting of survey questions has been described as consisting primarily of "the study of meaning" (Gerber and Wellens, 1997), and empirical studies by Presser and Blair (1994) and Rothgeb et al. (2001) have suggested that the majority of problems detected through the pretesting of common survey questions may involve failures of comprehension.

Ericsson and Simon's reviews did not strongly emphasize comprehension processes (Whitney and Budd, 1996). Rather, the psychological laboratory experiments they described tended to follow mainstream academic psychological practice in investigating tasks that were constrained and arguably artificial (Neisser and Winograd, 1998), but also well defined and generally well comprehended. Specifically, these tasks tended to emphasize concept formation and problem solving (Bickart and Felcher, 1996); a representative example is the *Tower of Hanoi puzzle*, a nonverbal task that involves moving a series of rings between several pegs according to defined rules of movement, in order to achieve a specified outcome (move all the rings from the first to the last peg, one ring at a time—a larger ring can never rest on a smaller one). Where previously learned tasks were studied, they tended to be of activities such as chess playing, which again represent a fairly well defined problem-solving domain (Austin and Delaney, 1998). Overall, the focus of the prototype verbal-report investigation has normally been on problem-solving steps, not on comprehension of the materials presented.

Since Ericsson and Simon's initial work, verbal report methods have been extended to the realm of comprehension, specifically in the area of understanding of written text (Magliano and Graesser, 1991; Trabasso and Suh, 1993). A special issue of the journal *Discourse Processing* has described the use of think-aloud techniques to investigate comprehension of written passages, inference making, use of spatial modeling, and individual differences (Whitney and Budd, 1996). However, much of this research seems to have been done without explicit consideration of criteria such as those listed in Tables 2.1 and 2.2, and even its proponents are somewhat cautionary in extending verbal reporting to this realm (Whitney and Budd, 1996).

Note that Ericsson and Simon (1993, 1998) were somewhat cautious in advocating the application of think-aloud methods to the general arena of comprehension processes, as the requisite cognitive operations may be somewhat different from those involved in problem solving. In particular, they suggested that the usefulness of verbal report methods may be limited in cases in *which the investigator and research participant fail to share an identical mental representation of the task*. Significantly, it is such a lack of a shared representation that aptly describes the comprehension and communication problems we normally detect when testing survey questions. It therefore remains especially unclear whether the assessment of comprehension of survey questions is effectively targeted through verbal report methods. Ongoing research related to text comprehension is potentially informative. Note, however, that the comprehension of text passages may differ somewhat from that of survey questions, as the tasks are somewhat divergent. In particular, the former emphasizes the understanding of declarative information (i.e., that stated as factual within the context of the presented passage) and the inferences the reader makes between a series of related statements, whereas the latter explicitly focuses on responses to individually presented interrogatives (i.e., answering a series of discrete direct queries).

Differences in Memory System Involvement Most studies reviewed by Ericsson and Simon involved strong STM involvement, hence satisfying one of their key requirements for veridical verbal reporting (retrieval from STM rather than LTM). For example, to the extent that Tower of Hanoi–variety problems are novel from a subject's point of view, their solution depends primarily on the operation within short-term cognitive workspace, or *working memory* (Baddeley, 1986), as opposed to retrieval of solutions from long-term storage (i.e., the solution is literally "worked out" by a cognitive processor or workspace rather than being drawn from permanent storage).

Some aspects of survey item processing would also appear to involve strong working memory involvement. Reports involving complex retrieval phenomena, such as those studied by Loftus, probably involve working memory contributions (although with the strong caveat that a good deal of this information originates in LTM as opposed to STM). Further, at least one aspect of survey item comprehension—the understanding of a long and complex question—would seem to require some type of temporary cognitive storage area, such as working memory. This aspect of processing seems to have much in common with text comprehension, which is now studied routinely through verbal report methods (Whitney and Budd, 1996). From this perspective, survey item processing appears to satisfy at least some requirements of veridical self-reporting.

However, survey item comprehension may be multidimensional and subsumes a further key process—the comprehension of particular items within the question (e.g., "medical specialist," "long-term cancer care")—which arguably requires qualitatively different memory-oriented cognitive operations. First, it is not clear that comprehension of individual terms involves significant short-term working memory involvement, but would rather seem to require direct retrieval from LTM. More specifically, drawing from the distinction between *episodic* and *semantic memory* (Tulving, 1984), whereas verbal reports reflecting the solution of Tower of Hanoi–type problems may involve heavy use of episodic memory (literally, recall of episodes involved in solving the puzzle), verbal reports relating to term comprehension are more clearly associated with semantic memory (the supposed repository of the language lexicon). That is, the probe "What does the term 'medical specialist' mean to you?" arguably requires information retrieval from LTM semantic storage rather than from a temporary STM store. In sum, term comprehension as an expression of semantic memory content may depend on memory systems fundamentally different from those typically operative in investigations of verbal reporting.

Differences in Social Context A third feature that distinguishes at least some survey question environments from the classic Ericsson–Simon domain is the degree of social interaction involved. Especially in their revised book, Ericsson and Simon (1993) strongly advocated minimizing interviewer involvement in task solution; for example, they caution that rather than telling the subject to "Tell me what you are thinking," the experimenter should avoid explicit reference to himself or herself as an involved agent (through the use of "me"), and instead, simply state: "Keep talking." Further, the experimenter should go so far as to locate physically behind the subject, out of direct view.[2] Diametrically opposed to this view, survey researchers have consistently characterized the survey-response task as an explicit form of social interaction between interviewer and respondent (Cannell and Kahn, 1968; Schaeffer, 1991a; Schaeffer and Maynard, 1996; Schuman and Presser, 1981), and it follows that cognitive interviewing would naturally reflect this social element, especially to the extent that the interviewer is engaged in active probing behaviors. Ericsson and Simon (1998) addressed this issue by explicitly distinguishing between inner speech and social speech, arguing that tasks involving the latter may be more reactive in nature and perhaps not well addressed through think-aloud methods in particular. Task differences

[2]Note that Ericsson and Simon's view in this regard is somewhat extreme among practitioners of think-aloud. Bolton and Bronkhorst (1996) describe an alternative approach, involving testing of survey questions, that promotes direct involves interviewer involvement (e.g., the interviewer is allowed to utter the phrase "tell me ...").

with respect to social interaction may have implications concerning the degree to which cognitive interviewers should intrude (as through probing), face physically, and otherwise interact with a subject. By extension, this argument suggests an important distinction between the cognitive testing of interviewer- and self-administered questionnaires. The latter clearly are more similar to the tasks that Ericsson and Simon envisioned, given that these largely remove the element of social interaction. Hence, testing of self-administered instruments might similarly benefit from verbal reporting approaches that minimize such interaction. Although researchers appear to recognize the potential need for use of divergent approaches as a function of administration mode (DeMaio and Rothgeb, 1996; Chapter 15 this volume), they have not yet addressed the theoretical underpinnings that may influence these procedural differences.

In summary, there are several major ways in which the "classic" tasks reviewed by Ericsson–Simon diverge from the task of answering survey question, whether the question being tested is attitudinal ("Would you say that your health in general is excellent, very good, good, fair, or poor?") or autobiographical ("Did you do any work at all LAST WEEK, not including work around the house?"). Virtually by definition, the survey response requires the involvement of a wide set of background experiences, memories, and social dynamics. Further, these differences may be associated with variation in the procedures that are appropriate for the study of cognitive processes. For example, Wilson et al. (1996) pose the concern that particular survey question varieties, specifically those requiring respondents to report *why* they possess particular attitudes, involve so much self-reflection and interpretation that they may not be amenable to unbiased verbal report procedures (see Nisbett and Wilson, 1977, for a similar but more general argument). Further, based on their finding that subjects had difficulty using think-aloud to verbalize responses to survey questions where direct retrieval from memory was used, Bickart and Felcher (1996) concluded that think-aloud is most useful where recall involves a nontrivial strategy (similar to the Tower of Hanoi task), but is of limited use under conditions of direct information retrieval.

2.2.4. Summary: Applicability of Ericsson–Simon Theory to Survey Questions

Some survey-oriented cognitive processes—especially related to comprehension of long and complex questions, nontrivial forms of information retrieval, and other reconstruction of memories—appear to involve strong elements of problem solving, working memory, STM involvement, and other features summarized in Table 2.1, and may fit well within the Ericsson–Simon rubric. However, justification for the *general* use of verbal report methods to target multiple cognitive processes associated with survey response processes, especially term comprehension, is more difficult to justify on the basis of their theory, as these processes simply do not appear to satisfy the implicit preconditions. Further, the large degree of social interaction inherently involved in interviewer-administered survey modes violates a key assumption set forth by Ericsson and Simon. This is not to say that the Ericsson–Simon framework could not be extended to subsume all cognitive processes relevant to the survey response process. Rather, at this point, CASM researchers have failed to take this step.

2.2.5. Use of Think-Aloud Procedures versus Verbal Probing

The remaining critical issue relating to Ericsson and Simon's theory concerns the question of whether their orientation, to the extent that it does apply (say, to the subset of processes that involve problem solving and STM memory retrieval), has direct implications concerning the specific practices that cognitive interviewers should select. Of particular relevance is the think-aloud method, simply because this procedure has been so closely tied to their work in the minds of CASM researchers, and because a fundamental decision for cognitive interviewers concerns the choice of think-aloud versus verbal probing (Blair and Presser, 1993; Conrad and Blair, 1996, 2001; Chapter 3, this volume). Willis (1994, 1999) describes these procedures in detail, along with the supposed advantages and disadvantages of each. However, to date the discussion has lacked a clear focus on exactly *why* we are using them, other than to suggest that adherents to Ericsson–Simon would naturally favor think-aloud.

Further, despite the regard with which think-aloud is held in the historical literature concerning cognitive interviewing, several authors have documented an evolution away from its use (Conrad and Blair, 1996; Forsyth, 1990; Willis, 1994, 1999). In particular, the migration toward an increased use of targeted verbal probing by interviewers [which Forsyth and Lessler (1991) label "targeted attention filters"] is seen as representing a divergence from the original Ericsson–Simon model (Tourangeau et al., 2000, p. 326). A reasonable question is therefore whether we should consider adhering more closely to those roots.

What did Ericsson and Simon say about directed probing versus think-aloud? I suggest that the close association between Ericsson and Simon's underlying theory, and think-aloud as resulting practice, is less direct than is generally recognized. Again, the ongoing focus on think-aloud is based largely on Loftus's selection of that procedure rather than on a strict interpretation of Ericsson and Simon's original works (1980, 1984). Ericsson and Simon did, especially in their revised (1993) book, emphasize forms of "pure" think-aloud, and criticized biasing and leading forms of probing. However, they also left the door open to the possibility that probes can be used effectively. In particular, Ericsson and Simon (1980) encompassed "the thinking aloud method *and other procedures using verbalization*" (italics mine) (1980, p. 217), and included interviewer-based verbal probes such as confidence judgments.

Concerning the use of targeted probing, they concluded that (1) to the extent that verbal probes are used, these should generally be asked immediately subsequent to the point at which the subject completes the task, as relevant information will be available in STM; and (2) use of delayed retrospective probing (i.e., after the interview has been conducted) can be effective in reproducing the relevant cognitions only if clear and specific probes are used, and if the probing is done immediately subsequent to the task (in their words, "the last trial": Ericsson and Simon, 1980, p. 243). At least in the original formulation (upon which survey-oriented cognitive interviewing procedures were based), Ericsson and Simon therefore advocated both timeliness and specificity of verbal probing—not solely the unguided concurrent think-aloud. Consistent with this, concerning the issue of probe timing, Shaft (1997) reported in a study of computer code comprehension that probe questions that were delayed for extended durations prior to administration were relatively ineffective in studying strategy use. Further, within the survey pretesting domain, Foddy (1998) evaluated probe specificity and reported that specific retrospective verbal probe questions were more effective than general questions.

In reaction to research findings since 1984 indicating that some types of probe questions cause nontrivial reactivity, Ericsson and Simon (1993) did later restrict their view of probing, mainly favoring an "ideal" procedure incorporating strict use of think-aloud. For example, Ericsson (2002) has argued that instructions to eyewitnesses involving explicit direction in precisely how to report nonverbal stimuli such as facial features were responsible for the *verbal overshadowing effect* described by Schooler and Engstler-Schooler (1990), in which extensive eyewitness verbalization of suspect characteristics was found to induce reactivity by reducing the accuracy of subsequent suspect recognition. Ericsson and Simon did not at any point conclude that probing is by nature biasing, however. What matters, rather, is how probing is done and whether it invokes level 3 verbalizations that contaminate cognitive processes. Interestingly, they noted that even within studies in which level 3 verbalizations were induced by the use of probing techniques they disfavored, the magnitude of the effects of such probing were generally small, and in their own words, "many of these deviations do not lead to reliable differences and their effects must be small compared to most other factors manipulated experimentally" (1993, p. 2xi).

Finally, Ericsson (2002) has recently reiterated that with respect to reporting on a single event, "the least reactive method to assess participants' memory of the experience is to instruct them to give a retrospective report"—again, a form of probing. Of particular significance to the cognitive interviewing field, Ericsson and Simon (1993) also observed that where probes were found to affect performance in studies reviewed, this was almost always in the direction of improvement relative to a silent control. Consistent with this view, Willis (1994, 1999) concluded that although the use of probing may enhance subject performance in answering survey questions and therefore lead to an underestimate of task difficulty within the field survey environment, the results of cognitive testing are generally characterized by a litany of problems. That is, we have no difficulty detecting problems with survey questions in the cognitive lab, and it is possible that artifacts due to the use of probing do not pose severe difficulties.

When should think-aloud and probing be used? Acknowledging that our current procedures have drifted from pure think-aloud, should practitioners retrench and eliminate their use of targeted verbal probing? Such an extreme step does not seem to be indicated. Again, recognizing the differences between the task conditions posed by survey questions and those induced by the classic Ericsson–Simon tasks, it may be that probing is much more effective for addressing some subtasks but that think-aloud is preferable for studying others. In particular, unguided forms of think-aloud seem especially well suited for subtasks that involve strong problem-solving features (e.g., retrieval strategy). Such features are well represented across a range of common survey questions (see Bickart and Felcher, 1996, for a review), and this supports the continued use of think-aloud methods.

Targeted verbal probing, on the other hand, may be more effective for the study of issues such as level of subject background knowledge (Shaft, 1997), and especially question term comprehension, which may not produce an easily verbalized representation within STM, does not therefore lead to the spontaneous reporting of useful information, and in turn requires more active forms of investigation. As a concrete example: In cognitive testing of a question involving the term *dental sealant* (a protective plastic coating placed on children's teeth), it is conceivable that an extensive think-aloud fails to reveal what the respondent assumes (incorrectly) a dental sealant to be. Rather, it is only through subsequent directed probing ("What, to you, is a dental sealant?") that comprehension of the term is externalized. Of course, such a result might have been produced via classic think-aloud ("Let's see—I didn't have any cavities, so I must not have gotten any kind of dental sealant"). It is not the case that think-aloud *cannot* be applied to study question comprehension, but rather, that this may be insufficient in the absence of further probing which *investigates that which is assumed but left unstated* (see Kuusela and Paul, 2000, for a discussion of the general possibility of omission in think-aloud protocols). I propose that the choice between think-aloud and verbal probing techniques is not best viewed as an either–or argument, and consistent with this view, current practitioners of cognitive interviewing have come to utilize both pure think-aloud and targeted verbal probing. What remains unresolved is the specification of the *conditions under which each is most useful*, especially from a theory-driven perspective.

To this end, we might consider more specifically the conditions under which think-aloud and probing procedures are likely to induce reactivity, and the seriousness of the resultant contamination. Consider the laboratory subject asked to answer a question (e.g., "Would you say that your health in general is excellent...poor?") Concurrent think-aloud could possibly contaminate ongoing cognitive processing of the question, due to imposition of the requirement to externalize thought processes. Therefore, as Ericsson (2002) points out, verbal reports given *after* the respondent answers the question are the least reactive; immediate retrospective probing cannot interfere with a response that it follows. However, a potential disadvantage is that the immediate retrospective report, even if not a contaminant of the processing of the tested question itself, might produce a biased self-report of the mental processes involved in answering that question. As such, we must distinguish between reactivity associated with *answering the question* from that associated with *reflection upon that question*, and decide which type of reactivity we are striving to avoid.

Only by sorting out issues related to our testing objectives will we be in a position to consider meaningfully the think-aloud versus probing conundrum. Further, to the extent that a fundamental goal of pretesting is to evaluate the usefulness of a survey question in potentially providing accurate information, and not solely to determine the subject's thoughts when answering it, the issue of reactivity takes on a different light. Some forms of reactivity (e.g., when probing induces the subject to correct a response that was based on a now-resolved misinterpretation) may be viewed as a desirable outcome; in this case, assessing the accuracy of survey performance is unrelated to the theoretical issue of reactivity effects due merely to verbalizing one's thoughts (K. Ericsson, personal communication).

Further, an even broader viewpoint—that viewing the questionnaire as a whole rather than the individual item—provides an additional perspective on reactivity. If we accept that reactivity may affect cognitive processing not only of the *current* target survey question, but as well, *later* questions, we must also consider the cumulative effects of verbal reporting over the course of a cognitive interview. Most seriously, a series of immediate retrospective verbal reports, as prompted by interviewer probes, may modify the respondent's behavior cumulatively, especially for a series of similar survey questions (e.g., leading to more effortful processing due to the subject's recognition that his or her answers are likely to be followed by interviewer probes that demand an explanation for those answers). Such an outcome has been observed in the problem-solving literature (summarized by Austin and Delaney, 1998), in which it is found that instructing novices presented the Tower of Hanoi task to verbalize reasons for their moves improves performance. Apart from reverting to think-aloud, this tendency could be avoided by engaging in delayed retrospective probing (after the interview), accepting that we have now lost access to STM traces that may be vital for veridical verbal reporting. Or it could be the case that the natural shifting between topics occurring within a survey questionnaire negates reactivity (i.e., the intensive immediate retrospective probing of 50 diet items may produce more cumulative reactivity effects than does probing of five items on each of 10 different survey topics).

The costs and benefits of these various approaches would seem to be an empirical issue (Ericsson, 2002). A logical empirical test of reactivity might be a comparison between the data distributions of answers to tested survey items produced by subjects in three independent groups: (1) one engaging in think-aloud, (2) one subjected to immediate retrospective verbal probing, and (3) one engaged in an unprobed control (i.e, normal survey) situation but with post-questionnaire retrospective probing. A preliminary study involving dietary reporting by 10 subjects (Davis and DeMaio, 1993) revealed no strong effect of the use of think-aloud versus silent control; an extension of this research would include both more subjects and the full design specified above. Even then, such a study goes only partway toward determining whether varieties of probing differ in effectiveness; for example, Conrad and Blair (2001) have advocated a hybrid form of probing (*conditional probing*) that is contingent on a set of defined behaviors exhibited through prior think-aloud. Such procedures may decrease variance in interviewer behavior; an open question is whether this increases the effectiveness of probing for problem detection. The conduct of further experimentation within this realm should contribute not only to the practice of cognitive interviewing but also to ongoing evaluation of the Ericsson–Simon perspective on the utility of verbal reports as applied to the specific task of answering survey questions.

2.3 IN FURTHER SEARCH OF THEORY: TASK ANALYSIS OF THE SURVEY RESPONSE PROCESS

I next consider a separate trend in the development of cognitive theory, to determine whether this provides a more explicit theoretical basis for evaluating and tailoring our cognitive interviewing techniques. Although the term is novel, I refer to this as *task analysis theory*. I have throughout this chapter relied on Tourangeau's (1984) four-stage task analytic model for descriptive purposes; that is, by parsing the survey response process into its generally accepted steps. However, in original form this model is not in itself a cognitive theory, but rather, a dissection of the response process, without explicit attention to the identification of underlying cognitive structures or hypothetical constructs, or to the development of a general model that applies beyond the domain of survey response. In this section I consider the extent to which this model has been further developed into a theory that serves to drive cognitive interviewing practices.

2.3.1 History of the Task Analysis Approach

Although the precise history of the task analysis model is somewhat unclear, and the various developments have little direct relationship to one another, I present this in Table 2.3 as a loose chronology. Interestingly, the rudiments of a task analysis cognitive model as applied to the survey response process existed well prior to the heyday of cognitive psychology, within Lansing et al.'s (1961) consumer finance survey work at the University of Illinois. Lansing et al. applied the terms *theory* and *memory error* predominantly and included the equivalent of comprehension and retrieval processes, although consistent with the times, they did not cite the term *cognition*. This work has unfortunately been forgotten with passage of the years.[3] Considerably more influential has been the 1984 Tourangeau model of the survey response process; this has been cited very widely and serves as the inspiration for several extensions that attempt to propose an elaborated system of cognitive processes that may serve as the theoretical underpinning of cognitive interviewing practice (see Table 2.3).

The common feature of these elaborated approaches which serves to differentiate them from the Ericsson and Simon approach—apart from their limited

[3]Of particular interest is the early, novel use of procedures such as the conduct of intensive interviews in a psychological laboratory, and the *postmortem interview*, which would today be referred to as *respondent debriefing*.

Table 2.3 Task Analysis-Based Models of the Survey Response Process

Lansing et al. (1961)
Motivational factors
Failure of communication (i.e., comprehension/communication)
Inaccessibility of the information to the respondent (i.e., retrieval)
Cannell et al. (1981)
Comprehension
Decision/retrieval/organization
Response evaluation
Response output
Martin (1983)
Imputing meaning to the question
Searching for relevant information
Formulating a judgment
Tourangeau (1984)
Comprehension
Retrieval
Judgment
Response
Since 1984, various extensions/modifications of the Tourangeau model: Beatty (undated); Conrad and Blair (1996); Jobe and Herrmann (1996); Sudman et al. (1996, p. 58)

"boutique" application to the specific domain of survey question answering as opposed to cognition in general—is the focus on *processing stage* as opposed to cognitive *structure*. Whereas Ericsson–Simon emphasized the hypothesized structural location (LTM versus STM) in which memories are stored during processing of the task, task analysis approaches are concerned with the processing of information at various stages of completion of the survey response (i.e., comprehension, retrieval, decision, response).

2.3.2 Ramifications of Adopting Task Analysis Theory

Does the task analysis approach, as an alternative to Ericsson–Simon, provide theoretical guidance in the selection or evaluation of cognitive interviewing procedures? To the extent that distinguishable cognitive processes have been emphasized (e.g., comprehension, retrieval, etc.), this does focus the cognitive interviewer toward procedures that target those processes directly. Given that cognitive probe questions can be scripted that apply to each process (there are comprehension probes, recall probes, and so on—see Willis, 1994, 1999), this explains why cognitive interviewers have migrated toward the use of targeted probes and away from purer forms of the think-aloud procedure. Beyond this, however, the variants listed in Table 2.3 differ primarily in detail (Jobe and Herrmann, 1996); none have been developed into a theory that invokes substantially more sophistication than that implied by the Tourangeau four-stage model.

Thus, it is difficult to label any of these as a true "theory" of cognition, akin to Ericsson and Simon (1980).

Does the task analysis model help us do cognitive interviewing? Perhaps as a result of limited theoretical development, the guidance provided by task analysis theory is very general. Despite its breadth with respect to range of cognitive processes, four-stage or other task-analytic models are intended as a general model of the survey response process (where they have been used very successful), as opposed to a theory of cognitive interviewing per se. Therefore, beyond providing a compendium of component cognitive processes, this theory provides limited guidance in exactly *how* to induce verbal reporting given particular varieties of question content or structure. As such, reminiscent of earlier cited comments by Conrad and Blair (1996) concerning paucity of appropriate theory, the model is mainly a skeletal framework that only begins to incorporate cognition—note especially that the early Lansing et al. (1961) model reflected a task analysis approach that was devoid of cognitive theory (as well as nomenclature), but instead, represents a commonsense approach to parsing the survey response process. Hence, the term *cognitive probe technique* could be viewed as technical jargon that embodies only a peripheral connection to theory, and Royston (1989) may be vindicated in her view that heavy application of verbal probing be described not as a *cognitive interview* but simply as an *intensive interview*.

Consistent with this view, Gerber and Wellens (1997) have pointed out that the cognitive probes advocated by Willis (1994) emanating from the four-stage model are suspiciously similar to those developed many years prior to the cognitive revolution in psychology by Cantril (1944). Further, the most detailed manuals to date describing cognitive interviewing techniques (Willis, 1994, 1999) present cognitive interviewing more as an acquired skill, akin to clinical psychological counseling, than as the application of techniques that are informed by a well-armed theory. Finally, Forsyth and Lessler (1991) described a variety of cognitive techniques, such as card sorting and use of vignettes, that extend beyond the usual complement of probing behaviors. However, they noted that current theory again provides virtually no guidance in the use of these techniques.

Overall, it is difficult to make a strong claim that the task analysis model became the wellspring on which our adaptation of verbal probes was based because it represented a breakthrough in the application of cognitive theory. Rather, it simply appeared to have provided a reasonable and practical framework for the application of a set of activities ("verbal probes") that had the appeal of being logical and straightforward, and that passed the commonsense test.

How can the task analysis model be developed further? Several authors have made the point that the task analysis model should be considered the inception of our cognitive theoretical focus, but that further progress will require significant development. Forsyth and Lessler (1991) relied on Ericsson and Simon (1980) as a point of departure and proposed several hypotheses concerning how the cognitive features presented by specified types of question-answering tasks may influence the varieties of interviewing techniques that should be applied by practitioners. In particular, they divided the relevant conceptual domain into dimensions related to task timing (concurrent versus delayed tasks) and degree of attention control (unrestricted, such as through pure think-aloud, as opposed to directed, through targeted probing). Generally, they argued that attention to these dimensions would address the otherwise superficial, ad hoc application of our techniques.

Similarly, over the past decade, several authors have emphasized the need for increased development of existing cognitive theory as a guide to practice and evaluation (Blair and Presser, 1993; Conrad et al., 2000; Tourangeau et al., 2000). To facilitate movement in this direction, Willis et al. (1999a) suggested that cognitive interviewers routinely describe and classify (1) their precise techniques, (2) the nature of the cognitive problems they observe, and (3) the nature of the question features that are associated with these problems (e.g., the extent to which retrieval problems are produced by questions having reference periods of varying types). Such a database of results may give rise to hypotheses that inform a coherent theory.

2.4 CONCLUSIONS

To the extent that we view cognitive interviewing as truly cognitive in orientation, it may be vital to develop a strong theory that provides precise guidance. I have examined two theoretical orientations that apply to cognitive interviewing and conclude that both are in current form somewhat limited in directing these practices. To rectify this situation, I recommend the following.

1. *Theoretical development should be assisted by cognitive interviewers.* Theory and practice can develop in iterative form. For example, if it is found repeatedly that comprehension problems are best diagnosed by the application of explicit comprehension-oriented probes that spur discussion of a respondent's interpretation of key terms, this would suggest that comprehension is a robust and stable process, not unduly sensitive to the evils of retrospection or to distortion induced through imposition of level 3 verbalization. As such, a key practical issue may not be so much as "What is the subject (now) thinking?" as implicit in the Ericsson–Simon view, but rather, "What does the person think?" from a more general, long-term perspective (see Groves, 1996). That is, some cognitions could be fragile and transient, exist in STM only temporarily, and must be captured through means that are sensitive to this. Others may be enduring, persistent, and best assessed through extensive probing techniques (such as those used by Belson, 1981) with relatively less concern about STM trace duration or probing-induced reactivity. General arguments about whether verbal probing methods are leading, biasing, or reactive may therefore be misguided, as this could be the case for some survey-relevant cognitive processes, or some question varieties, but not for others.

2. *Applications of theory from other disciplines are promising but will also require further development.* Our focus on "cognitive" theory may be too limiting (Gerber, 1999). A distinctly different theoretical approach to cognitive interviewing—the ethnographic approach—derives from the field of anthropology and focuses heavily on cultural variables. Similarly, the CASM movement has made little use of sociological or psychosociological theory. Schaeffer and Maynard (1996) have emphasized the interactional nature of the survey response process (i.e., between interviewer and respondent), and challenge the notion, inherent in our adoption of Ericsson–Simon, that survey-question cognition is individualistic in nature. Finally, the evaluation of survey questions implicitly involves an infusion of psycholinguistics, and Sudman et al. (1996) have made explicit use of linguistic theory (e.g., application of Gricean norms of communication) in influencing questionnaire design. Similar developments may serve to drive cognitive interviewing practice. However, theories that are useful for this purpose will probably not be established simply by recognizing that related disciplines provide a unique perspective. Rather, I advocate a review of the relevant literature related to applicable anthropological, social–psychological, or linguistic theory, followed by specific development in application to cognitive interviewing techniques.

3. *A caveat: We should apply theory only to the extent that this proves to be useful.* In a paper entitled "Are We Overloading Memory?" Underwood (1972) castigated researchers for applying too heavy a load of theoretical constructs to human memory, that is, creating a multitude of hypothesized cognitive operations and constructs that create complexity and confusion, as opposed to clarity. Similarly, we must be vigilant in avoiding the application of theory where this does not lead to clear enhancements of our pretesting activities. A counterargument to the thrust of this chapter is that Tourangeau's task analysis model has been extremely useful, and in fact provides as much structure as we need. We might simply work within this framework, and especially given the real-world complexities of questionnaire design, the multitude of trade-offs to be considered, and logistical and cost considerations such as interviewer and subject burden, additional theorizing may be a distinctly secondary concern. Further, practitioners might arguably be best off when guided by intuition, experience, flexibility, and subjectivity, and it is conceivable that further applications of theory result only in heavy-handed attempts to wrap a body of otherwise useful procedures with an illusory sheen of theoretical sophistication.

That said, I do not view such a perspective as optimal. Rather, we should anticipate that attention to theory will serve to provide insights suggesting that various components of the "cognitive toolbox"—think-aloud, probing, free-sort procedures, retrospective debriefing, vignettes, and so on—may be more or less useful under specifiable conditions that are not best determined simply through a trial-and-error process or researcher preference. Our current challenge is to further develop the core set of theories that will guide us in making these distinctions. Hopefully, this chapter will both spur practitioners in the direction of developing theory, as well as indicating the types of directions that appear to be indicated.

ACKNOWLEDGMENTS

The author thanks K. Anders Ericsson, Roger Tourangeau, and Betsy Martin for their very helpful suggestions.

CHAPTER 3

The Dynamics of Cognitive Interviewing

Paul Beatty
National Center for Health Statistics

3.1 INTRODUCTION

Over the last two decades, cognitive interviewing has become firmly established as a tool for evaluating survey questionnaires. The development of this methodology was one of several noteworthy consequences of groundbreaking discussions between survey researchers and cognitive psychologists in the mid-1980s aimed at understanding and reducing errors deriving from survey questions (Jabine et al., 1984). The first laboratory dedicated to this practice appeared at the National Center for Health Statistics (NCHS), with others following shortly thereafter at the U.S. Bureau of the Census, U.S. Bureau of Labor Statistics, and ultimately other academic and commercial research organizations (Sirken and Schechter, 1999).

The fact that cognitive interviewing is widely practiced is indisputable, but it is not always completely clear what that practice entails. As a working definition, I will suggest that cognitive interviewing is the practice of administering a survey questionnaire while collecting additional verbal information about the survey responses; this additional information is used to evaluate the quality of the response or to help determine whether the question is generating the sort of information that its author intends. Most would probably agree that cognitive interviewing includes either think-alouds (in which a research participant is asked to verbalize thoughts while answering questions), or interviewer probes (in which an interviewer asks follow-up questions about how the participant interpreted or

Methods for Testing and Evaluating Survey Questionnaires, Edited by Stanley Presser, Jennifer M. Rothgeb, Mick P. Couper, Judith T. Lessler, Elizabeth Martin, Jean Martin, and Eleanor Singer
ISBN 0-471-45841-4

answered questions), or both (Willis et al., 1999a). Several notable overviews have outlined the general parameters of the method. Forsyth and Lessler (1991) created a taxonomy of laboratory methods in use at that time, putting a stronger emphasis on think-alouds than on probing. Willis's (1994) overview of practices at NCHS reversed that emphasis and provided a number of examples of probes used to identify questionnaire flaws. DeMaio and Rothgeb's (1996) overview of methods at the U.S. Bureau of the Census struck a balance between the two, while extending the methods to field settings and post-questionnaire debriefings.

All of these overviews are useful, and each probably contributed to the proliferation of cognitive interviewing across institutions. Yet all were necessarily limited in their specificity. Within the general parameters established by these overviews, there is considerable room for interpretation regarding how to do these interviews, when and how to probe, what to say, what not to say, and how often to say it, among other particulars. Furthermore, cognitive interviewing became established in different institutions somewhat independently, with actual practices reflecting the backgrounds and preferences of the researchers in each (Beatty, 1997). Beyond the basic parameters, there does not appear to be any universally accepted standard of how cognitive interviews should be conducted. Actual practices may vary across laboratories as well as interviewers.

There are good reasons to look at actual cognitive interviewing practice in more depth. One is to foster discussions about "best practices." Discussions about what cognitive interviewers *should* do will benefit from an understanding of what they have actually done in practice, and whether there is a relationship between interviewer behavior and conclusions that are drawn. Another is to foster continuing methodological research on pretesting methods. Work started by Presser and Blair (1994) comparing the benefits of various methods should be continued but will require an up-to-date understanding of current practices. Also, questionnaire designers who rely on cognitive interviews would benefit from a greater understanding of the relationship between what interviewers do and what they conclude about question quality.

The primary goal of this chapter is to explore what actually happened in one set of cognitive interviews: what the interviewers actually did, how the content of the interview affected what they did (and how their behavior affected the subsequent content of the interview), and ultimately, how the dynamics of the cognitive interview affected the conclusions that were drawn. The investigation also opens up questions about what specific probing strategies cognitive interviewers employ, and whether they do anything other than probe per se. It should be acknowledged that the investigation is limited to one cognitive laboratory (and indeed, is based primarily on one cognitive interviewing project evaluating one particular type of survey question). Rather than generalizing to all cognitive interviews conducted on questionnaires anywhere, the objective here is to explore the content and dynamics of one interviewing project in depth. It is hoped that these findings will lead others to investigate the dynamics of cognitive interviewing practices in other laboratories.

3.2 BACKGROUND

This methodological investigation was prompted by a fairly typical questionnaire evaluation project at the National Center for Health Statistics (NCHS). The staff of the Questionnaire Design Research Laboratory (QDRL) had been asked to conduct cognitive interviews to test modules from the Behavioral Risk Factor Surveillance System (BRFSS). The BRFSS is a health-monitoring survey of noninstitutionalized adults, conducted by telephone at the state level on a continuous basis. A subset of the BRFSS questionnaire focused on health-related quality of life (QoL). Each question asked the respondent to indicate the number of days in the past 30 days that a particular health judgment applied to them (see Table 3.1).

As is the case for many surveys designed to track change over time, the authors of the QoL items had a strong interest in maintaining question wordings that had already been fielded. However, they felt that cognitive testing could be useful to improve their understanding of what the QoL questions were actually measuring, which presumably could strengthen the analytic uses of their data. At the same time, they did not completely rule out modifying the questions based on cognitive testing results.

3.2.1 Interviewing Methods

The testing methodology was typical of NCHS cognitive interviewing projects in recent years. Interviews were conducted with a total of 18 participants. Twelve of these participants were recruited from newspaper advertisements which offered $30 cash to participate in a one-hour interview on health topics. An additional six were recruited from a local senior citizens' center, because BRFSS staff had a particular interest in the experiences of older people. The total pool of participants was evenly divided among males and females. Ages ranged from 21 to 94, with a median age of 61. Education levels of the participants ranged from fourth grade to college graduates, with a median level of twelfth grade. Interviews were conducted individually by four NCHS staff members, all of whom had performed cognitive interviews on numerous questionnaire design projects. Interviewer experience with cognitive interviews ranged from 18 months to over six years. The interviews were conducted either at NCHS or at the senior citizens' center.

Per the usual practice at NCHS, the project leader provided interviewers with some suggested scripted probes, which were written based on a review of the questionnaire and evaluation of some issues that might be fruitfully explored. For example, some commonly suggested probes included: "How did you decide on that number of days?" "Describe for me the illnesses and injuries that you included in your answer." "Did you have any difficulty deciding whether days were 'good' or 'not good'?" Interviewers were instructed to rely on such probes as guidelines but were not required to use all of them, to use them exactly as written, or to be limited by them. Rather, interviewers were to probe based on the content of the interview and initial participant responses, using their judgment to determine the most useful lines of inquiry. Interviewers were also encouraged to add general unscripted probes about the meaning of responses or specific

Table 3.1 Quality of Life Questions from the BRFSS

(The "30-day" questions are numbers 2 to 4 and 10 to 14.)

1. Would you say that in general your health is excellent, very good, good, fair, or poor?
2. Now thinking about your physical health, which includes physical illness and injury, for how many days during the past 30 days was your physical health not good?
3. Now thinking about your mental health, which includes stress, depression, and problems with emotions, for how many days during the past 30 days was your mental health not good?
4. During the past 30 days, for about how many days did poor physical or mental health keep you from doing your usual activities, such as self-care, work, or recreation?
5. Are you limited in any way in any activities because of any impairment or health problem?
6. What is the major impairment or health problem that limits your activities?
7. For how long have your activities been limited because of your major impairment or health problem?
8. Because of any impairment or health problem, do you need the help of other persons with your *personal care* needs, such as eating, bathing, dressing, or getting around the house?
9. Because of any impairment or health problem, do you need the help of other persons in handling your *routine* needs, such as everyday household chores, doing necessary business, shopping, or getting around for other purposes?
10. During the past 30 days, for about how many days did pain make it hard for you to do your usual activities, such as self-care, work, or recreation?
11. During the past 30 days, for about how many days have you felt sad, blue, or depressed?
12. During the past 30 days, for about how many days have you felt worried, tense, or anxious?
13. During the past 30 days, for about how many days have you felt you did not get enough rest or sleep?
14. During the past 30 days, for about how many days have you felt very healthy and full of energy?

probes to follow up on issues that emerged during discussions with participants. In short, interviewers were given a sort of common denominator of issues to explore, but were allowed considerable discretion. Probes were administered concurrently (i.e., immediately following individual survey questions, as opposed to retrospectively, at the end of the questionnaire).[1]

[1]It is worth noting that other researchers, such as Forsyth and Lessler (1991), have used the term *concurrent* to refer to verbal reports during the question-answering process, and *retrospective* to refer to verbal reports anytime after the question has been answered.

One of the major conclusions of the cognitive interviewing was that it is difficult to respond to the QoL questions in the format requested (i.e., a number between zero and 30 days). Although some participants provided such responses in a straightforward manner, many others did not. Sometimes numeric responses were obtained only after extensive probing. Other participants never provided numeric responses to some of the questions, either rejecting the premise of the question ("I can't put it in days") or by responding to requests for a quantitative response with additional narrative information. Also, some responses that seemed straightforward became less clear after probing. For example, one participant claimed that she experienced pain "like every day," but follow-up probing revealed that actually "in the last couple of weeks I haven't noticed it as much." It was clear that participants were able to describe each aspect of health mentioned in the questions, but providing quantitative responses proved to be difficult. Furthermore, cognitive interviewers working on this project reached a clear consensus on this point. [For additional background on this cognitive testing, see Beatty and Schechter (1998).]

Although BRFSS researchers admitted that these questions posed a difficult response task, they expressed surprise at the reported magnitude of the problem. One reason for this disbelief was that the questions had been fielded successfully, reportedly with low levels of item nonresponse. The questions also appeared to have considerable analytic utility (Centers for Disease Control and Prevention, 1995). As an alternative, they asked whether it was possible that cognitive interviewing created the *appearance* of questionnaire problems—after all, cognitive interviewers tend to encourage narrative discussions, and interviewers are permitted to depart from standardized interviewing techniques. Could it be that cognitive interviewing found "problems" that would not exist under actual survey conditions? They also asked whether any evidence was available about the magnitude of these alleged questionnaire problems.

Although the staff members who conducted these interviews were confident in their methodology and findings, all had to admit that the evidence was far from systematic. Furthermore, the possibility that the conversational tone of the interaction discouraged participants from providing codable responses could not be dismissed out of hand. This challenge to defend cognitive interviewing methodology launched the analyses described in the remainder of this chapter.

3.3 DOES COGNITIVE INTERVIEWER BEHAVIOR AFFECT PARTICIPANT REPORTS AND CONCLUSIONS DRAWN?

In evaluation of the BRFSS module, NCHS cognitive interviewers observed that some participants failed to provide codable responses to the QoL items, even after repeated prompts. Based on qualitative analysis of the interviews, NCHS staff members concluded that the metric *days* was often the problem—participants could not express attributes of their health in these terms. These findings stood in contrast with the fact that the questions had been fielded successfully.

There were several possible explanations for this discrepancy. First, it is possible that some respondents in the field actually failed to answer the questions but that interviewers "glossed over" response imprecision. For example, imprecise answers ("it might have been the whole time") could have been accepted and recorded quantitatively ("30 days"). Such behavior goes against recommended practices of standardized interviewing (e.g., Fowler and Mangione, 1990; Guenzel et al., 1983) because it is based on interviewer interpretation rather than clear responses—however, such behavior would not be apparent in statistical tabulations. A second possibility is that interviewers actually succeeded in eliciting clear responses, but doing so required an inordinate amount of probing. This could still point to inadequacies of the questions, as research has suggested that interviewer errors increase when questions require extensive probing (Mangione et al., 1992). A third possibility is that problems answering these questions were unique to cognitive interviews. If true, this could indicate that the questions worked well but that the evaluation methodology of cognitive interviewing was flawed.

In subsequent monitoring of actual BRFSS telephone interviews, NCHS staff noted few instances of interviewer error, but did observe several instances in which interviewers needed to probe extensively before obtaining responses to these questions. This suggested that the second possibility discussed above was plausible. Unfortunately, resources were unavailable to record and behavior code the interviews, which would allow systematic analysis of interviewer and respondent behavior while answering (Fowler and Cannell, 1996). However, the cognitive interviews themselves had been recorded, allowing for more detailed analysis of whether there appeared to be a relationship between interviewer probing and the quality of responses (initially explored in Beatty et al., 1996). The cognitive interviews were therefore transcribed, and two NCHS staff members developed a coding plan to (1) identify the actual response to each survey question, (2) rate how closely this response matched the *expected* format of response (called *precision*), and (3) characterize the nature of probing that preceded each response.

3.3.1 Coding the Response to the Question

Identifying the actual response to a question in a cognitive interview is not always straightforward. In a survey interview, the interviewer moves to the next question as soon as the answer has been provided. In a cognitive interview, the response to the question may be both preceded and followed by discussion. In some cases participants may change their answers, and at times they may not provide a clear answer at all. In this study, if the participant gave one clear response to the question, it was clear that this response should be accepted as "the answer." But if multiple responses were given, the "best" response was defined as the answer that adhered most closely to the response format desired. That is, precise figures (such as "three days") were considered better than closed ranges ("three to five days"), which were considered better than open-ended ones ("at least three days"). If the participant gave multiple responses of equal quality, the first one given was recorded. Sometimes no acceptable responses were given, such as when participants rejected the premise of the question ("I can't tell you in days") or the best answer given was so vague that no quantitative response could reasonably be inferred ("I have these minor problems, but it's more of an inconvenience"; "I think it's most of the time"). Such instances were coded as nonquantifiable.

3.3.2 Coding the Precision of the Response

The precision of the best answer was then evaluated according to how closely it conformed to the response format. As noted earlier, the NCHS interviewers had suggested to BRFSS staff that imprecise answers might indicate a questionnaire problem. (Note that *precise* answers are not necessarily accurate or free of cognitive problems. However, unwillingness or inability to respond within the provided parameters of the question is one possible indication of trouble.) One objective of this analysis was to qualify the actual level of imprecision in responses. Imprecision was coded on a four-point scale (summarized in Table 3.2).

3.3.3 Coding the Probing That Preceded the Response

Another coding system was designed to record interviewer probes that appeared before the best response to the question. Because the study focused on how interviewer behavior could affect participant responses, probes that were administered following this best response were ignored.

Table 3.2 Codes for Level of Response Precision

Code 1: The answer required virtually no rounding, judgment, or interpretation from the coder. Examples:
Precise quantities: ("30 days"; "4 days")
Certain colloquialisms: "Every day" was accepted as 30; "never" was accepted as zero.

Code 2: The answer required minimal interpretation from the coder. Examples:
Moderately qualified responses: ("Probably every day"; "I think a day or so").
Narrow range of days: ("Six or eight days")—the midpoint was recorded as the response.
Fractions: ("Half of the days")—the response was recorded as 15 days.
Some imprecision in responding where the meaning could reasonably be inferred: (e.g., "I have no problems" after a series of "zero" answers).

Code 3: The answer required considerable interpretation from the coder.
Ranges of more than three days: ("16 to 20 days")—the midpoint was recorded as the response.
Anchored but qualified responses: ("More than 15 days")—the anchor point was accepted as the response.

Code 4: The answer could not be coded: ("I can't put the answer in days"; "For a while I was in horrible pain").

A distinction was made between two major categories of probes. Cognitive interviews generally include probes that are designed to get information beyond the specific answers to the survey questions. I refer to these as *elaborating probes*. Examples include "Tell me what you were thinking about while answering" or "How would you describe your health in the last 30 days?" All of the probes suggested in the interviewing protocol were of this type. In contrast, a *reorienting probe* asks a participant to refocus on the task of providing an answer to the question within the response format provided. For example, "So how many days would that be?" or "Which of the answers you gave is closer?" are both reorienting probes. These are similar to probes that survey interviewers often employ to obtain codable responses. Cognitive interviewers at NCHS were not specifically trained or instructed to use these probes, but they nevertheless used them frequently—as was necessary to keep participants focused on the specific questions being evaluated.

The interviewer probing that was administered prior to the "best answer" was coded using the following categories:

Code N: No probes appeared before the response.
Code R: One or more reorienting probes were used, but no elaborating probes.
Code E: One or more elaborating probes were used, but no reorienting probes.
Code B: Both: at least one of each type of probe (reorienting and elaborating) was used.

For example, consider the following interview excerpt:

Interviewer: Now thinking about your mental health, which includes stress, depression, and problems with emotions, for how many days during the past 30 days was your mental health not good?

Participant: Not good? My mental health has always been good.

Interviewer: So would you say zero?

Participant: No, not necessarily, because I do get stressful at times.

Interviewer: What kind of stress?

Participant: Well, the number one thing is I'm very impatient ... [discusses this for awhile].

Interviewer: So if you had to come up with a number of days in the past 30 where your mental health was not good, which could include stress, depression, emotional problems, or anything like that, what number would you pick?

Participant: I'd pick overall three to five days.

(Nothing else resembling a response was subsequently given in the discussion.)

In this example, "three to five days" was the best response, which was coded as "four days." The precision of the response was given code 2. The probing was given code B because the interviewer used both elaborating probes ("what kind of stress?") and reorienting probes ("so if you had to come up with a number...").

I created this coding scheme in collaboration with another NCHS staff member. We used it to code responses from 17 cognitive interviews independently (one of the original 18 had not been tape recorded). At various times during the coding process, the rules needed to be expanded to account for circumstances that did not seem to fit the coding scheme as developed at that point. Primarily for that reason, reliability statistics were not calculated during the coding. Instead, after working independently, we met, identified all discrepancies, and resolved them either by referring to existing rules or by agreeing on a new coding rule. Thus, the codes used in analysis represent complete agreement between two coders, and the reconciliation was completed before any quantitative analysis was conducted.

It should be noted that I was one of the interviewers on this project (whereas the other coder was not). More generally, it would have been preferable if there were no overlap between those who conducted interviews, developed and applied the codes, and analyzed the data. However, each of these tasks was conducted independently (i.e., the initial interviews were completed before the coding scheme was developed, and the coding was completed before any analysis was conducted) in an effort to maintain the integrity of each phase.

3.3.4 Results

These data provide a measure of the imprecision present in participants' answers to the QoL questions. Overall, 36.3% of all responses qualified as "precise" (code 1) and 23.0% of responses were completely uncodable (code 4). The remaining 40.7% of responses contained some degree of imprecision, but a codable answer was deemed salvageable. Interestingly, rates of imprecision varied considerably across questions, as shown in Table 3.3. Responses to Q11 (about depression) and Q14 (about being "healthy and full of energy") were relatively imprecise compared to Q10 (about difficulties caused by pain) and Q3 (about mental health) responses. Transcripts suggested that many participants had trouble answering Q14 because they thought that being "healthy" and being "full of energy" warranted different answers. Interestingly, responses to Q3 were much more precise than responses to Q11, even though the questions both deal with mental health issues. The conceptual overlap might have confused participants about the purpose of the latter question (Q11), leading participants to qualify or explain their answers there. For Q10, there were few precision problems because most subjects simply answered zero—most did not experience such pain.

Although this analysis shows how precision varies across questions, it does not yet address the issue of how probing relates to this imprecision. Table 3.4 provides a brief summary of the distribution of all probing activity for the battery of QoL questions. Some elaborating probes were used prior to 31.1% of all responses (adding E + B). Thus, this type of probing was common and might have encouraged some digressive behavior from participants. However, some reorienting probes were used prior to 30.3% of responses (adding R + B). Although

Table 3.3 Comparison of Precision Across 30-Day Subjective Health Questions

Question: "During the past 30 days, how many days have you . . ."	Responses with:	
	No/Minor Precision Problems (Codes 1 and 2)	Major Precision Problems (Codes 3 and 4)
Q11. Felt sad, blue, or depressed	9	8
Q14. Felt healthy and full of energy	10	7
Q13. Felt that you did not get enough rest or sleep	11	6
Q4. Had poor physical or mental health that kept you from usual activities	11	5
Q2. Felt that your physical health was not good	12	5
Q12. Felt worried, tense, or anxious	13	4
Q10. Experienced pain that made it hard for you for you to do usual activities	13	4
Q3. Felt that your mental health was not good	14	3

Table 3.4 Probing: All Questions About 30-Days Subjective Health Measures

Probing	Responses
N: None	55.6%
R: Reorienting only	13.3%
E: Elaborating only	14.1%
B: Both reorienting and elaborating	17.0%
	100% ($n = 135$)

interviewers often encouraged elaboration, there were also plenty of instances in which they encouraged participants to answer the question in codable terms.

The next analysis considers how probing relates to the precision of responses. It seemed likely that precise responses would be more likely to follow reorienting probes, and imprecise responses would be more likely to follow elaborating probes. Results in Table 3.5 suggest that this is correct: The mean precision when reorienting probes alone were used is 1.78. However, when only elaborating probes were used, the mean precision is 3.68. When both types of probes

Table 3.5 Precision of Responses to All Questions, by Type of Probe Used Before Response

Precision Code	No Probing	Only Reorienting Before Response	Both Used	Only Elaborating Before Response
1	33	9	1	1
2	34	6	8	1
3	4	1	5	1
4	4	2	9	16
	$n = 75$	$n = 18$	$n = 23$	$n = 19$
Mean precision	1.72	1.78	2.96	3.68

were used, the mean precision falls in the middle, at 2.96. Note that no probing preceded 75 responses; these had a mean precision of 1.72. It seems likely that reorienting probes encouraged subjects to answer the question in the specified format, whereas elaborating probes encouraged discussion at the expense of answering within the format specified.

3.3.5 Supplemental Interviewing

An important question remains: Was this response imprecision attributable to interviewer behavior or to attributes of the question? To further explore the relationship between probing and responding, 20 supplemental interviews were conducted. This second round of interviews used all of the interviewers from the earlier round (and one additional interviewer) and the same questionnaire. However, the interviewing procedure was modified: Interviewers were trained about the differences between reorienting and elaborating probes, and instructed to use *only* reorienting probes while administering the questionnaire. Thus, interviewers were instructed to push participants to provide codable responses, in a manner similar to standard survey interviewing. Elaborating probes were to be used only in a debriefing after completing the questionnaire.

Interviews were transcribed and coded using procedures identical to those in the preceding round. An initial check of the data confirmed that interviewers did probe as instructed, successfully avoiding elaborating probes (only one elaborating probe appeared throughout all 20 interviews). Reorienting probes were used prior to 30.4% of responses (almost identical to the incidence of reorienting probing in the preceding round), and no probing was done prior to 69.0% of responses (compared to 55.6% in the previous round).

The purpose of these interviews was to see if eliminating all discussion-oriented probes also eliminated the imprecision of responses. It seemed reasonable that some imprecision would remain if the response difficulties were attributable to question characteristics rather than to the style of cognitive interviewing. Yet surprisingly, the elimination of elaborating probes also eliminated most imprecise responses. Table 3.6 compares overall precision codes for the

Table 3.6 Precision of All Responses Compared Across First and Second Interview Rounds

Precision	Round 1	Round 2
1 (precise)	36.3%	82.3%
2	32.6%	14.6%
3	8.1%	0.0%
4 (uncodable)	23.0%	3.2%
	100%	100%
	(n = 135)	(n = 158)
	$X^2 = 71.65$, df = 3, $p < 0.001$	

first and second rounds of interviewing. When interviewers attempted to get a response through reorienting probes, they were generally successful at doing so. Not only were 82.3% of responses fully precise, but almost all of the remaining response imprecision was very minor (code 2). Only 3.2% of responses were uncodable, as opposed to 23.0% of responses in the first round.

However, these results need to be tempered by results from post-questionnaire debriefings, in which many participants revealed misgivings about their answers. Several complained that "days" were inadequate to describe how often they felt unhealthy, depressed, and so on, or expressed other dissatisfaction with their answers. Some admitted that they did not give their answers much thought. Also, some admitted that they gave numerical responses because interviewers clearly expected them to do so, and that these numerical answers did not allow them to express their reservations about the accuracy of their responses.

It is also important to note that participants in the second round were recruited entirely through a general advertisement. Consequently, they were younger than participants in the first round, with a median age of 48 (as opposed to a median age of 61 in round 1). Two of 20 participants in round 2 were over 65, as opposed to 8/18 in round 1. It seems reasonable that these questions are more difficult for people with more complex health portraits, and health complications may rise with age. Thus, the participants from the two rounds are not completely comparable, and we would expect that response imprecision would be reduced in this second round. However, the almost total *absence* of response imprecision in round 2 is surprising.

Taken altogether, several conclusions seem reasonable. First, it is unlikely that the appearance of problems with the QoL questions was entirely attributable to cognitive interviewer behavior. In fact, it is more likely that standardized interviews, which are geared toward obtaining quantitative responses, actually *suppress* the appearance of problems. Participants expressed frustration with the questions verbally in both rounds of interviews, and the actual imprecision of responses was quantifiable in round 1. Furthermore, the interviews highlighted the *reason* for many response difficulties: often, health could not be broken down into discrete "days," which were either "good" or "not good." Such a classification of days was especially difficult for participants with intermittent conditions or multiple health problems.

However, there does appear to be a relationship between what interviewers do in the laboratory and what participants report. It is possible that some laboratory participants did not provide codable answers because interviewers expressed greater interest in discussion than quantitative answers per se. Accordingly, if cognitive interviewing is used to evaluate participants' ability to answer questions and the quality of their responses, it seems advisable that interviewers should try to get participants to answer survey questions in the desired response format before using elaborating probes. If they do not, it may be difficult to judge whether participants *could not* answer questions, or whether they simply *did not*.

Furthermore, in evaluating cognitive interview results, response imprecision alone might not suffice as reasonable evidence of a problem with a question. Cognitive interviews are more conversational than survey interviews by nature. However, *relative* levels of precision across questions might be a useful measure of how well questions are working—after all, questions that require more discussion and qualifications of responses are likely to be the ones that are most difficult to answer.

3.4 CATEGORIZING INTERVIEWER BEHAVIOR IN THE COGNITIVE LABORATORY

Some paradigms of cognitive interviewing assume a relatively inactive role for the interviewer: The interviewer basically trains the participant to "think out loud" and engages in little discussion beyond that, other than reminding participants to think out loud while answering the questions [see Forsyth and Lessler, (1991) and Willis (1994) for more detailed descriptions of this paradigm]. Cognitive interviewing at NCHS over the past decade has been based on a more active conceptualization of the interviewer—for example, DeMaio and Rothgeb (1996) and Willis and Schechter (1997) describe cognitive interviewing approaches that combine probing and think-alouds. Recent conceptualizations of cognitive interviewing are often centered on the idea of probing (Willis et al., 1999a).

The literature on cognitive interviewing provides numerous examples of probes that are considered useful. Yet this literature leaves unresolved many specifics regarding ideal interviewer behavior. For example, should interviewers seek specific responses to probes that address cognitive processes explicitly, or should they attempt to generate broader discussions—and if so, how should they do this? How much should interviewers probe beyond those suggested in advance as relevant to the questionnaire evaluation? Should interviewers decide when the amount of probing has been sufficient? Are there particular types of probes that should be used liberally, and others that should be avoided altogether?

We cannot answer all of these questions definitively here, but can take some first steps toward answering them by proposing a coding scheme for what cognitive interviewers actually do, and applying it to the interviews discussed earlier. As before, the goal here is not to make definitive statements about how cognitive interviewing is done across all laboratories and all projects, but rather to look in depth at interviewer behavior in one project, considering the consequences and implications of conducting interviews in that manner.

3.4.1 Methods

Transcripts from the interviews in round 1 of the previous analysis were used for this one as well. (Transcripts from the second round were not used because those interviews were atypical of standard practice at NCHS.) As before, two NCHS staff members developed the coding scheme of interviewer behavior. It was clear that the categories of probes used in the previous study (reorienting and elaborating) were incomplete—numerous interviewer utterances could not be classified as either, and useful distinctions could be made within these categories. The coding scheme for this study was therefore constructed from scratch. It consisted of five general categories:

1. *Cognitive probes*: probes used to understand interpretation of terms, computational processes, information that was considered when answering, and level of difficulty encountered while thinking of an answer were included in this category. Some examples of these probes include:
   - "What were you thinking about while answering that?"
   - "How did you come up with your answer?"
   - "What does [term] mean to you?"
   - "Did you know the answer, or did you have to estimate?"
2. *Confirmatory probes*: probes that ask subjects whether the information provided so far is correct. One form of confirmatory probe is *mirroring*—repeating back all or part of what a subject said verbatim as an implicit request for confirmation.
   - "So, for the last 30 days you were unhealthy for only one day?"
   - Participant: "My health is pretty terrible right now." Interviewer: "Pretty terrible?"
3. *Expansive probes*: probes designed to obtain additional details and narrative information. (Some of the "elaborating probes" from the previous study fell into this category, but sometimes they could also be classified in the categories above.) For example:
   - "Tell me more about your arthritis—does that make it hard for you to get around?"
   - "When did that [event] happen?"
   - "Were you sick the entire day, or only part of it?"
4. *Functional remarks*: probes that redirect the subject back to the original survey question by repeating it or clarifying some aspect of it. (These should include all reorienting probes from the previous study, as well as responses to requests for information from participants.)

- Repeat of survey question: "And how many days did you feel that way?"
- Clarifications: "Yes, I'm talking about how you felt *in the last 30 days*."

5. *Feedback*: interviewer behaviors that are neither probes nor functional remarks, including traditional survey feedback designed to reinforce desirable response behavior, and also reactions to the actual substance of subjects' answers.
   - Traditional: "Thanks, that's just the sort of information I'm looking for."
   - Conversational: "I know what you mean, I feel the same way."

Transcripts from the discussions following nine questions from 17 interviews were included in the study.[2] Each distinct utterance made by interviewers following the initial reading of the questions was coded. In some cases, interviewers made two utterances in succession; when this happened, each utterance was coded separately. As in the previous study, coders worked independently; they met afterward to identify all discrepancies, which were resolved either by existing rules or, when necessary, by creating new rules to account for utterances that did not fit in the existing coding scheme.

3.4.2 Results

The four cognitive interviewers made a total of 610 distinct utterances (not counting initial question readings) over the course of 17 interviews. Table 3.7 shows the distribution of all interviewer utterances pooled together, as well as utterances broken down across questions. The "total" column shows the distribution of all cognitive interviewer utterances across all questions. Note that cognitive probes account for a relatively modest amount of total interviewer behavior. This is particularly striking: Given its name, one might assume that the purpose of cognitive interviewing is to explicitly explore cognitive processes. Also note that

Table 3.7 Cognitive Interviewer Behavior by Question

	Q1	Q2	Q3	Q4	Q10	Q11	Q12	Q13	Q14	Total	Avg. Uses per Interview
Cognitive	13	13	6	3	4	2	3	0	10	54	3.2
Confirm	29	36	29	12	9	20	14	26	35	210	12.4
Expansive	21	26	29	14	10	11	21	40	26	198	11.6
Functional	6	13	12	4	4	5	7	5	5	61	3.6
Feedback	7	16	17	11	1	9	10	9	7	87	5.1
Total	76	104	93	44	28	47	55	80	83	610	35.9
Avg. utterances per question	4.5	6.1	5.5	2.6	1.6	2.8	3.2	4.7	4.9	35.9	

[2]The nine questions include the eight QoL questions from the previous study, plus a general question on health status: "In general, would you say your health is excellent, very good, good, fair, or poor?"

interviewers employed this type of probe more for some questions than for others. For example, these probes appear much more for Q1 (overall health status) and Q2 (days of physical health not good) than for Q11 (days sad, blue, or depressed); there was no probing of this type at all associated with Q13 (days not getting enough rest or sleep). This is true despite the fact that the interviewer guide usually suggested several probes of this type for each question (e.g., "How did you come up with that number of days?" or "What illnesses or injuries were you thinking about while answering?").

More commonly, interviewers relied on expansive probes and confirmatory probes. The average interview in this study included 11.6 expansive probes and 12.4 confirmatory probes, as opposed to 3.2 cognitive probes. Transcripts show that these probes generally emerged from the interaction with the participant. For example, when participants revealed incomplete details about some aspect of their health, interviewers often followed up with expansive probes such as "tell me more about that." Sometimes these expansive probes were more specific, asking participants to provide additional details about previous statements ("what kind of illness was that?") Other expansive probes were geared toward understanding problems providing responses in the expected answer format. For example, rather than probing "how are you figuring out your answer to that," interviewers often asked participants to explain their health in their own words. This sort of probing turned out to be useful in the analysis phase, highlighting mismatches between the concept "days of good health" and participants' circumstances.

Confirmatory probes were even more common than expansive probes. In practice, the interviewer often simply repeated some of the participants' own words. For example, this exchange followed Q1, about general health status:

Participant: I'd say "good." Not "very good," no.

Interviewer: Not "very good"?

Participant: Because there are so many things. I have a lot of medical conditions like high cholesterol, like a little asthma, allergies. Considerations and worries ....

This sort of probing often served as an implicit request for more information and had a similar effect as an expansive probe—although the fact that a confirmatory probe relies on participants' own words distinguishes it from an expansive probe in practice. Other confirmatory probes were follow-ups on specific details (e.g., "so are you saying two days?"), and these also tended to generate narrative explanations.

The remaining behaviors—functional remarks and feedback—were employed less frequently than expansive or confirmatory probes, but were still relatively common. Functional remarks were generally geared toward returning the conversation to the actual survey questions. As for feedback, some instances were clearly "task-oriented," letting participants know that the type of information they provided was consistent with interview objectives ("that's very helpful the way you told me those things.") General conversational remarks also fell into this category. Some were neutral, but others could be seen as biasing—and clearly would have been unacceptable in a survey interview ("I'm glad to hear about that"; "I have the same problem").

Note also that the volume of interviewer activity varies considerably across questions. For example, there are almost four times as many interviewer utterances following Q2 than following Q10. It appears from transcripts that the amount of probing was largely driven by the relevance of the question to the respondent. Q10 is about pain that makes it difficult to do usual activities, but few participants had experienced such pain recently. Some questions also dealt with ground that had been covered thoroughly earlier—for example, the subject matter of Q11 and Q12 partially overlapped with that of Q3. Thus, probing on the latter questions may have seemed redundant.

The behavior tabulations, supplemented by transcripts, suggest a fairly dynamic interviewing process. While the interviewers were given suggestions for probing, a substantial majority of what they actually said was unscripted. The interviewer guide suggested a total of 12 cognitive scripted probes and nine expansive scripted probes; Table 3.7 reveals that interviewers tended to use far fewer cognitive probes (3.2) and somewhat more expansive probes (11.6, although according to transcripts, not necessarily the ones that were scripted). The interviewer guide contained no suggestions regarding confirmatory probes, functional remarks, or feedback, although all were used commonly.

Transcripts also suggest that many utterances were specifically chosen by interviewers to follow up on information provided by participants. Furthermore, it seems that a substantial amount of the behavior documented here falls outside the realm of what is usually thought of as cognitive interviewer behavior. Most examples of probing provided by DeMaio and Rothgeb (1996), Forsyth and Lessler (1991), and Willis (1994), among others, are of the "cognitive" variety, with some "expansive" probes as well. Interviewers seem to have employed these other behaviors on their own initiative, and they may serve important functions within cognitive interviews, but unfortunately, guidance on the most effective and appropriate use of confirmatory probing, feedback, and functional remarks does not appear to be available.

Finally, Table 3.8 considers differences across interviewers. The figures it contains are average times that each type of utterance was used per interview by three project interviewers.[3] Overall behavior varies little across interviewers, both in overall quantity and by specific categories. A few noteworthy differences are visible: Interviewer 1 used confirmatory probing much more often than did the other interviewers, perhaps instead of expansive probing. Also, interviewer 3 was more likely than the other interviewers to use cognitive probes (although still only 5.5 times per nine questions), and more likely to use feedback. But on the whole, the similarities across interviewers are more striking than the differences—only

[3]Four interviewers participated in the project, but one of the interviewers conducted only two interviews, one of which was not tape recorded and transcribed. His one interview was dropped from this analysis.

Table 3.8 Mean Number of Each Utterance Type per Interview, by Interviewer

	Interviewer				
	1	2	3	F^a	Significance[b]
Cognitive	2.3	3.0	5.5	3.63	$p < 0.06$
Confirm	17.3	8.8	9.3	5.32	$p < 0.02$
Expansive	10.1	14.0	13.0	1.29	n.s.
Functional	3.6	3.8	3.5	0.01	n.s.
Feedback	4.0	5.0	7.5	0.86	n.s.
Utterances per interview	37.3	34.6	38.8	0.12	n.s.
Interviews	7	5	4		
Total utterances	261	173	155		

[a]Computed through one-way analysis of variance.
[b]n.s., not significant.

the use of confirmatory probes was significantly different across interviewers at the 0.05 level.

3.5 DISCUSSION AND IMPLICATIONS FOR COGNITIVE INTERVIEWING PRACTICE

Although the studies described here focus on interviews based on one questionnaire module conducted by one group of interviewers, the results reveal important gaps in our knowledge of what actually happens in cognitive interviews. Most existing descriptions focus on certain types of probes, but we have seen that interviewer behavior is potentially broader (and could be broader still when considering the activities in other laboratories). Understanding what cognitive interviewers actually do, and the impact of their behavior on the conclusions that are drawn, is the first step toward developing better standards of practice.

Such standards are important, because there is some evidence that cognitive interviewing technique can affect the conclusions that are drawn. The first analysis sheds some light on how participant behavior may follow from interviewer cues. Conversely, the second analysis shows how interviewer behavior (at least at NCHS) may follow from what *participants* say. Both analyses point to practical implications.

For example, results from the first analysis suggest that if interviewers are trying to assess respondent difficulty answering questions, they should actually try to get participants to provide a codable answer. This may seem obvious, but NCHS interviewers did not always do this—and as a result, it was unclear whether participants *could* not answer or *did* not answer. The former would point to a problem with the questionnaire, whereas the latter might simply be a product of interviewer behavior. Probing about interpretations of questions and the broader meaning of the response (i.e., elaborating probes) is also an important component of cognitive interviewing, but such probing should be conducted after assessing whether the question poses a reasonable response task. As we saw in this study, apparent problems providing a "precise" response were much more likely to follow elaborating probes than reorienting probes. Although a solid case can be made that the format of the questions was responsible for these problems, the evidence would have been less ambiguous if interviewers had consistently pursued a codable response first. Cognitive interview practice might be improved if interviewers were trained in how to recognize the difference between reorienting and elaborating probes and when to employ each most effectively.

Results from the second analysis suggest that guidelines for cognitive interviewers need to cover a wider range of behaviors than have usually been considered. While the available cognitive interviewer guidance focuses on generating insights into cognitive processes, much interviewer probing seemed more oriented toward generating more general explanations of respondent circumstances and how those circumstances fit the parameters of the questions. Presumably, such probing can be done well or poorly. It would be useful for researchers to evaluate more fully what sorts of probes provide useful information and which do not, and to provide more training to cognitive interviewers to help them probe effectively. Interviews also included feedback and conversational remarks. Some of these remarks appeared to be benign chitchat, but others were potentially biasing. Providing guidance to cognitive interviewers regarding such behavior could also be valuable. It does not appear that practitioners of cognitive interviewing have developed such guidance, but creating and disseminating standards of best practices would clearly be valuable. Such guidelines should reflect the full range of probing and other behaviors that cognitive interviewers seem to engage in (keeping in mind that interviewers from other laboratories may engage in additional behaviors that have not been considered here). The potential for interviewers to bias findings is greatest where guidelines for behaviors are not well documented and where researchers do not train interviewers to perform those behaviors adequately and consistently.

Guidelines, however, are quite different from rigid standardization of cognitive interviewer behavior. Some researchers have called for reduction in interviewer variation: For example, Tucker (1997) argues that cognitive interviewer variation limits the scientific usefulness of the method, and Conrad and Blair (1996) propose analytic strategies for cognitive interview data that would require highly consistent behavior across interviews. Both proposals suggest a model of cognitive interviewing in which the creative contributions of interviewers are minimized, not unlike survey interviews. In such models, interviewers are instruments of a measurement process. The cognitive interviews examined here point toward a different paradigm, marked by an active interviewing style, where probes are selected based on specific issues that arise in discussions. Such a model puts a much greater burden on the interviewer, who not only collects data but doubles as an investigator, making important decisions about probing in a largely unscripted manner.

The potential advantage of emergent unscripted probing—where probes emerge from the content of the interview—is that questionnaire issues not anticipated in advance can be explored in a focused manner. If interviewer behavior is constrained to follow only scripted probes, interviews can yield only data about issues that have been anticipated in advance. If interviewer behavior is further constrained against probing (i.e., limited to encouraging "thinking aloud"), the interview will only cover topics that the participant thinks are most relevant.

In contrast, an interviewer with a thorough understanding of questionnaire design problems and good listening skills can actively explore issues that emerge while responding. Observant cognitive interviewers watch for contradictions and inconsistencies in responding. Recently, the following question was tested at NCHS: "In the past 12 months, how many times have you seen or talked on the telephone about your physical, emotional, or mental health with a family doctor or general practitioner?" One participant answered "zero" easily, but other answers suggested that he might have received medical care recently. This hunch led the interviewer to ask the participant about general patterns of medical utilization, which revealed that the person had in fact been to a doctor's office less than two months earlier. The interviewer reread the question—but the participant again answered "zero." The interviewer then asked the participant to reconcile this response with his report of going to the doctor's office two months ago; the participant replied that his answer was accurate because the survey question was about *talking on the phone*. The word "seen" had been lost within the verbiage of the question. In subsequent interviews, interviewers followed up "zero" responses with specific probes about visits to the doctor and found that this mistake was commonly made.

It would have been difficult to anticipate this problem and prescript the interviewer's follow-up probing. The active follow-up by the interviewer was especially fruitful: Not only did it identify a response error but also provided insight into the particular causes of the problem. Suggestions for revisions could therefore be based on specific information that emerged from this exchange.

Nevertheless, some may argue against this type of probing because it is not based on recognizable "cognitive" principles, nor does it appear to meet criteria of "science." In response to the former issue, it may be worthwhile to consider Gerber and Wellens's (1997) assertion that cognitive interviewing as currently practiced has more to do with understanding the meanings of questions and responses than cognitive processes per se. The actual emphasis on expansive and confirmatory probing in NCHS interviews seems consistent with a methodology that is more focused on exploring meanings in a qualitative sense than "cognition" in a scientific sense. Fortunately, an extensive literature on qualitative inquiry into meanings is available, although apparently it has not yet been tapped to provide input into cognitive interviewing methodology. Some potentially useful ideas could be gleaned from Strauss and Corbin's (1990) discussions of drawing conclusions inductively from textual data, from Kirk and Miller's (1986) overview of how the concepts of reliability and validity might be applied to qualitative research, and from Holstein and Gubrium's (1995) proposal for an "active" interviewing paradigm that adapts to the circumstances of the research participant. Cognitive interviewing fills a unique methodological niche—it has qualitative elements, although it is clearly linked to the process of providing better statistical measures—so not all of these new ideas may be fully applicable. At the same time, they are certain to contain some useful guidance about how to explore meaning from a participant's perspective. It may be that progress in cognitive interviewing methodology will come not from imposing a scientific paradigm onto it but by adapting standards of rigorous qualitative inquiry.

3.6 CONCLUSIONS

The findings here are based on interviews conducted at one institution, and primarily for one interviewing project. This chapter is not intended to draw conclusions about "the way" that cognitive interviews are conducted, but hopefully it will stimulate discussion about ways that interviewer behavior could affect what participants say and what conclusions are drawn. To the extent that we understand and can give guidance about what behaviors are optimal and acceptable, we can minimize the risk of problems with this methodology. More generally, the purpose of this chapter is to show the variety of behavior that may be taking place in cognitive interviews. Some of these findings may stimulate discussion about desirable and undesirable interviewing practices and encourage other researchers to examine interviewer behavior in other laboratories. It would be extremely useful to perform similar analyses about the dynamics of cognitive interviewing in other organizations and on projects involving different survey questions.

Methodological research on cognitive interviewing needs to be based on a thorough understanding of what actually happens in cognitive interviews. For example, comparisons of how well cognitive interviews fare next to other pretesting methods should be based on actual practices; comparisons based on idealized examples that did not reflect actual practices have limited utility. Thus, it is critical to continue to study what cognitive interviewers actually do.

I have suggested that some of the most useful findings from cognitive interviews are due to adaptive, investigative skills of individual interviewers, who improvise probing based on specific interview content. This idea may not be universally accepted, but if it is borne out by time and additional research, it has definite implications regarding the selection and training of cognitive interviewers. If interviewers are to listen actively for points to follow up on and select probes accordingly, thorough training in questionnaire design and response errors associated with questionnaires will be essential. Such interviewers would be methodological experts, with interviewing skills that might evolve over a long period of time, rather than research assistants who could be quickly trained to follow interviewing protocols. Qualitative interviewing experience, along with skills at adaptive listening, might become increasingly desirable attributes of cognitive interviewers.

Discussions about the role and ideal skills of cognitive interviewers will undoubtedly continue for some time. Hopefully, such discussions will be guided by further explorations of what cognitive interviewers do in various laboratories, marked by dialogue with an expanded circle of researchers interested in developing this interviewing methodology.

CHAPTER 4

Data Quality in Cognitive Interviews: The Case of Verbal Reports

Frederick G. Conrad
University of Michigan

Johnny Blair
Abt Associates

4.1 INTRODUCTION

Cognitive interviewing has been used extensively for over 15 years to pretest questionnaires. Practitioners use cognitive interview findings to decide whether or not to revise questions in many important survey instruments and, often, how to revise them. Considering the weight given to the results of cognitive interviews, evaluations of the method have been surprisingly few. The majority of studies that have been conducted compare a single version of cognitive interviewing to other pretest methods [e.g., Fowler and Roman, 1992; Lessler et al., 1989; Presser and Blair, 1994; Rothgeb et al., 2001; see Tourangeau et al. (2000, pp. 331–332) for a review]. Although a valuable beginning, the results of such studies can apply only to the particular cognitive interviewing technique investigated.

Cognitive interviewing is actually a generic designation for a variety of loosely related procedures. Different practitioners combine these procedures in various ways to produce alternative cognitive interview techniques (Willis et al., 1999a). Because almost no research has been published that compares the various techniques, we know little more about their performance than we did when

Methods for Testing and Evaluating Survey Questionnaires, Edited by Stanley Presser, Jennifer M. Rothgeb, Mick P. Couper, Judith T. Lessler, Elizabeth Martin, Jean Martin, and Eleanor Singer
ISBN 0-471-45841-4

cognitive interviewing was introduced. Perhaps one reason there has not been much progress in this area is the lack of an overarching approach to such research. Although Willis et al. (1999a) raise fundamental issues for the evaluation of cognitive interviewing, most empirical studies have been carried out as the opportunities present themselves rather than by design.

In this chapter we propose a research agenda to specify some aspects of cognitive interviewing that need inquiry and provide some suggestions as to how that inquiry might be undertaken. The goal of such an agenda is not to produce prescriptions for the practice of cognitive interviewing, although such prescriptions may come out of the research that we are recommending. Our point is that evaluation of cognitive interviewing is in such an early stage that progress will be most likely if the issues warranting study and the approach to this work follow a single framework. In the second section of this chapter, we specify such an agenda. In the third section, we report a case study that begins to address some of the issues on the agenda—those that concern the quality of verbal reports in cognitive interviews.

4.1.1 Characteristics of a Research Agenda

We propose that research on cognitive interviewing should be both empirical and theoretically grounded. In particular, methodological researchers will ask certain questions only if they consider the theory about how particular cognitive interview techniques should perform. For example, a technique that depends mainly on think-aloud procedures will not be good at detecting difficulties and errors in recall because it is generally known that information about retrieval processes is not available to respondents (e.g., Ericcson and Simon, 1993). Second, we propose that examining the *components* that comprise particular cognitive interviewing techniques will provide more information about why particular techniques perform as they do than will examining the technique as a whole. How do different types of instructions to respondents affect the outcome of cognitive interviewing? How does the information elicited by generic think-aloud procedures, in which the interviewer plays a relatively passive role, compare with the information provided in response to direct probes or in respondent paraphrasing (Willis et al., 1999a)? Are some respondent tasks better than others for explaining why a problem occurred or suggesting how to repair the question? And so on.

The common thread that connects cognitive interview techniques is that they all produce verbal reports. A useful research agenda, then, should include, but not be limited to, an assessment of the verbal reports produced by different techniques. We turn now to the theory of verbal reports and their use in survey pretesting.

4.1.2 Verbal Report Techniques

The lynchpin of cognitive interviewing is people's ability to provide verbal reports about their thinking when answering questions (Lessler et al., 1989;

Tourangeau et al., 2000; Willis et al., 1999). The reports are elicited in different ways by interviewers and provided in different ways by respondents. For example, interviewers may simply ask respondents to report their thinking or may ask specific probe questions. These probes may be crafted prior to the interview or in response to respondent reports (Willis et al., 1999a). Respondents may provide the reports concurrently (i.e., while formulating an answer) or retrospectively (i.e., after providing an answer). In all these cases, the assumption of practitioners is that respondents are able to accurately verbalize some of the thinking that underlies their answers.

Ericcson and Simon (1993) pioneered the modern use of verbal reports, providing a theoretical account of how the method works, including its limitations. They have been concerned primarily with thinking aloud in which laboratory participants report on their thinking in a relatively undirected way (i.e., the experimenter may prompt the participant to keep talking but does not probe with substantive questions). The key component of Ericcson and Simon's theory is that people can only report on thinking to which they have access. By this view, they should be able to report accurately on processes that engage working memory in a series of discrete mental steps. For example, when planning a move in chess, people must anticipate a series of moves and consequences, temporarily storing each one in working memory. In the survey response context, answering a behavioral frequency question by recalling specific events and counting each one involves similar temporary storage and thus should be amenable to verbal report techniques (e.g., Bickart and Felcher, 1996; Conrad et al., 1998). Respondents are less able to report on questions that do not have this character, such as answering "why" questions (Wilson et al., 1996) or those soliciting a preference (Wilson and Schooler, 1991).

Despite the popularity of verbal report data for studying high-level cognition, the method has been controversial. The crux of the controversy surrounds the observation that thinking aloud can degrade the process about which people are reporting, a reactive effect (e.g., Russo et al., 1989; Schooler et al., 1993). Although in most reactive situations, thinking aloud degrades the task being studied, it has also been shown to improve its performance (Russo et al., 1989). If verbal reports degrade survey respondents' question-answering performance in cognitive interviews, this could give the appearance of response difficulties where, in the absence of verbal reports, there actually are none. If verbal reports improve the question-answering process, they could mask problems that would otherwise be present. Taken together, these findings suggest that verbal reports are fragile sources of data, sometimes valid but sometimes not, sometimes independent of the process being reported but sometimes not.

In most studies that have evaluated cognitive interviewing, the typical measures are the number of problems and the type of problems. Although these measures may help practitioners choose among different pretesting methods for particular purposes, they do not help assess the *quality of the information* about problems produced in cognitive interviews. Implicit in much of this research is the assumption that verbal reports—the primary information about problems—are of high quality (i.e., veridical reflections of response problems). But these reports have seldom been treated as data in themselves and evaluated accordingly. This differs from how such reports are treated by cognitive psychologists (e.g., Ericsson and Simon, 1993), where the quality of verbal reports is routinely evaluated, for example, by computing agreement among coders.

Although Willis and Schechter (1997) did test the quality of cognitive interview results directly, their focus was not on the verbal reports themselves, but on the use to which those reports were put. The authors measured the accuracy of cognitive interview results for five questions in predicting problems and the absence of problems in several field tests. The authors assessed these predictions by comparing the response distributions of two samples of respondents. One sample was presented originally worded questions and the other was presented questions revised on the basis of cognitive interviews. The predictions were confirmed for four of the five questions and partially for the fifth. Although verbal reports were, presumably, the raw data that informed the revision process, it is impossible to know how much the revisions also benefited from the skill of the designer(s). We advocate disentangling these issues and in Section 4.3 present a case study that does this by addressing just the quality of verbal reports. The point for now is that verbal reports are data whose quality can be assessed just as the quality of data at other points in the survey process can be evaluated (e.g., Dippo, 1997; Esposito and Rothgeb, 1997).

4.2 RESEARCH AGENDA

In this section we sketch the broad features of a research agenda (see Table 4.1). In general, we advocate conducting stand-alone experiments that directly compare different cognitive interviewing techniques. To compare techniques, they must each be well defined and have reasonably well specified objectives. By defining techniques in terms of their components (e.g., the type of probes permitted by the interviewer, the instructions to the respondent, the types of tasks required of the respondent, etc.) it is easier to design experiments that can explain differences in outcome between techniques. If only one component differs between techniques, the explanation for different results is likely to lie here. Existing techniques may differ on just one or two components and can be compared as is; alternatively, a technique can be constructed by substituting a new component for one in an existing technique. Being clear on the objectives of a technique is important for assessing its success. For example, if the point is to detect as many potential problems as possible, one would apply a different set of criteria than if the point is to detect only the most severe problems.[1]

Problem detection and problem repair are the essential objectives of pretesting. A cognitive interview technique is efficient if it is cost-effective in identifying

[1]In the latter example, one would have to operationalize the measure of severity, which includes more than merely counting the number of problems.

Table 4.1 Research Agenda for the Evaluation of Cognitive Interview Techniques

1. Methodology for comparing techniques
   a. Define each technique, including its components (e.g., instructions, respondent tasks, probing)
   b. Specify each technique's objectives
      (1) Problem detection
      (2) Problem repair
   c. Design and conduct stand-alone experiments that compare techniques
      (1) Compare existing techniques
      (2) Create alternative techniques by varying one or more components
2. Data preparation
   a. Code verbal reports
   b. Transcribe and code interviewer–respondent interactions
3. Criteria for comparing techniques
   a. Problem detection
      (1) Number of problems
      (2) Types of problems
      (3) Quality of verbal report data: reliability; validity
      (4) Thresholds for problem acceptance
   b. Question repair
      (1) Reason for problem
      (2) Situations when the problem occurs
      (3) Effectiveness of question revision

question problems and providing useful information for their revision. Since detection and repair can be assessed in multiple ways, one concern is the choice of appropriate measures. Where possible, we advocate the use of multiple measures. Certainly for problem detection, one might count numbers of problems found, the nature of the detected problems, how reliable a technique is in uncovering problems, and to what extent those problems are valid (i.e., occur in field settings, adversely affect respondent answers, etc.). Other measures may also be useful.

The process of problem repair is harder to assess, since its effectiveness depends greatly on the skills of the personnel doing the revision. Certainly, question repair is more likely to succeed if it is based on information about why a problem occurred and the conditions under which it is most likely to occur. The main criterion in assessing question repair is whether or not a problem recurs in subsequent testing, either with cognitive interviews in the laboratory or in a field setting using other problem detection techniques.

In the remainder of Section 4.2, we provide more detail about the type of research that we believe is necessary in order to understand and, ultimately, improve cognitive interviewing techniques. In certain instances, we note that both the techniques studied and the methods used to study them may differ from common practice. This is a natural consequence of *studying* as opposed to *using* a method.

4.2.1 Defining a Cognitive Interview Technique

Despite variation in current practice (e.g., Blair and Presser, 1993; Willis et al., 1999a), cognitive interview techniques share a few basic features that need to be covered in a definition. The interviews involve a one-on-one interaction between an interviewer and a respondent. There may be a protocol for interviewers to follow, and if there is, the degree to which interviewers are expected to adhere to it may vary. Prior to the interview, the interviewer typically gives the respondent a general description of what will happen in the interaction; and the interviewer has decided how she will conduct the interview. What actually takes place in the interview may deviate from these intentions.

In the interview, respondents answer questions and produce other verbal reports. The interviewer asks the questions, probes, and follows up selectively on what the respondent says. Additionally, the interviewer can clarify or amend instructions to the respondent. Finally, the interviewer can summarize her understanding of a verbal report. The substance of the interview is a discourse between interviewer and respondent, which is typically summarized either in notes or recorded verbatim.

A definition should answer three basic questions:

1. What are the interviewer and respondent instructed to do?
2. What are the data from the cognitive interview?
3. How are these data analyzed?

Interviewer and Respondent Instructions Interviewers can instruct respondents to report their thinking either concurrently or retrospectively, or respondents can be instructed simply to answer the questions and respond to probes. Sometimes the respondent completes the interview without comment or interruption, after which the interviewer probes or asks other debriefing questions.

The degree to which the interviewer takes an active role can vary. The most passive role is to read the questions, and possibly prompt the respondent to keep thinking aloud, leaving it to him to produce verbal reports in accordance with the initial instructions. A more active role can involve myriad behaviors. For example, interviewers can administer scripted probes prepared prior to the interview which can be general and apply to any question (e.g., "In your own words what does this question mean to you?"). Probes can also be question-specific, such as "What do you think the phrase 'health care professional' means in this question?" Alternatively, unscripted probes can be improvised during the interview in response to something the respondent says, or based on an idea that occurs to the interviewer. Combinations of these options are also used.

Information about respondent instructions can be obtained from a written script of instructions, the transcript of the interviewer's introduction to the session, or the interview transcript. Since instructions may be changed or supplemented during the interview, an examination of the interview transcript is recommended.

Cognitive Interview Data Cognitive interviews produce verbal reports and we consider these to be the raw data for later analysis. The response processes that these reports describe are affected by the instructions to respondents and by the interviewer behaviors. As far as we are aware, little attention has been given to what actually constitutes verbal reports in cognitive interviews. Verbal reports certainly include respondent think-aloud statements and answers to probes (Willis et al., 1999a). They probably, but not necessarily, include other respondent remarks. But should verbal reports include interviewer statements, such as recapitulation by the interviewer of his or her understanding about what a respondent said, or the interviewer's summary of a potential problem? When the interviewer takes a very passive role, this is not an issue. The more active the interviewer role, the more ambiguity there is about what comprises the "verbal reports."

The definition of verbal reports is not typically a concern in everyday *use* of cognitive interviews and, in fact, may not be essential. As Tourangeau et al. (2000), concurring with Conrad and Blair (1996), note: "The conclusions drawn from cognitive interviews are only loosely constrained by the actual data they produce" (p. 333). However, in *methodological research*, such definition is essential; otherwise, it is hard to know what information is produced in the interviews and what information was produced in the accompanying processes (e.g., discussion by the research team).

From our perspective, methodological research on cognitive interview techniques benefits from analyses that separate the process of eliciting verbal reports (data collection) from their interpretation (data analysis). This approach permits assessing how well suited a technique is to each task. Again we note that this recommendation may depart from common pretest procedure, but our goal is to promote evaluation of the method, not make prescriptions for its practice. Eliciting verbal reports, interpreting them, and finally, applying those interpretations to question revision each involves different processes. A careful analysis of a technique will isolate one process from the others. It may well be that two techniques use identical processes to elicit reports (e.g., classical think aloud procedures). But interpretation of those reports may differ (e.g., one technique may rely on interviewer notes or even recollections of what took place in the interview, while the second method employs some sort of detailed analysis of interview transcripts).

Similarly, in one technique, question revision might be based primarily on the interviewer's judgment about the reason that a question caused problems for some respondents; another technique might rely more on follow-up probes to try to get respondents to articulate reasons for problems they appeared to have. Clearly, there are many possible approaches available to the cognitive interviewer/analyst. It seems prudent, whenever possible, to disentangle the effects of such alternative approaches.

Data Analysis From our perspective, there are at least two stages involved in analyzing verbal reports for methodological purposes. First, the verbal reports have to be interpreted and, where problems exist, coded into problem categories. If the verbal reports are not classified in some way, evaluation is restricted to the verbatim content of the reports. We note that the coding of problems may not be necessary, or even useful, in practice, but it is necessary for evaluation purposes. Second, the coded reports are counted and the tallies for techniques are compared.

The methodologist should design the specific codes to distinguish and capture the types of information of interest in the particular study. Our focus below is on problems, so we have primarily coded problems. Other methodologists might wish to study the connection between verbal reports and possible solutions to problems, so they might use codes for potential repairs. The coding scheme should be exhaustive and well defined. That is, it should be possible to assign any problem (or whatever the topic of interest) to a code. We advocate assigning a problem to one and only one code. Clearly, a particular verbal report may indicate multiple problems—and these should be coded individually—but a single problem should be uniquely classified so that it is tallied just once.

4.2.2 Identifying What the Technique Is Intended to Accomplish

The cognitive interview goals used by practitioners vary considerably (Blair and Presser, 1993; Willis et al., 1999a; Tourangeau et al., 2000). In addition to generally uncovering question flaws, some practitioners may want to use the method primarily to confirm their intuitions about possible question problems, while other practitioners may seek information from cognitive interviews to aid problem repair. Still others may wish to determine whether questions are problematic for subgroups of respondents with certain demographic or behavioral characteristics.

These purposes and most others depend on problem identification. If a technique is weak on problem identification, it will not (at least on its own) provide much value in achieving these other goals. A minimum quality requirement is for cognitive interviews to produce data that credibly identify problems. We take problem identification to be the logical starting point for research to evaluate cognitive interviewing techniques.

4.2.3 Measuring the Success of a Technique

Classifying and Counting Problems Problem classification and problem counting are closely related. First, if the goal is to count unique problems or recurrences of the same problem, some description of the type of problem is required to distinguish one problem from another. Second, if the classification scheme is designed to be exhaustive, the list of classes themselves will aid in identifying a verbal report's evidence of a problem. Thus, it is unsurprising that many of the attempts at formal analysis of cognitive interview techniques involve problem classification schemes. [See Tourangeau et al. (2000, pp. 327–328) for a discussion of many of the coding schemes that have been used.]

Thresholds for Problem Acceptance A problem report can be accepted based solely on the judgment of the interviewer. But this is not the only possible criterion. In summarizing "best practices" in cognitive interviewing, Snijkers (2002) notes that frequently the interviewer and someone who observed the interview meet to discuss the interview results. But he does not address how possible differences in their assessments are resolved. The implication is that two listeners can lead to better problem detection than just one.

A minimum requirement for treating a verbal report as evidence of a problem is that at least one interviewer/analyst concludes that a *problem* exists. However, it may be of methodological interest to examine the impact of higher thresholds (i.e., agreement among more than one judge). One issue on the proposed research agenda is to examine how the number and types of problems change when different amounts of agreement between analysts are required in order to accept the evidence.

Reliability and Validity in Problem Detection Whether agreement is measured between two or more judges, it is a clear way to assess the reliability of problem detection. Do different judges reviewing the same verbal report agree on whether or not it suggests a problem? If they agree that a problem exists, do they agree on the type of problem? Another way to assess the reliability of a technique is to compare the problems found on multiple administrations (or tests) of the method. By this view, a cognitive interviewing technique is successful to the extent that it turns up similar problems across different interviews.

To help explain why particular techniques are more or less reliable in either sense, one can examine the interaction between interviewers and respondents. For example, it may be that different types of probes lead to answers that are more or less reliably interpreted.

Similarly, validity can be defined in several ways, each of which entails a different view of what makes a detected problem "real." In one view, problems are real only if they lead demonstrably to incorrect data in field data collection. By extension, a reported problem is valid the more probable its occurrence in any given interview. A problem will rarely affect all respondents; most will affect only some portion of respondents. But if a potential problem detected in a cognitive interview does not affect any respondents in field administration of the questionnaire, it cannot be considered valid. Yet another sense of problem validity is severity—the size of the measurement error produced by the problem.

4.2.4 Stand-Alone Experiments

Methods research on cognitive interviews has been of two types: "piggybacking" the evaluation of a technique onto a production survey (e.g., Willis and Schechter, 1997) or conducting a stand-alone experiment (e.g., Presser and Blair, 1994), which can be a laboratory experiment and/or a field experiment. We favor the latter approach in order to exercise control over the variables of interest. In stand-alone experiments, the researcher can often determine the sample size. As has been advocated in the evaluation of usability testing methods in human–computer interaction (see Gray and Salzman, 1998, pp. 243–244), we endorse using larger samples than are typical in production pretesting. In the case study described below, eight interviewers each conducted five cognitive interviews for a total of 40 interviews, probably four or more times the number that is typical in production use of the method. This was sufficient for us to carry out some analyses but not all; the ideal sample size really depends on the questions that one is asking. Larger samples not only increase the power of the subsequent analysis, but also provide better estimation of the probability that respondents will experience a particular problem.

Although we recognize the central role of cognitive interview data in problem repair, we think evaluations of cognitive interview techniques should separately assess the quality of the data produced by the techniques and the use of those data in question revision.

4.3 CASE STUDY

In the following section we describe a study in which we examined two variants of cognitive interviewing. The study is one attempt to gather information about several, although by no means all, of the items on the research agenda presented in Section 4.2. We report the study here primarily to illustrate the type of evaluation we advocate rather than as the final word on either of the versions we examined or on cognitive interviewing in general. In particular, the study is an example of a stand-alone experiment carried out for methodological rather than survey production purposes.

The study produced thousands of observations of quantitative and qualitative data. Based on these data, the study illustrates the use of agreement measures to assess the interpretation of verbal reports, although it does not make use of validity measures. In addition, it illustrates the use of interaction analysis to explore the types of probes used and their relation to the types of problems identified. Finally, the study evaluates two versions of cognitive interviewing to test the effects of varying particular features of the general method rather than to evaluate these versions per se. One could vary the features in other ways or vary other features.

Because this was a methodological study and not a production application of cognitive interviewing, certain aspects of the way we used the method may depart from its use in production settings. For example, the number of interviews (40 total) may be larger than is common and the way that problems were reported (written reports for each interview and codes in a problem taxonomy) may differ from what seems to be typical (written reports that summarize across interviews).

We recruited eight interviewers and asked each to conduct five cognitive interviews with a questionnaire constructed from several draft instruments created by clients of the Survey Research Center at the University of Maryland. The questionnaire contained 49 substantive questions, about half of which were factual and half opinion questions. The topics included nutrition, health care, AIDS, general social issues, and computer use. The interviewers were told that they were participating in a methodological study sponsored by a federal agency.

Four of the interviewers were experienced practitioners of cognitive interviewing. Each had more than five years of experience at different organizations within the federal government and private-sector survey communities. Three of the four had doctoral degrees in psychology. This level of education seems to us to be typical of experienced practitioners of cognitive interviewing. This group conducted cognitive interviews using whatever method they ordinarily use. We refer to the procedures they used as *conventional* cognitive interviewing.

The remaining four interviewers were less experienced with pretesting questionnaires, although all four worked in survey organizations in either the academic, commercial, or federal sector. Two of the four had some experience with production cognitive interviewing, and the other two had been exposed to the theory behind the method and had completed relevant class exercises as part of their master's-level training in survey methodology. In contrast to the conventional cognitive interviews, three of whom held doctoral degrees, three of these interviewers held only bachelor's degrees and one held a master's degree. This level of experience and education seemed typical to us of junior staff in survey research centers—staff members that typically do not conduct cognitive interviews. This group of interviewers was trained to use a version of cognitive interviewing in which the types of probes and the circumstances of their use were restricted. One consequence of restricting the set of probing conditions, relative to conventional cognitive interviewing, was to simplify the interviewers' task by reducing the amount of experience and judgment required to know when and how to probe. We refer to this as the *conditional probe technique*.

Ideally, we would have crossed the factors of technique and experience/education to disentangle any effects of one factor on the other. This would have meant that in addition to the two groups that differed on both factors, we would have instructed inexperienced interviewers to carry out conventional cognitive interviews and trained experienced interviewers to follow the conditional probe technique, thus creating a 2×2 design. However, it was not feasible to test these additional groups. First, it seemed unlikely to us that experienced cognitive interviewers could "unlearn" their regular technique, developed over many years, for the duration of the study.[2] Interference from the old technique on the new one would have made the results hard to interpret. Moreover, they would not have been highly experienced with this particular technique.[3] Second, because there is no industry-wide consensus about what constitutes cognitive interviewing, we were not able to define the conventional method well enough to train inexperienced interviewers in its use. Even if we had been able to train them in

[2]We acknowledge that in some organizations, experienced cognitive interviewers are asked routinely to modify their practice. However, the success with which they do this is an empirical question about which we have little relevant data.

[3]Another approach to increasing the expertise of interviewers using the conditional probe technique would have been to give inexperienced interviewers substantial practice with the conditional probe technique prior to the study. However, it would have been impractical to give them several years of experience, which is what would have been required to match their experience to that of the conventional interviewers.

conventional cognitive interviewing, they would not have had the education or years of experience that the conventional cognitive interviewers had.

4.3.1 Cognitive Interviewing Technique(s)

Instructions to Interviewers Because of the lack of consensus about conventional practice, we asked the experienced practitioners to conduct cognitive interviews as they ordinarily do, allowing them to define "the method," both the procedure for conducting the interviews and the criteria for what constituted a problem. We did not provide them with an interview protocol or instructions for respondents, and we did not require that they examine the questionnaire ahead of time—if they did not ordinarily do so. We asked them to prepare written reports of problems in each question of each interview. We did not define *problem* for them but instructed them to use whatever criteria they used in ordinary practice. We required reporting at this level (individual questions on individual interviews) in order to compare the interviewers' judgments about each particular verbal report to those of coders listening to the same verbal report.[4] According to their written reports and a subsequent debriefing, these interviewers used a combination of scripted and unscripted probes to explore potential problems. Two of the four indicated that over the course of the interviews, they were less likely to probe already discovered problems than novel ones. Although we treat this as a single "conventional" method, we recognize that each interviewer might approach the data collection task somewhat differently.

The conditional probe interviewers were introduced to the technique in a two-day training session. They were instructed in both how to conduct the interviews and in how to classify problems identified in the interviews. For the interviewing technique, they were instructed to solicit concurrent verbal reports from respondents and to focus their probing on behavioral evidence of problems in those reports.[5] They were instructed to intervene only when the respondents' verbal reports corresponded to a generic pattern indicating a potential problem (e.g., an explicit statement of difficulty, or indirect indications such as a prolonged silence or disfluent speech). When such a condition was met, interviewers were instructed to probe by describing the respondent behavior that suggested the possibility of a problem (e.g., "You took some time to answer; can you tell me why?"). Other than probing under these conditions, the interviewers were not to play an active role. The interviewers practiced the technique in mock interviews with each other. The instructor (one of the authors) provided feedback to the interviewers and determined when all four had grasped the essentials of the probing technique.

[4]One interviewer indicated that this reporting procedure departed from her more typical practice of summarizing across interviews. The departure concerned her because it treated each interview in isolation; she indicated that she typically varied her interviews on the basis of what she had learned in previous ones.

[5]The written materials used to train the conditional probe interviewers are available from the authors.

The restrictions on probing were motivated by ideas and findings in the psychological literature on verbal report methods. First, people can only provide accurate verbal reports about the content of their working memory (Ericsson and Simon, 1993) and thus may sometimes legitimately have nothing to report. This is especially likely when respondents retrieve from long-term memory the information on which their answers are based. This type of retrieval usually occurs automatically (e.g., Shiffrin and Schneider, 1977) in the sense that people do not actively control the process and thus are not aware of its details. Probing respondents for more details when none are available could lead to rationalized or embellished reports. Under these circumstances, the conditional probe interviewers were instructed to do nothing.[6] However, if a respondent is able to report something and the report contains a hint of a problem, then presumably, reportable information exists which the respondent has not clearly articulated. In this case, probing should clarify the initial report without encouraging respondents to embellish it.

A further impetus to experiment with restricted probing was the set of findings about reactivity mentioned in the introduction (i.e., the observation that thinking aloud can distort the process about which respondents are reporting). Reactive effects seem particularly likely when the response task is difficult because pressing respondents to report may demand mental resources that would otherwise have been devoted to responding. By instructing interviewers to remain silent when respondents give no suggestion of problems, we believed the interviewers would be less likely to contribute to reactive effects.

In addition to practice with the probing technique, the conditional probe interviewers were taught to identify problems through (1) an introduction to well-known types of problems (e.g., double-barreled questions), (2) discussion of 12 detailed definitions for each category in a problem taxonomy they would later use, and (3) practice identifying problems in audiotaped mock cognitive interviews. The mock interviews illustrated respondent behaviors for which probing was and was not appropriate (see Conrad et al., 2000). The instructor provided feedback, including pointing out missed problems and determined when the interviewers had grasped the problem definitions.

Instructions to Respondents The conventional cognitive interviewers did not indicate a priori how they typically instruct respondents, so we relied on the interview transcripts to determine what they actually did. The transcripts showed substantial variation in wording and content between these interviewers. Two of the conventional interviewers said that although the questions would be administered as in a "real" interview, they were not as interested in the answers as in how the respondent came up with their answers. Two interviewers mentioned that respondents should tell the interviewer about any difficulties encountered. Three of the interviewers also gave some variant of think-aloud instructions.

[6]Note that the inability to report on a process does not mean that it is free of problems. It simply means that we do not have any data about the process.

All four conventional cognitive interviewers said that the purpose of the interview was to learn about comprehension problems before the survey went into the field.

The conditional probe interviewers were trained to provide think-aloud instructions to respondents that closely followed those of Ericsson and Simon (1993, p. 376), but they were not given an exact script to present to respondents. In general, they were instructed to encourage respondents to report everything that passed through their heads while answering each question and to do so without planning what to say after thinking. The interview transcripts confirmed that all four interviewers did this reasonably consistently.

Data from the Cognitive Interviews Each of the eight interviewers (four conventional, four conditional probe) conducted five interviews. All 40 interviews (20 per type of cognitive interviewing technique) were audio recorded. The conventional cognitive interviewers each wrote a narrative report listing the problems they identified in each administration of each question. We used this reporting format, instead of summarized reports, to measure agreement between interviewers and other analysts about the presence of problems in particular administrations of a question. Problems identified in the written reports were later classified into a taxonomy of problems (see Conrad and Blair, 1996) in what was essentially a transcription task. The main types of problems are lexical (primarily issues of word meaning), logical (both logical connectives such as "and" and "or" as well as presupposition), temporal (primarily reference periods), and computational (a residual category including problems with memory and mental arithmetic). Two transcribers independently mapped the written problem reports to the problem taxonomy and then worked out any discrepancies together.[7] Both transcribers had been introduced to cognitive interviewing in a graduate survey methodology course, but neither had conducted cognitive interviews. They were given written definitions of the problem categories and an oral introduction to the taxonomy. The exercise was discussed with both transcribers until they seemed competent to carry out the task.

The conditional probe interviewers directly classified any problems they detected for a given administration of a question into the problem categories in the taxonomy. They were required to choose a single category for a particular problem but could code more than one problem per question. They were provided with written definitions of the problem categories, an oral introduction, and were given coding practice until they were competent users of the taxonomy.

The exact rationale for this problem taxonomy (see Conrad and Blair, 1996, for a discussion of the rationale) was not relevant to its use in the current study. Other problem taxonomies certainly exist (e.g., Forsyth et al., 1992) and would have

[7]We did not compute any sort of reliability measure for the transcription task, primarily because it did not involve judgment about the nature of problems (this had already been done by the interviewers in their reports) but was a matter of translating the interviewers' descriptions to those in the taxonomy. In addition, because the transcribers worked jointly to resolve differences, their work was not independent.

been appropriate here. The point is that one needs some set of problem categories in order to tally the frequency of problems. For example, one cannot count two verbal reports as illustrating the same problem without somehow categorizing those reports.

In addition to the interviewers' own judgments about the presence of problems, four coders coded the presence of problems in all 40 interviews using the same problem taxonomy. They participated in the same training as did the conditional probe interviewers on the definition of problems (not on the interviewing technique) and had classroom exposure to cognitive interviewing. This made it possible to measure agreement between each interviewer and each of the four coders as well as agreement between each pair of coders.

We use agreement measures here to assess the quality of the data on which suggested revisions are based. Coding agreement in the current study measures the quality of the information that serves as *input* to the decision about revision. If two coders do not agree on the presence of a problem or its identity, the respondent's verbal report is ambiguous and thus less definitive about the need for revision than one would hope. Of course, in practice, researchers mull over the results of cognitive interviews before deciding whether to revise particular questions and how to revise them, but their decision making is constrained by the information produced in the cognitive interview (i.e., it's hard to make a good decision based on murky information).

4.3.2 Key Findings

Number and Type of Problems Because cognitive interviews have been found to be sensitive primarily to problems concerning comprehension (e.g., Presser and Blair, 1994), we would expect more lexical and logical than temporal and computational problems in the current study. This is in fact what we observed. Over all 40 cognitive interviews, interviewers identified 0.13 lexical and 0.11 logical problem per question versus 0.02 temporal and 0.04 computational problem per question. This serves as a general check that procedures were relatively similar to those used in other studies (if not in actual practice).

None of these patterns differed substantially between the two types of cognitive interviews. However, conventional cognitive interviewers reported 1.5 times as many potential problems as did the conditional probe interviewers. If more problems are detected with one technique than another, this could indicate that the larger number refers entirely to actual problems, and the smaller number reflects missed problems. Alternatively, the larger number could include reported problems that are not actually problems ("false alarms"), and the smaller number would therefore be the more accurate one. The truth could be somewhere in between if one technique promotes false alarms and the other tends to miss problems. This could be disentangled most clearly if we had a validity measure (i.e., an objective measure of whether a report was in fact a problem for that respondent). In the absence of such a measure, reliability provides some indication of how much stock to place in reports of problems.

Agreement Measures Problem detection was surprisingly unreliable when measured by agreement between interviewers and coders about the presence of problems in the *same* verbal report. That is, when an interviewer and coder listened to the same verbal report, they often reached different conclusions about whether or not it indicated the presence of a problem. The average kappa score for all interviewer–coder pairs for the judgment that a particular question administration did or did not indicate a problem was only 0.31 ("fair" agreement according to Everitt and Haye, 1992). This low agreement rate cannot be attributed to the complexity of the coding system since the judgment about the presence or absence of a problem did not involve the specific categories in the coding system. In fact, agreement on the particular problem category in the taxonomy for those cases where an interviewer and coder agreed there was a problem was reliably higher, 0.43 ("moderate" agreement according to Everitt and Haye, 1992) than their agreement that a problem simply was or was not present. However, even this score is disturbingly low, considering, again, that the interviewers and coders were interpreting the same verbal reports and considering that problem reports in cognitive interviews are used to justify changes to questions in influential surveys. At the very least, these low agreement scores suggest that verbal reports—the raw data from cognitive interviews—are often ambiguous.

Although agreement is low, it is reliably higher for conditional probe than for conventional cognitive interviews. For the simple judgment about whether or not a problem was indicated by a particular verbal report, the kappa score for conditional probe interviews is 0.38 but only 0.24 for conventional cognitive interviews, a reliable difference (see Table 4.2). When there is agreement that a verbal report indicates a problem, agreement about the particular problem category shows a nonsignificant advantage for the conditional probe interviews, kappa (κ) = 0.47, over the conventional interviews, $\kappa = 0.39$.

One might argue that the general advantage for the conditional probe interviews is due to the greater similarity in experience and training between coders and conditional probe interviewers than between coders and conventional cognitive interviewers. However, if that were the case, the kappa scores would be lower for pairs of conventional cognitive interviewers and coders than for pairs of coders. But this was not the case. Average kappas were statistically equivalent for interviewer–coder and coder–coder pairs interpreting the conventional cognitive interviews (see Table 4.3 for intercoder agreement scores).

Table 4.2 Average κ Values for Interviewer–Coder Pairs

	Conventional Cognitive Interviews	Conditional Probe Interviews	Difference
Is there a problem?	0.24	0.38	$p = 0.001$
If so, what type?	0.39	0.47	Not significant

Table 4.3 Average κ Values for Coder–Coder Pairs

	Conventional Cognitive Interviews	Conditional Probe Interviews	Difference
Is there a problem?	0.27	0.36	$p = 0.077$
If so, what type?	0.36	0.43	Not significant

Irrespective of interview type, one might expect interviewer–coder agreement to be lower than coder–coder agreement because interviewers had more information available than did coders: audiotaped interviews in the case of coders versus audiotapes as well as interview notes and memories by the interviewers. Interviewers may have taken into account nonverbal information, such as respondents' facial expressions and gestures, not available to the coders. However, these differences in available information did not affect agreement. There was no difference in average kappas for interviewer–coder pairs and coder–coder pairs.

Pairwise agreement scores indicate clearly that these verbal reports were difficult for different listeners to interpret in the same way. A further indication that these verbal reports are inherently ambiguous is evident when we raise the threshold for accepting problems. In particular, the number of problems identified by at least one coder (although possibly more) is 0.42 problem per question. The number identified by at least two coders was only 0.09 problem per question. This figure drops even further, to 0.02 problem per question, when the threshold is at least three coders; and practically no problems, 0.004 problem per question, are detected by all four coders.

Interviewer–Respondent Interaction It is possible that the low agreement in interpreting verbal reports can be understood by examining the interaction between interviewers and respondents, since it is this interaction that produces the reports. At the very least, examining this interaction should help to document what actually happens in cognitive interviews. To address this, all 40 interviews were transcribed, and each conversational turn was assigned a code to reflect its role in the interaction. These interaction codes should not be confused with problem codes: The interaction codes were assigned to each statement in the interview, whereas problem codes were assigned to each question in the interview. The particular interaction codes for respondent turns included, among other things, potential indications of a problem (e.g., long pauses, disfluencies, and changed answers) and explicit respondent descriptions of a problem. Interviewer turns were coded, among other things, as probes about a problem that was expressed, at least potentially, in an earlier respondent turn and probes about a problem that was not expressed, even potentially, in a prior respondent turn.

Over all the interviews, conventional cognitive interviewers probed 4.2 times as often as did conditional probe interviewers. However, conditional probe interviewers probed about an explicit respondent utterance that potentially or explicitly indicated a problem 4.8 times as often (61% versus 13%) as conventional cognitive interviewers. In the following example, a conventional probe interviewer asks about the respondent's silence.[8]

I: Transportation of high-level radioactive wastes means that state and local governments may have to spend money on things like emergency preparedness. Do you think that the federal government should reimburse the states and localities for costs like these?

R: Um: . I don't want to say no, because I think ultimately states:, should handle their own business, unless it's a federal issue ...

I: I noticed you paused there for a second after I, I asked you that. Was there something unclear in *that question*?

R: *Um yeah* I would, unless I wasn't hearing the question, completely clearly, I, I wasn't sure WHO was producing the waste. Which that may have been the case I was thinking back to what the question asked, what you know who, who was producing the waste whether it was federal government or state.

I: Alright.

In contrast, in the following exchange, a conventional cognitive interviewer probes about the meaning of several terms in the question without any indication from the respondent that these terms are misinterpreted or are causing difficulty.

I: In general, do you think the police in your neighborhood treat cases of domestic violence seriously enough when they are called, or do you think the police are more likely to treat domestic violence as just a family matter?

R: [sighs] I don't, I haven't had much experience with the police in my area and I don't know anybody who's been abused, so I don't know how it's handled in my area.

I: Okay.

R: I know in some areas it's handled, differently, so.

I: Okay and how, how do you define "differently"?

R: I have heard of people getting slapped on the wrist for domestic violence and just let off the hook saying like "don't do this again." And I've also heard of cases where people have gone to jail for quite awhile as a result of domestic abuse.

[8]In the transcribed excerpts, overlapping speech is enclosed in asterisks. A period between two spaces (.) represents a pause. A colon within a word indicates a lengthened sound. A hyphen at the end of a word ("that-") indicates that the word was cut off. Question marks indicate rising intonation, and utterance-final periods indicate falling or flat intonation, regardless of whether the utterance is grammatically a question or an assertion. Upper case indicates vocal stress.

I: Okay. So tell me in your own words what you think THAT question is asking.

R: It basically seemed to be asking, um, "do the cops take it seriously or, or do they take it kind as leniently and none of their business."

I: Okay, so as a family matter, um, as, as "just a family matter," what does "just a family matter" mean, to you?

R: It seems to me as something that, they don't think is in their authority to handle and also that it's not severe enough that it would warrant their involvement.

I: Okay.

Exchanges like the second, which focused on possible meaning-related problems, were quite frequent in the conventional cognitive interviews but relatively rare in the conditional probe interviews. In particular, 36% of conventional cognitive interviewers' probes concerned respondents' understanding of specific terms when respondents had not given any verbal evidence of misunderstanding these terms. Conditional probe interviewers administered this type of probe under these circumstances only 5% of the time that they probed.

Such differences in the type of interaction could be related to differences in levels of agreement for the two types of interviews. When interviewers probe about a particular respondent utterance to determine if it indicates a problem, the respondent's reply to the probe should lead to a relatively clear-cut problem judgment; the initial utterance either did or did not indicate a problem. However, probes that are not clearly tied to something the respondent said earlier may produce less definitive results. Suppose that the interviewer asks the respondent what a particular term means even though the respondent has not indicated any confusion up until this point. If the respondent's answer to this probe indicates possible confusion, it may be hard for listeners to evaluate this. Has the interviewer uncovered an actual problem, or introduced one? For example, the respondent may have understood the question well enough in context to answer the question accurately but not well enough to provide the relatively formal definition that the probe requests. Different listeners may hear such an exchange differently, which would lead to low agreement.

4.3.3 Conclusions from Case Study

We have presented this study primarily as an example of the type of evaluation research that we advocate in Section 4.2. One methodological lesson from the case study is that even though it was hard to find four experienced cognitive interviewers, future studies should involve even more interviewers. Without a larger numbers of interviewers, idiosyncratic practices are a threat to the generalizations that one can draw confidently. Another methodological issue concerned the lack of consensus about what current practice involves. We tried to overcome this by allowing the traditional interviewers to follow their ordinary practice, but it would have been preferable to know ahead of time that they were following a single representative approach.

The first substantive conclusion is that the overall low agreement scores suggest that verbal reports in cognitive interviews (even when interviewers are constrained in their probing) often lend themselves to different interpretations. This is potentially of great concern when we consider that based on cognitive interviews, designers change and decide not to change the content of questionnaires in major surveys that produce high-profile statistics. If the information on which those decisions are based is inherently ambiguous, the decisions will be compromised, no matter how thoughtfully considered.

Similarly, the number of problems that are reported in a particular application of cognitive interviewing is greatly reduced by requiring identification by more than one analyst. This suggests that the number of problems produced in particular pretests can lead to different conclusions, depending on what threshold is used.

Finally, conventional cognitive interviewers report more problems than conditional probe interviewers, but they agree less often with coders about the presence of problems than do conditional probe interviewers. Conventional cognitive interviewers may be erring on the side of including questionable problems as actual problems, a strategy that may reduce the risk of missing actual problems but may also introduce new problems by leading to changes in questions that are not actually problematic.

4.4 FUTURE WORK

In cognitive interviewing, a deceptively simple set of procedures—asking respondents to report what they are thinking while answering survey questions—sets in motion a complex series of mental and social processes that have gone largely unstudied. Yet the effectiveness of the method is certain to rest on the nature of these underlying processes. We have proposed an agenda for research that compares these processes between techniques. Our case study begins to address the processes involved in producing and interpreting verbal reports. Subsequent research on this topic might include reliability—in both senses mentioned above—across different types of questions and different types of probes. In addition, validity of problems found in cognitive interviews has received very little attention, in part because its measurement is elusive. We have suggested several possible measures, although none is perfect. If it were known to what degree potential problems identified in cognitive interviews really are problems for respondents, this would enable practitioners to use cognitive interview results more wisely in revising questionnaires.

Similarly, little is known about the degree to which revising questions in response to cognitive interview results actually prevents problems from recurring. This needs to be evaluated in laboratory as well as field settings. But more fundamentally, the revision process is a largely creative enterprise that may vary widely depending on who is involved and in what organization they work. By better examining how designers use the information from cognitive interviews to reword questions and redesign questionnaires, it becomes more feasible to codify their practices and disseminate this to students.

Finally, the type of research that we are proposing would be facilitated by certain procedural changes by practitioners. Most significant is greater vigilance by practitioners in defining their cognitive interviewing techniques. Although a definition does not guarantee that the technique is actually used in a particular way, it provides a starting point for evaluation research. Techniques that are clearly defined make it possible to identify and then evaluate their key aspects, and this increases the chances that the results are relevant and useful to practitioners.

ACKNOWLEDGMENTS

We thank Ed Blair, David Cantor, and Greg Claxton for their advice and assistance.

CHAPTER 5

Do Different Cognitive Interview Techniques Produce Different Results?

Theresa J. DeMaio and Ashley Landreth
U.S. Bureau of the Census

5.1 INTRODUCTION

In recent years, cognitive interviews have become widely used as a method of pretesting questionnaires in the federal government and survey organizations, and have become accepted as a survey methodological tool. A criticism of the method, however, is that there is no standard definition of what a cognitive interview is and no set of standardized interview practices. Differences in the application of the method range from whether respondents are specifically instructed at the beginning of the interview to think out loud or whether interviewers simply probe the thoughts and answers of the respondents, to whether interviews are tape-recorded, and if so, whether the tapes are ever used.

Documentation of the procedural details for conducting cognitive interviews has been sparse. How exactly do different organizations conduct cognitive interviews? Do the procedural differences affect the problems that are identified and the recommendations for questionnaire changes that result from the interviews? This last question goes to the purpose of conducting cognitive interviews in the first place: to identify and eliminate problems in the questionnaire and improve the quality of the data collected.

Systematic research on the cognitive interview method has been even rarer. In this chapter we conduct a systematic investigation to evaluate alternative methods

Methods for Testing and Evaluating Survey Questionnaires, Edited by Stanley Presser, Jennifer M. Rothgeb, Mick P. Couper, Judith T. Lessler, Elizabeth Martin, Jean Martin, and Eleanor Singer
ISBN 0-471-45841-4

of conducting cognitive interviews. This is not a strictly controlled experiment, as we have chosen to evaluate three "packages" of methods that reflect actual differences in interviewing practices among survey research organizations. This is not optimal from a theoretical point of view, since it does not permit evaluation of any individual procedural elements. It can, however, provide useful information from an applied perspective about whether the procedural differences taken as a whole affect the thoroughness of the questionnaire evaluation and the quality of the resulting recommendations. The value of this experiment, loosely controlled though it is, lies in the documentation of whether the differences that abound in the cognitive interviewing method matter.

The chapter consists of five additional sections. First, we present a review of the literature pertaining to the cognitive interviewing method. Next, we discuss the research objectives of our experiment. Then we describe the research methods, including the three experimental treatments and supplemental data collection. Finally, we present the research results, followed by a discussion of the implications of the research and a concluding section.

5.2 BACKGROUND

As cognitive testing of questionnaires has evolved, the cognitive interview has become a basic method in the repertoire of pretesting techniques. The goal of cognitive interviews is to understand the thought processes used to answer survey questions and to use this knowledge to find better ways of constructing, formulating, and asking survey questions (Forsyth and Lessler, 1991). Respondents are asked to think out loud as they answer survey questions or to answer probing questions either after they answer the survey question or at the end of the interview.

However, evolution of the cognitive interview has occurred in a rather haphazard manner, with little consistency across organizations (Willis et al., 1999a). Relatively few attempts have been made by survey organizations to document the procedures used in cognitive interviews [only DeMaio et al., (1993) and Willis (1994) exist, to our knowledge]. Short courses in cognitive interviewing methods have been conducted at professional association meetings (see, e.g., Willis, 1999), but the adoption of any particular interviewing method cannot be assumed from this fact. Research papers using the method focus on substantive research results with little description of the methodology provided (Blair and Presser, 1993).

Variations are known to occur in basic details of the method. For example, reporting results from a survey of organizations that develop or test questionnaires, Blair and Presser (1993) note differences in the number of cognitive interviews conducted in pretesting projects, the number of interviewers who work on a single project, whether the interviews are conducted by survey interviewers or professional staff, whether the interviews are tape-recorded, whether tape recordings are reviewed after the interview and by whom, and whether the think-aloud method is used. Research manuals produced by U.S. federal statistical agencies (DeMaio et al., 1993; Willis, 1994) document use of professional research staff to conduct cognitive interviews, while statistical agencies in other countries vary between having cognitive interviews conducted by professional research staff (Smith, 2000; Snijkers, 1997) and specially trained survey interviewers (Cosenza, 2001; Henningsson, 2001). Cognitive interviews have become widely used but include a number of different options: think-alouds in conjunction with verbal probes, think-alouds without probes, or probes without think-aloud instructions.

Several investigations have been conducted to evaluate cognitive interviews compared with other pretesting methods (Presser and Blair, 1994; Rothgeb et al., 2001; Willis et al., 1999b), but the specific implementation of this method was not the focus in these investigations. These research efforts have barely scratched the surface, and there have been several calls (Blair and Presser, 1993; Tucker, 1997; Willis et al., 1999a) for research involving adequate, comparable, and replicable implementation of cognitive interview procedures.

We know of no investigations that involve alternative implementations of the entire cognitive interview method. From an applied perspective, such an effort has merit because a systematic evaluation of the different methods currently used by survey organizations can provide information about the ability of the methods to detect questionnaire problems. Evidence about successful methods might encourage their adoption by other organizations, thus decreasing the diverse range of methods currently in use.

5.3 RESEARCH OBJECTIVES

Toward the end of supplementing the holes in the research outlined above, we aim to address three research objectives in this chapter:

1. Do different cognitive interviewing methods identify different numbers and types of questionnaire problems?
2. Do different cognitive interviewing methods have differing abilities to identify problems predicted by expert reviewers?
3. Do different cognitive interviewing methods produce different questionnaire recommendations?

5.4 RESEARCH METHODS

To begin answering the research questions outlined above, three research teams from three survey research organizations, practicing three disparate cognitive interview methods, were asked to cognitively pretest a household survey on recycling. The resulting questionnaire problems and revision recommendations, documented by each research team in a report, were compared with one another. Then the results were compared against a more objective evaluation tool, created from expert review findings of the questionnaire by three independent evaluators.

We focused on the total number and type of problems identified, whether the problems were identified through cognitive interviews or questionnaire design expertise alone, and how frequently the teams and the experts found the same problems in the same questions. Additional data were used to provide a better understanding of the results, and included an evaluation of the interview protocols and interviewers' probing behavior and a performance evaluation of the revised questionnaires through additional cognitive interviewing.

5.4.1 Experimental Treatments

The experimental treatments reflect three different packages of procedures, currently practiced by survey research organizations, for pretesting questionnaires using cognitive interviews. Although the differences across implementations of the cognitive interview method are of primary interest in this research, the results must be viewed as part of a larger "package" (e.g., the cognitive interview plus recruiting procedures, interviewer experience and education level, organizational culture, etc.). The three approaches, labeled teams A, B, and C, were selected because each team's implementation of the method included elements that might be expected to affect the cognitive interviewing results. These include (1) the type and training of personnel who conduct interviews—field interviewers versus professional researchers; (2) the role of survey researchers in the overall process; (3) the types of data collected and the degree to which they are reviewed; and (4) facets of the interview protocol developed for the interviews. Although the teams were not selected based on the nature of the institutions employing the researchers, they came from a variety of survey research organizations: academic, governmental, and private.

Table 5.1 describes many characteristics of the teams' implementations of the cognitive interview method.[1] Differences of particular interest for this research include (1) team A uses field interviewers for cognitive interviews, while teams B and C use professional researchers; (2) team C alone explicitly instructed respondents to use the think-aloud procedure; (3) teams A and C review audiotaped interviews, while team B reviews notes taken during the interviews, when preparing a summary of identified problems; and (4) reports are prepared by the interviewer/researcher for teams B and C, whereas researchers debrief interviewers prior to formulating a report for team A.

In the summer of 2001, each research team received general information about the questionnaire and was provided the following guidelines for conducting the cognitive interviews: (1) two interviewers should conduct and audiotape five interviews each between June and September 2001 for a total of 10 interviews per research team; (2) one interviewer should be more experienced and one should

[1]Details of the implementation of the method came from telephone debriefing sessions conducted separately with each research team in December 2001. Debriefing sessions elicited information on the research team's project preparation, procedures conducting the interviews, data analysis steps, formulation of recommendations, and miscellaneous information such as the researchers' survey methodology backgrounds and experience levels.

Table 5.1 Summary of Differences in Implementation of Cognitive Interview Methods

Cognitive Interview Aspects	Team A	Team B	Team C
Who conducted interviews?	Field interviewers	Professional researchers	Professional researchers
No. of interviewers	3	2	2
Interview location	Respondent's home	Research offices	Respondent's home and research offices
Respondent payment	$40	$40	$30
Recruiting considerations: targeting certain types of respondents	None (no specific geographic or subject matter characteristics)	Geographic and demographic variation (proxy for different types of recycling/trash disposal)	Recyclers, nonrecyclers, landfill users, apartment dumpster users, curbside trash pickup households
Recruiting method	Database of people who have answered ads	Volunteers from prior projects, flyers	Personal networks, database of volunteers
Who developed the protocol?	Researchers, not interviewers	Researchers/ interviewers	Researchers/ interviewers
Think-aloud procedure stated?	No	No	Yes
Listened to audiotaped interviews?	Yes	No	Yes
Method and type of data summarized	Interviewers summarize audiotapes, debriefed by researchers	Researchers review notes taken during interview, compile list of all problems	Researchers summarize audiotapes, compile list of all problems
Report authors	Researchers, not interviewers	Researchers/ interviewers	Researchers/ interviewers
Level of interviewer experience/training	5+ years conducting household interviews; 2 years experience in cognitive interviews; some college	$2\frac{1}{2}$–12 years of experience in cognitive interviews; MA in social science	3–8 years of experience in cognitive interviews; MA in social science
Average interview length	48 minutes	45 minutes	40 minutes

be less experienced in cognitive interviewing; (3) the questionnaire should be tested "as is"; (4) research teams would not have access, as would be typical, to a survey sponsor for guidance and/or feedback; (5) the interviews should use typical procedures to the extent that the experimental environment would allow; and (6) all cognitive interview findings should be reported in a question-by-question format, with a revised questionnaire attached.

5.4.2 The Questionnaire

Each research team used the same questionnaire, a paper version of a preexisting general population computer-assisted phone survey on recycling containing 48 items,[2] with an expected interview time of 10 to 15 minutes. The questionnaire asked about trash removal practices in households, recycling practices, attitudes about recycling, and opinions on alternative recycling strategies. This particular survey was selected for several reasons: (1) a general population survey decreases the possibility of bias due to recruiting respondents with special characteristics or behaviors; (2) the survey topic was not known to be an area of substantive expertise for any of the research teams; (3) it contains both factual and opinion questions; (4) it had been fielded previously;[3] and (5) there seemed to be room for improvement in the design of the questionnaire.

5.4.3 Evaluating the Data

Comparing Teams' Cognitive Interviewing Results To quantify and analyze the cognitive interview results, the authors applied a questionnaire appraisal coding scheme to problems described in the teams' reports. Many coding schemes exist (see, e.g., Blair et al., 1991; Bolton, 1993; Conrad and Blair, 1996; Lessler and Forsyth, 1996; Presser and Blair, 1994; Rothgeb et al., 2001), and most are based on a four-stage response process model. The scheme used for the current experiment was developed by Lessler and Forsyth (1996) and refined by Rothgeb et al., (2001) and was selected because of its fine-grained approach to identifying questionnaire problems.[4] As shown in Table 5.2, it contains 28 unique problem types, organized into five general categories, two of which have three subdivisions, corresponding to a response process model (Tourangeau, 1984; Tourangeau et al., 2000). Although the problems presented in each team's written report were coded in one of the 29 categories,[5] our analyses collapse the 29 categories into

[2]Copies of the questionnaire are available from the authors on request.

[3]This telephone survey was implemented in August 1999 by the University of Delaware Survey Research Center.

[4]Minor alterations were made to the scheme used by Rothgeb et al. (2001). Mainly, the *interviewer difficulties* category was separated from the *comprehension* category, since we deemed this a separate phenomenon from the process of respondent comprehension. Also, a code for *multidimensional response set* (code 28) was added.

[5]The teams' reports were coded in their entirety under blind conditions by both authors, with a good level of intercoder agreement (80.8%). Differences in the application of the problem-type codes were reconciled by a postcoding coder review.

Table 5.2 Questionnaire Appraisal Coding Scheme

Interviewer Difficulties	Comprehension	Retrieval	Judgment	Response Selection
IR difficulties 1 Inaccurate instruction 2 Complicated instruction 3 Difficult for interviewer to administer	Question content 4 Vague/unclear question 5 Complex topic 6 Topic carried over from earlier question 7 Undefined/vague term Question structure 8 Transition needed 9 Unclear respondent instruction 10 Question too long 11 Complex/awkward syntax 12 Erroneous assumption 13 Several questions Reference period 14 Period carried over from earlier question 15 Undefined period 16 Unanchored/rolling period	Retrieval from memory 17 Shortage of memory cues 18 High detail required or information unavailable 19 Long recall or reference period	Judgment and evaluation 20 Complex estimation, difficult mental calculation required 21 Potentially sensitive or desirability bias	Response terminology 22 Undefined term 23 Vague term Response units 24 Responses use wrong or mismatching units 27 Unclear to respondent what response options are 28 Multidimensional response set Response structure 25 Overlapping categories 26 Missing response categories

either the five broadest categories or the nine somewhat finer categories. The question number where the problem occurred was also noted.

In addition, the reports indicated that some questionnaire problems were identified by questionnaire design expertise alone, whereas others were clearly evidenced in cognitive interviews, and we coded the source of information.[6] (The process used by teams to translate respondent input into specification of question problems was out of the scope of this project.) Whether the documented problems were addressed by the questionnaire revision recommendations was also coded.

An illustration is provided to demonstrate how the coding scheme was applied. We present the raw data for an open-ended question (presented below), which asked how often the household recycles, and required the interviewer to field code the responses.

EXAMPLE QUESTION

When was the last time you (took any of these materials to a recycling center/put materials out to be collected by the service)?

(1) Within the past week
(2) Two or three weeks ago
(3) More than one month ago
(4) Don't remember

Table 5.3 presents the problems contained in each team's report and how they were coded. Problems were reported by at least one team in three of the five broadest categories. While six problems were reported, none of these were the same problems. In one case, the same code was applied, however. Teams A and C both noted problems with vague or undefined terms. Team A noted a problem with the definition of a "recycling center," and team C noted ambiguity with the term "you"—it was unclear whether it referred just to the respondent or to all household members. Teams A and C both found problems with the response categories, but they focused on different aspects of the categories.

Comparing Teams' Results with Expert Review Findings In the fall of 2001, three survey methodologists were enlisted to conduct expert reviews of the recycling questionnaire. These survey methodologists, from three different federal agencies, were recruited based on their experience with questionnaire evaluation and their availability to complete the task in the time frame required.[7]

[6] Although researchers were not given explicit instructions about including problems identified only through their own expertise, all the teams included such problems and clearly described them as such. In coding this variable, we relied on the way the problem was presented in the report. Where the problem was identified both in the interviews and through the expertise of the research teams, we gave precedence to the cognitive interviews. Where the problem's source was unclear, this was also noted.

[7] Although one of the expert reviewers belonged to the same organization as one of the teams, the expert was not involved in any aspect of this research besides providing the questionnaire review. Neither author performed expert reviews. Each reviewer had 10 or more years of experience in questionnaire design, cognitive interview methods, and/or survey interview process research.

Table 5.3 Illustration of Problem Codes Applied to Question About Materials Used for Recycling

Problem Codes	Description of Problems in Report	Team A	Team B	Team C
Interviewer difficulty				
Difficult for interviewer to administer (code 3)	Not clear whether response categories should be read to respondent	×		
Comprehension				
Undefined/vague term (code 7)	Meaning of "recycling center" unclear	×		
Undefined/vague term (code 7)	Meaning of "you" unclear			×
Erroneous assumption (code 12)	Both situations (curbside recycling, drop-off) may apply, question allows for only one		×	
Response selection				
Missing response categories (code 26)	No category for "one month ago"			×
Unclear to respondent what response options are (code 28)	Respondents don't know list of precoded answers	×		

Expert reviewers were provided with a questionnaire and a paper form for noting and describing potential problems as they occurred across the questionnaire in a question-by-question format. Each form was then coded independently by both authors, by applying the previously mentioned coding scheme in the same manner as for the teams' reports.[8] Thus, it was possible to determine the number and types of problems that experts found in the questionnaire and the extent of agreement among them regarding where and how the questionnaire was flawed.

Comparing Revised Questionnaires Each team produced a revised questionnaire based on the problems identified and recommendations proposed. To evaluate the effectiveness of the revised questionnaires, each version was subjected to another round of 10 cognitive interviews (referred to here as *confirmatory interviews*). The original questionnaire was also included, as a benchmark measure for comparison with the revised questionnaire versions.

[8]The questionnaire problems documented by the expert review forms were coded by both authors, with 76.3% intercoder agreement.

Four professional cognitive interviewers from an independent survey research organization, unaffiliated with any of the teams' organizations, conducted confirmatory cognitive interviews using the four questionnaires. All interviewers had comparable interviewing experience, and each pretested one questionnaire version. Cognitive interviewers were instructed not to discuss the results with one another, to prevent contamination across the versions. After conducting the interviews, interviewers were instructed to generate one report for each version, including (1) a record of the questions where problems occurred, (2) a description of the nature of each problem, and (3) a count of the number of times that each problem occurred over 10 interviews.

5.4.4 Evaluating Interview Protocols and Administration

At the outset of this research, we suspected that certain facets of the interview protocol and its administration might differ across research teams, especially as used by survey researchers versus field interviewers. But because our objective was to use cognitive interview methods as they are *typically* implemented, protocol-related restrictions were not imposed. Since the research design called for each team to develop its own protocol, information was gathered regarding the relative similarities of the protocols and the ways in which they were administered. Major differences in these aspects may affect the conclusions drawn from the team's results.

To this end, the contents of the interview protocol and interviewers' probing behavior during the interviews were analyzed. A question-by-question evaluation of the protocols determined the total number of scripted probes[9] and the nature of those probes (i.e., concurrent versus retrospective[10] and general versus specific probes[11]). Then, six audiotaped interviews were sampled from each team, for a total of 18 interviews, and coded by the authors.[12] Interviewers' behavior was also analyzed with regard to the number of scripted and unscripted probes administered during the interviews.[13]

[9] *Scripted probes* are defined as those included in the protocol. They did not have to be read verbatim as long as the intent of the probe was maintained.

[10] *Concurrent probes* are defined as probes administered right after the question was posed during the interview; *retrospective probes* were probes administered at the conclusion of the interview that were about particular items. A category for *general debriefing probes* was also included, which referred to probes administered at the end of the interview that did not refer to any specific survey question.

[11] *General probes* are defined as those that could be used to probe any survey question (e.g., "What did you think that question was asking?" and "How did you come up with your answer?"). In contrast, *specific probes* could not be used to probe other survey questions because they referred to specific aspects of the question (e.g., "What is the difference between a drop-off center and a recycling center?" and "Do you think that these two questions are asking the same thing?").

[12] The authors coded the same six interviews for reliability, two from each team, with an intercoder agreement statistic of 89%.

[13] *Unscripted probes* are defined as impromptu probes not present in the protocol but rather those that the interviewer initiated to get additional information about the respondent's question interpretation, response formulation, or problems with the question. This does not include interviewers' confirmatory comments, where they sought confirmation by restating information that they believed respondents had already provided.

5.4.5 Data Analysis

Coded data from the cognitive interview reports and the expert reviews were merged and matched by problem type and question number to facilitate comparison. Data were entered into a Microsoft Excel document and statistical analysis was conducted using SAS. Significance tests consist of chi-square statistics.[14]

5.4.6 Limitations

1. *Ability to differentiate effects of the methods.* By imposing few artificial constraints on the teams' implementations of the cognitive interview method, the comparisons are embedded in cognitive interview "packages." This makes it impossible to identify the effects of particular differences. Furthermore, each method is represented here by a single effort from a single team, and other teams using similar implementations of these methods may differ in subtle ways. Nevertheless, this research sheds light on the strengths and possible weaknesses of these commonly used implementations of the cognitive interview, which is necessary for exploring a frequently used, unstandardized, and inadequately documented survey development procedure.

2. *Use of expert reviews as an evaluation method.* That expert reviews typically uncover far more questionnaire problems than cognitive interviews is well documented (Presser and Blair, 1994). One reason for this is that expert opinions cast a wider net over potential problems that interviewers and respondents may encounter than may actually surface during cognitive interviews. Thus, we did not expect that the cognitive interviews would uncover all the problems identified by the experts. Nevertheless, the expert results are a convenient benchmark against which to gauge the effectiveness of the cognitive interview implementation.

3. *Use of cognitive interviews as an evaluation tool.* Confirmatory cognitive interviews conducted with the original and revised questionnaires have some potential for confounding, since the evaluation method is not independent of the experimental manipulation. The procedures used in the evaluation interviews most closely matched those used by team B—interviewers took notes on problems identified in the interviews and used the notes to prepare the report. Due to an oversight, the confirmatory interviews were not tape-recorded.

4. *Lack of interaction with a survey sponsor.* There was no interaction with a survey sponsor, which typically serves as a means of obtaining detailed information about the question objectives. All three research teams mentioned this during the debriefing session as a factor that may have affected their work. This may have affected the recommendations made in the revised questionnaires, since the amount of perceived liberty available regarding the structure of the questionnaire may have differed by research team.

[14]We recognize that statistical tests are not strictly appropriate, since the data do not arise from a probability sample. Rather, we use them for heuristic purposes.

5.5 RESULTS

Do different cognitive interviewing methods identify different numbers and types of questionnaire problems? Table 5.4 presents the number of problems identified by research teams, in total and in the cognitive interviews alone. Teams A and C identified similar numbers of problems, while team B found 30% fewer. *Comprehension* issues constitute the most common problem type for all three teams. This includes problems identified during the interviews themselves as well as by the prior review conducted by the researchers and is consistent with previous research on cognitive interviewing (Presser and Blair, 1994; Rothgeb et al., 2001). Problems with *response selection* were the second-most-frequent type of problem identified by all the teams.

Problems related to *interviewer difficulties* exhibited wide variation among the teams. While teams A and C documented many of these problems, team B found only three. The significant difference overall in the types of problems identified by the research teams was due entirely to this difference in interviewer difficulty problems.

Table 5.4 Distribution of Problem Types by Source and Research Teams

	Problem Source					
	Team A		Team B		Team C	
Problem Type	Total ($n = 106$)	C.I.[a] ($n = 51$)	Total ($n = 60$)	C.I. ($n = 32$)	Total ($n = 84$)[b]	C.I. ($n = 53$)
Interviewer difficulties	24	1	3	1	16	11
Subtotal	24	1	3	1	16	11
Comprehension						
Question content	28	22	16	13	20	16
Question structure	15	14	21	8	21	8
Reference period	4	0	1	0	0	0
Subtotal	47	36	38	21	41	24
Retrieval	4	4	5	4	3	3
Subtotal	4	4	5	4	3	3
Judgment	2	2	2	1	0	0
Subtotal	2	2	2	1	0	0
Response selection						
Response terminology	2	2	1	1	0	0
Response units	17	4	5	2	11	7
Response structure	10	2	6	2	13	8
Subtotal	29	8	12	5	24	15

[a] Evidence for problem found in cognitive interview (C.I.).
[b] Total n includes two problems for which a problem source (i.e., C.I. or expertise) could not be identified.

Essentially the same results are obtained in comparing the distribution of problem types identified during the cognitive interviews alone (shown in the "C.I." column). Team C identified significantly more interviewer difficulty problems based on the cognitive interviews than did either team A or B. The low prevalence of interviewer problems for team A occurred because the researchers identified and repaired the bulk of these problems before they finalized the questionnaire for testing. The low prevalence for team B may result from a lack of attention to these problems by the researchers, and team C's high rate of interviewer difficulty problems runs counter to previous research (Presser and Blair, 1994) in which the number of interviewer problems identified in cognitive interviews was virtually nil.

Another way to evaluate the teams is by the number of problematic questions they identified rather than the number of problems they found. Team A identified 27 of the 48 questions, team B identified 20 questions, and team C identified 28 questions (data not shown). These differences are not statistically significant ($X^2 = 3.17$, $df = 2$, $p = 0.20$).

Although there is no difference either in the number of problematic questions identified or in the number of problems found (except for interviewer difficulty), there is evidence that the teams were not reporting the same problems at the same point in the questionnaire. The agreement rate among the teams was only 34%.[15] This contrasts with a much higher level of agreement, 81%,[16] for the teams in identifying any problems in those questions.

Do different cognitive interviewing methods have differing abilities to identify problems predicted by expert reviewers? To answer the second research question, we compared the results of the expert reviews with the findings in the teams' reports. We conducted a question-by-question review of problems found by the experts in the original questionnaire to create a set of 39 "defective" questions—that is, two or more experts described at least one of the same general problem types (the five broadest categories) for a given question.[17]

The top half of Table 5.5 shows that the teams found fewer problematic questions in the questionnaire than the experts did. Overall, however, there were no differences among the teams in identifying the questions judged defective by the experts. The bottom half of the table shows the extent to which the teams

[15]Agreement statistic was generated by dividing the total occurrences of cases where two or more teams agreed on problem type and question number by the total number of mutually exclusive problem types across all teams ($n = 94$).

[16]Agreement statistic was generated by dividing the total occurrences of cases where two or more teams agreed that a question had any type of problem by the total number of problematic questions identified across all teams ($n = 36$).

[17]Initially, we hoped to use the evaluations as an objective "standard" to evaluate the quality and quantity of the problems identified by the three teams. However, in comparing the problems identified by each of the three experts, the level of agreement among the experts regarding both the number and types of problems identified was quite low, only 21%. As a result, we switched our focus to problem questions rather than the problems themselves. For this approach, the agreement rate was much higher, 81%.

Table 5.5 Number of Defective Questions Identified by Experts and Research Teams in Cognitive Interviews

	Team A	Team B	Team C
Number of defective questions[a]			
Identified by team and experts[b]	24	18	26
Identified by experts but not team	15	21	13
Total	39	39	39
Number of defective questions identified			
Where $\geq$ 1 problem matched experts[c,d]	18	16	25
Where no problems matched experts	6	2	1
Total	24	18	26

[a] 39 out of 48 questions were identified by experts as defective (i.e., at least two experts identified at least one problem for a particular question and agreed on its general nature; collapsing the 26 problem types into the five response stage categories).
[b] No significant difference in teams' ability to identify defective questions ($X^2 = 3.65$, df = 2, $p = 0.16$).
[c] *Matched* means teams agreed with experts on the nature of at least one problem in these questions.
[d] Difference in teams' ability to find problems noted by experts approaches significance ($X^2 = 4.96$, df = 2, $p = 0.08$); team A vs. C: $X^2 = 4.64$, df = 1, $p = 0.03$.

identified the same types of problems recognized by the experts in the defective questions. The criteria for agreement were very broad—at least one of the problem types (as defined by the five broadest categories) identified by the experts and the teams had to be the same. In most cases, there was some agreement between problems identified by the experts and the teams for the same question. Some of the reasons for the teams' identification of defective questions were different from the reasons specified by the experts, and the variation across teams approaches significance. The main contributor to this result is that team A found more problems that did not match the experts than did team C. Teams B and C did not differ significantly in this regard, although team C had almost perfect agreement with the experts.

Table 5.6 focuses on problems rather than on defective questions. It presents information about the number of problems identified by the teams and the overlap with those identified by the experts. The top half of the table shows that the great majority of problems identified by the teams were in the questions judged defective by the experts. Very few problems were identified with other questions, and the differences among the teams in this regard were not significant.

The bottom half of the table demonstrates that although the teams and the experts identified problems in the same questions, the nature of the problems differed in half or more of the cases. Teams differed in their ability to identify problems found by experts in the defective questions,[18] and this difference approaches significance. Team C differs significantly from team B, and the

[18] A total of 65 problems were found by experts in the 39 defective questions.

Table 5.6 Number of Problems Identified for Defective Questions by Experts and Research Teams in Cognitive Interviews[a]

	Team A	Team B	Team C
Number of problems identified			
In defective questions[b]	48	30	50
In all other questions	3	2	3
Total	51	32	53
Number of problems identified in defective questions			
By teams and experts[c,d]	21	19	31
By experts only	44	46	34
Total	65	65	65

[a] 39 out of 48 questions were identified by experts as defective—they identified at least one problem and agreed on the nature of the problem.
[b] Difference in teams' ability to identify defective questions not significant ($X^2 = 0.01$, df = 2, $p = 0.99$).
[c] Teams agreed with experts on the type of problem.
[d] Difference in teams' ability to identify same problems as experts in defective questions approaches significance ($X^2 = 15.49$, df = 2, $p = 0.06$); team C vs. B: $X^2 = 4.68$, df = 1, $p = 0.03$; team C vs. A: $X^2 = 3.21$, df = 1, $p = 0.07$.

differences between teams A and C are close to significant. Overall, team C did a better job of identifying problems that were also found by the experts.

Do interview protocol and probing behavior influence the results? We explored whether differences in the teams' protocols might have influenced the results by reviewing the content of each team's protocol and listening to a sample of the audiotaped interviews to determine how the protocols were administered.[19] This revealed significant differences in the types of probes administered by the teams. This is in marked contrast to the previous results, which show very few differences across teams.

Table 5.7 presents summary information about the types and timing of probes as they occurred during the cognitive interviews.[20] There was a significant difference between the teams' administration of concurrent and debriefing probes: Team A made greater use of debriefing probes than did teams B and C. There was also a significant difference in the teams' administration of scripted and unscripted probes. Team B used significantly fewer unscripted probes in the sampled interviews than did either team A or C. Although team A used more unscripted probes than team C, this difference was not significant. There was

[19] A review of the protocols shows that team A included 68, team B included 107, and team C included 42 probes. Comparison of the protocols across teams shows pretty much the same pattern as the administration of the interviews. We focus on the protocols as administered.

[20] Although these results are presented as averages, the analysis was performed on the teams' total numbers of probes administered in the interview samples. During six interviews each, team A administered 354 probes, team B administered 344 probes, and team C administered 233 probes.

Table 5.7 Average Number and Types of Probes Administered by Research Teams in Cognitive Interviews[a]

	Number of Probes Administered per Interview		
Probe type	Team A	Team B	Team C
Concurrent[b]	54.8	55.2	37.8
Debriefing[b]	4.2	2.1	1.0
Total	59.0	57.3	38.8
Scripted[c]	40.1	46.3	23.9
Unscripted[c]	18.9	11.0	14.9
Total	59.0	57.3	38.8
General[d]	13.2	8.5	12.0
Specific[d]	45.8	48.8	26.8
Total	59.0	57.3	38.8

[a]Averages for number and types were derived from a sample of six interviews from each research team.
[b]Difference in teams' use of concurrent and debriefing probes is significant ($X^2 = 7.37$, df = 2, $p = 0.03$); team A vs. B: $X^2 = 3.65$, df = 1, $p = 0.06$; team A vs. C: $X^2 = 5.67$, df = 1, $p = 0.02$.
[c]Difference in teams' use of scripted and unscripted probes is significant ($X^2 = 27.23$, df = 2, $p = <0.0001$); team A vs. B: $X^2 = 14.84$, df = 1, $p = 0.0001$; team B vs. C: $X^2 = 25.55$, df = 1, $p = <0.0001$.
[d]Difference in teams' use of specific and general probes is significant ($X^2 = 21.26$, df = 2, $p = <0.0001$); team A vs. B: $X^2 = 6.46$, df = 1, $p = 0.01$; team B vs. C: $X^2 = 21.40$, df = 1, $p = <0.0001$; team A vs. C: $X^2 = 5.42$, df = 1, $p = 0.02$.

also a significant difference in the teams' use of general and specific probes. Although two-thirds or more of each team's probes were specific, team B relied on them most heavily. Team A seemed to have the most varied implementation: It used more debriefing probes than did either of the other teams; along with team C, it included more unscripted probes than did team B; and it used more general probes than did team B.

Comparing the number of probes used by the teams with the number of problems the teams identified provides evidence about the efficiency of probing strategies. Team C identified as many problems in the cognitive interviews as did team A (and more than team B), despite administering an average of 20 fewer probes per interview.

Do different cognitive interviewing methods produce different questionnaire recommendations? Our final research question deals with the teams' recommendations for revising the questionnaire. In the revised questionnaires, the teams addressed almost all the problems they had identified.[21] Rewording questions was the most frequent means of addressing problems and was used heavily by all three teams. We subjected the original and revised questionnaires to an additional series of confirmatory cognitive interviews, which provides some indication of each version's performance.

[21]Teams A, B, and C addressed 96, 98, and 95% of the problems, respectively (data not shown).

Table 5.8 shows that team A's revised questionnaire had the greatest number of *problem incidences* compared to the other three questionnaires (first row). But in terms of the number of *questions found to be problematic* (second row), team A's questionnaire was the only one that was significantly better than the original questionnaire (although the teams' revised questionnaires did not differ significantly from one another on this measure).

Table 5.9 presents the number of times that questions were problematic across the original and revised questionnaires in the confirmatory interviews. Pairwise comparisons suggest that team A's questionnaire performed differently from team B's and C's, and team C's questionnaire performed differently than the original. In particular, team A had the largest number of questions with four or more problems, and team C the largest number of questions with only one problem.

Table 5.8 Summary Information About Problems Found Through Confirmatory Interviews by Questionnaire Version

	Questionnaire Version			
Measure of Problems	Original ($n = 48$)[a]	Team A ($n = 60$)	Team B ($n = 61$)	Team C ($n = 61$)
Total incidences of problems	76	77	60	70
Number of questions found problematic[b]	28	19	27	34

[a]Total number of questions per questionnaire version.
[b]No significant difference in problematic questions found by teams ($X^2 = 2.86$, df $= 2$, $p = 0.24$); only team A differs significantly from original—team A vs. original: $X^2 = 7.71$, df $= 1$, $p = 0.006$.

Table 5.9 Number of Questions with Single and Multiple Problems Found Through Confirmatory Cognitive Interviews by Questionnaire Version

	Questionnaire Version[a]			
Number of Questions Containing:	Original ($n = 48$)[b]	Team A ($n = 60$)	Team B ($n = 61$)	Team C ($n = 61$)
1 problem	6	4	12	18[c]
2 or 3 problems	16	6	13	12
≥4 problems	6	9[d]	2	4

[a]Difference in performance of questionnaire versions significant ($X^2 = 19.07$, df $= 6$, $p = 0.004$); team A vs. B: $X^2 = 9.94$, df $= 2$, $p = 0.007$; team A vs. C: $X^2 = 9.33$, df $= 2$, $p = 0.009$; team C vs. original: $X^2 = 6.45$, df $= 2$, $p = 0.04$.
[b]Total number of questions in questionnaire.
[c]Difference in performance of versions collapsed to 1 to 3 and 4 or more problems is significant; team A vs. B: $X^2 = 9.79$, df $= 1$, $p = 0.02$; team A vs. C: $X^2 = 8.35$, df $= 1$, $p = 0.004$.
[d]Difference in performance of versions collapsed to 1 and more than 1 problem is significant; team A vs. C: $X^2 = 5.11$, df $= 1$, $p = 0.02$; team C vs. original: $X^2 = 6.43$, df $= 1$, $p = 0.01$.

Hence, the results of the evaluation are mixed. Team A's revised questionnaire contained significantly fewer problematic questions than did the original version, but contained more problematic instances and questions that were frequently problematic. Team B had the lowest number of problematic instances but did not significantly outperform the other teams or the original in the number of problematic questions. Team B did, however, have significantly fewer frequently problematic questions than did team A. Finally, team C's revised questionnaire was the only version to differ significantly from the original on the number of questions with rare problems.

5.6 DISCUSSION

Which team did best? There is no clear answer to this question. Teams A and C performed similarly in their ability to identify problems. Team C was more likely to identify problems predicted by the experts. Team A's revised questionnaire had the fewest problematic questions, although it also had more "high-frequency" problem questions. In contrast, team C's revised questionnaire had more rarely problematic questions. Although no team clearly performed best across the different measures, teams A and C clearly outperformed team B.

In interpreting these results, we consider the characteristics of the teams. Teams A and C were highly divergent in terms of their interviewing personnel. However, they shared the fact that interviewers listened to the audiotapes as a standard part of problem identification. Given the similarity between the performance of these teams, one conclusion is that this additional time spent with the data is well worth it. Team B, whose interviewers relied only on notes taken during the interviews, were much less likely to identify problems.

The relative similarity of team A's and C's performance, however, may obscure differences in other aspects of the methods used. The process of protocol development, for example, was very different for the two teams. For team A, the researchers went through a very careful and explicit procedure to document for the field interviewers what the objectives of the probing were for each question, and relied heavily on scripted probes. Team C, on the other hand, used a much more general approach to designing the protocol, since the protocol designers were themselves the interviewers. They used fewer scripted probes and presumably relied on their knowledge of the response process to guide their probing. Thus, these differences in the design and administration of the protocol may not be critical to the overall performance of the team (e.g., the number and type of probes found).

The teams also differed in use of the think-aloud method. Team C instructed respondents to express all their thoughts as they interpreted and answered the survey questions and provided reminders during the interview. Teams A and B did not provide such instruction, but rather, conducted standardized survey interviews with probing. The similarity of team A's and C's performance would seem to suggest that the think-aloud method is not beneficial in eliciting problems. That would be too strong a statement, however, because our review of the tapes revealed that all three teams did, in fact, get respondents to think out loud. Perhaps the act of probing suggests to respondents that their thoughts are important, and this elicits spontaneous verbalization of thoughts without explicit instruction. Thus, we conclude that think-aloud instructions may not be critical to cognitive interviews.

Finally, we note differences among the teams in who conducted the interviews. Cognitive interviews for both teams B and C were conducted by professional survey researchers with similar levels of experience, and these two teams performed very differently. In contrast, teams A and C, which used quite different interviewing staffs, performed similarly. This suggests that the level of experience of the interviewing staff may not be critical as long as professional researchers are closely associated with the field staff who conduct the interviews.

5.7 CONCLUSION

These results suggest that the extra time and effort associated with listening to tapes of cognitive interviews have a big payoff in identifying respondent problems. Regardless of who listens to the tapes, the added exposure to the thoughts and comments of respondents can supply further insight into or clarification of the response process. This provides some evidence that a more rigorous review of the data may result in a greater understanding of questionnaire problems.

The results also suggest that properly trained field staff, working in conjunction with experienced survey methodologists, can identify equal numbers of questionnaire problems as professional survey researchers. Here the guidance is not so clear-cut. Although we know that the same problems were not identified across teams in the same questions, we were unable to determine the specific nature of the differences. This causes us some concern, since more information about these differences might shed a different light on the teams' performances.

The dramatic differences observed in the administration of the protocols suggest that this is a promising area for future research. To some extent the differences can be explained by the dissimilarities in interviewing staffs—the protocol had to be spelled out in more detail for the field interviewers, while the researcher/interviewers may not need as structured a protocol. This is not a blanket statement, however, since the two teams of interviewer/researchers performed quite differently in this regard. Variations in the length of the protocols, the number of probes administered, and the types of probes used demonstrate that there are differences in the efficiency of probing strategies, and this too warrants more investigation.

ACKNOWLEDGMENTS

The authors gratefully acknowledge the anonymous contributions of the survey researchers, field interviewers, and expert reviewers who participated in the research. We also acknowledge Elizabeth Martin, Jennifer Rothgeb, and Stanley Presser, who provided insightful comments on previous drafts of the chapter.

In this chapter we report the results of research and analysis undertaken by U.S. Bureau of the Census staff. It has undergone a Census Bureau review more limited in scope than that given to official Census Bureau publications. This report is released to inform interested parties of ongoing research and to encourage discussion of work in progress. The views expressed are attributable to the authors and do not necessarily reflect those of the Census Bureau.

CHAPTER 6

Evaluating Survey Questions by Analyzing Patterns of Behavior Codes and Question–Answer Sequences: A Diagnostic Approach

Johannes van der Zouwen and Johannes H. Smit
Free University

6.1 EVALUATION OF QUESTIONS USING BEHAVIOR CODING AND SEQUENCE ANALYSIS

The idea of using what is said during an interview as a basis for evaluating the adequacy of the questions of a questionnaire is already over 30 years old (Fowler and Cannell, 1996). Marquis and Cannell's (1969) study of interviewer–respondent interaction in the Urban Employment Survey was specifically aimed at an assessment of task-related behavior of interviewers and respondents. However, the authors found that some questions evoked much more inadequate behavior than others. Thus, the results of behavior coding could be used for diagnosing specific question problems. "The list of incorrectly asked questions revealed several items which contained parenthetical phrases, others which contained difficult syntax, and still others were extremely cumbersome to handle in verbal form" (Cannell et al., 1977, p. 27).

This first application of behavior coding to the evaluation of a questionnaire was followed by quite a few others (see Table 6.1). In line with the development of survey practice, the earlier paper-and-pencil interviews (PAPI) were succeeded

Methods for Testing and Evaluating Survey Questionnaires, Edited by Stanley Presser, Jennifer M. Rothgeb, Mick P. Couper, Judith T. Lessler, Elizabeth Martin, Jean Martin, and Eleanor Singer
ISBN 0-471-45841-4

Table 6.1 Overview of Studies Using Behavior Coding for the Evaluation of Questions

(1) Study Year and Author(s)	(2) Mode of Interview[a]	(3) Procedure[b]	(4) Coding[c]	(5) Criteria for Problem Question
1969, Marquis and Cannell	PAPI	AT	ALL	<85% correct asked <85% adequate answers
1979, Morton-Williams	PAPI	AT	SEL	<85 correct asked <85% adequate answers >20% secondary activity
1982, Brenner	PAPI	AT	ALL	
1985, Prüfer and Rexroth	PAPI	AT	ALL	% ideal sequences
1987, Sykes and Morton-Williams, Study 1	PAPI	AT	SEL	(mean +5)% problem indicators
1991, Oksenberg et al.	PAPI	L	SEL	% respondent problem % no adequate answer
1992, Sykes and Collins	PAPI	AT	SEL	% straightforward sequences
1992, Esposito et al.	CATI	L	ALL	<90% adequate answers
1992, Fowler	CATI	AT		>15% requests for clarification >15% inadequate answers
1994, Cahalan et al.	CATI	AT	SEL	<85% of 'asked and answered only' sequences
1995, van der Zouwen and Dijkstra	PAPI	TR	ALL	>60% nonparadigmatic sequences
1996, Bates and Good	CAPI	AT	ALL	>15% incorrect asked >10% question omitted
1997, Dykema et al.	PAPI	AT	SEL	Probability of accurate answer
1999, Hess et al.	TI	AT	SEL	<85% correct asked <85% adequate answer Reliability of answers
2000, Comijs et al.	PAPI	TR	ALL	% questions skipped % unusable answers
2002, van der Zouwen and Dijkstra	CATI	AT TR	ALL	>60% nonparadigmatic sequences
Present study	CAPI	TR	SEL	% paradigmatic−% inadequate sequences

[a]PAPI, paper and pencil interview; CATI, computer-assisted telephone interview; TI, telephone interview; CAPI, computer-assisted personal interview.
[b]AT, behavior coding from audiotape of the interview; L, live coding (during the interview); TR, coding from transcripts of the audiotape.
[c]ALL, all behaviors of interviewer and respondent are coded; SEL, only a selection of these behaviors is coded.

by computer-assisted telephone interviews (CATI) or computer-assisted personal interviews (CAPI). Most interviews are coded from audiotapes (AT); live coding (L) permits only a rough coding scheme because otherwise the coders cannot keep up with the interview. Only recently have some researchers based their coding of behaviors and question–answer sequences on transcripts of the tapes (TR) or on a combination of transcripts and tapes (AT&TR).

For studies aimed primarily at the interaction in interviews, coding of virtually all (verbal) behaviors (ALL) is necessary. For the evaluation of questions, only behaviors that deviate from those prescribed by the paradigmatic sequence have to be coded (SEL).[1] The number of different codes for the interviewer and for the respondent ranges from only a few [e.g., in the study of Dykema et al. (1997)] to dozens, as in the second study of Prüfer and Rexroth (1985).

The first researchers who coded not only separate behaviors but entire sequences were Prüfer and Rexroth (1985). They used a coding scheme with only three types of sequences: *ideal sequences, adequate but not ideal sequences*, and *sequences not resulting in an adequate response*.[2] Seven years later Sykes and Collins (1992) published a coding scheme for behavior coding on the basis of which sequences could be classified into six different types, and since then sequence classification has gained popularity.

The criteria for the identification of problematic questions are shown in the last column of Table 6.1. If only frequency tables of behavior codes are available, the criterion is a certain percentage of interviews in which the question is (not) asked correctly or answered adequately. If information is also available about the quality of the entire question–answer sequence, the criterion is a certain percentage of sequences that are not *ideal* (Prüfer and Rexroth, 1985), not *straightforward* (Sykes and Collins, 1992)[3] or not *paradigmatic* (van der Zouwen and Dijkstra, 1995). If, as in the study by Dykema et al. (1997), external information is available by which the answers given can be checked, the probability that the answer to a specific question is accurate can be used as an identifier for problematic questions as well. In the study by Hess et al. (1999), low test–retest reliability of the responses was used as another identifier of problematic questions.

The most common method for diagnosing the problematic character of a question is by means of the *code pattern*, the frequency with which specific codes are given to behaviors following that question. Other methods are follow-up interviews, analyses of the notes that coders made during the coding process, debriefings of interviewers and respondents, application of other procedures for pretesting, and recently, content analysis of transcripts of problematic parts of the interview.

[1]Using terminology coined by Schaeffer and Maynard (1996, p. 71), we call a sequence *paradigmatic* if the interviewer poses the question about the same as worded in the questionnaire, the respondent chooses one of the response alternatives, and the interviewer records the corresponding code.

[2]Prüfer and Rexroth (1985, p. 29) define an *ideal sequence* as a sequence in which the interviewer reads the question literally, the respondent answers adequately, and no other activities occur between the question and the answer.

[3]Sykes and Collins (1992, p. 286) describe straightforward sequences as those sequences that do not depart from the stimulus-response model of the interview.

Because the studies mentioned in Table 6.1 are in chronological order, it is possible to discern some developments over the last 30 years: (1) behavior coding is supplemented with classification of question–answer sequences, and as a result, (2) the percentage of nonparadigmatic sequences becomes an identifier of problem questions; and (3) the slowly increasing use of transcripts makes possible content analysis of these transcripts as a diagnostic method. The present study, characterized in the last row of Table 6.1, can be conceived as a further continuation of these three developments: behavior coding of transcripts has been supplemented with the analysis of question–answer sequences. The coding scheme we used is described in the next section.

6.2 SCHEME FOR BEHAVIOR CODING

6.2.1 Phase 1: Selection and Formulation of the Question

The question–answer process for a particular question starts with a decision by the interviewer as to whether to pose that question to this particular respondent. Not posing the question to a respondent where this actually is required (i.e., a wrong skip, abbreviated as WS) means that a sequence will not start at all (see arrow 1 in Figure 6.1). The omission of that sequence will, at least in principle,

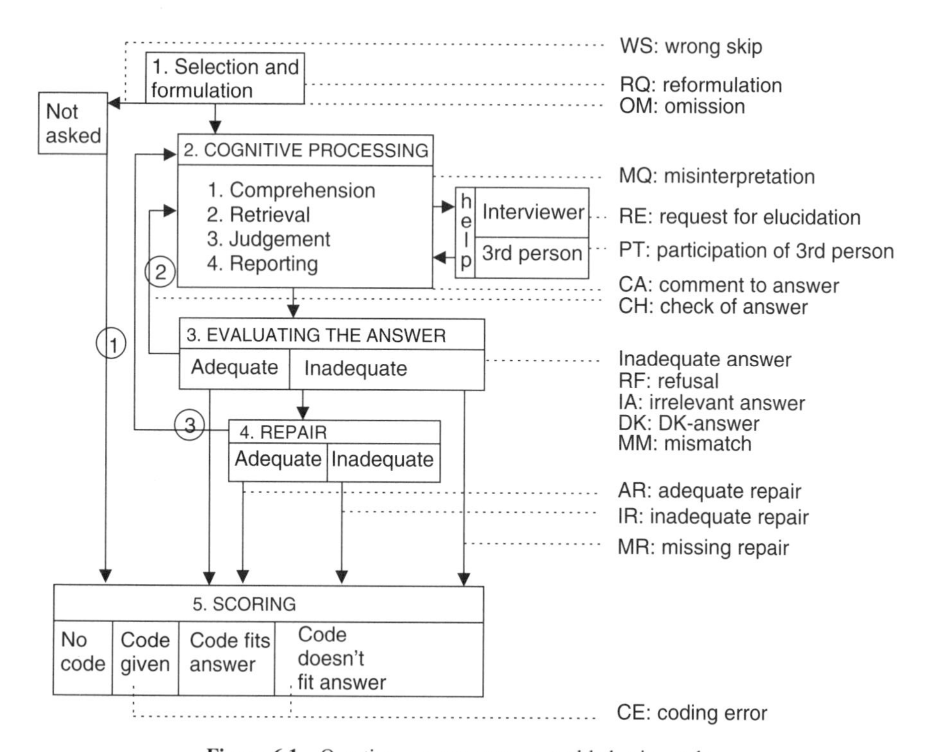

Figure 6.1 Question–answer process and behavior codes.

lead to a blank in the data matrix. However, in quite a few cases the interviewer assigns a response code to the respondent; this code is in all likelihood based on a guess by the interviewer. Such a coding error (CE) will of course lead to data of questionable quality.

It is possible that the interviewer poses the question in a way that deviates substantially from the text of the questionnaire. Such a reformulation (RQ) means that the respondent will answer a question different from the one intended, with the possibility that irrelevant or noncomparable information will be provided. Another deviation here is the omission (OM) of (part of) the response alternatives, or of the explanation provided with the question. A potential effect is that respondents give answers that are different from the ones they would have given if all alternatives and the explanation had been presented.

One of the reasons that things go wrong in the phase of formulating a question is that respondents interrupt the interviewers while they are reading the question. It turns out to be difficult to decide on the basis of the transcripts whether it was the respondent who interrupted or the interviewer who, by leaving a pause, gave the respondent the impression that the question was already finished. We therefore code only whether the question as it is eventually posed to the respondent contains errors of reformulation (RQ) or omission (OM).

6.2.2 Phase 2: Cognitive Processing

As soon as the interviewer has presented the text of the question and the response alternatives, a particular cognitive process begins within the respondent, aimed at finding an answer to the question. In line with Tourangeau and Rasinski (see Snijkers, 2002), it is assumed that this cognitive process consists of four steps:

1. Interpretation and comprehension of the question
2. Retrieval of information pertinent to answering the question
3. Integration and evaluation of the retrieved information
4. Selection of a response category

This cognitive process is not directly observable by others. However, during an interview, parts of this cognitive process may become indirectly observable. This happens if help is sought from the interviewer or a third person present during the interview or if respondents think aloud, indicating how they understand the question or which part of the question causes them trouble. These interactions and comments are very useful for assessing questions.

If the respondent asks for a repetition of a question or for an elucidation of a question, code RE applies. If a third person interferes with the response process (e.g., by providing the requested information), the code PT is given.

If respondents paraphrase the question, and by doing so show that they have misunderstood it, the code MQ applies. If in that case the interviewer does not supply the correct meaning, the misinterpretation may result in an answer that does not pertain to the actual question.

Sometimes the respondent mentions the reasons for the choice of a particular response category. If that happens, the behavior code CA (comment on answer) is given.

6.2.3 Phase 3: Evaluating the Answer

As soon as the respondent gives an answer, the interviewer has to evaluate it. For closed-ended questions to be adequate, the response has to be one of the response categories offered. However, an adequate response is not always a correct one. If interviewers are in doubt whether an answer is correct, they may decide to check (CH) the response and pose the question again (arrow 2). Usually, an adequate response, followed by a short reaction by the interviewer, forms the end of the audible part of the question–answer sequence.

We distinguish four types of inadequate responses to the question:

1. Refusal to give an answer (RF)
2. An answer that is irrelevant to the topic of the question (IA)
3. A "don't know" response (DK)
4. A mismatch answer (i.e., a relevant answer, but not one of the response categories) (MM)

6.2.4 Phase 4: Repair

If respondents give an inadequate answer, interviewers are expected to try, in collaboration with the respondent, to repair the initial response in such a way that it becomes adequate (see arrow 3). Interviewing skills (e.g., the skill to probe in a nonleading manner) play a crucial role in this. If the response is "adequately and sufficiently repaired," the sequence is coded (AR).

However, the repair activity may be inadequate, for example, if interviewers, in reaction to a mismatch answer, suggestively present only one of the response alternatives: "So you are satisfied with your income?" Here the remedy may be worse than the disease. It is also possible for respondents to stick to their initial reaction, for example by refusing to answer the question. In all these cases the sequence is "inadequately or insufficiently repaired" (code IR).

It may also happen that interviewers, after hearing an inadequate answer, do not attempt any repair activity at all but use the inadequate answer to guess what a respondent's adequate answer would have been, and code it accordingly. A sequence with necessary repair activity missing receives the behavior code "missing repair" (MR).

6.2.5 Phase 5: Coding the Response

If coding errors occur relatively often with a particular question, this may indicate that the response alternatives are unclear. The set of response codes is usually larger than the set of response options offered to the respondent; it may contain codes for refusals and don't know responses. These responses need not cause trouble in the coding phase. But if the respondent gives an irrelevant answer or a mismatch answer, and this is not repaired, no appropriate response code is available. If the interviewer then enters a response code, this can only occur on the basis of guessing.

6.3 THE DATASET

6.3.1 Sample

The data used to illustrate the diagnostic approach advocated in this chapter were collected as part of the Longitudinal Aging Study of Amsterdam (LASA), a study of the antecedents and consequences of changes in autonomy and well-being among older adults (see Beekman et al., 1995). In 1992 a representative sample of 3805 adults aged 55 to 85, stratified for age and gender, was drawn from the population registers of 11 municipalities in the Netherlands. The data for the present study were collected by CAPI in 1998. At that time 1770 participants were still in the study and capable of completing a face-to-face interview of approximately two hours.

6.3.2 Interviewers

Interviewers were recruited from the area the respondents lived in. New interviewers received a five-day training on interviewing techniques. The training included reading assignments on interviewer techniques, videotape exercises, pilot testing of instruments, and a complete interview with an elderly person. Interviewers who had worked on earlier waves of the study received 2.5 days of additional training before the survey to refresh their interviewing techniques and instruct them in new procedures and instruments. The interviewers were mostly female (95%), between 25 and 60 years old, and had a medium level of education (10 years). All interviews were audiotaped for review by the fieldwork staff in order to give supportive feedback on interviewer behavior.

6.3.3 Questions

To illustrate the diagnostic approach, we selected eight questions about the income of respondents and their partners. We know that questions about income are often problematic, because respondents either don't have the information or are not willing to answer the questions (Moore et al., 2000). The set of eight income questions is presented in Table 6.2. The set contains five questions about the income of the respondent and his or her partner (Q1 to Q5), two attitude questions about their income (Q6 and Q7), and one about expectations for future income (Q8). The question formats are also different: two are simple yes/no questions (Q1 and Q2), three are simple multiple-choice questions (Q6 to Q8),

Table 6.2 The Eight Questions About Income

Q1: Do you have an income of your own? We mean an income from work, but also from payments like pension, AOW,[a] or dividends that are coming to you in your own name. 1, no; 2, yes.

[If the respondent has a partner[b]]

Q2: Does your partner have an income of his/her own? We mean an income from work, but also from payments such as pension, AOW, or dividends coming in your partner's name. 1, no; 2, yes.

[If the respondent has no partner]

Q3a: Could you indicate, by means of this card, in which category your *net* income falls? Please mention the number of the category that applies.

[If the respondent has a partner]

Q3b: Could you indicate, by means of this card, in which category the joint *net* income of you and your partner falls? Please mention the number of the category that applies.
N.B. The card contains 12 income categories (excluding vacation allowance) ranging from 1000 to 1250 guilders per month (12,000 to 15,000 guilders per year) to 5001 guilders per month or more (60,001 guilders per year or more).

[If the respondent has no partner]

Q4a: Did your income decrease in the last three years?

[If the respondent has a partner]

Q4b: Did your income decrease in the last three years? Do *not* count the income of your partner.
0, don't know; 1, no; 2, yes, less than 100 guilders net per month; 3, yes, 100 to 200 guilders net per month; 4, yes, 200 to 300 guilders net per month; 5, yes, 300 to 400 guilders net per month; 6, yes, 400 to 500 guilders net per month; 7, yes, more than 500 guilders net per month.

[For respondents reporting a decrease in their income]

Q5: Could you indicate at what moment this decrease occurred?
Month: . . . Year: 19. . .

[If the respondent has no partner]

Q6a: Are you satisfied with your income level?

[If the respondent has a partner]

Q6b: Are you satisfied with your income level? We mean the joint income of you and your partner. 1, dissatisfied; 2, a little dissatisfied; 3, neither satisfied nor dissatisfied; 4, a little satisfied; 5, satisfied.

Q7: Are you satisfied with the standard of living you attain on your income?
1, dissatisfied; 2, a little dissatisfied; 3, neither satisfied nor dissatisfied; 4, a little satisfied; 5, satisfied.

Q8: Do you expect that in the next two years your income will 1, decrease?; 2, remain the same?; 3, increase?

[a]The acronym AOW used in questions Q1 and Q2 stands for *Algemene Ouderdoms Wet*, the Dutch Old Age Pensions Law, providing an income for everyone 65 years and older.
[b]If the respondent is married, the word *partner* is automatically replaced in the questionnaire by *husband* or *wife*.

and the other three questions have a slightly more complex format. Question Q3 is the only one in which a show card has to be presented by the interviewer.

6.4 EVALUATION OF THE QUESTIONS

6.4.1 Transcribing and Coding the Question–Answer Sequences

Those parts of the tapes that contained the question–answer sequences related to the income questions were transcribed, using very simple rules. The transcript gives information about the speaker (interviewer, respondent, or third person) and what he or she literally said.

For each of the eight questions we used a coding sheet, formatted as a spreadsheet with the 15 behavior codes shown in Section 6.2 as columns, and the interviews to which the transcripts belonged as rows. A total of 201 interviews was coded.[4]

For practical reasons, the coding was done by the authors. The coding turned out to be neither very difficult nor very time consuming, because the coding scheme only requires putting a "1" in the corresponding cell of the coding sheet if a particular deviation from the paradigmatic sequence occurs. All other behaviors receive a "0" by default. Moreover, the coding scheme runs parallel with the phases of the sequence. Most of the 15 codes belong to only one phase of the sequence, meaning that at each phase of the sequence only a few codes have to be considered. The average sequence could be coded in less than 1 minute.

6.4.2 Overall Distribution of Behavior Codes

Table 6.3 shows the relative distribution of the behavior codes for each separate question and for all eight questions taken together. As an arbitrarily chosen indication of a relatively high frequency of occurrence, all percentages for a particular question that are more than 1.5 times higher than for all questions together are printed in boldface type. From the last column it appears that in the phase of *selecting and formulating,* omission of (part of) the set of response alternatives or of the explanation of the question (OM) is the most common behavior code (in 42.1% of all 1358 sequences). In the *cognitive processing* phase we see that quite often (18.6%) respondents give comments (CA). When the initial *answer is evaluated*, the most likely reason for an inadequate answer is a mismatch (MM) between the answer given and the response alternatives (in 27.3% of cases). Over one-third of the initial responses given require *repair* by the interviewer. But less than half of these to-be-repaired responses are adequately repaired (AR). Finally, we were surprised by the high percentage (13.5%) of *coding* errors (CEs).

[4]In order to obtain a sample with enough variation regarding the interviewers and respondent characteristics, we randomly selected 240 interviews, of which 201 had audiotapes with sufficient sound quality for making transcriptions.

Table 6.3 Distribution of Behavior Codes and Sequence Types for the Eight Questions Separately and for All Questions Combined (Percent)

Code	Q1	Q2	Q3	Q4	Q5	Q6	Q7	Q8	Total
Phase 1: Selection and Formulation									
Wrong skip (WS)	0.0	**2.4**	1.5	0.5	**6.1**	0.0	**2.5**	0.0	1.0
Reformulation of the question (RQ)	2.5	0.8	**15.5**	4.5	**18.2**	0.5	6.0	4.5	5.5
Omission of response alternatives or explanation (OM)	55.6	**83.1**	0.5	32.8	0.0	**69.7**	**75.5**	0.0	42.1
Phase 2: Cognitive Processing									
Misunderstanding of the question (MQ)	0.0	0.0	2.5	**3.5**	**12.1**	0.0	0.0	**3.5**	1.7
Request for repetition or elucidation of the question (RE)	0.5	0.0	**28.5**	1.0	0.0	0.0	2.0	1.0	4.9
Participation of third person (PT)	0.5	0.8	**8.0**	2.5	3.0	2.5	1.5	1.0	2.5
Comment on answer (CA)	4.5	0.0	**26.5**	24.4	15.2	20.4	16.0	**31.8**	18.6
Phase 3: Evaluating the Answer									
Check of response (CH)	3.0	4.8	**12.5**	3.0	0.0	1.0	1.0	1.5	3.7
Refusal to answer (RF)	0.0	0.0	**2.0**	0.0	0.0	0.0	0.0	0.0	0.3
Irrelevant answer (IA)	1.0	0.0	0.5	**2.0**	0.0	1.0	0.5	0.0	0.7
Don't know (DK)	0.0	0.0	**16.5**	6.0	**12.1**	0.5	0.0	**15.9**	6.0
Mismatch answer (MM)	1.0	0.8	29.0	14.4	**42.4**	**57.7**	**64.5**	10.9	27.3
Phase 4: Repair									
Adequately and sufficiently repaired (AR)	1.5	0.0	**31.0**	10.4	24.2	**26.4**	21.5	14.4	16.1
Inadequately or insufficiently repaired (IR)	0.5	0.8	**13.0**	6.0	**24.2**	11.9	11.0	9.5	8.3
Missing repair (MR)	0.0	0.0	4.0	6.5	12.1	**24.9**	**34.5**	7.4	11.7
Phase 5: Coding									
Coding error (CE)	1.5	3.2	8.0	7.5	**27.3**	**27.4**	**35.5**	5.0	13.5
Type of sequence									
Paradigmatic sequence (PAS)	37.9	16.9	23.0	42.8	24.2	22.4	15.0	**47.3**	29.8
Problematic sequence (PRS)	4.5	0.0	**52.5**	17.9	30.3	30.8	23.5	33.3	25.0
Inadequate sequence (INS)	57.6	**80.6**	24.5	39.3	45.5	46.8	61.5	19.4	45.1
%PAS − %INS	−19.7	−63.7	−1.5	3.5	−21.3	−24.4	−46.5	27.9	−15.3
n	198	124	200	201	33	201	200	201	1358

6.4.3 Code Patterns

When we look at the frequency distribution of the codes for each question separately, it is obvious that they are very different. For questions Q1 and Q2, omission (OM) is relatively frequent. The text of the question begins with the query "Do you have an income of your own?" followed by a definition of personal income. In the majority of cases, the interviewer did not read this definition. Many respondents give an answer immediately after hearing the first part, leaving the interviewer no opportunity to present the definition. It should be noted that if the interviewer had already provided the definition of income with Q1, skipping the definition while reading Q2 is not coded as an omission. Questions Q1 and Q2 could be improved by sticking to the principle formulated by Fowler (1995, p. 86): "If definitions are to be given, they should be given before the question itself is asked."

The code "wrong skip" (WS) is also found frequently with Q2. These wrong skips occur because the series of income questions starts with a question about respondents' own income. A respondent does not know that the income of his or her partner will be asked for in a separate question. A number of respondents react to the question about their own income (Q1) by giving information about the household income (e.g., "I have a small pension and my husband, too"). The interviewer is then faced with the tricky task of asking again for information already supplied by the respondent. Thus, the question is often skipped, and the information given previously is used to code an answer to Q2, without asking the question itself.

The codes that are found relatively frequently with question Q3 (income level) belong to the phase of *cognitive processing* of the question (RE, PT, and CA). This indicates that the question is not clear to respondents and is difficult to answer. From the high frequency of RQ it appears that the question is not well written, because interviewers tend to change the *formulation* of the question. Moreover, from the high frequency of codes involving *cognitive processing* and *repair* (AR, IR), it appears that the help of an interviewer is necessary for obtaining adequate answers to this question. This type of income question is often found in self-administered questionnaires where no interviewer is available to help the respondent.

The code pattern for question Q4 (income decrease) consists of the codes for misunderstanding the question (MQ) and for irrelevant answers (IA). This means that some respondents answer a different question than the one intended by the investigator. A careful inspection of sequences with codes MQ and IA is necessary to identify the actual problem with this question.

From the code pattern for question Q5 (date of income decrease) (RQ, MQ, DK, MM, IR, and CE), it is apparent that this question has a number of problems. However, because Q5 is asked only of those who give a positive answer to Q4, the number of sequences evaluated is low ($N = 33$), and conclusions should thus be drawn with caution. The frequent occurrence of code RQ indicates that there is something wrong with the question text. Code MQ points at misunderstandings of the question, and the combination of codes MM and CE indicates that there is a problem with the response categories. Finally, the frequency of inadequate repair (IR) codes shows that problems with this question are not easily solved by interviewers.

Questions Q6 and Q7 have roughly the same code pattern (OM, MM, MR, and CE): omission of the response categories, initial answers that do not fit the response categories, missing repair by the interviewer, and checking of an answer box by the interviewer not based on information provided by the respondent. From further inspection of the sequences, it appears that these questions are not written correctly. Both Q6 and Q7 begin with "Are you satisfied with. . .?", suggesting a dichotomous answer format (yes/no). However, the question has five response categories. This mismatch between question format and response categories leads to many nonparadigmatic sequences. The solution to these problems is to rewrite the questions to make it clear that the response should be a choice from a set of response categories (e.g., "*How* satisfied are you with. . .?" Is that dissatisfied, a little dissatisfied, . . .").

The frequent occurrence of wrong skips (WSs) with question Q7 can be explained by the same mechanism mentioned above regarding Q1 and Q2: The answer to Q6 often includes information the interviewer can use for coding an answer to Q7 without actually asking the question itself.

The code pattern for question Q8 (MQ, CA, and DK) indicates, with MQ, the changes that respondents make in the text of the question while answering it. The need that respondents feel here to motivate the choice of a response alternative (CA) and the large number of don't know answers (DK) show that respondents find this a difficult question to answer, although not much is wrong with the question as it is written.

6.4.4 Codes as Qualitative Indicators

The 15 behavior codes above were used for the generation and exploration of question–specific code patterns. Furthermore, four of the 15 behavior codes can be conceived as indicators for a qualitative content analysis of transcripts of sequences, in search of additional information to be used for the evaluation and improvement of questions and for the improvement of interviewer training.

Code RQ, *reformulation of the question by the interviewer,* may inform us about difficult question wording. A closer inspection of the transcripts for Q3 shows, for example, that there is a tendency to omit the word *net* from the question text and by doing so to change the meaning of the original question. It also appears that reformulation of the question is often done by incorporating previously given information into the text of the question. For example: "So you expect that your income will decrease?" (Q8), or "So you have a pension?" (Q1).

Code MQ, *misunderstanding of the question by the respondent*, points at sequences in which respondents start to answer a question that is different from the one intended (e.g., "hope" instead of "expect" in Q8). Because interviewers are confronted with this type of respondent behavior during their fieldwork, interviewer instructions based on information from the transcripts may be helpful in standardizing their reaction to this type of misunderstanding.

Requests for elucidation by the respondent (REs) belong to two different types. First, there may be a problem with hearing (parts of) the question: "Did you say...?". Another, less frequent type is asking for an explanation of a particular word used in the question. Here it is not the frequency of the code that counts; sometimes just by a remark made in only one interview, a shortcoming of a question becomes apparent. For example, a respondent reacting to Q7 ("Are you satisfied with the standard of living you can attain on your income?") asked the interviewer "What do you actually mean by 'standard of living'? I cannot do now what I used to do when I was young." Not an easy question to answer for an interviewer or for the researcher. Actually, when we confronted the researcher with this remark, it appeared that even the researcher had no clear view of the concept "standard of living."

Comment on the answer (CA) by the respondent can be used to investigate why a certain answer was given. By doing so, we get a glimpse of the cognitive process the respondent goes through while formulating the answer. Two types of comments can be found in sequences with code CA. First, a word in the question is understood by the respondent in a way not intended by the researcher. Another type of CA is the motivation of "don't know" answers. For example, with question Q5 ("Could you indicate at what moment this decrease [in income] occurred?") we found a relatively high frequency of "don't know" answers. Inspection of the transcripts showed that a substantial number of respondents answered with: "I don't know, it decreased gradually," indicating that the response format (giving month and year of the decline) is probably not the best one for this question.

6.4.5 Evaluation of the Entire Sequence

On the basis of the 15 behavior codes, codes for the entire sequence can be constructed. We distinguish three types of question–answer sequences. First, we have the paradigmatic sequence (PAS). This is the sequence intended by the researcher. Second, there are sequences in which deviations from the paradigmatic sequence occur, but these deviations are adequately and sufficiently repaired (AR). There was a problem in the sequence, but the problem is solved correctly. For that reason we call these sequences *problematic sequences* (PRSs). Other sequences belonging to this type have only "harmless" sequence codes [e.g., RE (request for elucidation), CH (check), and CA (comment on the answer given)]. All other sequences are called inadequate sequences (INSs). In these sequences significant deviations of the paradigmatic sequence occur, but they are not repaired, or at least not adequately.

To assess the intercoder reliability of the classification of sequence types, 276 randomly selected sequences were coded independently by both coders. On the basis of the behavior codes assigned by each coder, the sequences were classified as being paradigmatic (PAS), problematic (PRS), or inadequate (INS). The degree of agreement between the two classifications of the sequences, expressed as Cohen's kappa values, ranged from 0.62 for Q4 to 0.92 for Q7, with an average of 0.76.

Table 6.3 shows that the eight questions differ not only with respect to their code patterns but also with respect to the proportion of paradigmatic, problematic, and inadequate sequences. A rough measure of the quality of the questions is the difference between the percentage of paradigmatic and inadequate sequences (%PAS – %INS). Question Q8 performs relatively well. The topic of the question is difficult, but the wording seems adequate. Questions Q2, Q5, Q6, and Q7 turn out to be very problematic, leading to far more inadequate than paradigmatic sequences. They are badly in need of improvement.

Question Q3 (income level) requires much activity by the interviewer, and the quality of the response is thus strongly dependent on his or her competence. Question Q4 is problematic only for respondents who have recently experienced a decrease in their income. The others could simply answer "no" without having to retrieve the magnitude of the decrease.

6.5 EFFECTS OF THE INTERVIEWER AND OF CHARACTERISTICS OF RESPONDENTS

The quality of a sequence is the joint product of the adequacy of the question, the competence of the interviewer, and characteristics of the respondent. There is always a chance that the occurrence of codes indicating potential problems with a question is "produced" by a small and selective group of interviewers and respondents—with, for example, a different task orientation regarding that question. For that reason we examined whether certain interviewers, and respondents with specific characteristics, influenced the distribution of codes classifying the entire sequence (codes PAS, PRS, INS) for each question.

6.5.1 Interviewers

The 201 interviews were administered by 36 interviewers (who completed from 1 to 13 interviews). To reduce the number of empty cells, we selected interviews done by interviewers who had administered at least seven interviews (13 interviewers and 128 respondents). The influence of the interviewer on the distribution of PAS, PRS, and INS for the separate questions was investigated with chi-square statistics. Because of the relatively large number of low-frequency cells, we used the likelihood ratio test as a test for significance. For questions Q1 and Q2 the number of sequences classified as PRS was very low (9 and 0, respectively), and these sequences were left out of the analyses to minimize the occurrence of empty cells.

6.5.2 Respondents

Using the same approach, we examined the following respondent characteristics in relation to the overall classification of the sequence: gender, age, education, and cognitive functioning. Age was dichotomized around the mean (and median) age of respondents in the sample: young–old (<75) and old–old (≥75). Education was divided into three categories: low (up to 8 years of education), medium (9 to 12 years of education), and high (13 years or more). Cognitive functioning, measured with the Mini Mental State Examination (MMSE; Folstein et al., 1975), was dichotomized into impaired (MMSE score below 24 points) and not impaired (MMSE score of 24 points or more).

Because of the large number of statistical tests (eight tests for the interviewer and 32 for the tests of the influence of respondent characteristics), the significance level of the likelihood ratio test had to be adjusted. We used the method described by Holm (1979), which is an adaptation of the classical Bonferroni correction.

6.5.3 Results

We found a significant relationship between interviewer and sequence type for questions Q1, Q2, Q6, and Q7. We have seen that questions Q1 and Q2 are problematic because a definition of the core concept "personal income" is given after the question itself is asked. Some interviewers are better able than others to counter this formulation defect. Questions Q6 and Q7 turned out to be problematic because the five response categories do not match the (yes/no) question. Again, it appears that some interviewers are better able to handle the troublesome situations caused by this wording defect. In our opinion it was the flaw in the question wording that initiated systematic differences in interviewer behavior, and therefore these interviewer effects could be interpreted as specifications of the impact that the adequacy of the question wording has on the quality of the resulting sequences.

The four respondent characteristics generally appeared not to have a significant effect on the distribution of the sequences over the three sequence types. There is only one exception: With question Q5, age has a significant effect on the distribution. Of the 10 sequences stemming from respondents 75 years or over, nine (90%) were inadequate (INS), compared with 26% of sequences stemming from younger respondents.

These results indicate that the outcome of the diagnostic approach is partly affected by the interviewer, but hardly at all by the respondent (although this conclusion might be different for a sample including younger respondents).[5] Good interviewers make only few mistakes when asking the questions, and they often adequately repair inadequate answers initially given by respondents, thus leading to a low frequency of INS sequences. But even the best interviewers cannot mask wording defects. If these defects lead respondents to give inadequate answers, they reduce the proportion of paradigmatic sequences, thus reducing our measure of quality.

[5]An interesting research question is whether the format of the question affects the sensitivity of the diagnostic approach as a testing device. To answer that question, one needs a greater variety of questions (with a greater variety of formats) than the eight questions included in the present study.

6.6 COMPARISON WITH OTHER METHODS FOR QUESTION EVALUATION

6.6.1 Do Different Methods Lead to Different Evaluations?

From evaluation of the eight income questions using the diagnostic approach, it appeared that each of these questions is in some way problematic. Even the least problematic, question Q8, resulted in nearly one-fifth of inadequate question–answer sequences (Table 6.3). Thus, we will not compare the various methods by seeing what questions they identify as problematic, but by looking at the rank order of questions they produce with regard to *how* problematic they are.

Behavior Coding We first compared the assessment of the questions by means of the diagnostic approach (DA), as presented in column (2) of Table 6.4, with that using "classical" behavior coding (BC). We reconstructed the assessment by BC by selecting from our database all information regarding behavior of interviewer and respondent during the first part of the question–answer sequence (first-order exchange). For the description of interviewer behavior we used the codes WS (wrong skip), RQ (reformulation of the question), and OM (omission of response alternatives or explanations). For respondent behavior we used the data regarding RF (refusal to answer), IA (irrelevant answer), DK (don't know response), and MM (mismatch between response given and answer categories). We looked at all 1358 question–answer sequences to see whether any of these inadequate behaviors had occurred and computed the percentage of inadequate behaviors for each of the eight questions. The results are presented in column (3) of Table 6.4. It is quite clear that all questions failed to meet the common criterion of less than 15% inadequate interviewer and respondent behaviors.

The rank ordering of the eight questions with respect to how problematic they are is about the same for BC as for DA (Spearman's rank-order correlation rho = 0.95). The evaluation of question Q7 is even more negative in BC than in DA. On the other hand, the very negative evaluation of question Q2 is less pronounced with BC than it is with DA.

Expert Panel It is common practice to pretest a questionnaire before it is used in the field by asking a panel of experts to evaluate the questions. To see how well such an ex-ante evaluation of questions would correspond with our field-based evaluation, we asked five senior researchers of the Department of Social Research Methodology of the Vrije Universiteit, all experts in survey methodology, to evaluate the relevant part of the questionnaire. They were asked to indicate whether they expected a particular question to cause problems for interviewers or respondents, and if so, to explain the nature of that problem. They were also asked to rank-order the questions with respect to the expected number of errors made by interviewers and respondents. The rank orderings of the questions by the experts were aggregated, and the resulting numbers are presented in column (4) of Table 6.4.

Table 6.4 Assessment of the Problematic Character of the Questions by Various Methods[a]

(1) Question	(2) DA[b]	(3) BC[c]	(4) EP[d]	(5) QAS[e]	(6) TDS[f]	(7) SQP[g]
Q1: Do you have an income of your own?	−19.7	**59.6%**	8.7	1	3	0.80
Q2: Does your partner have an income of his/her own?	**−63.7**	**85.5%**	7.9	1	3	0.80
Q3a: In which category does your *net* income fall?	−1.5 (Q3a/Q3b)	53.2% (Q3a/Q3b)	7.0	4	5	**0.58**
Q3b: In which category does the joint *net* income of you and your partner fall?			**4.9**	5	5	**0.58**
Q4a: Did your income decrease in the last three years?	+3.5 (Q4a/Q4b)	50.7% (Q4a/Q4b)	5.7	**13**	**6**	**0.61**
Q4b: Did your income decrease in the last three years? (Do *not* count the income of your partner)			**4.5**	**14**	**6**	**0.61**
Q5: When did this decrease occur? (month/year)	**−21.3**	57.6%	6.0	5	4	**0.72**
Q6a: Are you satisfied with your income level?	**−24.4** (Q6a/Q6b)	**73.1%** (Q6a/Q6b)	6.0	**7**	**6**	0.97
Q6b: Are you satisfied with your income level? (We mean the joint income of you and your partner).			**4.4**	**9**	**6**	0.97
Q7: Are you satisfied with the standard of living you attain on your income?	**−46.5**	**86.5%**	**5.0**	**9**	**7**	0.94
Q8: Do you expect that your income will decrease, remain the same, or increase?	+27.9	27.9%	6.1	2	3	0.92
All questions together	−15.3	60.9%	—	6.4	4.9	0.77
n (sequences or questions)	1358	1358	11	11	11	11

[a] Scores in boldface type indicate the most problematic half of the questions.
[b] DA, diagnostic approach: percentage paradigmatic sequences minus percentage inadequate sequences. Range: −100 to +100.
[c] BC, classical behavior coding: percentage sequences with inadequate behavior of interviewer (codes WS, RQ, OM) or respondent (RF, IA, DK, MM) in the first-order exchange. Range: 0% to 100%.
[d] EP, expert panel: rank order of number of errors made by interviewer and respondent as expected by an expert panel. Range: 1 to 11.
[e] QAS, questionnaire appraisal system: number of potential flaws in the question. Range: 0 to 26 points.
[f] TDS, task difficulty score: number of question characteristics making the question difficult to answer. Range: 0 to 11 points.
[g] SQP, survey quality predictor: expected validity of the information collected by this question. Range: 0.00 to 1.00.

If we compare the rank ordering of the questions by the expert panel (EP) with the evaluation of the questions using DA, we see striking differences. Question Q2, the question that proved to be the most problematic according to DA, was rated as one of the least problematic by the experts. On the other hand, our experts were very critical of question Q4b, whereas according to DA, this question did not cause much trouble. The differences between the evaluations based on DA and the experts' opinions are reflected in a slightly negative rank-order correlation coefficient (rho = −0.07). The differences between the outcomes of classical behavior coding and EP are about the same: rho = −0.06. Even for experienced survey researchers, it seems hard to predict what problems will actually show up during the question–answer process.

Questionnaire Appraisal System (QAS) The Research Triangle Institute has developed a guided checklist-based means of identifying potential flaws in survey questions: "For each survey question to be evaluated, the researcher completes a QAS form that leads the user to consider specific characteristics of the question and the researcher decides whether the item may be problematic with respect to that characteristic" (Rothgeb et al., 2001; see also Chapter 25, this volume). The QAS form consists of 26 checks: for example "Question is not fully scripted and therefore difficult to read," "Respondent may not remember the information asked for," and so on. The present authors completed independent QAS forms for each of the income questions. Their judgments were highly similar. By counting the number of checks answered with "yes" we could assign a QAS score to each question. These scores are presented in column (5) of Table 6.4. Both versions of question Q4 are evaluated by QAS as problematic, and to a lesser degree, so were questions Q6b and Q7. The QAS-based evaluation does not correspond at all with the results of DA (or BC for that matter). This is reflected in a negative rank-order correlation coefficient (rho = −0.16). Again, an ex-ante evaluation seems to be quite different from an evaluation done after the data are collected.

Task Difficulty Score (TDS) Another method for ex-ante evaluation of a questionnaire was developed by van der Zouwen (van der Zouwen and Dijkstra, 2002). Like the QAS method, it asks for each question whether one or more (of 11) potentially problematic question characteristics is present. For example: "The question relates to a 'hypothetical' or future situation" or "The question structure does not indicate when respondent's turn begins." The outcomes of the evaluation with TDS are presented in column (6). As the rank ordering of the questions by QAS and TDS is about equal (rho = +0.91), it is not surprising that the correspondence between the evaluations by DA and TDS is close to zero (rho = +0.17).

Survey Quality Predictor (SQP) Saris and his colleagues (see Chapter 14, this volume) have developed another system for the ex-ante evaluation of survey questions. SQP predicts the reliability and the validity of survey questions on the basis of (mostly formal) question characteristics, such as question length and number of response categories. The effects of these characteristics on the validity and reliability, and thereby on the quality, of the responses are derived from metaanalyses of dozens of multitrait, multimethod studies.[6]

As can be seen in the last column of Table 6.4, the SQP scores for the estimated quality of the eight income questions are very high, with the exception of questions Q3 and Q4. The positive SQP estimations contrast strongly with the negative evaluations of the questions by the diagnostic approach. The range of evaluations of the various questions is far wider in the diagnostic approach. Moreover, the rank-order correlation between SQP evaluations and those of the diagnostic approach is moderately negative (rho $= -0.50$). The SQP rankings also do not correspond with those of the other ex-ante evaluation methods ($-0.13 <$ rho $< +0.03$).[7]

We can summarize the data presented in Table 6.4 as follows:

1. There is a very strong relationship between evaluations based on classical behavior coding and the diagnostic approach (rho $= 0.95$);
2. There is also a strong relationship between three methods of ex-ante evaluation of questions: EP, QAS, and TDS ($0.75 <$ rho < 0.91);
3. The relation between the three ex-ante evaluation methods on the one hand, and DA or BC on the other, is close to zero ($-0.16 <$ rho $< +0.17$);
4. The ranking by SQP does not correspond with either DA/BC or with the other ex-ante methods.

Thus, the two evaluations of the questions based on observation of what actually goes on (or rather: goes wrong) in interviews lead to results that differ markedly from evaluations based on the text of the questionnaire only [for a similar conclusion based on other data, see van der Zouwen and Dijkstra (2002)].

6.6.2 Do Different Methods Identify the Same Problems and Flaws in Problematic Questions?

To answer this question, we compare the observed problems identified by the diagnostic approach with potential problems mentioned by our expert panel. We asked the members of the panel to indicate, for each question, whether they expected it to cause a problem for the interviewer and/or the respondent and to explain the problem as specifically as possible. This request resulted in a list of potential problems. To this list we added the problems as identified by DA. The result was a list of 25 potential problems. Of these 25 potential problems, 15 were detected by members of the panel. Fourteen problems were detected by DA. Only

[6]In a multitrait, multimethod (MTMM) study, the same latent traits or variables are measured using different data collection methods. For example, different attitudes are each measured using questions with category scales, number scales, and graphical line scales. The MTMM design is used for estimating the validity and reliability of data collected with each of these methods.

[7]The SQP scores are computed with the version described in van der Zouwen et al. (2001).

four of the problems on the list of 25 (16%) were detected by both methods. The remaining 21 were unique to a particular method. The experts pointed out what, according to their experience, *could go wrong* with using this questionnaire. The field testing by BC and DA signaled what *actually went wrong*.

6.7 CONCLUSIONS

6.7.1 Difference between Ex-ante Evaluations and Field Testing of Questionnaires

The data presented in Section 6.6 emphasize the differences between the outcomes of ex-ante pretesting (using EP, QAS, TDS, or SQP) and field testing (using BC or DA) of questionnaires. But these results do not provide an easy answer to the obvious question of which method is better, because the ex-ante methods base their evaluations only on the text of the questionnaire, whereas field testing is based not only on the questionnaire but also on recordings of the interviews conducted with this questionnaire. To use a metaphor borrowed from meteorology: The ex-ante pretesting differs from field testing as next week's weather *forecast* differs from today's weather *report*. Unfortunately, but understandably, these are seldom identical. The best practice for testing questionnaires would seem to be first to use ex-ante methods for testing early versions of the questionnaire, and after necessary improvements are made, to evaluate the draft questionnaire using field-testing methods.

6.7.2 Advantages of the Proposed Extension of Behavior Coding

The diagnostic approach presented in this chapter can be conceived as an extension of classical behavior coding. In our opinion this extension has three advantages. First, DA does not restrict itself (as BC does) to information about the first phases of the question–answer sequences. It also uses *information about the later phases* (i.e., repair and coding) and the errors made in these phases that could harm data quality. The information about the necessity and effectiveness of repair activities can be used to train interviewers to deal more effectively with respondents' inadequate initial answers. Information about the prevalence, and the character, of coding errors can be used to improve the way that response codes are presented.

Second, DA makes use of the *content* of what is said by interviewers and respondents when they are giving comments on, and explanations of, their questions and answers. Analysis of these texts is a rich source of information about the actual "performance" of the questionnaire: how questions are interpreted by respondents, how they reach the final choice of a response category, and so on.

Third, DA permits an analysis of aspects of the *interaction* between respondent and interviewer, that is, certain combinations of actions by both actors in the interview. A systematic study of these interactions may lead to a questionnaire that is easier to administer and to answer. For example, in question Q3 the interviewer is assumed to read the text of the question, but at the same time to hand over a show card with a set of rather complicated response categories (i.e., 12 income classes, each on a monthly and a yearly base). The respondent is assumed to listen to the interviewer but at the same time to inspect the show card, to understand what the interviewer is actually saying. The poor ergonomics of this question design leads to awkward interactions between interviewer and respondent, resulting in a decrease in paradigmatic sequences and an increase in problematic sequences. This diagnosis of Q3, based on the interaction between interviewer and respondent, could be used to improve the question.

6.7.3 Additional Costs of the Diagnostic Approach

These advantages of DA come with additional costs, especially those of transcribing the relevant parts of the interviews. For the analysis reported here, the extra costs of the transcription amount to 640 euros (or U.S. dollars). The costs of both classical behavioral coding and the diagnostic approach are proportional to the number of interviews analyzed. In this illustration we used a sample with $n = 201$. Would we have arrived at about the same evaluation of the questions if we had used a sample half this size? To answer that practical question, raised by Zukerberg et al. (1995), we drew a sample of $n = 100$ from the sample of 201 sequences. For this smaller sample we computed the code patterns for each question and the difference between the percentage of paradigmatic and inadequate sequences. The result was a table similar to Table 6.3. The code patterns were identical, except for question Q5, where codes for MQ were missing, probably because we now had only 17 sequences for Q5. The ranking of the questions by the quality of the sequences was also almost identical. Only questions Q5 and Q6 changed places, resulting in a Spearman's rho of 0.98. This means that we would have received about the same results with a sample half as large.

We do not recommend reducing the sample size below $n = 100$, because then the results of the statistical analyses will become less reliable, and conditional questions such as Q5 will lead to too few sequences. Moreover, the diagnostic approach is based on the nonparadigmatic sequences, especially sequences with behavior codes RQ, MQ, RE, and CA. Because these four codes appear in less than one-third of all sequences, we would end up with too few sequences for a reliable content analysis.

6.7.4 Future Research

The diagnostic approach aims at the evaluation of the quality of the questions of the questionnaire by looking at the sequences produced by interviewers and respondents when administering the questions. Nonparadigmatic sequences often proved to be informative about the question–answer process. By studying those sequences, flaws in the questions could be identified and suggestions for improvement made.

A limitation of the present study is that we did not have an external criterion for evaluating use of the diagnostic approach with respect to the improvement of data quality. Future research into the validity of the diagnostic approach requires a design in which (1) the diagnostic approach is used to identify potential flaws in a question, then (2) the question is revised based on the DA information, and (3) the diagnostic approach is applied again to new sequences. Such a design should preferably be developed around data for which (4) external validating information is available. In other words, a key next step is to collect validating information in order to see whether the improvement of the *adequacy* of the questionnaire has indeed led to an improvement in the *accuracy* of the answers received. An important extension of such a design would be the incorporation of other forms of behavior coding (from tapes and from transcripts) in order to gain more insight into the benefits and costs of the different approaches.

ACKNOWLEDGMENTS

This chapter is based on data collected in the context of the Longitudinal Aging Study Amsterdam (LASA). This program is conducted at the Vrije Universiteit in Amsterdam and is supported by the Ministry of Health, Welfare, and Sports. The authors would like to thank Jaap Ruseler, for his corrections of the translation of the text of the eight income questions. We also thank members of the Department of Social Research Methodology of the Vrije Universiteit in Amsterdam for their willingness to act as members of the expert panel and for their comments on an earlier version of this chapter.

CHAPTER 7

Response Latency and (Para)Linguistic Expressions as Indicators of Response Error

Stasja Draisma and Wil Dijkstra
Free University

7.1 INTRODUCTION

Since the rise of the cognitive movement in survey research, substantial effort has been invested in the study of factors that determine the task burden of answering questions (Lessler and Forsyth, 1996; Sirken et al., 1999a; van der Zouwen, 2000). The difficulty of a question is assumed to be an important cause of response errors: The more difficult the question, the higher the probability of the occurrence of an incorrect answer. Ample evidence exists for the positive relationship between question difficulty and response error (e.g., Bless et al., 1992; Knäuper et al., 1997; Krosnick, 1991; Smith, 1982). However, researchers have used very different definitions and indicators of question difficulty.

A currently popular way to consider the difficulty of question answering is by means of insights from cognitive psychology and information processing theory (e.g., Tourangeau, 1984; Tourangeau et al., 2000). Among cognitive psychologists, the measurement of response latency is often used as an indicator of task difficulty [for a good overview of technical procedures, see Luce (1986)]. But despite the attention given to cognition in question answering, few authors have investigated the usefulness and validity of response latency (hereafter, RL) as a measurement device in survey research. The major exception is the work of John Bassili (Bassili, 1993, 1995, 1996; Bassili and Fletcher, 1991; Bassili and Krosnick, 2000; Bassili and Scott, 1996).

Methods for Testing and Evaluating Survey Questionnaires, Edited by Stanley Presser, Jennifer M. Rothgeb, Mick P. Couper, Judith T. Lessler, Elizabeth Martin, Jean Martin, and Eleanor Singer
ISBN 0-471-45841-4

In what way might response latency, cognitive processing, and question difficulty be related? We believe that the time it takes a respondent to answer a question reflects the processing that is necessary to arrive at an answer. A difficult question needs more processing and hence results in a long response latency. Difficult questions may also cause *uncertainty* about the correctness of an answer. Schaeffer and Thomson (1992) distinguish *state uncertainty* (respondents having doubts about the "true answer," for example, as a result of memory problems) and *task uncertainty* (the standardized format of a question making it difficult for respondents to express their "true state" within the range of alternatives offered). The respondent's experienced uncertainty is thus a result of the interaction of the question format and uncertainty about his or her true state. Response latencies, verbal behavior, and paralinguistic behavior may all serve as useful indicators of uncertainty in question answering.

The aim of this study is to explore the relationships of such indicators to measurement error. Specifically, we address the following questions:

1. What is the relation between response latency and response errors?
2. Is the perceived difficulty of questions related to response errors?
3. Are response latencies related to the perceived difficulty of questions?
4. Are response latencies and response error related to other paralinguistic and linguistic indicators of uncertainty?

7.2 USE OF RESPONSE LATENCY IN SURVEY RESEARCH

Response latencies have been used primarily as a performance measure in experimental settings in cognitive psychology. Aaker et al. (1980) and LaBarbera and MacLachlan (1979) were among the first to introduce response latency in a surveylike environment such as telephone interviews. They assessed the contribution of RL to interpreting paired comparisons in a conventional marketing research study. It was assumed that the faster a choice was made between two brands, the stronger the preference for the chosen brand. Following this line of research, MacLachlan et al. (1979) found that incorrect answers to knowledge questions about the companies that manufactured certain products took considerably more time than correct answers (about 3 seconds for incorrect answers, versus $1\frac{1}{2}$ seconds for correct answers). They concluded that response latencies longer than a certain threshold might be based on "helpful guessing" instead of actual knowledge by the respondent.

The most important work on response latencies and (survey) question answering is that of John Bassili and his co-workers. They used RL as an indicator of several concepts, for example, attitude stability and question difficulty, and as a predictor of actual behavior. Bassili (1993) found that response latency was a better predictor of discrepancies between voting intentions and actual voting behavior than a verbal measure of "certainty" (a question about the finality of the voter's intentions). In another study, Bassili and Fletcher (1991) demonstrated orderliness in the latencies of various types of questions, ranging from very brief latencies for simple factual questions to long ones for complex attitudinal ones. More complex questions, especially the ones flawed in construction, seemed to take more time to answer (Bassili, 1996). Bassili and Scott (1996) also used response latency as an index of question problems. Poorly formulated questions resulted in longer RLs than repaired versions of the same question.

In social psychological research, RLs have been used to study the accessibility, strength, and stability of attitudes. Respondents who hold stable attitudes tend to react faster to questions than respondents with unstable, noncrystallized attitudes. The stronger, more accessible, and more stable the attitude, the more certain a respondent is about the answer alternative selected (Bassili and Krosnick, 2000; Fazio, 1990; Johnson et al., 1999).

In sum, several studies suggest that response latency in survey research might be useful both as an indicator of attitude strength, stability, and accessibility, and as a signal of problems in question formulation. We believe that the common factor in these studies is *question difficulty*. The more stable an attitude or the stronger a preference, the less difficult a question concerning the attitude or preference will be. Such difficulty expresses itself in a state of uncertainty, which in turn causes longer response latencies. In general, the outcomes of the studies discussed indicate that RL may be a good indicator of respondent certainty; that is, the longer the RL, the less certain a respondent is and the higher the probability of obtaining an incorrect answer.

An additional relevant issue is the well-known phenomenon of a speed–accuracy trade-off. It has often been found that the faster subjects react, the more errors they are prone to make in cognitive tasks, for instance in adding or subtracting numbers. In answering survey questions, however, we believe that the speed–accuracy trade-off does not apply in this way. Instead, the longer it takes the respondent to answer a question, the greater the chance of obtaining an incorrect answer. For example, MacLachlan et al. (1979) found shorter response latencies for correct answers to knowledge questions than for incorrect answers. In question answering, a longer latency signifies that the respondent has to invest in more cognitive processing: for example, as a result of a greater demand on memory. Stated otherwise, the more difficult a question is, the more cognitive processing is required, and thus the longer the RL will be. Further, the more difficult a question is, the higher the probability of response error. Such difficulty will result in uncertainty, which in turn influences RL. Thus, we expect that long RLs in question answering indicate a poor performance on such an item, which can be assessed by the number of response errors.

7.2.1 Measuring Response Latency in Question Answering

Response latencies are ordinarily measured by, or with the assistance of, the interviewer. In computer-assisted interviews, interviewers may be asked to start and stop an internal computer clock by pressing a key on the keyboard. Bassili, for instance, instructed interviewers to press a key immediately after reading the last word of the question and press a key a second time at the moment they judged that the respondent gave an answer. A voice onset key may also be used for the answer onset. Bassili and Fletcher (1991) used a voice key connected to the computer that signaled the first sound emitted by the respondent and immediately activated the computer clock [see also Fletcher (2000)]. In older studies, more complicated methods were used [e.g., MacLachlan et al. (1979) used a voice-operated relay system in which the interviewer had to press a pedal at the offset of the question reading. At the start of a verbal utterance by the respondent, the watch was stopped by the voice-operated relay system. A special device was used to register the durations.]

Use of the interviewer to measure RL has several drawbacks. First, the burden for the interviewer is increased, and this may result in less reliable and valid measurement, as well as having detrimental effects on the quality of the interview data. Second, it is almost impossible for an interviewer to determine at what particular moment the respondent initiates an adequate answer to a question. The start of an adequate answer cannot be perceived validly *during* a conversation but must be established afterward. Quite often, respondents make several introductory verbal utterances during their search for an answer (e.g. "Aah, let me see, I think people in my neighborhood do not like the current president. But I still think they do vote Republican, so I would say, yes"). Many utterances made by the respondent before a definite answer is provided form an essential part of the latencies between question and answer. Often, such utterances provide no evidence that the respondent has arrived at an answer yet, but simply indicate task processing.

In assessing RLs, most authors only take *paradigmatic* question–answer sequences into account (Schaeffer and Maynard, 1996), those in which there are no utterances between the question as posed by the interviewer (as worded in the questionnaire) and the respondent's provision of an acceptable answer. But many question–answer sequences are not paradigmatic, and include verbal utterances not directly related to the answer itself, such as repetitions, think-aloud reflections on the content of the question, requests for clarification, and so on. In the measurement of RL, question–answer sequences in which such utterances occur are quite often discarded as data. Yet it can be argued that the unit of analysis of RL measurements is formed by a sequence of several utterances by the respondent and sometimes by the interviewer as well. In between beginning and ending, *turns* can be distinguished which consist of all utterances a speaker produces between the utterances of other speakers (Dijkstra, 1999; Schegloff, 1996). Extensive use of conversation analysis in recent years (Houtkoop-Steenstra, 2002; Mathiowetz, 1998; Schaeffer, 1991a; Schober and Conrad, 2002) has shown time and again that the specific reactions of interviewer and respondent to each other's utterances may have beneficial effects on the validity of the answers provided. Sequences of several utterances may also be informative about processing, difficulty of the task, and respondent's (un)certainty.

We believe that it is useful to measure RLs for such nonparadigmatic sequences, in which a respondent needs several utterances to arrive at an answer. Take, for instance, the sequence:

Interviewer: "Did you donate more than ten dollar to the Red Cross?"
Respondent: "Now let me see. I usually pay 15 dollars to such funds. . . ."
Respondent: "Now in this case I am not really sure."
Respondent: "But I think I did."

In this sequence, we have two verbal events before the actual answer is provided. Yet these two utterances are extremely informative about the cognitive processing of the respondent.

Measuring RL becomes more difficult when the respondent provides several comments that have no direct relation to the question, or when the interviewer interjects explanatory comments. The time consumed by the interviewer's comments is no indication of the respondent's cognitive processes in arriving at an answer, but they provide the respondent with extra information about the question. It would thus be wiser to refer to such events as 'listening time.'

In short, we recommend assessing RLs not only in straightforward paradigmatic sequences of utterances, but also in longer sequences in which the respondent adds remarks that are relevant to the cognitive process of finding an answer. Sequences in which the interviewer interferes in the process should be discarded, since they add spurious listening time to the RLs.

A final consideration for the measurement of RL as applied in the survey field concerns the removal of outliers. Some of the procedures described in the literature seem to suffer from massive information loss, especially if we are interested in problematic aspects of questions and the cognitive processes involved in question answering. Johnson et al. (1999, p. 13), for instance, not only removed 18% of their data where the interviewer recorded the response time incorrectly, but also excluded extreme outliers in response latencies. Fletcher (2000) removed the data of respondents who answered before the end of the question reading, yielding negative RLs. It can be argued that negative RLs do have meaning: They indicate that respondents start their information processing during the presentation of the question. Schegloff (2002) would call this a *terminal overlap*: One speaker starts talking because of a perception that the previous speaker is finishing his or her turn. In surveys, the question is sometimes so easy that the respondent has already selected the answer alternative before the interviewer finishes reading the entire question. Moreover, the decision to start measuring the latency at the end of the question reading may even seem counterintuitive: Why would respondent processing wait until then? Especially with lists of questions of the same form in a questionnaire, negative RLs can be expected and interpreted in a meaningful way. On the other hand, positive RLs do not always reflect the time it took a respondent to arrive at an answer. Some respondents may wait until the interviewer finishes reading the question for the sake of politeness, even though they have already selected an answer alternative.

To reduce noise in latency data, Fazio (1990) recommends special instructions, practice trials, and a within-subjects design with filler trials to handle the sometimes immense variability of RLs between subjects. However, such precautions are only feasible in truly experimental conditions in laboratory research. It is difficult to imagine how all these recommendations could be followed in conventional large-scale survey research.

7.2.2 Other Indicators of Respondent Uncertainty

Other indicators of respondent uncertainty may also provide information about the quality of respondent answers. Mathiowetz (1998), for example, found that expressions such as "I think," "I'm not sure," and so on, were more often used by uncertain respondents than by certain respondents in response to frequency questions about health services utilization.

Verbal as well as nonverbal respondent utterances are useful indicators of uncertainty and task difficulty. It is now generally agreed that paralinguistic verbal behaviors such as "Oh well, I'm not so sure ...," or "Uhm ..." may express uncertainty. But nonverbal paralinguistic behaviors such as prosody and speech rates may also be indicators of uncertainty. A special type of paralinguistic behavior is the pause (Sacks et al., 1974). *Filled pauses* are formed by expressions such as "uhm", "pff," and so on. A *silent pause* is a hesitation in speech within a turn that is not filled with any speech sounds. *Gaps* between turns in conversations may express cognitive effort and can be used explicitly in RL measurement. In paradigmatic interaction sequences (*Interviewer*: "Are you older than 40 years of age ..."—gap—*Respondent*: "No") the entire RL actually consists of the duration of the gap.

Survey methodologists often point to the usefulness of an external evaluation of question difficulty (see, e.g., Graesser et al., 1999; Lessler and Forsyth, 1996; Chapter 6, this volume). Draisma (2000) let a panel of experts judge questions with respect to the respondent's "expected" certainty about the correctness of the answer alternative chosen. Not much is known about respondents' own evaluations of the task difficulty of survey questions. Yet it would be interesting

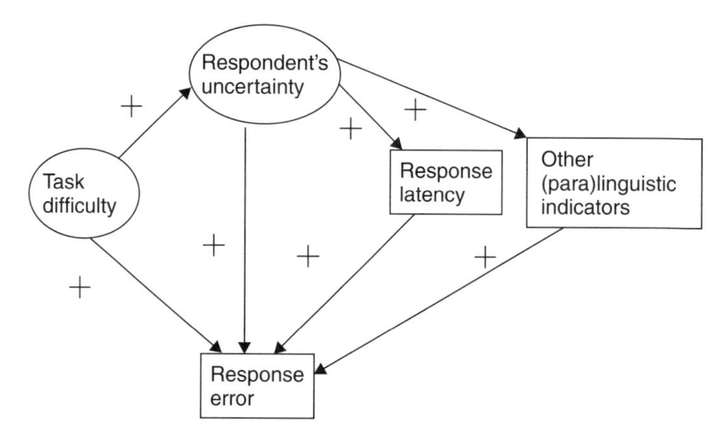

Figure 7.1 Relationships between several indicators of task difficulty.

to find out whether their question evaluations can be used in the same way as those obtained from so-called "experts," especially since problems with expert judgments are often mentioned.

The relationships among the various indicators of task difficulty and uncertainty are depicted in Figure 7.1. In the figure, variables that can be observed directly are given in squares and latent nonobservable variables are depicted in circles. As shown, the variables "task difficulty" and "respondent uncertainty" are assumed to exert direct as well as indirect effects on the dependent variable "response error." Nevertheless, in the present study the direct effects of these two variables cannot be established empirically.

7.3 DESIGN OF THE STUDY

To study measurement error in binary factual questions, a validation study was carried out. Individual true scores could be determined for several questions by using information from the records of an environmental organization, whose members were interviewed by telephone. Thus, we were able to determine measurement error in the form of correct or incorrect answers given to the questions. Question–answer sequences of the telephone survey were used to assess RL, response error, difficulty of questions, and uncertainty of respondents.

Two datasets involving different samples were available from telephone interviews of members of the environmental organization *Milieudefensie*. The topic of the telephone survey was environmental issues. Questions referred to environmental issues, details of membership in the organization, and acquaintance with campaigns held by the organization. Information obtained from the organization provided two ways of establishing the validity of respondents' answers. First, organizational records were used to determine the accuracy of respondents' answers about length of membership and amount of membership fee paid. Second, we added questions to which the same answer was true for the entire sample. For example, questions were posed about acquaintance with campaigns organized by Milieudefensie. However, the set of questions included some about nonexistent campaigns. For such "fake" questions, no one should have answered "yes." In addition, three other questions were asked with a uniform true score distribution. One question concerned the use of a nonexistent environmentally friendly potato. Two other questions asked whether a product had been used longer than the actual period of its total existence (energy-saving lamps and an environmentally friendly washing liquid). For these questions only the answer "no" was considered correct. All these questions could be answered with a "yes," a "no," or a nonsubstantive answer. Appendix 7A gives the exact questions for the various datasets, together with the percentages of correct answers and the number of respondents.

The target questions were embedded in a questionnaire with filler questions, which also contained attitudinal questions about environmental issues. Two samples of respondents were drawn; half a year after the first interviews (dataset 1),

Table 7.1 Number of Respondents, Questions, and Question–Answer Sequences

Dataset	Number of Respondents	Number of Questions	Number of Q-A Sequences
1	146	11	1261
2	145	11	1468

a new sample was interviewed (dataset 2). Table 7.1 contains the numbers of respondents and questions that were usable for the present study. It should be noted that there is considerable overlap between the questions used in the two samples. Yet differences exist for questions concerning membership duration (numbers 2, 3, and 13), questions about the size of the membership fee paid [as the interval between the payment and the time at which the question was posed varies (numbers 4, 5, 15, and 16)] and a question about a nonexistent brand of ecologically friendly potatoes (number 14).

At the end of the interview, respondents were asked to give a judgment using an 11-point scale on a selection of questions posed in the interview itself. The judgments concerned the degree of certainty the respondent had about the correctness of the answer alternative chosen and the importance of the question (topic).

7.3.1 Coding Procedures

The telephone interviews were taped, digitized, transcribed, and coded with the use of the computer program SEQUENCE VIEWER (Dijkstra, 2002). Five coders processed the data (transcription, coding, RL registration). Interactions between interviewer and respondent were divided into separate *events*, that is, meaningful separate utterances. Onset and offset time were established in the following way. While listening to the audiofile, coders pressed a button at appropriate moments to register the onset and offset times of all the separate events, which were already transcribed into SEQUENCE VIEWER. For convenience, we present the results in seconds, but it should be noted that the results and analyses are based on RLs measured with a precision of 0.1 second. Note also that an event is not the same as the well-known concept *turn* in conversation analysis. An event corresponds to a single *utterance*, so several events are possible in one turn, as shown in the example in Table 7.2. Events 2 and 3 together make up one respondent turn. (The example is a special case in which the respondent changes his answer from "Don't know" to "No.") The resulting datasets consist of collections of sequences similar in appearance to the sequence in Table 7.2.

For a sample of 124 events, intercoder reliability among three different coders was calculated for the difference between the onset time of an event and the offset time of the preceding event. The average correlation between the coded latencies was 0.97, suggesting that the task performed by the coders was highly reliable.

Table 7.2 Question–Answer Sequence with Five Events

Event	Utterances	Onset Time	Offset Time
1	*I:* Do you use energy-saving lamps longer than five years?	0	4.8
2	*R:* Oh, I don't know [silence]	5.1	6.1
3	*R:* Uhm, but let's say no.	6.4	7.2
4	*I:* No?	7.3	7.5
5	*I:* O.K.	7.7	8.0

7.3.2 Procedures for Assessing Response Latency

The SEQUENCE VIEWER program allows for various operationalizations of response latency. First, the user can choose between the onset time and the offset time of a question. One may argue that processing starts during asking of the question. On the other hand, adequate processing may start only at the moment at which the question is finished. Similarly, one may argue that either the onset time or the offset time of an answer should be used. Second, one must decide which answer should be used for calculating the response latency. Quite often, the respondent repeats the answer. Sometimes, the respondent even changes the answer. Third, one must decide what constitutes an answer. It is not uncommon for respondents, in deciding on the correct answer, to give their reasons. In the coding process, such reasons are viewed as different from the answer itself.

For the sake of comparability, we chose an operationalization that is most consistent with those used by other authors. RL is determined from the offset of the question until the onset of a definite answer. Sequences in which the respondent interacts with the interviewer (e.g., in the form of a request for clarification of the question) are discarded in our analyses. For sequences without interviewer interference, the proportion of incorrect answers is 0.16, whereas sequences with interviewer interference generate 0.28 incorrect answers. By discarding the problematic sequences we lose 393 sequences in both datasets, or 14% of the total number of sequences. Finally, negative RLs were recoded to RLs with zero values (approximately 0.6% of the data used).

7.3.3 Indicators of Uncertainty

In Table 7.2 the respondent changes an initial nonsubstantive answer into a substantive one. In such sequences, the respondent is said to *switch* answers. RL is assessed using event 3 as the answer. In our data, in nearly 7% of sequences the respondent made one or more switches. In most of these cases the respondent changed a nonsubstantive answer into a substantive one, as in Table 7.2. This change is a direct indication of uncertainty. In our analysis, the second event would count as an instance of linguistic uncertainty. Yet the second event also takes time, increasing RL. In other cases a "yes" answer is changed to a "no," or vice versa. In such cases the final answer is taken as definite.

Table 7.3 Question–Answer Sequence with Four Events

	Onset	Offset
I: Have you been a member of Milieudefensie for more than six years?	0	4
R: I ehm ... I might, I think I am.	5	9
R: Yes, no, I am quite sure of that.	11	14
I: Yes.	15	16

Table 7.3 illustrates some other indicators of uncertainty. The second event consists of a linguistic as well as a paralinguistic expression of uncertainty: the expression "ehm" is taken as a paralinguistic indicator, and the subsequent phrases of words ("might," "I think") are seen as verbal indicators. In our analyses, we used such expressions of doubt (or hedge words) as verbal indicators of uncertainty. Instances of such expressions were "I believe," "I think," "probably," "I suppose." As nonverbal paralinguistic indicators of uncertainty we used speech rate and number of words, calculated over the event in which the definite answer is given. This was done to investigate the relation between response latency and paralinguistic and linguistic indicators of uncertainty.

7.4 RESULTS

7.4.1 Bivariate Analyses

In this section we explore the relations between respondent uncertainty, response latency, and response error shown in Figure 7.1. We begin by analyzing the bivariate relations. Subsequently, the results of a multivariate logistic regression analysis, with response error as the dependent variable, are presented.

RL and Response Error The relation between RL and response error was examined for all the sequences in the two datasets for which information about the true score was available (11 questions in both datasets; see Appendix 7A). Only sequences in which no interviewer events were present were analyzed. The average RL for these sequences in dataset 1 was 1.11 seconds, and in dataset 2, 1.34 seconds.

Table 7.4 presents the results of a one-way analysis of variance with RL as dependent variable and the correctness of the answer as the grouping factor. A t-test was also performed for the difference between correct and incorrect answers only. As can be seen, there are large differences, in the direction expected, between latencies for correct, incorrect, and nonsubstantive answers. The longer the RL, the higher the probability of an incorrect answer. The longest RLs are found for nonsubstantive answers. This may be due to the number of events, or verbal utterances, present in a sequence between the offset of the question reading and the onset of the answer. "Don't know" (DK) answers express uncertainty in themselves, and this final answer may result from considerable deliberation.

Table 7.4 Response Latencies in Seconds for Correct, Incorrect, and NonSubstantive Answers

Dataset[a]	RL for Correct Answers	RL for Incorrect Answers	RL for Nonsubstantive Answers	F	t
1	0.96 ($n = 635$)	1.47 ($n = 114$)	2.58 ($n = 40$)	18.24[b]	2.95[b]
2	1.14 ($n = 726$)	1.60 ($n = 147$)	3.40 ($n = 52$)	39.73[b]	3.03[b]

[a]Dataset 1: 789 sequences, 11 questions, 146 respondents; dataset 2: 925 sequences 11 questions, 145 respondents.
[b]$p < 0.01$.

Indeed, the rank order of sequences for both datasets, in terms of the average number of events, was: nonsubstantive, incorrect, and correct. A separate test of the difference between correct and incorrect answers confirms that they differ significantly.

We inspected the pattern of RLs for the individual questions (see Appendix 7A). The shortest RL was found for the objectively easiest question: "Are you a member of [the organization] Milieudefensie?" All respondents were actually members and had received a letter that introduced the interview. The question was used as a type of base-rate question. Two questions about the size of the membership fee paid (numbers 4 and 5 in dataset 1, numbers 15 and 16 in dataset 2) posed tasks that differed in difficulty since the length of their recall period differed. The question about the earlier payment generated the larger RL. Of the set of four questions about fake campaigns in both datasets, the answers to the campaign "get rid of the dung [fertilizer]" generated the longest average RL. At the time the interviews were conducted, the Dutch media paid considerable attention to the effects of the discharge of dung into the environment. This media attention may have confused respondents.

Another interesting result is the finding that the shortest RLs are found for sequences in which the true answer is "yes" and the answer given is also "yes" ($p < 0.01$). This is not completely explained by the answers to the very easy base-rate question. Of the 22 questions in both datasets, 13 have a true score distribution of 100% "no" (the fake campaigns, fake potato brand, duration of product use). The probability of obtaining significantly shorter average latencies for these questions is larger, yet RLs for yes–yes answers are much shorter (average $RL_{yes-yes} = 0.52$, $n = 520$; average $RL_{no-no} = 1.34$, $n = 841$).

Perceived Difficulty of Questions and Response Error Next, we examined the relationship between perceived question difficulty, response error, and RL (Table 7.5). Respondents judged three characteristics of some of the questions posed to them: certainty about the correctness of their answer, difficulty of the question, and importance of the question topic. For the first dataset, two questions were judged by respondents on certainty, but only one question could be used in the analysis because all the answers to the other question were correct. For the second dataset, two questions were evaluated.

Table 7.5 Response Error and Respondents' Certainty About the Correctness of Their Answers

Dataset[a]	Average Certainty Evaluation for Correct Answers	Average Certainty Evaluation for Incorrect Answers	Average Certainty Evaluation for Nonsubstantive Answers	F	t
1	8.93 ($n = 87$)	8.07 ($n = 28$)	4.42 ($n = 19$)	27.34^b	-1.94^c
2	8.82 ($n = 172$)	7.63 ($n = 59$)	5.74 ($n = 27$)	26.01^b	4.03^b

[a] Dataset 1, contribution size question (no. 5), 136 respondents; dataset 2, contribution size question (no. 15), 119 respondents, and membership duration question (no. 13), 139 respondents.
[b] $p < 0.01$.
[c] $p < 0.05$.

Judgments about difficulty and importance were unrelated to response error, but evaluations of certainty were significantly related to error in the expected direction. Lower values for certainty were found for incorrect answers than for correct ones, and the lowest values were found for nonsubstantive answers. Examination of the relationship between respondent evaluations of correctness and RL showed approximately the same results. The correlations between certainty and RL had values of $r = -0.26$, $p < 0.01$ for dataset 1 and $r = -0.41$, $p < 0.01$, for dataset 2, indicating that the more certain a respondent is, the shorter the response latencies. No interpretable relations were found between RL and either judged question difficulty or importance of the topic. It is apparently easier to evaluate the correctness of the answer given to a question than to evaluate the characteristics of the questions themselves.

Linguistic Indicators of Uncertainty For every sequence, we coded whether the respondent uttered a *doubt word* (e.g., "I think," "I believe"). Differences between sequences with and without doubt words were calculated for correct and incorrect responses. Table 7.6 shows that doubt words are more likely to be expressed with incorrect than with correct answers.

Table 7.6 Verbal Expressions of Doubt and Response Accuracy[a]

Dataset		Correct	Incorrect	Nonsubstantive	Total
1	No doubt	79%	14%	7%	100% ($n = 797$)
	Doubt	67%	26%	7%	100% ($n = 168$)
			$\chi^2 = 14.91^b$		
2	No doubt	77%	13%	9%	100% ($n = 952$)
	Doubt	60%	35%	5%	100% ($n = 190$)
			$\chi^2 = 48.48^b$		

[a] From here on, the results of explicit t-tests of data without the nonsubstantive answers are no longer mentioned. Except when explicitly noted, they generated the same results as those already described.
[b] $p < 0.01$.

Paralinguistic Indicators of Uncertainty Four paralinguistic indicators of uncertainty were used: (1) switches of answer alternatives in a sequence, (2) speech rate, (3) number of words used, and (4) filled pauses.

Switches Sequences in which the respondent switches one or more times from one answer to another are distinguished from sequences in which the initial answer remains the final answer. (See Figure 7.1, in which the respondent switches from "Don't know" to "No.") Table 7.7 shows that respondents who do not switch their answers are more likely to give correct answers than are those who do switch. Thus, switches can be considered as an indicator of question difficulty or response error.

Speech Rate Speech rate is defined as the number of words uttered per second in an answer to a question. In using speech rate, one must control for the total number of words used in the answer to a (closed) question. In a substantial proportion of the answers given to our questions, only one word ("yes" or "no") is used. It is, of course, not surprising that "I don't know" answers have the fastest speech rate, since three words are required for such answers. We performed one-way analyses of variance for several subsets of sequences: those in which one or more words were used (dataset 1: $F = 49.13$, $p < 0.01$), those in which two or more words were used (dataset 1: $F = 15.12$, $p < 0.01$), and those in which three or more words were used (dataset 1: $F = 10.47$, $p < 0.01$). Although these analyses yielded significant F values, when we performed separate analyses with and without nonsubstantive answers, it became apparent that the effects found for speech rate are explained completely by "don't know" answers. If these are left out of the analysis, the effect of speech rate disappears. We still find an effect for sequences in which the answer consists of more than three words, but this is not very meaningful, since in more than half of these sequences the answer consists of only two words. Thus, although at first sight it is tempting to conclude that an increase in speech rate indicates an increase in response error, our analyses indicate that at least for simple closed questions with "yes" or "no" response alternatives, this conclusion is not warranted.

Number of Words Used As an answer to a binary closed question, the alternatives "yes," "no," or "don't know" would be sufficient. Nevertheless, many

Table 7.7 Switches in Answer Alternatives and Response Accuracy

Dataset		Correct	Incorrect	Nonsubstantive	Total
1	No switch	79%	15%	6%	100% ($n = 893$)
	Switch	54%	29%	17%	100% ($n = 72$)
			$\chi^2 = 25.55$		
2	No switch	76%	16%	8%	100% ($n = 1056$)
	Switch	48%	29%	23%	100% ($n = 85$)
			$\chi^2 = 36.83^a$		

[a] $p < 0.01$.

answers contain more words (e.g., "I pay more than 25 NLG to Milieudefensie"). When nonsubstantive answers are left out of the analysis (because of the inherently larger number of words), the effect of number of words on response error is significant (dataset 1: $t = 3.93$, $p < 0.01$; dataset 2: $t = 5.35$, $p < 0.01$). We considered whether these results were affected by the association between expressions of doubt (e.g., "I believe no") and incorrect answers. When the analyses were repeated for only sequences without doubt expressions, the results were still significant: The more words used, the higher the probability of an incorrect answer

Filled Pauses We examined whether incorrect answers are disproportionately preceded by filled pauses, "hemming" expressions such as "ehm," "pff," and "oeh." However, an important danger with this kind of analysis is circularity in reasoning: A longer RL increases the probability of the use of such expressions, and the use of a hemming expression increases the RL. Thus, we analyzed only sequences between 1 and 5 seconds in length. We found no link between filled pauses and error.

Of the four paralinguistic indicators, only switches of answer alternatives and number of words can be interpreted unambiguously. The larger the number of switches in answers and the more words used in answers to questions, the higher the probability of obtaining an incorrect answer.

7.4.2 Multivariate Analyses

To investigate whether RL contributes anything over and above what we would know from other indicators, multiple logistic regression was performed for the entire set of indicators, taking response error as the dependent variable. In these analyses, only correct and incorrect answers were considered; nonsubstantive answers were omitted. A (natural) logarithmic transformation was applied to the response latencies. Only variables that generated significant results in the bivariate analyses were considered for the multiple logistic regression. Table 7.8 shows the results of a logistic regression analysis in which RL and verbal expressions of

Table 7.8 Effect of RL and Doubt on Response Accuracy

Model A	Variable	*B* Coefficient	Standard Error
Dataset 1	Constant	2.18^{a}	0.18
($n = 749$)	RL	-0.19^{b}	0.08
	Doubt	-0.69^{a}	0.26
Dataset 2	Constant	2.18^{a}	0.18
($n = 873$)	RL	-0.18^{b}	0.08
	Doubt	-1.07^{a}	0.23

[a] $p < 0.01$.
[b] $p < 0.05$.

Table 7.9 Effects of Verbal Expressions of Doubt and Paralinguistic Indicators of Uncertainty on Response Accuracy

	Variable	*B* Coefficient	Standard Error
Dataset 1	Constant	1.89[a]	0.12
($n = 901$)	Doubt	−0.52[b]	0.22
	Switches	−0.68[b]	0.30
	Number of words	−0.04[b]	0.02
Dataset 2	Constant	1.90[a]	0.11
($n = 1043$)	Doubt	−0.95[a]	0.20
	Switches	−0.53	0.29
	Number of words	0.05[b]	0.02

[a] $p < 0.01$.
[b] $p < 0.05$.

doubt are used as predictors of response error. The effects of both doubt and RL are significant in this simple model. We next added the paralinguistic indicators "number of words in the answer" as well as "switches of answer alternatives." Neither variable reached significance in either dataset.

We also estimated a model in which respondent judgments of certainty, and their evaluation of question importance, were used together with RL to predict response error. No significant effects were found for these variables in either dataset, but this may be due to the very small number of cases for which respondent evaluations were available.

Finally, a model without RL was considered, in which the verbal doubt indicator and the two paralinguistic indicators were investigated. The results are shown in Table 7.9. It is evident that we can use verbal expressions of doubt together with the paralinguistic indicators to predict response error. However, a model that substitutes RL for doubt also provides highly significant results ($p < 0.01$), except for switches. At this point it is not clear whether RL or doubt is the best predictor of response error. They both behave as good predictors, together with the number of words used in the answer.

7.5 CONCLUSIONS

Our study demonstrates a link between response latency and response error: The longer the latencies, the higher the probability of an incorrect (and nonsubstantive) answer. This confirms Bassili's contention that "RL's may provide a powerful tool for screening survey questions during questionnaire development.... It is reasonable to expect that questions that suffer from problems identified by behavior coding require longer latencies" (1996, p. 332).

But we also found that respondent certainty judgments and (para)linguistic indicators, such as the number of words uttered, doubt words, and switches in answer alternatives, were related to response error. Thus, these, too, could be used as indicators of question difficulty in the pretesting phases of questionnaire development. Indeed, the results of our multivariate analyses suggest that the expression of doubt is as good a predictor of response error as is RL. Recording expressions of doubt is, of course, less tedious than the exact measurement of latencies for all elements of question–answer sequences. Whether our findings will hold in general—for instance, with different types of questions—is an important topic for future research.

Appendix 7A Question Wording, Percent Correct/Incorrect, RLs, and Number of Respondents Answering a Question

	Percent Correct	Percent Incorrect	Percent Don't Know	RL (seconds)	*n*
Dataset 1					
1. Are you member of Milieudefensie?	100	—	—	0.22	121
2. Have you been a member of Milieudefensie for more than 6 years?	89	6	5	0.64	101
3. Have you been a member of Milieudefensie for more than 1 year?	71	20	9	0.85	35
4. Did you pay a subscription of more than 25 NLG to Milieudefensie during the most recent subscription drive? This refers to the drive that was held just recently.	71	18	10	0.97	138
5. Did you pay a subscription of more than 24 NLG to Milieudefensie during the preceding subscription drive, held during the winter? (question posed half a year after the subscription drive)	21	64	15	1.61	136
6. Have you used the biological washing-up liquid ECOVER for more than 6 years?	83	9	8	1.43	61
7. Have you used the energy-saving lamps for more than 5 years?	60	35	5	1.33	98
8. Have you read about the "Get rid of the dung" campaign?	72	23	5	2.32	64

Appendix 7A (*continued*)

	Percent Correct	Percent Incorrect	Percent Don't Know	RL (seconds)	*n*
9. Have you read about the "Buy unsprayed lettuce" campaign?	80	18	2	1.08	60
10. Have you read about the "On your bicycle" campaign?	68	24	8	1.74	76
11. Have you read about the "Buy secondhand" campaign?	89	9	2	1.04	75
Dataset 2					
12. Are you a member of Milieudefensie?	97	2	1	0.20	138
13. Have you been a member of Milieudefensie for less than 3 years?	69	24	7	1.34	119
14. Do you buy potatoes of the brand Artemis?	82	3	14	1.50	141
15. Did you pay a subscription of more than 25 NLG to Milieudefensie during the most recent subscription drive in the summer? (half a year later)	64	22	14	1.45	141
16. Did you pay a subscription of more than 24 NLG to Milieudefensie during the preceding subscription drive, held during the winter? (question posed a year later)	56	28	16	1.54	143
17. Have you used the energy saving lamps for more than 5 years?	66	26	8	1.46	97
18. Have you used the biological washing-up liquid ECOVER for more than 6 years?	65	26	9	1.31	78
19. Have you read about the "Get rid of the dung" campaign?	76	21	3	2.54	66
20. Have you read about the "On your bicycle" campaign?	85	11	4	2.27	75
21. Have you read about the "Buy unsprayed lettuce" campaign?	73	21	6	1.38	67
22. Have you read about the "Buy secondhand" campaign?	89	9	2	1.20	77

CHAPTER 8

Vignettes and Respondent Debriefing for Questionnaire Design and Evaluation

Elizabeth Martin
U.S. Bureau of the Census

8.1 INTRODUCTION

Recent decades have seen theoretical and empirical advances in understanding the cognitive sources of measurement errors introduced by failures of comprehension or retrieval. In this chapter we describe how two methods, vignettes and respondent debriefing questions, can be used to identify measurement problems and craft and test questionnaire designs to address them. The focus is on their application in field-based tests of interviewer-administered questionnaires, although they also are used in laboratory and qualitative studies (the latter use is discussed) and with other types of questionnaires, such as self-administered ones. The chapter draws on research (much of it hitherto unpublished) conducted for the redesign of several Census Bureau surveys as well as other studies. In Section 8.2 we describe the use of vignettes for questionnaire design, drawing on research on problems of comprehension in the Current Population Survey (CPS). In Section 8.3 we describe respondent debriefing questions, drawing on research undertaken to redesign instruments for the National Crime Victimization Survey (NCVS) to reduce recall and reporting problems.

Methods for Testing and Evaluating Survey Questionnaires, Edited by Stanley Presser, Jennifer M. Rothgeb, Mick P. Couper, Judith T. Lessler, Elizabeth Martin, Jean Martin, and Eleanor Singer
ISBN 0-471-45841-4

8.2 VIGNETTES

Vignettes have a long history in qualitative and quantitative research on social judgments, going back (at least) to Piaget's (1932/1965) use of "story situations" to investigate moral reasoning in children. Piaget offers an important rationale for using vignettes, as well as the main methodological question about their validity: "... while pure observation is the only sure method, it allows for the acquisition of no more than a small number of fragmentary facts. ... Let us therefore make the best of it and ... analyse, not the child's actual decisions nor even his memory of his actions, but the way he evaluates a given piece of conduct. ... We shall only be able to describe [it] ... by means of a story, obviously a very indirect method. To ask a child to say what he thinks about actions that are merely told to him—can this have the least connection with child morality?" (Piaget, 1932/1965, pp. 112–113).

In other words, the use of vignettes permits an investigator to gather data that could not otherwise be collected at all, or only for a small number of cases, but the question of whether evaluations of hypothetical situations relate to judgments in real life remains an issue. Piaget himself adopts a pragmatic approach to validity when he states that "... any method that leads to constant results is interesting, and only the meaning of the results is a matter for discussion" (1932/1965, p. 114).

Vignettes are brief stories or scenarios that describe hypothetical characters or situations to which a respondent is asked to react. Because they portray hypothetical situations, they offer a less threatening way to explore sensitive subjects (Finch, 1987). Their specificity allows contextual influences on judgments to be examined. To preserve realism, qualitative researchers may create vignettes based on actual situations reported to them. These are then used to stimulate open-ended discussions with respondents to explore their reasoning and judgments. Quantitative researchers more often construct vignettes by systematically manipulating features in different vignettes, which are administered in a controlled, experimental evaluation of the factors that affect respondents' judgments. This strategy was devised by Rossi and his colleagues, who labeled their vignettes *factorial objects* to capture this approach to vignette design, in which cells in a formal experimental design were represented by different vignettes (Rossi and Anderson, 1982). Respondents in quantitative or qualitative studies may be asked to perform a task, such as ranking, rating, or sorting vignettes into categories, or projecting themselves into a vignette situation, to imagine what a vignette character would or should do or feel.

Although vignettes typically contain detailed descriptive information, they may vary in degree of elaboration, as suggested by several illustrative vignettes used in studies of normative judgments:

1. "Armed street holdup stealing $25 in cash" (from Rossi et al.'s 1974 study of the seriousness of crimes).
2. "Cindy M., a freshman, often had occasion to talk to Gary T., a single 65-year-old professor. She went to his office after class. She seemed

worried and asked him about grades. He remarked that she was making good progress in class. He reached out and straightened her hair. He said that he could substantially improve her grade if she cooperated" (from Rossi and Anderson's 1982 study of judgments of sexual harassment).

In both studies, vignette features [the amount stolen in (1); and social setting, prior relationship, male physical acts, and five other dimensions in (2)] were manipulated systematically in order to evaluate their effects on judgments.

The use of vignettes as a methodological tool for designing and evaluating questionnaires is more recent. [They were first used to evaluate alternative instruments for the redesign of the National Crime Victimization Survey (Biderman et al., 1986).] This chapter describes the use of vignettes to:

- Explore a conceptual domain
- Test consistency of respondents' interpretations with survey definitions, diagnose question-wording problems, and assess uniformity of meanings
- Evaluate the effects of alternative questionnaires on interpretations of survey concepts
- Analyze the dimensionality of survey concepts

8.2.1 Exploring Concepts

This use of vignettes is similar to substantive investigations of conceptual domains, although more focused on survey concepts and the implications for framing survey questions. A good example is Gerber's (1990, 1994) exploratory research employing ethnographic interviews to examine how people think about residence, the language they use to describe it, and the factors they take into account in deciding how to report it in surveys. Gerber initially conducted open-ended interviews with 25 informants, many with tenuous living situations that made residence determination difficult. (Respondents were recruited from a homeless shelter and a church.) She gathered descriptive information about informants' residence patterns and elicited the terms they used to describe them. The following excerpt from an interview illustrates the distinction that one informant made between "living" and "staying" and several of the criteria (e.g., intentions, location of belongings) she invoked to explain her situation:

A: "I'm just a friend of hers. I lost my apartment in December. ... That's why I said I'm staying there, cause I'm not living there. I'm doing everything I can to find a way out of there.

Q: So you're not living there. ...

A: Well, you would say I'm living there, I been there since December, but I'm just saying it's not mine ... But I live there, I bathe there, I sleep there, I dress there, my clothes are there—not everything I own. Most of my things I got out of storage and took to my mother's, but basically everything I have to live with since December is there. As a matter of fact, it's packed up at the door. Because I'm trying to get out. ..." (Gerber, 1990, pp. 15–16).

Gerber (1990) used the situations reported by the first set of informants to construct vignettes that were the focus of a second set of interviews. The excerpt above was simplified into the following vignette: "Mary asked her friend Helen if she could stay with her for a few days while she looked for a place of her own. It has been five months since then. Mary's suitcases are still packed and are at the front door. Should Helen count Mary as usually living there?"

All the vignettes described ambiguous living situations and were used to elicit informants' calculations of residence. According to Gerber, "In making the judgment about a complex or ambiguous case, informants revealed what elements of the situation were important to them and what sort of logic they followed in arriving at a decision. In the course of the interview, I would vary the circumstances somewhat in order to follow out these trains of logic. ..." (1990, pp. 5–6). For example, she probed to determine if informants' answers to the vignette above would change for stays of longer or shorter duration. As illustrated in the excerpt above, when life circumstances were complex, her informants used various criteria to determine where someone lives. Respondents' calculations may lead them to omit marginal residents who should be listed on a census roster or include those who should not be. Gerber (1994) interviewed additional respondents using an expanded set of vignettes in a follow-up study to identify appropriate terminology for census roster questions. Several features of her use of vignettes are worth noting.

First, vignettes were culled from ethnographic sources to present respondents with living situations they might actually encounter. As Gerber notes, "By providing respondents with situations they recognize as 'real,' we were able to tap into the expectations and reactions which they would have in similar social circumstances. This increases our confidence that the way respondents reasoned during our interviews is similar to the judgments they make in reporting rosters in survey situations" (1994, p. 4).

Second, Gerber (1994) took care to use neutral vocabulary in the vignettes, to elicit the vocabulary that respondents naturally use to describe residence situations. For example, vignette characters were described as sleeping in a certain place or spending time with a particular person, rather than as "living" or "staying" there. She also attempted to create an entirely neutral probe that would elicit residence terms without actually using any ("What would you call the time X spends with Y?"), but respondents did not understand the probe. Therefore, the common term *live* was introduced early in the interview, to train respondents in the task. Other, less common terms used in the census rostering process were introduced using structured probes (e.g., "Is X a member of Y's household?") later in the interview to avoid biasing answers.

Third, the ambiguity of the vignette situations stimulated respondents to think through and articulate the criteria they would apply to decide where a person should be considered to live. Altering the details of a vignette in unstructured follow-up probes helped clarify respondents' reasoning, as illustrated in the following interview excerpt:

A: Well, it seemed to me that if you had said he ate his meals and slept there, then I would consider that he lived there.

Q: ... if we said he eats at his wife's house, but he always sleeps at his mother's ...

A: I'd say that's a weird arrangement.

Q: That's weird, but would you say that changed where he lived?

A: Well, if he slept at his mother's, I would consider that he lived at his mother's. On a permanent basis ... if he just slept there occasionally, I would not consider that he lived there ..." (Gerber, 1994).

By separating eating and sleeping (and other circumstantial details), Gerber was able to develop a more nuanced understanding of which factors influenced the answer given.

Fourth, the tasks involving vignettes were readily understood, even by respondents without much education or fluency in English. Respondents often treated the task as a puzzle or game, and only one interview (of 37) in Gerber's 1994 study had to be terminated because the respondent did not understand the task. However, focusing on hypothetical situations influenced responses to a subsequent request for factual information, which elicited a number of obviously fabricated answers.

8.2.2 Testing Interpretations of Question Intent and Diagnosing Question Wording Problems

It is well known that small changes in question wording can substantially affect responses (see, e.g., Schuman and Presser, 1981), presumably by affecting respondents' interpretations of question meaning. Despite their sensitivity to wording changes, respondents commonly misinterpret the intended meanings of survey questions (Belson, 1981). Vignettes provide a tool for investigating the effects of question wording and context on interpretations of survey concepts, as illustrated by research conducted for the 1994 redesign of the Current Population Survey (CPS), the source of official U.S. unemployment estimates.

The CPS questions used ordinary words (such as "work" and "looking for work") with technical survey meanings that were not communicated to respondents. This situation arose because concepts had been refined over the years but the questionnaire had remained largely unchanged since the 1940s. (The pertinent questions about work were, "What were you doing most of *last week* [working, keeping house, going to school] or something else?" and if the respondent did not report working, "Did you do any work at all *last week*, not counting work around the house? [*Note:* If farm or business operator in household, ask about unpaid work.]" The question about looking for work asked, "Has NAME been looking for

Table 8.1 CPS Respondents' and Interviewers' Classifications of Vignette Situations

Vignette	Percent "Yes"[a]	
	Respondents	Interviewers
I asked you a question about *working* last week. Now, I'm going to read a list of examples. After each example, please tell me whether or not you think the person should be reported as *working* last week.		
(1) Last week, Susan only did volunteer work at a local hospital. Do you think she should be reported as *working* last week?	38% (1973)	4% (1458)
(2) Last week, Amy spent 20 hours at home doing the accounting for her husband's business. She did not receive a paycheck. Do you think she should be reported as *working* last week?	50% (1977)	83% (1324)
(3) Sam spent 2 hours last week painting a friend's house and was given 20 dollars. Do you think he should be reported as *working* last week?	64% (1976)	93% (1395)
(4) Last week, Sarah cleaned and painted the back room of her house in preparation for setting up an antique shop there. Do you think she should be reported as *working* last week?	59% (1949)	66% (1348)
Please tell me whether or not each of the following activities should be reported as *looking for work*.		
(5) During the past 4 weeks, George has occasionally looked at newspaper ads. He hasn't yet found any jobs in which he's interested. Do you think he should report that he is *looking for work*?	36% (1122)	37% (1413)

Source: Campanelli et al. (1989b).

[a]Correct answers are "yes" to vignettes 2, 3, and 4 and "no" to 1 and 5. Missing data are excluded from calculations. *n*'s are given in parentheses. Vignettes were asked in the order shown.

work during the past 4 weeks?" and [if yes] "What has ... been doing in the last 4 weeks to find work?")

A series of vignettes (shown in Table 8.1) portraying irregular employment situations was administered to about 2300 respondents in a computer-assisted telephone debriefing interview (CATI) conducted in 1988, immediately after a final CPS interview (Campanelli et al., 1989a). Also shown in Table 8.1 are interviewers' responses to the same vignettes, administered as part of a "knowledge of concepts" test conducted with the CPS field staff (Campanelli et al., 1989b). The results revealed common misunderstandings and suggested that the intended meanings of "work" and "looking for work" were not communicated by the questions as worded. For example, only half of respondents correctly interpreted unpaid work in a family business (vignette 2) as work, with many interviewers (17%) also classifying it incorrectly. Results for vignettes 2 and 4 suggested that the phrase "not counting work around the house" might have led respondents (and some interviewers) to exclude legitimate work activities that took place in the home.

The misinterpretations shown in Table 8.1 were likely to lead to reporting error, although the vignette results did not, in themselves, provide direct evidence about its magnitude. The vignettes were asked of people who had been asked a pertinent question in the main survey, for whom a vignette situation may or may not have been relevant. Reporting error would result if a respondent whose actual situation a vignette describes misinterprets how it should be reported. Campanelli et al. (1989b) found that in one case respondents for whom a vignette situation was relevant were more likely than others to classify it correctly, and in another, less.

However, the combined use of respondent and interviewer classifications of vignettes supported inferences about the probable impact of misinterpretations on the data (as discussed by Campanelli et al., 1991). For example, the situation portrayed in vignette 5 is problematic because many (over a third) interviewers as well as respondents erroneously considered this passive job search to be "looking for work." Overly broad interpretations would be expected to lead respondents erroneously to report passive job searches and interviewers erroneously to accept them. This inference was consistent with evidence of high rates of misclassification of passive job searches (Fracasso, 1989; Martin, 1987). The results suggested that the wording of the question led both respondents and interviewers (including highly experienced ones) to misinterpret its intent. The problem could not be overcome by additional interviewer training or experience, but required rewording the question to better communicate its intent.

Discrepancies between intended and actual interpretations of work were more serious because they were correlated with age, with older respondents generally defining work much too narrowly and younger ones too broadly (Campanelli et al., 1989a). The correlation with age was consistent with a suspected underreporting of teenage work activities by older proxy respondents (such as parents) and underreporting of work activities by retirees.

Thus, the vignettes confirmed that key questions did not adequately communicate the intended meaning of important concepts. The results supported the need to revise question wordings and helped identify probable sources of misunderstanding.

Several points should be noted about the use of vignettes in this application. First, administering the vignettes as part of a CATI supplement made it possible to tailor the vignettes to be asked of people who had also been asked the target question in the main survey. Contextual validity was preserved, since the vignettes captured respondents' interpretations in the context of an actual survey interview rather than in a laboratory setting from which generalization is less certain.

Second, administering the vignettes to probability samples made it possible to generalize about differences between groups in their interpretations of key concepts. This is not possible with convenience samples.

Third, administering vignettes to both interviewers and respondents yielded more information than either study alone would have provided. Situations that were poorly understood by both seemed most vulnerable to reporting error and pointed to a need for questionnaire revision and improved interviewer training to address them.

Fourth, response rates for the supplement were high and item nonresponse rates for the vignettes were low (less than 3% per item) (Campanelli et al., 1991). As in the qualitative study, the vignette task does not appear to be overly difficult for respondents or to lead to high rates of "don't know" answers in general samples.

Fifth, the vignettes did not provide direct evidence about the magnitude of reporting errors. Rather, they provided feedback about misinterpretations of key survey concepts held by respondents who were asked a target question; misinterpretations may or may not result in error. One would expect a misinterpretation to result in error when the situation portrayed in the vignette applies, but this remains an inference.

8.2.3 Evaluating the Effects of Alternative Questionnaires on Interpretations

The CPS redesign also illustrates the use of vignettes to evaluate whether questionnaire revisions bring respondents' interpretations more in line with question intent. The CPS instrument went through several iterations of revision and testing, with a final split-sample comparison of old and proposed new questionnaires in a large national RDD sample in 1991. Vignettes were administered to one in 10 respondents after a final interview, in order to test whether interpretations were more standardized and consistent with CPS definitions under the new questionnaire. The last two columns of Table 8.2 present the 1991 vignette results by questionnaire version; selected 1988 results from Table 8.1 are included in the first column for comparison.

Several methodological differences between the 1988 and 1991 vignette studies should be noted. Vignette 1 was reworded slightly to avoid using the word *work* and perhaps biasing respondents' classifications. (As it turns out, the same fraction classified vignette 1 as "work" in the first and second columns.) Second, the introduction was revised to repeat key portions of the target question, to ensure that classifications were contextualized by the question that respondents had been asked earlier. Similarly, the wording of follow-up questions more closely mirrored the target question to reinforce the effect of question wording. Third, another work vignette (not shown) was added before vignette 1 in 1991, while a vignette that had preceded vignette 4 in 1988 was dropped (the latter is discussed below). Vignette responses were probed ("Why would you consider/not consider that person to be working?"). Finally, the gender of the subject person was manipulated experimentally to examine whether men's and women's work activities were viewed differently. A random half of respondents received a vignette with a female name, and the other half received a male

Table 8.2 Classifications of Vignettes Following New and Old CPS Questionnaires in Two Surveys

	Percent "Yes"[a]		
Vignette	Old Q'aire 1988	Old Q'aire 1991	New Q'aire 1991
Earlier I asked you a question about working. Now I want you to tell me how you would answer that question for each of the persons in the following imaginary work situations. Would you report her/him as ... *(Old q'aire)*: working last week, not counting work around the house? *(New q'aire)*: working for pay (or profit) last week?[b]			
(1) Last week, Susan/Al put in 20 hours of volunteer service at a local hospital.	38%	37%[c]	4%
(2) Last week, Amy/Joe spent 20 hours at home doing the accounting for her husband's/his wife's business. She/he did not receive a paycheck.	50%	46%[c]	29%
(3) Sam/Diane spent 2 hours last week painting a friend's house and was given 20 dollars.	64%	61%[c]	71%
(4) Last week, Sarah/Jeff cleaned and painted the back room of her/his house in preparation for setting up an antique shop there.	59%	47%[d]	42%
Total n asked work vignettes	1980	305	319

Source: Martin and Polivka (1995).

[a]Correct answers are "no" to 1 and "yes" to 2, 3, and 4.

[b]The parenthetical "or profit" was used after vignettes 2 and 4, to which it was applicable.

[c]Difference between versions in 1991 significant at $p < 0.05$.

[d]Difference between years (old q'aire only) significant at $p < 0.05$.

version (these results are discussed below). Results in Table 8.2 combine male and female versions.[1]

Results in Table 8.2 support several conclusions about using vignettes to compare question wording and context effects. First, except for one vignette, the old questionnaire evoked the same classifications in both the 1988 and 1991 surveys. Results in the first and second columns are similar despite the survey differences described above. The replicability of vignette results in independent surveys using the same questionnaire suggests that they reliably capture the effects of questionnaire context and wording on interpretations. In addition,

[1]Classifications of vignette 2 were affected significantly by subject gender. "Amy" was more likely than "Joe" to be considered as working (for the Amy version, 54% and 34% answered "yes" following old and new questionnaire versions, respectively).

the response structure remained stable. Martin and Polivka (1995) fit log linear models to joint distributions of responses to the work vignettes in the 1988 and 1991 surveys (following the old questionnaire) and found that the same model described associations among items in both years. The one significant difference was a drop from 1988 to 1991 in "yes" responses to vignette 4, which may be due to a contrast effect. In 1988, vignette 4 immediately followed another that described donating blood for money, which few respondents considered work (it was dropped in 1991). The contrast to selling blood may have made setting up an antique shop seem more like "work." In addition to being sensitive to the context created by questions in the main survey, vignette responses also may be vulnerable to context effects created by the order in which they are asked.

Second, the questionnaire modifications partially succeeded in bringing respondents' interpretations more in line with CPS, but also may have led to some new misinterpretations. Responses to the "why" probe confirmed that the meaning of work in the old question was vague. Respondents gave more, and more various, reasons for their vignette responses, including the irrelevant consideration of location. In contrast, the revised wording focused their attention on payment. Comparison of the second and third columns in Table 8.2 shows that the revised wording reduced positive responses to several vignettes (1 and 2) that did not involve payment, and broadened respondents' interpretations to more often include casual paid labor (vignette 3). Unfortunately, the narrower focus on pay led some respondents to rely exclusively on present payment and to rule out some legitimate work activities not yet yielding pay or profit, such as those in vignette 2.

The experimental manipulation of gender also suggested that the new questionnaire created a more gender-neutral interpretation of work. The meaning of work in the old questionnaire was vulnerable to gender bias. Vignette 1 (and other "helping" vignettes not shown) were more likely to be classified as work if the subject was female, and male respondents were more sensitive to gender than female respondents. In other words, "helping is 'women's work,' if you ask men" (Martin et al., 1995, p. 43). The focus of the revised question wording on "pay or profit" eliminated the effects of both respondent and subject gender on classification of "helping" vignettes (Martin et al., 1993). Thus, the new questionnaire elicited more gender-neutral interpretations of helping activities as well as reducing the extent to which respondents thought they should be reported at all.

Did altered interpretations of "work" influence reporting under the new questionnaire? Evidence on this question is somewhat mixed. The expanded frame of reference indicated by results for vignette 3 should increase reporting of casual employment in the new questionnaire, and this prediction is borne out by evidence. A larger fraction of persons 16 to 19 years old (but not of older persons) were reported as working (Martin and Polivka, 1995), and there were significantly more reports of work activities involving a few hours. A slight gender bias due to underreporting of the number of female workers was eliminated. Thus, evidence from several sources suggested that the new questionnaire was more inclusive of casual labor.

In another situation, the new questionnaire narrowed the respondents' frame of reference too much, leading them to exclude unpaid work in a family business (vignette 2). Nevertheless, the new questionnaire elicited more, not fewer, reports of unpaid work in a family business (Polivka and Rothgeb, 1993), because a direct question about unpaid work in a family business was added. Respondents were no longer expected to understand that they should report it in response to a general question about working. Thus, the questionnaire solution was to add a specific question, rather than try to improve respondents' understanding of a complex concept.

Several conclusions about the method are suggested by this research. First, vignette classifications are highly sensitive to questionnaire context. Even relatively small samples provide useful feedback about the effect of questionnaire revisions on respondent interpretations. This is especially useful when (as is true for CPS as well as many other surveys) really enormous samples are required to detect actual reporting differences for specific, relatively rare situations. This conclusion is also consistent with research on the NCVS (Biderman et al., 1986).

Second, research to date suggests that vignettes are reasonably robust measures of context and question-wording effects on interpretations. Despite survey differences, similar results were obtained in a replication of the vignettes in two independent surveys using the same questionnaire. Additional research is needed to establish the conditions under which vignettes reliably measure context and wording effects, but results such as those in Table 8.2 are promising. An apparent contrast effect induced by the order of the vignettes in one survey provides a caution that exact replication of a vignette series is necessary to ensure comparable results. Caution is also warranted in using vignette results to make improvements in items; improved items need to be evaluated *in situ*, in the context of a revised questionnaire.

Third, the open-ended "why" probes proved useful in understanding respondents' reasoning and interpreting wording effects.

Thus, revising questions based on vignette results and retesting the revised questions using the same vignettes can tell a questionnaire designer whether or not the questionnaire revisions have addressed the problems of interpretation satisfactorily. However, as illustrated by the CPS example, a questionnaire revision may correct one misunderstanding but create a new one. This reinforces the need for several rounds of testing, to ensure that problems have been addressed satisfactorily and that new problems have not arisen (see also Esposito, 2002; Esposito et al., 1992).

8.2.4 Exploring the Dimensionality of Survey Concepts

By analyzing the joint distribution of responses to a set of vignettes rather than analyzing them one at a time, a questionnaire designer can assess the global effects of a questionnaire revision on the inclusiveness or exclusiveness of the underlying concept.

For example, Martin et al. (1991) applied the Rasch measurement model to the joint distribution of the work vignettes to examine whether a latent dimension of meaning accounted for response patterns, or alternatively, whether respondents applied different criteria to classify each vignette situation, with no unifying concept.[2] Their analysis suggested that both were true to some extent. The data were consistent with a latent dimension of inclusiveness, with respondents who held a strict interpretation of work at one end and respondents willing to include marginal activities at the other. Beyond a propensity to be inclusive (or not), responses to three pairs of vignettes were associated, suggesting that respondents applied similar criteria or rules to classify each pair (e.g., vignettes 2 and 4 were associated, perhaps because some respondents ruled them both out because they took place at home). One vignette (about donating blood for money) could not be scaled with the others and therefore did not belong in the same conceptual domain; it did not partake of the meaning of "work." Alternative scorings of the vignette responses, as "yes/no" and as "correct/incorrect," led to the conclusion that the dimension underlying respondents' classifications was inclusiveness, not correctness. This implied that vignette series should include a balance of items with both "yes" and "no" correct answers, to avoid confounding correctness and inclusiveness.

Ideally, if the Rasch model had fit (with no additional between-item associations beyond those attributable to the underlying latent dimension), a very simple scale formed by summing the number of "yes" responses could have been used as a practical tool to evaluate the inclusiveness of respondents' interpretations of work. [See Duncan (1984) for discussions of scaling and Rasch models.]

Martin and Polivka (1995) took this analysis one step further to assess the effects of the questionnaire revision on the structure of responses to the vignettes. A series of log-linear analyses showed that the revision affected respondents' interpretations of particular types of situations, and hence responses to particular vignettes and associations among them, but did not globally broaden or narrow their interpretations of work.

Thus, modeling vignette responses can yield insight into the underlying dimensions of meaning evoked by a survey questionnaire, whether a questionnaire defines global or particular meanings, and whether respondents use rules or heuristics to judge situations. When the goal is specificity, the failure of the Rasch model to describe vignette data adequately may be taken as evidence of improvement. When a questionnaire designer intends to measure a global construct, abstracted or generalized from particular situations, it would be desirable to find that the Rasch model fits and that respondents adopt global rather than specific rules for classifying different situations.

[2]The Rasch measurement model (Rasch, 1960/1980) treats each response as the product of two parameters: one unique to the item, the other unique to the person. When items are scored "yes/no," the person parameter represents an individual's latent tendency to interpret the concept of work inclusively or restrictively. The item parameter represents how difficult or easy an item is; in other words, how congruent the activity described in the vignette is with respondents' underlying concept of "work." The Rasch measurement model is useful for analyzing the dimensions underlying a set of dichotomous items [see Duncan (1984) for a detailed discussion]. Other models, such as factor analysis, are appropriate when response categories can be considered to form an interval scale.

8.3 RESPONDENT DEBRIEFING QUESTIONS

A second method is similar to vignettes in being administered following a field interview, but encompasses more diverse types of questions. Typically, respondents are told that the main interview is complete and then asked general probing questions or standardized, retrospective questions about their experience of the interview, how they answered or interpreted questions, and other topics.

The respondent debriefing method reinvents and adapts probing techniques that have been used for decades to examine question meaning in survey pretests or the survey itself. In 1944, Cantril and Fried conducted an intensive study of 40 respondents, using follow-up probes to identify specific misunderstandings of poll questions. For example, answers to the question "After the war is over, do you think people will have to work harder, about the same, or not so hard as before?" were probed by asking "When you said (harder/about the same/not so hard), were you thinking of people everywhere and in all walks of life (laborers, white-collar workers, farmers and businessmen) or did you have in mind one class or group of people in particular?" About half thought "people" meant everybody, a third interpreted the word as indicating a particular class, and a tenth didn't know what it referred to.

Belson (1981) relied on similar probes when he introduced the *question-testing method*, using what he called *double interviews* to identify problems of question understanding. Target questions were embedded in a questionnaire administered in personal interviews. A second, intensive interview was conducted the following day by another, specially trained interviewer. Respondents were paid for the second interview, which began with informal conversation about the previous day's interview, then (for each of seven target questions) reminded respondents of the question and their answer, and interrogated them extensively about how they understood the question and arrived at their answer. Belson notes that "the intensive interviewer was responsible also for probing for a full reconstruction, for challenging inconsistencies between the indications of the present evidence and the answer actually given 'yesterday,' and for keeping the respondent thinking of how she answered the question yesterday as distinct from her interpretation of it now" (1981, pp. 35–36). Belson evaluated specific misunderstandings of terms (e.g., "you," "usually") and developed hypotheses about sources of misunderstanding. He concluded: "There is simply no way in which standard piloting can be used reliably to reveal the many misunderstandings of respondents, many of them unsuspected by the respondent himself and not visible to the piloting interviewer. . . . Direct question testing is essential" (1981, pp. 390, 397).

The various applications of respondent debriefing or special probes have several methodological features in common. First, question meaning (or other response issue) is evaluated in the context of a real survey interview, typically conducted in the field. Second, respondent debriefing questions or special probes are standardized and asked after the main interview is complete, to avoid influencing responses to survey questions. [Other uses of special probes (e.g., Schuman, 1966) employ them as part of the interview, immediately after a question has been asked.] Third, the method frankly enlists respondents' help in improving a survey by inviting them "to assume a new role: to become informants rather than respondents" (Oksenberg et al., 1991, p. 357) by commenting and elaborating on their interview experiences. Fourth, studies that employ special probing methods conclude that misunderstandings are quite common but that respondents (and interviewers) are largely unaware of them.

Respondent debriefing studies vary considerably in scope, in the amount of time that elapses between original and debriefing interviews, and whether the debriefing interview takes place in the lab or in the field. Respondent debriefing questions used in pretesting may probe for:

- Interpretations of terminology, questions, or instructions
- Subjective reactions or thoughts during questioning
- Direct measures of missed or misreported information[3]

8.3.1 Meaning Probes

Probes to test interpretations of terminology or question intent are the most common form of debriefing question, and their usefulness for detecting misunderstandings is well documented (Belson, 1981; DeMaio, 1983a; DeMaio and Rothgeb, 1996; Oksenberg et al., 1991; Schuman, 1966). An illustration is drawn from an evaluation (Von Thurn, 1996) of interpretations of the term *regular school* in the following question: "Is ... attending or enrolled in regular school? (Regular school includes elementary school, high school, and schooling that leads to a college or professional school degree.)"

Reviewers doubted that the term was meaningful to respondents, even with the parenthetical definition, which was judged likely to be interrupted by respondents or skipped by interviewers. To test interpretations, several open- and closed-ended debriefing questions were administered in the field after completion of the interview: "Earlier I asked if ... is attending regular school. What does the phrase regular school mean to you in this question? Would a technical or vocational school be considered a regular school?"

Both open and closed probes confirmed that regular school was poorly understood, with closed probes providing more usable information. Responses to the open probe required coding, and many were too general or meaningless to be categorized as correct or incorrect (Von Thurn, 1996).

Oksenberg et al. (1991) found that comprehension probes, similar to the closed probes in the example above, were useful for revealing misinterpretations of key terms in survey questions. (An example from their research, following a question about consumption of red meat, was: "Would you include things like bacon,

[3]Another type, not discussed here, is a more general "debriefing" about a prior interview, which is not a pretest or evaluation of survey questions but may yield insights about questionnaire problems. For example, a large-scale reinterview was conducted after the 1980 census to learn more about the mail response process and perceptions of the census form (DeMaio, 1983b). In another example, Wobus and de la Puente (1995) conducted telephone debriefing interviews to learn respondents' reactions to receiving both English- and Spanish-language forms in the mail as part of a census test.

hot dogs, or lunch meats as red meat?"). Similar probes have been asked to test interpretations of reference periods, such as "the past 12 months" or "last week" (see Campanelli et al., 1989a; Hess and Singer, 1995; Moyer et al., 1997). Other types of probes also have proved useful for uncovering misinterpretations. Hess and Singer (1995) asked respondents in a field pretest to paraphrase survey questions about hunger ("Could you tell me in your own words what that question means to you?") and found that several complex questions were commonly misunderstood. However, Oksenberg et al. (1991) found that general "tell me more" probes and probes for direct reports of problems were not productive. (An example of the latter, following a question about illnesses "that kept you in bed for more than half of the day," was "How clear was it to you what to include as a half day in bed?"). Perhaps respondents were reluctant to admit or were unaware of their own misunderstandings. The authors noted that "respondents did not appear to doubt their own, often mistaken, interpretations" and concluded that "the particular strength of special probes lies in their ability to reveal problems that are not evident in interview behavior" (1991, pp. 358, 363). Other authors (DeMaio and Rothgeb, 1996; Morton-Williams and Sykes, 1984) reach similar conclusions. Research also shows that revising survey questions to correct problems revealed by special probes appeared to reduce misreporting (Fowler, 1992).

8.3.2 Thoughts and Subjective Reactions

Debriefing questions about respondents' thoughts or feelings have been used to address a variety of questionnaire issues.

Question Sensitivity Miller and Davis (1994) conducted a field pretest of potentially sensitive questions with 29 mothers of children whose fathers lived elsewhere. Of particular concern were questions about whether the child's paternity had been established. After the pretest interview, respondents were asked: "Were there any questions in this interview that you felt uncomfortable answering?" and about one in five expressed discomfort with the paternity questions. Similar questions have been asked to assess discomfort with a request for social security numbers in a mailed census form (Bates, 1992) and sensitivity of long-form census questions (Martin, 2001). Debriefing questions also may be asked about respondents' sensitivity to specific features of a survey's design. For example, in a debriefing conducted after adolescent respondents filled out a self-administered questionnaire that included sensitive questions, Hess et al. (1998a) learned that respondents would be more concerned about the privacy of their answers if survey questions were printed where others might read them than if the questions were administered using an audiocassette tape player.

Confidence Debriefing questions that ask respondents about their certainty or confidence in their answers have thus far not provided useful information about questionnaires (see, e.g., Oksenberg et al. 1991). Respondents typically express high levels of confidence and certainty, which appear to have little relationship to the correctness of their interpretations of survey questions. Campanelli et al. (1991) found no correlation between respondents' confidence and how well their classifications of various employment situations corresponded with CPS definitions. Moyer et al. (1997) found that respondents who misinterpreted the reference period for an income question were more confident of their answers than respondents who correctly interpreted the question (probably because they had reinterpreted the question to be one they could answer with confidence). Schaeffer and Dykema (Chapter 23, this volume) report that the accuracy of answers was unrelated to respondents' judgments about how certain they were of their answers, but Draisma and Djikstra (Chapter 7, this volume) report that respondents' use of "doubt words" was related to accuracy.

Mental Processes Questionnaire pretests do not usually employ retrospective debriefing questions to assess respondents' mental processes while answering survey questions, and it is not clear how fruitful this might turn out to be as a pretesting method. Oksenberg et al. (1991) had limited success asking respondents how they came up with their answers to survey questions. It may be difficult for respondents to recall what they were thinking while answering a prior question, especially if other questions have intervened, and survey interviewers may not be skilled at asking what are in effect "think-aloud" probes of the sort typically asked in cognitive interviews. On the other hand, Moyer et al. (1997) obtained useful information from a probe asking respondents how they came up with their answers to a survey question, as did Blair and Burton (1987) when investigating respondents' retrieval strategies (their debriefing question immediately followed the pertinent survey question, however).

An example of the use of debriefing questions to assess respondents' mental processes is drawn from research to redesign and test alternative screening questions for the National Crime Victimization Survey. The major goal of the redesign project was to reduce severe underreporting of victimizations that had been documented by record check studies (Biderman et al., 1986). The redesign of the screening questions was informed by a theory of cognitive barriers to recall and reporting of victimization incidents (see, e.g., Martin et al., 1986; Biderman, 1980a,b, 1981a,b; Loftus and Marburger, 1983; Sparks, 1982), and a screener designed to address the problems was tested against the standard screener in a split-panel CATI field experiment. Debriefing questions were asked after the interview to test whether the revised screener reduced hypothesized sources of retrieval and reporting problems, including failure of metamemory, recall interference, fatigue and negative response sets, mnemonic failure, and selective reporting.

Failure of metamemory Psychological research (e.g., Hart, 1965) had suggested that retrieval efforts are guided by an initial "feeling of knowing" that there is or is not something relevant available in memory to recall. If respondents conclude too quickly that they have nothing to report, they may fail to engage in a memory search. The experimental screener was designed to prevent respondents from committing themselves to a "nothing happened" response before screening. Debriefing item 1 (see Table 8.3) was intended to measure respondents' expectation, before screening started, that they would have something to report.

Recall interference Recall of one incident may block retrieval of additional incidents, because a respondent in effect keeps recalling the same incident (Roediger and Neely, 1982). Item 2 was intended to measure whether respondents experienced interference from an incident recalled previously.

Fatigue and negative response set Respondents may become annoyed and fatigued by a long list of screening questions to which the answer is almost

Table 8.3 Results of Debriefing Questions in an Experimental Comparison of Two Victimization Screener Designs

Debriefing Question	Percent "Yes"	
	Experimental (Short Cues) Screener	Traditional (Question-and-Answer) Design
(1) At the beginning, before I asked you any questions, did you think you would have any crimes to report?	18%	15%
(2) I (asked questions/gave examples) to help you remember crimes that might have happened to you. You told me about one incident. Did you find you were still thinking about that incident when we went back to the (examples/questions)? (Asked of respondents who reported one or more incidents)	57%	71%
(3) While I was asking you the (questions about/examples of) crimes that might have happened to you, did you lose track or have a hard time concentrating?	12%	8%
(4) Did you feel bored or impatient?	30%	25%
(5) Were you reminded of types of crime you hadn't already thought of on your own?	41%	26%
(6) Was there an incident you thought of that you didn't mention during the interview? I don't need details.	8.6%	4.6%
If yes: "Were you unsure whether it was the type of crime covered by the survey?"	20%	17%
"Was the incident a sensitive or embarrassing one?"	11%	33%
n	522	534

Source: Martin et al. (1986).

invariably "no." The redesigned screener employed short cues and reminders rather than questions to try to avoid these problems. Items 3 and 4 measured the subjective burden of the alternative formats.

Mnemonic failure Because many crime experiences are not salient events in memory, their recall requires contextual cues to aid in retrieval. The new screener employed extensive cues, including reminders of nonstereotypical crimes, to improve the mnemonic properties of the screener and improve respondents' understanding of the crime scope of the survey.[4] Item 5 asked whether respondents were reminded of types of crime they hadn't thought of.

Selective reporting The new screener adopted a "broad net" approach (Biderman, 1980b), encouraging respondents to report, and interviewers to accept, all incidents they thought of, even if they were unsure whether the incidents were covered by the survey. (Answers to follow-up questions determined if an incident was or was not in scope.) Item 6 provided a measure of whether respondents were withholding information about potentially in-scope incidents. By telling respondents in advance that they would not be asked to disclose anything about unreported incidents, we hoped to increase their willingness to acknowledge them. Respondents were also probed for reasons why an incident was not mentioned.

The results suggested that respondents were willing and apparently able to report their subjective reactions and thought processes during screening. The cognitive properties of the two instruments differed substantially by some measures, and the differences were consistent with objective evidence about screener performance. Item 1 suggested that few respondents (less than 20% in both screeners) said that they initially expected to have anything to report, which seemed to confirm the designers' concerns about possible failure of metamemory. (As it turned out, 37 and 48% of respondents in the traditional and experimental screeners, respectively, reported at least one incident.) According to item 2, experimental respondents were significantly less likely to persevere in thinking about an incident after the interviewer returned to the screening task, indicating less recall interference. This was consistent with the much higher rates of victimization reporting, especially of multiple incidents, produced by the experimental screener. However, most respondents in both screeners experienced interference from incidents recalled previously. Items 3 and 4 suggested that the experimental screener was more cognitively burdensome, with experimental respondents significantly more likely to say that they had a hard time concentrating or lost track during screening. This was consistent with its greater length (32 minutes, compared to 21 minutes for the traditional instrument). On the other hand, experimental respondents were much more likely to report being reminded of types of

[4]To test understanding of the crime scope of the survey, respondents were read vignettes and asked "whether you think it is the kind of crime we are interested in, in this survey." Six vignettes portrayed situations vulnerable to misreporting, such as domestic violence ("Jean and her husband got into an argument. He slapped her hard across the face and chipped her tooth. Do you think we would want Jean to mention this incident to us when we asked her about crimes that happened to her?").

crime they hadn't thought of (item 5), suggesting that the experimental screener was a more effective mnemonic aid, and consistent with this they reported many more victimizations. Finally, results of item 6 indicated less selective reporting in the new instrument, with fewer respondents reporting that they thought of an incident but failed to mention it. This was consistent with the elevated reporting of both in-scope and out-of-scope incidents. Although fewer reports were withheld in the new instrument, a significantly larger fraction of them pertained to "sensitive or embarrassing" incidents. Thus, the estimated fraction of respondents who withheld reports of sensitive incidents did not differ significantly between questionnaires.

Thus, debriefing questions derived from hypotheses about the cognitive sources of response errors appeared to yield meaningful information about the retrieval and reporting process. Although the results are suggestive, they do not demonstrate that the questions represent valid measures of the intermediate cognitive processes that were the intended target of the redesigned screening procedures. (For example, one might doubt whether respondents could accurately recall their expectations at the start of an interview, as item 1 in Table 8.3 asked them to do.)

Debriefing questions about respondents' mental processes have a natural connection with research on metacognition, or people's knowledge of (and presumably, ability to report about) their own memories and cognitive processes (see, e.g., Koriat et al., 2000, for a recent review). Cognitive psychologists have explored certain tasks (e.g., feeling-of-knowing and confidence judgments) similar to debriefing questions that are (or might be) asked of respondents. The experimental literature on metacognition may shed light on the validity of self-reports about cognitive processes, and help answer the question of what respondents can report about their cognitive states, especially their memory processes. Although a review of this literature is outside the scope of this chapter, it might be applied usefully to the design of respondent debriefing questions and other survey topics.

8.3.3 Direct Measures of Missed or Misreported Information

A third type of debriefing question probes for events or facts that a respondent failed to report or reported incorrectly in the main survey. In some cases, detailed debriefing questions may test whether questions in the main survey are eliciting reports consistent with survey definitions (e.g., Esposito, 2002; Fowler and Roman, 1992). For example, to evaluate how well a general question about income is performing, debriefing questions may probe whether a respondent reported net or gross income, and whether specific sources were included. If the question in the main survey is not obtaining the intended information, the question wording or response categories might be revised, or a question added. Sometimes the debriefing question itself may be moved into the survey to improve measurement.

This type of debriefing question may also be used as a direct measure of underreporting. In the final split-panel comparison of new and old CPS instruments, a subsample of respondents who had not reported any work activities in the main survey was probed to determine if they had neglected to report a few hours of work (Martin and Polivka, 1995). The probe was: "In addition to people who have regular jobs, we are also interested in people who may work only a few hours per week. *Last week* did NAME do any work at all, even for as little as one hour?"

Follow-up probes ("What kind of work did NAME do?" and "Did NAME get paid for the work?") were asked to screen out reports that were not legitimate work activities (about 80% of responses were bona fide). Between 2 and 3% of respondents who had not reported working in the survey did report bona fide work activities in debriefing, with no overall questionnaire difference. The age bias in underreporting casual labor was reduced but not eliminated in the new questionnaire. A problem with the missed work probe was that large samples were required to detect meaningful differences, so most of the questionnaire differences were not statistically reliable.

Item 6 in Table 8.3 also permitted direct examination of victimization underreporting, although it had the disadvantage that the characteristics of the unreported incidents were unknown. Because it was unknown whether the unreported incidents were in scope, the information could not be used to estimate the fraction of in-scope victimizations missed.

8.4 CONCLUSIONS

In the past, it has been necessary to approach questionnaire design and revision as a process of redesigning a "black box" whose output is evaluated but whose inner workings are poorly understood and often produce puzzling results. Respondent debriefing and vignettes do not eliminate all the surprises involved in questionnaire design and pretesting, but they can help a designer better understand and predict the nature and underpinnings of questionnaire effects. By shedding empirical light on the inner workings of a survey instrument, these methods help demystify the questionnaire design process and take us a step toward placing the design of survey measurements on a firmer (dare I say scientific?) footing. Below I summarize the advantages and disadvantages of vignettes and respondent debriefing questions and compare them with some other questionnaire pretesting methods.

An advantage of both vignettes and respondent debriefing questions is that they reveal hidden problems of meaning that respondents and interviewers may be unaware of and that do not necessarily result in interviewing difficulties. This advantage is shared by cognitive interviewing but not by pretesting methods that do not probe respondents' interpretations, such as behavior coding. A second advantage is that respondents appear able to step into the informant role and perform the tasks, and even appear willing to disclose their less-than-complete reporting of sensitive facts, as in the case of the NCVS debriefing question.

The methods are flexible and may be used in exploratory, qualitative studies, laboratory investigations and experiments, small field pretests, large-scale pilot studies, ongoing surveys, and split-panel field experiments. Indeed, the same set of debriefing and vignette questions may be carried forward from one stage of pretesting to the next to provide systematic, comparable measures at each stage. Behavior coding can be applied in both field and laboratory settings, but cognitive interviewing probably cannot be.

The methods yield useful information even when administered to relatively small samples. Their efficiency is increased because a respondent does not need to have experienced a specific situation in order to interpret how it should be reported. In contrast, very large samples are needed to measure actual reporting differences, especially for uncommon situations. Large samples also may be required for debriefing questions about actual events or behavior, such as missed work.

The methods are cheap when administered as supplements to an ongoing survey or a field pretest, because they involve no separate field contact. Most of the debriefing interviews (including vignettes) reported in this chapter took no more than 5 to 8 minutes of interviewing time. This cost advantage may be shared by cognitive interviewing, which can provide useful results with a small number of interviews, but not by behavior coding, which requires labor intensive coding.

When administered as a survey supplement, both methods preserve contextual validity because respondents are asked to interpret a term or concept, classify a vignette, and so on, in the context of the actual survey. This is less true of laboratory methods, such as cognitive interviewing.

When administered to probability samples, results are generalizable to the survey population, and group differences in question interpretation may be assessed meaningfully. Other pretesting methods, including cognitive interviewing and (in applications to date) behavior coding, are not generalizable because they are not sample-based.

Respondent debriefing questions and vignettes also share several disadvantages. With the exception of probes for missed or misreported information, they do not provide direct evidence of reporting error. They provide indirect evidence about questionnaire performance that is useful in conjunction with other performance indicators, such as reporting differences. They are most useful when their design is informed by substantive and methodological knowledge and theory. In the examples discussed in this chapter, advance knowledge was provided by prior ethnographic interviews, as in Gerber's research, by prior investigations of reporting problems, as in the CPS example, and by hypotheses derived from cognitive literature on recall, as in the NCVS example. In contrast, cognitive interviewing provides more opportunity for an interviewer to design probes flexibly and explore problems that emerge during an interview.

Although their results are plausible and consistent with other evidence, respondent debriefing and vignettes have not been evaluated rigorously. Evidence to date suggests that vignettes are sensitive and relatively robust measures of the context and wording effects of the particular questionnaire they follow, but more research is needed to evaluate this key assumption. And although it is reasonable to assume that misinterpretations (as measured by responses to debriefing questions or vignettes) are indicative of measurement error, the connection is indirect rather than direct, and needs additional investigation.

Issues involved in the design and implementation of vignettes and debriefings need further exploration. Evidence suggests that some types of debriefing questions are more meaningful and valid than others, and both vignettes and debriefing questions are probably affected by their own sources of error and bias. Research to date suggests that questions about respondent certainty or confidence elicit exaggerated reports that shed little light on question misunderstandings, because respondents seldom doubt their own idiosyncratic interpretations. Responses appear to be sensitive to vignette order, suggesting the need for careful replication when comparisons are made between surveys. The validity of debriefing questions about mental processes during an interview is subject to the same limitations that self-reports about cognitive processes are generally subject to, and this type of question has not been much explored in pretesting. The effects of delay between a survey question and a retrospective debriefing question have not been addressed, but ephemeral thoughts or reactions are likely to be quickly forgotten. Subjective reactions and interpretations may also be affected by questions that intervene between target questions and debriefing questions or vignettes.

Each method also offers some advantages not shared by the other. Because vignettes are posed in hypothetical terms removed from a respondent's own situation, this method is well suited for exploring sensitive or stigmatizing subjects (this type of application was not illustrated here). Vignettes offer advantages for exploring how respondents arrive at complex judgments that are influenced by social context, because situational factors can be varied among vignettes in qualitative or experimental research. Probes to determine why respondents classify vignettes as they do shed light on their reasoning. Vignettes can be used to examine particular problematic situations, test the match between respondents' interpretations and survey definitions, and assess the degree of standardization of meaning. Administering them to interviewers helps identify concepts and situations that are poorly understood and require additional training.

Respondent debriefing questions are more direct measures than vignettes and can provide information about a greater range of response problems, including direct measures of reporting error, although this requires larger samples than are needed for comprehension probes. Vignettes and respondent debriefing questions are useful in three phases of questionnaire development and pretesting. First, vignettes are useful for exploring respondents' understandings of terms and concepts before designing a questionnaire and can help designers design better questions and avoid wording pitfalls. A second use is to identify or verify problems of interpretation of existing questions in a survey. The statistical modeling of vignette responses may yield insights about changes or inconsistencies in underlying survey concepts, marrying methodological and substantive purposes. Respondent debriefing questions can elicit information about subjective reactions to the questions and feedback about unreported or misreported information. Information about which words and phrases are misunderstood and which types of situations are misreported can help the questionnaire designer address the problems by rewording questions, adding instructions or examples, and so on. (Alternatively, a designer might revise survey definitions to bring them in line with respondents' interpretations.)

In a third phase, vignettes and debriefing questions can be used to evaluate alternative questionnaires. Performance measures based on vignettes and debriefing questions can be used (together with actual reporting differences) to select the questionnaire that is best understood, least cognitively burdensome, or which yields other measurable improvements. By using these methods through a program of iterative design and pretesting, it is possible to gain much richer knowledge about the performance of questions and the nature of the errors affecting survey measurement of a phenomenon.

ACKNOWLEDGMENTS

This chapter reports the results of research and analysis undertaken by the U.S. Bureau of the Census staff. This report is released to inform interested parties of ongoing research and to encourage discussion of work in progress. The views expressed are those of the author and not necessarily those of the U.S. Census Bureau. I am grateful to the colleagues and friends who were collaborators on some of the research reported here, including Al Biderman (now deceased), Pam Campanelli, Bob Fay, Bob Groves, Jennifer Hess, Jay Matlin, Anne Polivka, Stanley Presser, Jennifer Rothgeb, Donny Rothwell (now deceased), and Paul Siegel.

CHAPTER 9

The Case for More Split-Sample Experiments in Developing Survey Instruments

Floyd Jackson Fowler, Jr.
University of Massachusetts–Boston

9.1 INTRODUCTION

The bridge that has been built in the past two decades between cognitive psychology and survey research has generated a more refined understanding of how people answer questions (e.g., Hippler et al., 1987; Jabine et al., 1984; Sirken et al., 1999a; Tourangeau, 1999). This, in turn, has provided an impetus for better question evaluation.

The past decade or so has seen three important improvements in the routine procedures that are used for presurvey evaluation of questions. First, some kind of cognitive testing is frequently used to help evaluate how questions are understood and what answers mean (DeMaio and Rothgeb, 1996; Forsyth and Lessler, 1991; Willis et al., 1999). Second, field pretests often are augmented with behavior coding, the systematic recording of how interviewers and respondents handle the question-and-answer process, as another way of identifying questions that are difficult to ask or to answer (Fowler and Cannell, 1996). Third, there has been some advance, although perhaps less developed and widely accepted, in systematic presurvey assessment of questions using fixed standards. Appraisal forms, such as those described by Lessler and Forsyth (1996), can be used to flag question characteristics that are likely to cause problems of one sort or another.

Methods for Testing and Evaluating Survey Questionnaires, Edited by Stanley Presser, Jennifer M. Rothgeb, Mick P. Couper, Judith T. Lessler, Elizabeth Martin, Jean Martin, and Eleanor Singer
ISBN 0-471-45841-4

There is no doubt that the use of these techniques has improved the quality of survey questions. Identifying and revising questions that are difficult to understand or to answer in a consistent way can only improve survey measurement. However, there are limits to all of these techniques. Specifically, while they identify features of questions that could potentially cause problems and affect data quality, these techniques provide no direct information about how data quality is affected. Moreover, when questions are revised to address problems that are identified, they provide no basis for knowing whether or not the quality of the data resulting from the alternative question will be better than, or different in any way from, the original question.

There are at least three reasons why some type of quantitative assessment of the effects of a revised question is important:

1. Sometimes "fixing a problem" (e.g., by defining a term or concept that can be misunderstood by respondents) makes a question worse from other perspectives. For example, adding detailed definitions of terms or concepts can make a question harder to administer and harder to understand for respondents. Thus, it is important to know whether the effects of the fix constitute an improvement over the original from the various perspectives from which questions must be evaluated.
2. In a similar way, one of the most important conservative forces for keeping problem questions is the desire to use questions that have been used in previous surveys. Reviewers often like the fact that an item has been used previously, whether or not it has been carefully evaluated. Researchers also value the potential to compare their results with those obtained by other researchers in other times and places. In that context, when "problems" are found in questions with a pedigree, how much the problems affect data quality and how improved versions of the question will affect estimates are important considerations.
3. Finally, even if the two issues above are not relevant, when a researcher "fixes a problem," it is highly valuable to know whether or not the new question produces data that are more likely to be valid or better by some other standard than the original.

One way to address these issues is to conduct a field experiment in which parallel questions are asked of comparable samples. By randomizing two versions of a question designed to achieve the same objective to two samples, researchers can see how the alternatives affect the distribution of answers obtained. In some cases, if the samples are large enough, there may also be the potential to assess other psychometric properties of the answers.

The purpose of this chapter is to illustrate the way that experimental tests can add to the presurvey question design process. Indeed, one could argue that the benefits of presurvey evaluations, such as cognitive testing, can be fully realized only when they are paired with experimental pilot studies prior to conducting a full-scale survey.

9.2 GAPS IN COGNITIVE TESTING RESULTS

Cognitive testing identifies problems that people have in understanding and answering questions. However, the samples tested typically are small and not designed to be statistically representative in any sense. In addition, cognitive procedures themselves, particularly think-aloud approaches, may have an effect on how respondents choose answers. Thus, once "problems" are identified through cognitive testing, there are three critical issues with respect to how much these problems will affect the survey results:

1. If a question is found to be inconsistently understood, what percentage of people actually have that misunderstanding?
2. What percentage of the time is the answer that people have to give actually affected by the ambiguity?
3. If a question poses a problem that makes it unclear how to answer, how in fact do respondents resolve those problems when they translate what they have to say into an answer?

To elaborate briefly on these generalizations, consider the following example: On days when you drink, how many drinks do you usually have?

Cognitive testing consistently shows that there are at least three problems with this question:

1. What constitutes a drink is not defined. Respondents could have different understandings about how many ounces of wine or ounces of liquor would constitute a drink. So one issue would be: What is the distribution of people's understandings of what constitutes a drink, and how widely do they vary?
2. The second issue is how much difference it makes how one counts drinks. With this particular question, a considerable amount of averaging and rounding is needed in any case. It is plausible that in the process of averaging and summarizing, a good conception of how many ounces constitutes a drink may have little effect on answers.
3. A further issue with this question is that it requires people to average and imposes an assumption of "regularity" on people that is not justified for some respondents. For example, if a person usually has three drinks on the weekend, but only one during the week, he has to decide whether to report the most common number or to provide some kind of weighted average. How respondents resolve that difference, how they translate what they have to say into an answer, will have a lot to do with how much this ambiguity affects the results.

It is to address these problems that experiments in pilot studies, following cognitive testing, can make a major contribution to assessing the significance of question problems identified in cognitive testing and decisions about whether or not, and how, to revise potentially problematic questions.

9.3 HOW TO USE FIELD EXPERIMENTS

The concept behind field experiments is simple. Two "comparable" samples of respondents are asked different versions of questions designed to achieve the same question objective. The questions can then be compared in up to three different ways:

1. The most important comparison often is simply to compare the response distributions. If the response distributions seem to be virtually identical, one may conclude that the two questions are virtually the same question.
2. It may also be possible to compare the validity of the two questions. To do that, there has to be some kind of a standard against which to assess answers, so that one can tell which answers are "more valid."
3. The "usability" of the two questions also can be compared. Often, cognitive testing identifies questions that are worded in an awkward or confusing way. One goal of revising questions may be to make them easier for respondents and interviewers to use, and field experiments can provide information about those features as well.

The basic design required is to randomize respondents to different versions of a test question. Depending on the circumstances, sometimes only a few questions are randomized; in other cases, two versions of a survey instrument are created, with two versions of a number of questions being tested (see, e.g., Chapter 10). Of course, if data collection is computer assisted, randomization of alternative versions of questions or survey instruments is particularly easy to accomplish.

How big do the samples need to be? There is never a completely satisfactory answer to a question like that; researchers are always trading off resources and time versus statistical precision. Table 9.1 is one approach to thinking about the answer. If the goal was to see whether or not distributions were different for alternative versions of a question, this table shows two standard errors of

Table 9.1 Standard Errors of Differences When Comparing Percentages That Are Around 50% for Various Sample Sizes

Two standard errors of differences[a] (%)	24	20	16	14	11
When comparing groups of these sizes	35	50	75	100	150

[a] Differences between percentages this big or larger would be statistically significant ($p < 0.05$).

differences based on various sample sizes, assuming that the optimal design is to have equal-sized samples for testing each version of a question. It can be seen that considerable precision is gained by going from 35 in each group to 75 in each group. Things improve more slowly with increased sample size after that.

These figures also may be too conservative. Most often when one is testing alternative versions of questions, there are good hypotheses about why the answers might differ and in what direction. In that context, if an experimental test provided evidence that a particular question wording had a predicted effect, one might not feel the need for adherence to a two-tailed 0.05 statistical standard for interpreting the results. For further discussion of sample size, see Chapter 11.

The other sampling issue is how representative the sample needs to be. Obviously, as with any pretest, researchers want to feel confident that the results are similar to those they will obtain in their full-scale survey. However, since creating statistics about populations is not the goal of such tests, cost-cutting measures to improve efficiency are likely to be appropriate. For example, when these studies are done by telephone, researchers often limit callbacks and efforts to convert reluctant respondents, which produces lower response rates, in an effort to save time and money. For the purposes at hand, as long as the samples answering alternative versions of the questions are randomized and reasonably representative of the survey population, nonresponse usually would not be expected to affect comparisons between alternative questions.

There are two ways in which researchers assess the validity of the answers to two alternative versions of questions. With respect to factual items, one legitimate approach is simply to look at distributions. Sometimes, cognitive testing has identified a "problem" that seems likely to affect the answers systematically; an alternative question is designed to fix an error in people's understanding of the intent of the question. If the new or revised question changes the distribution in the expected way, researchers may well be justified in concluding that the answers are more valid.

For example, a question about exercise was found in cognitive testing to be ambiguous about whether or not walking for exercise should be counted. A revised question was written that explicitly told respondents that it was all right to include walking for exercise. An experimental test showed that the explicit inclusion of walking increased the rate at which people reported that they exercised. In that case, the researchers seemed justified in concluding that the latter answers were more valid (Fowler, 1992).

The other approach to assessing validity is, of course, to correlate answers with some standard or some measure with which the answers are supposed to be correlated. Opportunities for that kind of assessment are not universal, but researchers should be on the lookout for ways that they can use studies of association to evaluate the quality of data they are getting.

Finally, one of the reasons for revising questions is to make them easier for respondents and interviewers to use. If instruments are self-administered, it is harder to get a quantitative measure of usability. However, if an instrument is interviewer administered, tape recording the interviews and behavior coding the way that interviewers ask questions and respondents answer them provides a systematic way of assessing the usability of two versions of a question.

9.4 SOME EXAMPLES

To illustrate the role that experimental testing can play to complement cognitive testing results, six examples from two different studies are reported below. In those studies, the initial phase of the project involved extensive cognitive testing of a set of items. Based on that testing, problems were identified and alternative versions of the questions, designed to accomplish the same objectives, were created. In one study, the field test was carried out doing telephone interviews with a national sample of households identified through random-digit dialing. A large pool of numbers was made available to interviewers, and interviewing was stopped when the target numbers were reached. Few callbacks were made to those who did not answer or who were reluctant. Respondents were randomized to either the original or alternative forms of survey questions. The sample sizes for the two sets of questions were 77 and 79 interviews.

The other test was carried out with samples of adults enrolled in a health plan in Washington State. The data reported here came from telephone interviews conducted with respondents who were assigned randomly to either the original or alternative questions. The resulting samples were 256 and 263 for the comparisons. In this case about 52% of the originally selected sample members were interviewed by phone.

As an additional way of evaluating questions, approximately the first 50 telephone interviews using each version of the questions were tape recorded, with respondent permission of course, and the interviewer and respondent behaviors during the interviews were coded. The key measures used in this chapter are the rates at which the questions were read exactly as worded by interviewers and the rates at which respondents asked for clarification of questions.

The first two examples both address the value of providing specific examples to help clarify the meaning of a general or abstract concept.

Original question 1: About how many months has it been since you last saw or talked to a dentist? Include all types of dentists, such as orthodontists, oral surgeons, or all other dental specialists, as well as dental hygienists.

Original question 2: During the past 30 days, on how many days did you do strenuous tasks in or around your home? By *strenuous task*, we mean things such as shoveling soil in a garden, chopping wood, major carpentry projects, cleaning the garage, scrubbing floors, or moving furniture.

Both of these questions provide fairly long and elaborate explanations of what the question covers. In both cases, these explanations are provided after the question itself is already asked. When these questions were tested, there was evidence that the explanations constituted overload, that they were not well attended to, and that the possibility existed that they were unnecessary. As a result, the two following alternatives were created:

Alternative to question 1: About how many months has it been since you last went to a dentist's office for any type of dental care?

Alternative to question 2: During the past 30 days, on how many days did you do any strenuous tasks in or around your home?

The results of the experimental test are reported in Tables 9.2 and 9.3. In Table 9.2 it can be seen that the results from the two versions of the dentist visits question are virtually interchangeable. Almost identical percentages of respondents said they had seen a dentist within six months and within one year. Apparently, the elaboration has no practical effect on the data.

In contrast, in Table 9.3 it can be seen that there is a great difference in the number of days that people report doing strenuous tasks in and around the home. Specifically, while the list of activities may have created a long question, and everyone may not have attended to all the details, in fact it substantially increased the number of days that people said that they engaged in these activities. The mean was 4.67 for the long version, only 2.72 for the short version. This difference is statistically significant ($p < 0.05$).

A final point is that one goal of the streamlined wording was to make the questions easier to administer. With respect to question 1, interviewers committed small or large reading errors about 10% of the time, and the reading of the question was interrupted another 10% of the time when the original question was asked. The simplified question was read much better; there were no major reading errors or interruptions observed in the taped interviews.

Table 9.2 Results of Dentist Visits Question

Original: About how many months has it been since you last saw or talked to a dentist? Include all types of dentists, such as orthodontists, oral surgeons, or all other dental specialists, as well as dental hygienists.

Alternative: About how many months has it been since you last went to a dentist office for any type of dental care?

Time Since Dental Care	Original (%)	Alternative (%)
6 months or less	60	57
More than 6 months but not more than 1 year	14	18
More than 1 year	26	25
Total	100 ($n = 77$)	100 ($n = 79$)

Table 9.3 Results of Strenuous Work at Home Question

Original: During the past 30 days, on how many days did you do strenuous tasks in or around your home? By strenuous tasks, we mean things such as shoveling soil in a garden, chopping wood, major carpentry projects, cleaning the garage, scrubbing floors, or moving furniture.

Alternative: During the past 30 days, on how many days did you do any strenuous tasks in or around your home?

Number of Days	Original (%)	Alternative (%)
0	31	41
1–5	34	37
6–10	13	10
11–20	16	7
30	5	4
Unable to do this type of activity	1	1
Total	100 ($n = 77$)	100 ($n = 79$)
Mean days	4.66	2.72
	$p < 0.05$	

In contrast, when we look at the behavior coding for question 2, reading errors, if anything, were higher in the alternative version (13% in the original, 22% for the revision). Interruptions (5%) were not very prominent in the original question, and requests for clarification were actually much higher in response to the simplified question (27%) than they were with the original question (15%).

Thus, with respect to question 1, the experimental evidence shows that the questions are virtually equivalent from the respondent point of view; the data are virtually the same. Moreover, if anything, the usability of the simpler question is higher. In contrast, with respect to question 2, the original question elicits considerably more reporting of strenuous activities, which seems highly likely to be more accurate and complete reporting. Moreover, if anything, the usability of the alternative question is worse, not better.

The next example is actually a short series of questions designed to identify people who have chronic health conditions:

Original question series 3:

a. Do you now have any physical or medical conditions that have lasted for at least 3 months?

b. (*If yes*) In the past 12 months, have you seen a doctor or other health provider more than twice for any of these conditions?

c. (*If yes to a*) Have you been taking prescription medication for at least 3 months for any of these conditions?

A condition was considered chronic if it had lasted for at least 3 months and had been cause for the respondent either to see a doctor more than twice or to take prescription medications for at least 3 months.

Obviously, this series is very complicated. It requires respondents to think about a number of issues all at once in order to answer the questions correctly. A particular issue that arose in the course of cognitive testing was whether or not people had a shared understanding of what was meant by a "physical or medical condition." Concerns about inconsistent understanding of this concept were particularly important because if a respondent said "no" to the first question, the second and third questions were not even asked.

An alternative set of questions was designed:

Alternative question series 3:

a. In the past 12 months, have you seen a doctor or other health provider 3 or more times for the same condition or problem?
b. (*If yes*) Is this a condition that has lasted for at least 3 months?
c. Do you now need or take medicine prescribed by a doctor?
d. (*If yes to c*) Is this to treat a condition that has lasted at least for 3 months?

The results of the testing of these two alternative series are presented in Table 9.4. It can be seen that there is a marked difference in the results. The

Table 9.4 Chronic Condition Identifiers by Questionnaire Version

Original (%)		Alternative (%)
a. Do you now have any physical or medical conditions that have lasted *for at least 3 months*? (*Women*: Do *not* include pregnancy.)		a. In the past 12 months, have you seen a doctor or other health provider three or more times for the same condition or problem?
b. In the last 12 months, have you *seen a doctor* or other health provider *more than twice* for any of these conditions?		b. Is this a condition that has lasted for at least 3 months? (Do *not* include pregnancy.)
c. Have you been taking prescription medicine for at least 3 months for any of these conditions?		c. Do you now need to take medicine prescribed by a doctor (other than birth control)?
		d. Is this to treat a condition that has *lasted for at least 3 months*? (Do not include pregnancy or menopause.)
Has Chronic Condition		
Yes	38	56
No	62	44
Total	100 ($n = 335$)	100 ($n = 347$)
	$p < 0.01$	

original question identified about 38% of respondents as having a chronic condition that met these criteria. The alternative series identified 56% of respondents as having a chronic condition that met these criteria. In this case, it is quite clear that starting with the behavioral implications of the question, either seeing a doctor more than twice or taking prescription medications, led to recall and reporting of many more conditions than when the series depended on respondents identifying and reporting a stand-alone "condition." It should be noted that the behavior coding showed no notable issues with the usability of either series of questions.

A fourth example also involves extraordinary cognitive complexity.

Original question 4: This question is about automobile injuries, including injuries from crashes, burns, and any other kinds of accidents. Have you ever had an injury because of your driving?

Cognitive testing suggested that there were a lot of issues for respondents to think about when dealing with this question. The question tells people what types of injuries to think about. The time frame is for the respondent's entire life. The cognitive testers felt that it was hard for respondents to keep all the elements of the question in mind when they tried to answer. As a result, an alternative was designed that broke the cognitive task into pieces.

Alternative question 4:

a. This question is about automobile injuries, including injuries from crashes, burns, and any other kind of accident. Have you ever had an injury while you were in a car?
b. (*If yes*) Were you ever the driver when you were injured?
c. (*If yes to a and b*) Were you ever injured because of your driving?

Table 9.5 shows that while 8% of the respondents were classified as "yes" in response to the original question, only 2% were identified as being "yes" in response to the alternative series. Based on the cognitive testing, it seemed most likely that respondents to the original question were overreporting because they were not attending to the last part of the question, which asked whether or not the injury was due to the respondent's faulty driving. It seems highly likely that the alternative series is collecting more valid data.

As a further note, based on the behavior coding, while there were no differences in requests for clarification, which were fewer than 4% for both versions, the original question produced more reading errors (18%) than the alternative question (12%). The other two questions in the alternative series had even fewer reading errors (fewer than 10%). Hence, the alternative, if anything, is also more usable.

The next example addresses the issue of how people go about forming answers. In the CAHPS survey, respondents are asked to summarize their experiences in getting health care across a reference period, usually 12 months.

Table 9.5 Results of Automobile Injury Question

Original: This question is about automobile injuries, including injuries from crashes, burns, and any other type of accident. Have you ever had an injury because of your driving?

Alternative:
a. This question is about automobile injuries, including injuries from crashes, burns, and any other kind of accident. Have you ever had an injury while you were in a car?
b. Were you ever the driver when you were injured?
c. Were you ever injured because of your driving?

Injured Because of Driving?	Original (%)		Alternative (%)
Yes	8		2
No	92		98
Total	100		100
	($n = 79$)		($n = 77$)
		$p < 0.05$	

Original question 5: In the last 12 months, how often did you get an appointment for regular or routine health care as soon as you wanted—always, usually, sometimes, never?

This response task was chosen in part based on cognitive testing that indicated that respondents were reluctant to choose negative alternatives when their experience had been mostly positive. Thus, giving them the alternative of saying that "usually" their experience met the standard was an attractive option to respondents. The downside of this response task, however, is the tension it produces when respondents have only one or two experiences that they are summarizing. If they had one experience (e.g., if they had made only one appointment), their answers should be restricted to "always" or "never." Although many respondents use those alternatives, some respondents with only one event use the middle categories, which clearly cannot be technically accurate. Either they are bringing in experiences from outside the reference period when they answer the question or they are using the middle categories to moderate their answers for reasons that technically should not be included in their answers.

A question that would eliminate this tension for those respondents with only one or a few visits would be the following:

Alternative question 5: In the last 12 months, were you *always* able to get an appointment as soon as you wanted?

Such a question should pose no problem for someone who only had one visit; the technically correct answer is clear.

Table 9.6 Results for Health Care Appointment Question

	Standard (%)	Alternative (%)
	In the last 12 months, how often did you get an appointment for *regular or routine* health care as soon as you wanted—always, usually, sometimes, never? (*"Always" recoded to "Yes"*)	In the last 12 months, were you always able to get an appointment as soon as you wanted?
Always Got Appointment		
Yes	47	66
No	53	34
Total	100 ($n = 261$)	100 ($n = 299$)
	$p < 0.001$	

Table 9.6 presents the results from the experiment. It can be seen that 66% of the people responding to the alternative question said that they were "always" able to get an appointment when they wanted, while only 47% of respondents answered "always" when given the four-category choice. On the one hand, psychometricians would not be surprised at this result; more categories usually spread out answers. On the other hand, the data are highly consistent with the results of cognitive testing, which suggested that respondents have trouble using the negative category when things are generally pretty good. Undoubtedly, some of the "yeses" to the alternative question would have turned out to be "usuallys" had people been offered the four-category response task.

The behavior coding indicates that interviewers read these questions equally well and respondents have no trouble answering them.

The final example also relates to how respondents go about answering questions when faced with ambiguity.

Original question 6: When riding in the back seat of a car, do you wear a seat belt all the time, most of the time, some of the time, once in a while, or never?

When this question was tested cognitively, it was quickly learned that there are some people who never ride in the back seat of cars. For such people, this question poses a dilemma. There are three ways they can answer the question. First, they can volunteer to the interviewer that the question does not apply to them. Second, they can generalize from their seat belt–wearing behavior when they are in the front seat, essentially saying that they would probably behave in the same way if they had been riding in the back seat. Such an answer does not describe actual behavior, but it does describe the type of person the respondent is from a seat belt–wearing point of view. Third, the respondent could give the technically correct answer "never." However, note that in essence this is answering a different question. Rather than saying that they ride in the back seat and never wear a seat belt, they are saying that they never ride in the back seat and hence they never ride wearing a seat belt in the back seat.

A simple alternative approach is to identify those people who ride in the back seat of the car in advance by asking the question.

Alternative question 6:

a. In the past year, have you ever ridden in the back seat of a car?
b. When you are riding in the back seat of a car, do you wear a seat belt all the time, most of time, some of the time, once in while, or never?

Table 9.7 shows that the results are quite different in several ways. Clearly, as we would expect from many other studies, respondents do not volunteer that questions do not apply to them very often. When people who do not ride in the back seat are explicitly identified, 20% fall in that category, compared with 8% who volunteered that response. Moreover, it is obvious that a large number of people were using the "never" response category to say they never rode in the back seat. The subsequent effect of that distortion is quite striking. If one looks at the data from the original question, one would conclude that nearly a quarter of the people riding in back seats are not wearing seat belts, possibly a significant public health problem. In contrast, when the alternative question is asked, only 4% of the people say that they do not wear seat belts when riding in the back seat—much less of a public health problem.

Table 9.7 Results for Back-Seat Seat Belt Question

Original: When riding in the back seat of a car, do you wear a seat belt all of the time, most of the time, some of the time, once in a while, or never?

Alternative:
a. In the past year, have you ever ridden in the back seat of a car?
b. When you are riding in the back seat of a car, do you wear a seat belt all of the time, most of the time, some of the time, once in a while, or never?

Wear Seat Belt in Back Seat	Original (%)		Alternative (%)
All of the time	30		42
Most of the time	17		16
Some of the time	13		8
Once in a while	8		10
Never	24		4
Don't ride in back seat	8		20
Total	100 ($n = 77$)		100 ($n = 79$)
		$p < 0.002$	

Behavior coding indicated that neither of these approaches to designing the questions posed any problems for interviewers or respondents.

9.5 CONCLUSIONS

Each of the six examples was found in cognitive testing to have some kinds of issues that might affect data quality. However, based on cognitive testing alone, the significance of the issues for survey estimates was uncertain, and whether or not alternative forms of the question would in fact improve the data was also unknown until further data were collected.

The first two questions, the one about dentists and the other about everyday exercise, pose fairly typical problems that survey researchers need to address. How important is it to provide definitions and examples of general concepts when asking people about fairly common things? Providing examples can help clarify what the question is asking about, and it can serve to remind people, jogging their memories of things they might otherwise forget. As the data show, the added examples with respect to going for dental care added little to the questions; in contrast, having examples of activities that people engage in that might count as exercise added a great deal to the questions and to the number of events reported.

Researchers are often tempted to pile multiple concepts and issues into a single question, leaving it to respondents to sort them out and figure out what they are really being asked. The question is, how such complexity affects the data. Example 3, about injuries that are the driver's fault, and example 6, about wearing seat belts in the back seat, are both examples of when question complexity led people to focus on only a part of the question and distorted answers. In these cases, the issues identified in cognitive testing mattered in ways that were important for data quality, and the results from the experiment made that clear.

Two other examples made similar points. We knew from cognitive testing that people were not always clear about what counted as a condition. It was difficult to tell how much difference that would make. It would be easy for researchers to say that people who have significant conditions probably will know it, will think of them as such, and will report them. The results from the field experiment show that clearly that is not the case. In the same way, the results from cognitive testing suggested that people were reluctant to use the negative category when things were mostly positive. The significance of this observation was shown clearly when the yes/no responses were compared to the always-to-never answers.

Cognitive testing can make a major contribution to the quality of survey data. If anyone doubts the importance of getting the questions right, they only need to look at the data from five of the six examples in this paper. The differences in estimates between the alternative questions tested in most cases exceed by manyfold the types of variability one would get from sampling error or nonresponse bias. Either 24% or 4% ride in the back seat and do not wear seat belts.

Either 55% or 36% have a chronic condition that has lasted for 3 months and for which they either take prescription medication or have seen a doctor more than twice. The important point is that these problems were identifiable and identified in cognitive testing. However, without a field experiment to go with it, no one could have predicted the size of the effect on survey estimates that would result from the poorly designed questions.

When a survey research project is designed, researchers have to decide how they are going to allocate resources. Increasingly, researchers now build into their budgets the time and resources needed to do some type of cognitive testing of their questions. However, a weak link in the process to date is that the guidelines for when questions need to be changed or fixed are not clear. Experimental testing provides an answer to the question of how much difference a poorly designed question will make. It also provides evidence about whether the revision is actually better than the original question?

We also know that the forces for keeping old questions are strong. All methodologists have worked with researchers who want to use old questions because they "worked before," even if what constitutes "working" was never very systematically evaluated. There also are researchers who legitimately want to replicate past work, which may mean using questions that now seem flawed. We have literally hundreds of questions in our survey research literature that were designed and used before current standards were in place: that we should use questions that respondents can understand in a consistent way and that they can answer. When such questions are found to be wanting, given our current perspectives, people understandably ask what the effect would be of asking a better designed question. Would data quality really be improved?

Experimental designs have been used for many years to examine the effects of question wording. For example, Schuman and Presser (1981) wrote a widely cited book based primarily on a series of split-sample wording experiments. However, while such designs have been used widely for methodological studies, they are not commonly used as part of routine presurvey question evaluation.

Experimental designs, such as those outlined in this chapter, are extraordinarily useful. Of course, it costs more money to do a pretest with 100 or 150 cases, randomized with two sets of questions, than it does to do a 25- or 35-interview pretest. However, in the context of a large survey budget, the cost of an extra 100 pretest interviews may be pretty small. In particular, when data collection is by telephone, mail, or the Web, the absolute costs for such experiments are not high. At the same time, as we have seen, the data from such experimental studies can make a very large contribution by helping researchers decide which of two alternative forms of questions should be asked and showing the implications of those decisions for their estimates. Although not every project will be able to include an experimental pilot study in its presurvey activities, there is a strong case to be made that they should be included more often than they are and that the result would be more informed decisions about question design issues. In short, pilot tests that include experimental designs are a key way to deliver the full potential of the cognitive testing.

ACKNOWLEDGMENTS

The research reported here was supported in part by a contract from the National Center for Health Statistics (NCHS) and a cooperative agreement between the Agency for Health Care Research and Quality and the Harvard Medical School, Department of Health Care Policy (HS-0925). The contribution to this work of Paul Beatty at NCHS is also gratefully acknowledged.

CHAPTER 10

Using Field Experiments to Improve Instrument Design: The SIPP Methods Panel Project

Jeffrey Moore, Joanne Pascale, Pat Doyle, Anna Chan, and Julia Klein Griffiths
U.S. Bureau of the Census

10.1 INTRODUCTION

In this chapter we discuss the use of field experimentation to improve questionnaire design. The fundamental characteristic of a field experiment, as with any scientific experiment, is the random assignment of respondents (sample cases) to alternative questionnaire versions (treatment groups) under controlled conditions of interview administration. The chapter draws heavily on our experience with the field experiments conducted in a U.S. Bureau of the Census research effort to develop and evaluate improvements to the Survey of Income and Program Participation (SIPP), through a project called the SIPP Methods Panel.

Although the overall project carries the "methods panel" nomenclature in its name, we distinguish our experiments from that particular method of questionnaire research. We would define a *methods panel* as a small panel of sample cases, run separately from but parallel to a "live" parent survey, which is available to researchers as a test vehicle for new question wordings and other methodological changes. The field experiments we describe do maintain some features of a panel—most notably, their repeat visits to sample households. But in the end, they are "just" field experiments, albeit with some unusual features: They are large in scale; their samples and procedures closely mirror the production survey; they

Methods for Testing and Evaluating Survey Questionnaires, Edited by Stanley Presser, Jennifer M. Rothgeb, Mick P. Couper, Judith T. Lessler, Elizabeth Martin, Jean Martin, and Eleanor Singer
ISBN 0-471-45841-4

use both a control group and a test group, with a comprehensive test instrument that implements all proposed instrument changes at once; and they make use of multiple outcome measures to assess instrument improvements, especially including close attention to field staff assessments.

10.2 BACKGROUND: SIPP DESIGN BASICS AND THE MOTIVATION FOR INSTRUMENT IMPROVEMENTS

SIPP is a nationally representative, interviewer-administered, longitudinal survey conducted by the U.S. Bureau of the Census. It provides data on income, wealth, and poverty in the United States, the dynamics of program participation, and the effects of government programs on families and individuals. Currently, a SIPP panel consists of nine waves (or rounds) of interviewing, with waves administered at four-month intervals. All SIPP interviews are conducted with a computer-assisted questionnaire; the first interview is administered in-person, subsequent interviews are often conducted via telephone. The SIPP core instrument (which contains the survey content that is repeated in every survey wave) is detailed, long, and complex, collecting information about household structure, labor force participation, income sources and amounts, educational attainment, school enrollment, and health insurance over the prior four-month period. A typical SIPP interview takes about 30 minutes per interviewed adult. See U.S. Bureau of the Census (2001c) for a more complete description of the SIPP program.

As with other government demographic surveys (Bates and Morgan, 2002), SIPP's nonresponse levels rose noticeably in the late 1990s (Eargle, 2000). Coincidentally, the survey's burdens[1] on respondents, and especially its unnecessary burdens, seemed to emerge with force following its transformation in 1996 from a paper-and-pencil questionnaire to an automated, computer-assisted instrument. Although few content changes were introduced with automation, to many reviewers the automated instrument suddenly appeared rife with design problems, and generally much more burdensome than it needed to be.

These twin concerns—nonresponse and questionnaire design problems—led the Census Bureau to implement a top-to-bottom review of the SIPP core instrument, seeking improvements to increase efficiency, reduce tedium, develop "friendlier" wording, simplify response tasks, and in general reduce perceived burden. We call these *interview process improvements*. In addition to the inherent benefits of a less burdensome interview, we hoped such improvements might also improve survey cooperation. The review also produced recommendations for improvements to two long-standing data quality problems—seam bias in month-to-month transitions, and high rates of nonresponse for income amount items—and recommendations for introducing new questions to address new

[1] An interview's *burden* is often narrowly defined as commensurate with its duration. We intend a broader meaning here, one that emphasizes the subjective burdens imposed by poorly worded, ill-defined, or unnecessary questions, general repetitiousness and inefficiency, and other evidence of lack of attention to the niceties of questionnaire design from the respondent's perspective.

data needs. The SIPP Methods Panel Project was created to implement these improvements in an experimental SIPP instrument and to evaluate their success in a series of split-panel field tests.

10.3 INSTRUMENT IMPROVEMENT EXAMPLES

10.3.1 Improving the Interview Process

Interview process improvements were of three main types:

1. Screening questions and related procedures, to avoid asking questions that are at best unlikely to be productive and at worst may be perceived as repetitive or insensitive. Two primary examples: Standard SIPP procedures require that all adults, even those in obviously wealthy households, be asked all of the instrument's welfare program participation questions. This contrasts with the new instrument, which asks the full battery of such questions only in low-income households, or households in which someone is known to receive such benefits. Another screening procedure saves asset-poor households from receiving the entire series of asset ownership questions, including a half-dozen questions about extremely rare asset types. Again, the standard SIPP procedure is to ask all asset questions in all interviews, whereas the new instrument administers the full battery only to those who report owning at least one common asset type. The obvious risk of such screening procedures is that too many respondents will be screened out and data quality will suffer; the risk of *not* implementing them is that the perceived inappropriate questions will negatively affect respondents' (and interviewers') attitudes about the survey, and thus their engagement in the interview task. Related changes were the elimination of questions about the race of biological children of parents whose race is known, and our greater reliance on interviewers' judgment about when to ask questions that might sound ridiculous in some circumstances (e.g., whether a person's mother is a household member).

2. Respondent-focused procedures, to make the interview less demanding, more sensitive to respondents' needs, and in general a more pleasant experience. A prime example of this category is the introduction of a "topic-based" format for reporting basic demographic characteristics.[2] Even more notable, however, is the new instrument's openness to respondent-defined periods for reporting wage/salary earnings and asset income. For example, the stated goal for earnings is to obtain monthly amounts, but to do that "we can . . . [also] work with biweekly paychecks, or hourly pay rates or annual income amounts. . . . What would be easiest for you?"

[2]SIPP is currently person-based, administering the entire sequence of questions for each person, one person at a time. In contrast, the topic-based approach asks each question of all persons before proceeding to the next question (e.g., "Is Mary currently married, widowed, divorced, separated, or has she never been married?" "How about John?" "And Susan . . .?"). An obvious benefit of this approach is its use of very brief cues after the full text has been administered once. See Moore (1996) and Moore and Moyer (1998a,b).

3. Improved instrument flow and more natural language, to make individual questions, and the entire question sequence, less stilted, less artificial, less jarring, and less repetitive. An example of this type of improvement is the use of optional restatements of the reference period [e.g., "(Since July 1st) did you receive any social security payments?"] throughout the survey's potentially long series of questions about transfer program receipt, to avoid constant, unnecessary (and no doubt highly annoying) repetitiveness.

The interview process improvements were motivated primarily by the desire to increase survey cooperation. But noncooperation decisions occur mostly within the first seconds of the interviewer–respondent interaction, well before the nature of the interview or the presence of questionnaire improvements could directly affect respondents' desire to participate (Groves and Couper, 1998). Why, then, would we expect that interview process differences might affect initial cooperation?

Research has occasionally found that instrument improvements of this type do reduce nonresponse (e.g., Hess et al., 2001; Moore and Moyer, 1998a,b). A likely cause, we suspect, is that even though potential respondents may not be aware of instrument design problems, interviewers are, and they, too, are troubled by badly worded questions, inefficient procedures, and other violations of conversational norms (Grice, 1975), which often result in "interactional troubles" during an interview (Briggs, 1986; Clark and Schober, 1992; Suchman and Jordan, 1990). It does not strain credulity to imagine that interviewers' greater eagerness to administer a better instrument compared to a worse one would lead to subtle differences in conversion efforts when faced with potential respondents who express reluctance to cooperate (Collins et al., 1988). This indirect influence path perhaps explains the paucity of evidence of instrument design impacts on initial cooperation. Attrition in subsequent interviews ought to change the calculus considerably, of course, since respondents' experiences with the first interview can inform their decision concerning cooperation with the second. This notion finds support in the literature [see Kalton et al. (1990) and Lepkowski and Couper (2002)].

10.3.2 Improving Data Quality

Reducing Item Nonresponse Like other income surveys, SIPP suffers high rates of nonresponse for some types of income. Asset income nonresponse rates, for example, can reach 40% or more. One of the new instrument procedures designed to reduce nonresponse has already been alluded to: easing the reporting task by allowing respondents to select the period over which the income is to be reported. Whereas standard SIPP procedures force all respondents to report monthly earnings and asset income as a four-month total,[3] the new instrument

[3]In fact, there is some flexibility in the current SIPP instrument—in addition to monthly amounts, earnings can be reported via hourly pay rate and hours worked, and in addition to four-month totals, asset income can be reported as an annual amount. These options are only revealed to interviewers, however, and not to respondents.

offers several additional options (e.g., for assets, the new instrument also allows monthly, quarterly, or annual reporting).

Other new procedures attack nonresponse directly. The new instrument expands the use of closed-ended follow-up questions to capture an asset income amount range in the event of an initial nonresponse to an amount question.[4] The standard SIPP instrument makes only limited use of nonresponse follow-ups, and the follow-ups focus on the balance or value of the asset. The new instrument's wave 2+ interviews include a more radical change across all income types—in the event of an initial nonresponse, the instrument presents a dependent follow-up question: "... last time you received [$$$] in food stamps. Does that still sound about right?" Currently, there are no circumstances under which SIPP feeds back a prior wave's amount report in the current interview wave.

Reducing Seam Bias As in other panel surveys, SIPP's estimates of month-to-month transitions (e.g., from being "on" a transfer program to being "off") are subject to seam bias.[5] We introduced several important new features to address this problem. For the sake of convenience, current SIPP procedures allow respondents to report events and circumstances that have occurred during the current (interview) month up to "today," the day of the interview, but in fact anything that happens outside of the preceding-four-calendar-month reference period, which ends at the end of the preceding calendar month, is largely ignored. The new procedures, in contrast, place extra emphasis on those interview month reports. Note that the interview month in one wave is the most distant month of the next wave's reference period. If a respondent reports a relevant event "this month" (e.g., receipt of food stamps), the next interview starts by confirming with the respondent their already-reported receipt of food stamps in the first month of the current wave, thus heading off a potentially mistaken report of a change on the seam.

Although Moore et al. (1996) report little success with similar techniques in an experimental paper-and-pencil SIPP instrument, this is probably due to complex procedures that required extensive "paper shuffling." Other researchers (e.g., Brown et al., 1998; Mathiowetz and McGonagle, 2000; Rips et al., 2002) have found some success in reducing seam bias problems using dependent interviewing procedures such as those proposed here.

10.3.3 Introducing New Content

The world inevitably changes over the course of a survey panel, often in ways that demand new questions. Midstream instrument changes in a complex, continuously-in-the-field survey are challenging—while quick patching in of new items is

[4]Note the similarities with the "unfolding brackets" techniques developed by Juster and colleagues [see, e.g., Heeringa et al. (1995), Juster and Smith (1997, 1998), Kennickell (1997)].

[5]*Seam bias* occurs when month-to-month transitions appear much more often in month-pairs that include the "seam" between survey waves compared to month-pairs that reside wholly within a single wave. Seam bias has been widely observed in SIPP [see Czajka (1983), Kalton and Miller (1991), Moore and Marquis (1989), and a variety of other surveys (e.g., Hill, 1987)].

sometimes possible, it is often difficult to effectively integrate the new and the old in a coherent way, and "on the fly" changes can produce unexpected impacts on instrument operations. The Methods Panel's test instrument thus acquired the task of proving in new questions developed to address new data needs; examples include questions about citizenship, participation in new types of benefit programs, non-English language use, and questions necessary to implement the Census Bureau's new respondent identification policy (RIP).[6] Note that this aspect of the project's mission is at clear cross purposes with its primary task of finding ways to make the survey shorter, less burdensome, and so on.

10.4 METHODS PANEL FIELD EXPERIMENTS

10.4.1 History and Motivation

The use of field experiments to evaluate comprehensive questionnaire design changes has a long history, going back at least to the 1950s work of Trussell and Elinson (1959). One of their concerns was similar to our concerns regarding SIPP—how to improve interview efficiency. The specific modification they tested was a "family-style" interviewing procedure for one part of their chronic illness and disability survey questionnaire ("Does anyone in this family have [condition X]?"), in contrast to the standard procedure, which asked individual-level questions of each household member.

Trussell and Elinson's main research focus, however, was not improving the interview process; they were concerned primarily with the impact of new methods on the quality of their key survey measurements (in this case, measurements of illness and disability). This focus has predominated in the field experiments conducted since Trussell and Elinson's day. The nature of the alternative methods under examination can take many different forms:

- Repairs of faulty procedures, known (or suspected) to cause measurement quality problems (e.g., Rosenthal and Hubble, 1993)
- Investigation of promising new procedural innovations (e.g., Tucker and Bennett, 1988)
- Modernization of survey administration mode: for example from personal visit interviewing to telephone interviewing (Hochstim, 1967; Rogers, 1976), or from paper-and-pencil to computer-assisted interviewing (Tucker et al., 1991)
- A combination of survey automation and content updating to reflect changed conditions and needs (e.g., Rothgeb et al., 1991; U.S. Department of Health and Human Services, 1997)

But the primary focus of virtually all such experiments has been measurement quality. In fact, in their otherwise firm defense of field experimentation, Tucker

[6]The RIP policy prohibits revealing one respondent's answers to another household member without first getting the respondent's permission to do so. The test instrument included new questions to obtain this permission in wave 1, and in wave 2, new procedures for responding appropriately to the prior wave's RIP response.

et al. (1998) devote no attention to using field experiments to address less-than-optimal questionnaire design from an interaction/burden standpoint, and we find almost no examples of such work in the literature.

Despite the dearth of precedent, we had little doubt that field experimentation was the best method for evaluating the success or failure of our attempted improvements. In large part this was because several of our redesign goals (i.e., reducing unit nonresponse and attrition, reducing income item nonresponse, and reducing seam bias) could only be evaluated via carefully obtained quantitative data. Although field experiments have some competitors when it comes to evaluating interview process improvements (e.g., behavior coding, debriefings, cognitive interviews), they are arguably still the best source of information for understanding how, under "live" interview conditions, interviewers and respondents react to new design features.

Other more practical considerations also played a role in the decision to use field experiments as our primary evaluation tool: the culture of the Census Bureau, in particular its demand for "hard" data to justify changes to its demographic survey programs [see, e.g., the agency's pretesting policy (U.S. Bureau of the Census, 1998)]; the need to evaluate the impact of the new procedures on the survey's processing systems, to anticipate any problem that might increase the likelihood of data release delays; and the need to avoid imposing additional tasks and difficulties on the already overburdened production survey staff. Finally, of course, there is the absolutely essential criterion for interview process improvements in a survey like SIPP: that no harm must be done to the survey's key estimates. Almost regardless of the extent of their benefits, evidence that the interview process improvements harmed data quality in a survey which is a major source of official government statistics (Citro and Michael, 1995) would trump any such benefits. Thus, full evaluation of the interview process improvements also required controlled experimentation to assess their impact on survey estimates.

10.4.2 Design of the Field Experiments

The most basic characteristic of the Methods Panel field experiments was their true experimental design, with random assignment of sample cases to test and control groups. Simply testing the new instrument alone, and comparing the results to current production SIPP outcomes and estimates would of course be cheaper and easier, but the "noise" in such comparisons (different timing, different staffs, different interview conditions, different processing systems, etc.) would render them very difficult to interpret. Other basic design features included a fully realized automated test instrument, with the entire array of improvement recommendations, to enable a complete assessment of its impacts on field procedures, processing systems, and survey estimates[7]; special procedures to quantify the

[7] Although we designed the experiments to assess the impact of instrument changes on survey estimates, we chose not to attempt a full set of postcollection processing adjustments (for item nonresponse and data anomalies) because our real interest was the influence of instrument design on responses. This choice placed some limits on the types of test/control comparisons we were able to make.

impact of the interview "process improvements"; an iterative design, to allow experimentation with different alternatives; and a longitudinal design for testing dependent interviewing.

Staffing issues also played an important design role. We chose to cut costs, and to increase face validity, by using experienced SIPP interviewers, already trained on SIPP goals and concepts and how to gain respondent cooperation. We had interviewers administer both the control and test instruments, so they could provide feedback on the relative merits of each, and to avoid confounding staffing differences with instrument treatments. The major considerations in sample size determination were the need to detect important treatment differences for key estimates while keeping the field operation within budget limits. Ultimately, the sample size permitted the detection of differences of 3 to 8 percentage points in nonresponse rates across the two instrument treatments, depending on item universes and baseline nonresponse rates. This level of precision translated to a sample size of approximately 2000 cases (i.e., 1000 per treatment) in each of our three experiments, which we label according to the year in which they were conducted: MP2000, MP2001, and MP2002. See Doyle et al. (2000) for a detailed description of the design of the field experiments.

In hindsight, several of our design decisions were questionable, or had unforeseen impacts. The decision to use SIPP-experienced interviewers, for example, did not fully anticipate the effects of interviewers' familiarity with the "old" instrument on their immediate ability to administer the new one confidently and efficiently. This problem was exacerbated by an imposed change in the implementation of the experiments: namely, the use of many more interviewers than had originally been planned. As a result, interviewers' workloads were often very small, about three test cases per month on average, affording scant opportunity for achieving real familiarity with the new instrument, especially in contrast with their great familiarity with the control instrument. Due to budget and timing constraints, our longitudinal "panels" consisted of only two waves, leaving us with weak data to assess attrition effects. Finally, we failed to include any procedures for assessing respondents' subjective evaluations of the two instruments, operating under the naive assumption that their objective cooperation/noncooperation behaviors would make such assessments unnecessary.

10.5 SELECTED RESULTS OF THE EXPERIMENTS

10.5.1 Evidence for an Improved Interview Process

Interviewer Evaluations We used debriefing questionnaires to assess interviewers' reactions to the control and test instruments. In general, interviewers reacted quite positively to most of the test instrument's interview process improvements, and increasingly so over the three experiments. With regard to the evaluation of specific experiments in the wave 1 instrument: In MP2000, interviewers rated the test instrument to be significantly superior to the control on 52% of the

67 individual debriefing questionnaire items (versus 34% for the test; 13% no difference); in MP2001 and MP2002 the comparable figures were 73% (15%; 11%) and 93% (3%; 4%), respectively. Debriefing items assessing the general performance of the two wave 1 instruments show an even more dramatic pattern: interviewers rated the test instrument superior on none (0%) of the 11 general evaluation items in MP2000 (the control instrument scored higher on all items), a pattern that completely reversed itself in MP2001 and MP2002, where the test instrument elicited higher ratings on 91% and 100% of the items, respectively. With regard to the new wave 2 instrument, introduced in MP2001, interviewers' attitudes were substantially favorable from the beginning. Interviewers also expressed highly positive attitudes toward most features of the new instrument design in several in-person debriefing sessions.

We attribute the emphatic shift toward more positive attitudes from MP2000 to MP2001 and MP2002 to both instrument refinements and increased interviewer familiarity. The latter notion finds some support in the debriefing questionnaire data, which reveal a tendency for interviewers with more SIPP experience to display a greater preference for the control instrument, as contrasted to those with less SIPP experience, who were more positively disposed to the test instrument.

Respondent Evaluations As noted, the first two experiments did not include any direct measurement of respondents' reactions to the two instruments. (We have corrected this oversight in the MP2002 test, the results of which are not yet available.) However, we do have evidence of respondents' preferences concerning the reporting period options for income amounts, as indicated by the choices they made during the interview. Table 10.1 shows that the vast majority of test instrument respondents chose to report their earnings income in a manner other than the monthly amounts prescribed by the control instrument. (Similar results are evident in respondents' asset income-reporting choices—data not shown.) Income is a difficult and highly sensitive subject for survey respondents (Moore et al., 2000), and reporting it is arguably *the* central SIPP task. We conclude from these results that adding other options for reporting income, and encouraging respondents to select whatever option best suits them, has substantially eased the SIPP response task.

Number of Questions Avoided We can also use the survey data themselves to calculate instrument efficiencies in terms of questions avoided as a result of the new screening procedures. For example, Griffiths (2001) reports that the "income screener" procedures designed to skip most assistance program questions in wealthy households resulted in a 51% reduction in the number of questions asked in that section of the SIPP interview; similarly, Moore (2001) finds an 18% reduction in asset ownership questions.

Timer Data The efficiencies incorporated into the new instrument did not reduce the duration of test instrument interviews, according to the instruments' automatic

Table 10.1 Percentage of Wage/Salary Earnings Reports Using Each of the Various Periodicity Reporting Options in the MP2001 Field Test, by Instrument Treatment

Periodicity Reporting Option	Wave 1		Wave 2	
	Control	Test	Control	Test
Monthly	83	16	86	18
Weekly/biweekly/ bimonthly	[a]	30	[a]	30
Hourly pay rate	5	10	2	9
Annual salary	12	36	12	36
Other; D/R[b]	[a]	8	[a]	6

[a]This option was not available in the control instrument.
[b]In the test instrument, these responses defaulted to "monthly" reporting.

timers. None of the individual sections showed a time gain, and in fact, the average test instrument interview length actually exceeded the control by about 3 minutes per interviewed adult. This is a somewhat surprising outcome given the objective evidence of fewer welfare program and asset ownership questions being asked in the test instrument, and the known time efficiencies of some of the test instrument's alternative procedures, such as the topic-based questions (Colosi, 2001). A likely explanation is that the efficiencies introduced into the test instrument were counterbalanced by other factors, including:

- The additional new content in the test instrument.
- Interviewers' unfamiliarity with the test instrument; in debriefing sessions we repeatedly heard interviewers claim that they could administer the regular SIPP interview "in my sleep," while the test instrument's new features required them to "think about every keystroke."
- The fact that the test instrument yielded more responses (less missing data).

Research also shows that interviewers sometimes take advantage of the additional time made available by more efficient interview procedures by spending more time in relaxed, rapport-building "side" conversation with respondents (see Zabel, 1994); to the extent that this happened during the interview, it may also have masked the expected timer results.

10.5.2 Effects of the Interview Process Improvements on Data Quality

Whatever their other benefits, the new procedures are of questionable value if they reduce data quality. Although the field experiments did not include any procedures that would permit direct assessment of data quality, we can apply some reasonable assumptions to the task: (1) that underreporting is the predominant form of reporting error in income surveys (see Moore et al., 2000), (2) that screening procedures of the type employed in the new instrument might be expected to exacerbate any underreporting problems, and (3) that comparability of estimates is indicative of comparable underlying data quality.

Welfare Program Participation Table 10.2 reveals a tendency in the first two field experiments for estimates of participation rates for some welfare programs to be significantly lower for the test instrument than for the control; we suspect that this tendency indicates increased underreporting as a result of the new screening procedures. The results are far from consistent, but there is some suggestion that benefit programs which target children tend to fare less well in the test treatment. This effect emerges most clearly for WIC, the Women, Infants, and Children Nutrition Program, which is known to have income eligibility cutoffs that are high relative to other programs. For the MP2002 field experiments we have introduced slight modifications to the screening procedures, which in essence will set higher income thresholds in households containing children.

Asset Ownership The asset ownership estimates, in contrast, offer no evidence that substituting a single global question about very rare assets in place of

Table 10.2 Reported Participation Rates for Need-Based Welfare Benefit Programs in the MP2000 and MP2001 Wave 1 Field Tests, by Instrument Treatment[a] (Weighted Percent)

	MP2000 Wave 1		MP2001 Wave 1	
Welfare Program Type	Control	Test	Control	Test
Person-level benefits				
SSI	2.3*	1.7*	2.2	2.0
Food stamps	3.8*	2.9*	4.4	4.5
WIC	5.6*	3.3*	4.9*	3.6*
"Cash" assistance	0.6	0.9	1.1	1.0
Other[b]	0.5*	1.7*	0.5*	2.2*
% of all adults reporting at least one person-level welfare program benefit	5.6	5.2	5.4	5.5
Household-level benefits				
Public housing	10.8*	5.8*	8.6	5.2
Rental assistance	4.4	2.9	4.7*	3.7*
Energy assistance	0.1*	0.6*	0.3*	1.2*
School lunch/breakfast	22.0	18.2	25.0*	18.6*
% of all households reporting at least one household-level welfare program benefit	9.9*	7.4*	10.9*	7.7*

[a] Asterisks denote mean estimates that differ significantly ($p < 0.10$) by instrument treatment.
[b] The "other" category of person-level benefit programs includes: general assistance or general relief, short-term cash assistance, pass-through assistance, child care assistance, transportation assistance, and other welfare assistance.

Table 10.3 Reported Asset Ownership Rates in the MP2000 and MP2001 Field Tests, by Instrument Treatment[a] (Percent)

	MP2000 Wave 1		MP2001 Wave 1	
Asset Type	Control	Test	Control	Test
Interest-earning checking accounts	29.9	28.2	27.4	32.6
Savings accounts	42.2	41.2	43.1	46.7
Money market deposit accounts	13.4	14.0	14.0	14.4
Certificates of deposit (CDs)	10.9	10.5	9.4	12.2
Mutual funds	15.3	14.5	15.0	16.6
Stocks	17.7	18.0	17.4	20.2
Municipal or corporate bonds[b]	2.2	1.8	2.2	1.9
U.S. government securities[b]	1.0	1.1	1.3	1.8
Mortgages[b]	0.5*	1.1*	0.7*	1.8*
Rental property[b]	4.6	3.8	4.3	5.3
Royalties[b]	0.2	0.5	0.9	0.7
Other[b]	1.9	1.6	1.4	1.4

[a]Asterisks denote mean estimates that differ significantly ($p < 0.10$) by instrument treatment.
[b]Detailed, individual questions about these asset types may not have been asked in the test instrument treatment, due to its screening procedures (see the text).

several individual questions causes any trouble in terms of data quality (see Table 10.3). Specifically, in neither MP2000 nor MP2001 do we see any evidence that the screening procedures resulted in a reduction of reported ownership for the "screened" asset types.

Income Reporting Because income reporting is so central to SIPP's goals, we examined income amount estimates for evidence of possible data quality problems, even though the test instrument did not include any new procedures that presented a clear risk to data quality. We found no such evidence; mean and median monthly amounts for household total income and for person total earnings did not differ significantly by instrument treatment for any month, nor were there any treatment differences in the proportions of the population reporting each type of income (earnings, assets, other).[8] See Doyle (2002) for details.

10.5.3 Effects of the Interview Process Improvements on Nonresponse/Attrition

The weight of the evidence suggests that we succeeded in improving the interview process and that the improvements met the necessary condition of doing

[8]Our analysis weights included simple corrections for unit nonresponse and for a slight discrepancy in weighted household counts across the two treatments, but not for item nonresponse. Ideally, we would have preferred a test/control comparison of total aggregated income, which would provide a somewhat more definitive assessment of the impact of the new instrument procedures on income reporting. However, differential item nonresponse across the two experimental group (see Section 10.5.4) prevents that comparison.

no harm to the survey's key estimates (the latter conclusion remains tentative, pending the results of the modified welfare program screening procedures). As noted earlier, however, improving the interview process was primarily a means to another goal, reducing unit nonresponse and attrition. Table 10.4 shows the wave 1 and 2 unit nonresponse rates for each experiment, by instrument treatment. For comparison purposes we also show the relevant rates for the 2000 and 2001 SIPP panels (there was no new SIPP panel in 2002). The estimates are constructed in accordance with standard practice (American Association for Public Opinion Research, 2000); they show the proportion of eligible, occupied housing units that failed to respond to the wave 1 interview, and the proportion of households containing wave 1 interviewed people that failed to respond to the wave 2 interview.[9] Table 10.4 also shows the refusal component of each of the nonresponse rates, the only nonresponse component likely to be affected by instrument design differences.

The results fail to display the hoped-for outcomes; none of the test/control differences in Table 10.4 is statistically significant. One possible explanation for the absence of positive effects is that instrument design differences exert such a weak impact on cooperation that they emerge only under certain circumstances, especially in an initial interview. The Moore and Moyer (1998a,b) and Hess et al. (2001) studies that found such impacts were telephone interviews with

Table 10.4 Unit Nonresponse Rates and their Refusal Components, by Wave, in the MP2000, MP2001, and MP2002 Field Tests, by Instrument Treatment[a] (Percent)

	Instrument Treatment Group		
Methods Panel Field Test	Control	Test	Production SIPP
MP2000			
Wave 1 unit nonresponse rate	14.8	17.2	11.3
(refusal rate)	(9.4)	(10.2)	(8.7)
MP2001			
Wave 1 unit nonresponse rate	15.2	16.4	13.3
(refusal rate)	(9.8)	(10.8)	(10.0)
Wave 2 unit nonresponse (attrition) rate	7.7	9.0	6.2
(refusal rate)	(2.9)	(3.8)	(3.7)
MP2002			
Wave 1 unit nonresponse rate	13.2	12.2	n/a
(refusal rate)	(8.5)	(7.3)	
Wave 2 unit nonresponse (attrition) rate	(data not yet available)		n/a
(refusal rate)			

[a] No control/test comparison is statistically significant ($p < 0.10$).

[9] Due to SIPP's "following rules," the wave 2 denominator is defined as eligible households at the time of the wave 2 interview containing at least one adult interviewed from a wave 1 interviewed household (see U.S. Bureau of the Census, 2001c).

substantially lower baseline cooperation rates than those of SIPP. Perhaps the much larger initially reluctant pools in those studies contained a much lower proportion of truly committed noncooperators than in the SIPP Methods Panel tests, thus rendering more of them more subject to influence by subtle differences in interviewers' behaviors.

Other factors to consider have been noted before: the test instrument's new content, whose additional burdens may have exerted a negative counterbalance to any interview process improvement effects; and interviewers' inexperience with the new questionnaire, which the literature consistently links with reduced response rates (e.g., Couper and Groves, 1992). Thus, the decision to use SIPP-experienced interviewers may have reduced the likelihood that we would see the desired nonresponse results, especially in wave 1. Another Methods Panel design shortcoming may also have reduced the likelihood of seeing positive results in wave 2. Resource and other constraints restricted us to only two interview waves. This severely limits our ability to detect the real impact of any interview process improvements on attrition over time, which would require several waves of data.

The Table 10.4 results do suggest a possible trend toward an increasingly positive impact of the new instrument design across the three field tests. This gives us some hope that, over the long run, the desired nonresponse and attrition benefits may yet emerge as interviewers become more familiar with the new instrument.

10.5.4 Reducing Income Item Nonresponse

As noted earlier, we introduced several new procedures to reduce nonresponse for SIPP's income amount items: the new "periodicity" options, the expanded use of closed-ended nonresponse follow-ups for asset income amounts, and the use of dependent income amount questions in wave 2, again as a nonresponse follow-up procedure. For each type of income, we computed a nonresponse ratio for each adult in which the numerator was the number of final "don't know" or "refused" nonresponses to an income amount question, and the denominator was

Table 10.5 Nonresponse Rates for Income Amount Items in the MP2000 (Wave 1) and MP2001 (Waves 1 and 2) Field Tests, by Instrument Treatment[a] (Percent)

	MP2000 (Wave 1 Only)		MP2001			
			Wave 1		Wave 2	
Income Type	Control	Test	Control	Test	Control	Test
Job/business earnings	0.22[b]	0.23[b]	0.17*	0.11*	0.24*	0.05*
Asset income before DK follow-ups	0.45*	0.40*	0.38*	0.34*	0.46*	0.23*
Asset income after DK follow-ups	0.31*	0.18*	0.24*	0.17*	0.29*	0.19*
Other unearned income	0.23[b]	0.20[b]	0.20[b]	0.17[b]	0.24*	0.13*

[a] Asterisks denote mean estimates that differ significantly ($p < 0.10$) by instrument treatment. Some test instrument procedures were first introduced later in the field experiment series.

[b] Comparisons where we would expect no nonresponse difference by instrument treatment, because in fact the control and test instrument procedures did not differ.

the total number of income amount questions administered during that person's interview. Table 10.5 summarizes those individual nonresponse ratios averaged across all adults, by income type and instrument treatment. Overall, wherever the two instruments' procedures differed, the test instrument yielded significantly lower item nonresponse than the control.

The Table 10.5 results suggest that there were beneficial effects of each new procedure. The benefits of increased reporting flexibility are evidenced in the significant test/control differences in wave 1 for earnings (MP2001 only—the periodicity options were absent in the MP2000 field test) and assets (see the "before DK follow-ups" row of Table 10.5). The additional impact of the enhanced nonresponse follow-up procedures in the test instrument is evidenced by the even more pronounced earnings nonresponse difference in wave 2, the appearance of such a difference for welfare program income in wave 2, and the across-the-board differences in the asset income "after DK follow-ups" results.

10.5.5 Improving Reports of Receipt Transitions: Seam Bias

Table 10.6 presents the overall seam bias results from the MP2001 field test summed across all 26 welfare program income sources for which (1) both instrument

Table 10.6 Preliminary Seam Bias Results for Welfare Program Participation Reports from the MP2001 Field Test, by Instrument Treatment[a]

Month-to-Month Participation Changes for Seam and Nonseam Month Pairs	Instrument Treatment Group	
	Control	Test
Data		
Total n—eligible cases	700	635
Cases excluded due to nonresponse for at least one seam month	119	62
Cases lost due to wave 2 test instrument error	—	29
n available for analysis	581	544
Results		
Number of participation changes reported at the wave 1/wave 2 seam	83	70
Number of reported nonseam participation changes	42	64
Ratio of seam:nonseam participation changes	1.98	1.09
[Expected seam:nonseam ratio, in the absence of any seam bias]	[0.17]	

[a]Eligible cases consist of people whose recipiency pattern was anything *except* all "no's" across all 8 months of the wave 1 and wave 2 reference periods, for any of the 26 programs which (a) captured monthly recipiency information, and (b) for which the test instrument included new dependent interviewing procedures (see the text). Results are preliminary due to an instrument error (since corrected) which resulted in the loss of some data in the test treatment. Data for most cases affected by the error were salvageable, and the salvage procedures were conservative (i.e., they tend to undercount nonseam changes in the test treatment); nevertheless, we believe that it is necessary to treat these results with some caution.

treatments collected monthly participation data, and (2) for which the test instrument implemented new dependent interviewing procedures to try to reduce seam bias. The table reveals clear evidence of reduced seam bias with the new procedures, although it is also clear that much bias still remains.

10.5.6 Introducing New Content to Address New Data Needs

As noted earlier, the Methods Panel's test instrument also served as a vehicle for proving in new SIPP content, although design limitations left us with rather weak bases for evaluating the new items. On the whole, however, the field experiments met accepted standards (U.S. Bureau of the Census, 1998) for providing sufficient evidence that the new items "work" and thus can be implemented in the production survey. The field test results showed acceptable levels of item nonresponse, reasonably positive interviewer evaluations of the new items and an absence of negative feedback, and the items also worked in the sense of being free of implementation "bugs" and of fitting in smoothly and naturally with existing SIPP content.

10.6 CONCLUSIONS

10.6.1 Evaluation Goals

The Methods Panel's field experiments provided extremely useful information concerning our efforts to improve the SIPP interview process. Of particular value were the interviewers' evaluations following their experience administering the test and control instruments side by side in the field. Interviewers' assessments could have been captured with less cost and effort, but no other evaluation method could have duplicated the combination of quantitative data, derived from "live" administration of the survey, under conditions of a successively greater interviewer familiarity with a successively refined test instrument. As noted, we regret not applying equal attention to respondents' reactions and evaluations.

Despite their great usefulness, we harbor no illusions that the field experiments (at least as we implemented them) were perfectly suited to the evaluation of interview process improvements. In particular, the test instrument bore two burdens not shared by the control instrument which undoubtedly affected some evaluation outcomes: additional questions in several new content areas, and a marked disadvantage in interviewer familiarity, especially at the outset of the field test sequence. With regard to the former issue, we simply lacked sufficient resources to add comparable new questions to the control instrument. In other circumstances, this would be a surmountable barrier.

Interviewer familiarity differences present a somewhat more thorny problem, at least in the context of a complex, ongoing survey like SIPP. A solution that did *not* appeal to us was to use new interviewers, without a strong foundation in basic SIPP concepts, and equally unfamiliar with both instruments. We feared that this would result in poorer performance administering the questionnaires and eliciting cooperation, and thus would seriously jeopardize confidence in generalizing our experimental results to a production survey setting. Under ideal circumstances, interviewers would receive sufficient training to render them as familiar and facile with the new instrument as they are with the old, but this is difficult to achieve in practice. In the end, we are comfortable with our conservatively biased design, in which the impact of increasing familiarity could be tracked, and in which positive interviewer evaluations of the test instrument achieve extra weight by dint of their having to overcome a familiarity disadvantage.

With regard to our need for an evaluation method that would convincingly assess the (non-) impact of the new procedures on important survey estimates, we quickly narrowed the set of viable options to one: field experimentation. If questionnaire design changes must pass the "do no harm to the estimates" litmus test, there is simply no other evaluation method that will suffice. We do note, however, that under some circumstances a simple examination of differences in estimates produced by different experimental treatment groups is not particularly informative. One of our improvement efforts involved substantial modifications to the survey's labor force questions, to clarify types of work arrangements and ensure appropriate detailed questions about those arrangements. The new procedures did seem to yield a small but significant increase in reported work activities—presumably, the sorts of small secondary or tertiary "jobs" that tend to be missed in surveys of earnings income (Moore et al., 2000). But they also produced a shift in the distribution of work categories, away from typical wage/salary jobs and toward less regular types of employment (Pascale, 2001). In part because of uncertainty over the meaning of the latter changes, the new procedures for this section of the questionnaire were not approved for use in the 2004 production survey.

Although the new instrument's procedures do not seem to have affected income reporting in any negative way, we noted earlier the problems in drawing a definitive conclusion due to its success in reducing income item nonresponse. The addition of a record check component to the field experiments would have overcome this problem, and we highly recommend this augmentation wherever possible. In the present case, we hope eventually to be able to use records to assess the quality of income reports for selected sources.

As with the assessment of survey estimate impacts, the effects of instrument design on unit nonresponse, item nonresponse, and seam bias could only be evaluated with data generated by field experimentation. The experiments offer strong evidence that our item nonresponse efforts met with great success and hopeful evidence of some reduction in seam bias as well. The fact that we cannot draw the same conclusion—yet—about unit nonresponse and attrition does not diminish the value of the field experiments as the appropriate evaluation method. It may mean that our expectation of a positive nonresponse impact due to instrument design was unrealistic or that some specific aspects of the design of our experiments were flawed, but the experimental approach itself remains unchallenged as the method of choice when the questionnaire improvement goals involve quantitative survey estimates and outcome measures.

10.6.2 Other Lessons Learned

The Methods Panel's field experiments were ambitious, challenging, and costly of both dollar and staff resources. Nevertheless, the Census Bureau has derived major benefits from these efforts, and we recommend them as valuable tools for questionnaire evaluation, particularly for ongoing, large-scale surveys, where it is crucial to pretest new content and assess impacts of instrument changes on key estimates and outcome measures.

We strongly endorse the iterative design of this project, which was fundamental to our ability to hone the questionnaire recommendations and to address the volume of the recommendations, and whose replicated measures provided additional weight for many important findings. The discipline imposed by the iterative design, essentially mimicking a production survey, was invaluable in keeping the project team focused on specific goals and realistic instrument design options. Especially since our goals were heavily oriented to improving the interviewing process, we also realized great benefits from heavily involving interviewers in the instrument evaluation. This gained important buy-in and engagement in the experiments, since interviewers could see real interest in their opinions and experiences, and real attempts to address their concerns.

As is always the case, practical considerations and resource constraints played a large role in the design of this project and limited our ability to assess the degree to which we attained some goals. Our sample size, while huge by instrument pretest standards, was not sufficient to measure all the impacts that we wanted to measure. For example, an attempt to reduce within-household undercoverage could not be assessed objectively because its impact affects only a small handful of people. Other evaluation methods, such as cognitive interviews, respondent debriefings, or reliability measures may be better suited for these types of measurement issues.

Our time constraints (i.e., the need to be ready for implementation in the 2004 SIPP panel) and our need to implement all of the myriad instrument changes in the test instrument (in order to examine impacts on key estimates and processing procedures) had a downside, which was that the amount of time between the field tests was quite short given the volume of instrument changes to be evaluated. This sometimes forced decisions to be made based on less in-depth analysis than would have been ideal and delayed implementation of some instrument improvements pending additional research, a prime example being the decision to delay implementation of changes to SIPP's labor force questions due to lack of time to fully understand the new instrument's differences for some estimates. Thus, we recommend making sure to build in sufficient time between experiment iterations to conduct a volume of research that is consistent with the volume of changes to be evaluated.

ACKNOWLEDGMENTS

In this chapter we report the results of research undertaken by U.S. Bureau of the Census staff. It has undergone a Census Bureau review more limited in scope to that given to official Census Bureau publications, and is released to inform interested parties of ongoing research and to encourage discussion of work in progress. The authors gratefully acknowledge the numerous and important contributions of our many Census Bureau colleagues on the Methods Panel project, especially Adele Alvey, Nancy Bates, Nancy Cioffi, Liz Griffin, Elaine Hock, Heather Holbert, Johanna Rupp, Tom Spaulding, Tim Stewart, and Ceci Villa. In addition to our project colleagues, we also thank Chuck Nelson for his helpful comments on an early version of this chapter.

CHAPTER 11

Experimental Design Considerations for Testing and Evaluating Questionnaires

Roger Tourangeau
University of Michigan

11.1 INTRODUCTION

Questionnaire design texts often begin by observing that writing survey questions is more of an art than a science, and sometimes a tone of frustration accompanies this observation. Converse and Presser's classic text (1986, p. 7) is quite explicit about the frustration: "Yet it is unrewarding to be told, always, that writing questions is simply an art. It is surely that, but there are also some guidelines that have emerged from the collective artistic experience and the collective research experience." The books and articles by Converse and Presser, Sudman and Bradburn (1982), and others (e.g., Dillman, 2000a; Krosnick and Fabrigar, 1997) offer useful compilations of these guidelines and, in some cases, summaries of the research evidence on which they are based. Despite these efforts, crafting survey questionnaires remains something less than a scientific enterprise.

There are several reasons for this. The people who write survey questionnaires are often unaware of the growing literature on questionnaire design, and in most survey organizations, the questionnaire design process emphasizes substantive rather than methodological issues; the key question for the survey designers is often not whether the questions will elicit accurate (or even meaningful) answers but whether they cover everything the clients believe they need without exceeding the allotted administration time.

Methods for Testing and Evaluating Survey Questionnaires, Edited by Stanley Presser, Jennifer M. Rothgeb, Mick P. Couper, Judith T. Lessler, Elizabeth Martin, Jean Martin, and Eleanor Singer
ISBN 0-471-45841-4

11.1.1 Reliance on Nonrigorous Methods

Another reason for the nonscientific character of the typical questionnaire design process is its reliance on low-cost, low-rigor methods for testing draft questions and questionnaires. For example, two of the most popular methods for testing questions—review of the questions by survey experts and cognitive interviewing—are both relatively quick and inexpensive. As a result, most of the questionnaire design texts recommend these or similar methods for pretesting draft questions (see, e.g., Converse and Presser, 1986, pp. 55–75, or Dillman, 2000a, pp. 140–147). My guess is that both are reasonably effective at detecting problems with draft questions. The problem is that both methods tend to be carried out in a nonrigorous, even subjective way. Most expert reviews simply elicit the *opinions* of the experts; in my experience, the review process makes no particular effort to assess consistency across the experts or to apply the same criteria to different questionnaires (for an exception, see Lessler and Forsyth, 1996, who propose a detailed system for coding the problems with questions). Similarly, in most cases, cognitive interviewers simply record their impressions of what went wrong with a given item and, based on these impressions, make recommendations about how to fix the items. Both procedures may be quite useful, but clearly they could be a lot more scientific.

It's helpful to frame this discussion of pretesting methods by drawing some basic distinctions among the types of research done in testing questionnaires:

- Studies that yield qualitative data (e.g., verbatim text) versus those that yield quantitative data (such as the marginal frequencies of responses to draft questions)
- Observational studies that simply try to record key variables in a more or less natural setting versus experimental studies that deliberately manipulate one or more variables, often in a laboratory
- Exploratory studies that attempt to develop hypotheses versus confirmatory studies that attempt to test them
- Applied studies that attempt to resolve questions about specific items in specific questionnaires (does this version of the question seem to work?) versus basic studies that attempt to resolve general issues relevant to many items in many questionnaires (when does decomposition improve the accuracy of the answers?)

Although this chapter focuses on hypothesis-testing experiments that gather quantitative data, it's not an argument for dropping or deemphasizing any of the other tools commonly used to develop and test questionnaires. (Questionnaires need all the help they can get!) It is an argument for applying all of the available tools with greater rigor. In addition, it underscores the importance of gathering hard evidence that these methods actually yield better questions.

11.1.2 Neglect of Statistical Considerations

The informal character of many questionnaire pretests stands in sharp contrast to the careful attention to statistical considerations in the design of the surveys themselves. Cognitive interviews are generally done with self-selected samples of volunteers; pilot tests are often done with haphazard samples drawn from one locality; interviewer debriefings are often done with a few interviewers who may have their own axes to grind. Such tests may be better than no tests at all (and they may be all that the budget for a study will permit), but they are hardly likely to yield conclusions that can be confidently generalized to other questionnaires or other samples.

Another statistical consideration that is generally neglected during the questionnaire development process is the reliability of the judgments made by experts, cognitive interviewers, or field interviewers about particular items. Apart from one study by Presser and Blair (1994), we have little evidence that different experts reach similar conclusions about an item or that one cognitive interviewer turns up the same problems with an item that another interviewer does. (Actually, the results reported by Presser and Blair are none too encouraging in this regard.) As survey researchers, we worry a lot about variation across interviewers, samples, and survey houses in the results of real surveys, but we tend to ignore similar sources of variation in the pretesting process. Part of the problem here is that procedures such as cognitive interviewing and expert reviews aren't very well defined, and there's tremendous variation in how they are done. Another part of the problem is that these methods can yield very rich qualitative data that are hard to reduce to a usable form. As a result, it's easy for different observers to reach different conclusions even from the same set of data.

11.1.3 Reliability and Validity of Questionnaire Responses

A final weakness with the typical questionnaire design process is that the standards used to select questions for particular surveys are not the ones we would really like to apply—whether the questions yield consistent and accurate answers. Most pretests simply do not yield the data needed to assess the validity or reliability of survey responses. Validity data are often a challenge to get, especially with attitude questions; with more objective items, some outside source of validation data (often, administrative records of some sort) is generally tapped and matched to the survey responses. For example, respondents' answers to a question about doctor visits are checked against HMO records (Mathiowetz and Dipko, 2000). Although some of the classic studies in the survey methods literature have used this method to come to conclusions about the relative accuracy of different data collection procedures (see, e.g., Cannell et al., 1977; Neter and Waksberg, 1964; Sudman and Bradburn, 1973), record checks and validity studies are still a rarity. Instead, questions are dropped or rewritten for a variety of other reasons—they are judged to be hard to understand or to administer, they yield highly skewed distributions of responses, they trigger lots of requests for clarification from the respondents, they take a long time for respondents to answer, and so on. It is quite possible that all these features of survey questions are, in fact, useful proxies for the validity and reliability of the answers (e.g., if respondents misunderstand a question, they are undoubtedly less likely to answer it correctly), but again, there's not much evidence for this assumption, in part because so few studies gather validity data.

The same could be said for reliability data. In principle, it's a lot easier to measure the reliability of survey responses than to assess their validity. All that is needed to measure reliability is reinterviews with some portion of the sample at a suitable time lag. Some surveys, notably the Current Population Survey, routinely conduct reinterviews (Forsman and Schreiner, 1991); even more surveys routinely recontact survey respondents to detect "curbstoning" (fabrication of data) by the interviewers. Unfortunately, reinterview data are rarely collected during the pretesting or questionnaire design process; in fact, apart from an intriguing paper by O'Muircheartaigh (1991), not much seems to be done with the reinterview data that are collected.

11.1.4 Statistical Considerations for Questionnaire Experiments

Even when questionnaire designers do set out to do experimental comparisons to inform the development of a questionnaire, they still may neglect such statistical issues as sampling error, power, and confounding of variables in the design of these experiments. Survey experiments are often carried in a few localities rather than nationally and they often have relatively small sample sizes.[1] For example, all of the early experiments comparing audio computer-assisted self-interviewing (ACASI) with more traditional means of data collection were based on relatively small samples from a single or a few localities. An initial study by O'Reilly and colleagues (1994) was based on a volunteer sample of 40 respondents from a single community. Similarly, a later study by Tourangeau and Smith (1996) was based on a total of 339 interviews (spread across three modes of data collection) in one urban county. A third study examined 500 women in five experimental conditions from localities in three states (Lessler et al., 1994; Mosher et al., 1994). The first national experiment comparing ACASI with another mode of data collection (a paper self-administered questionnaire) was subsequently reported by Turner and his colleagues (1998); aside from having a more representative sample, this study also had a much larger sample than its predecessors ($n = 1729$ completed interviews, 1361 of them via ACASI). Even this study was restricted to teenage boys. Finally, national studies of the general population were reported by Lessler et al. (1993) and Epstein et al. (2001).

The danger with small-scale studies is lack of power, the possibility they will fail to detect real differences. The danger with unrepresentative samples is,

[1]Our examples of survey experiments involve both mode comparisons and studies comparing different versions of questions; many experiments involve both types of comparisons. Fienberg and Tanur (1987, 1989) provide a general discussion of experiments embedded in surveys.

of course, unrepresentative results. It's easy to imagine that volunteers, big-city dwellers, or residents of three states might react differently to a new technology such as ACASI than ordinary survey respondents, residents of rural areas, or residents of other states. (Fortunately, the results of all the studies converged on the conclusion that ACASI preserved or enhanced the gains from self-administration.)

Another common issue for questionnaire design experiments involves the confounding of multiple design features. Confounding simply means that two or more variables are systematically correlated in the design, often perfectly correlated so that their effects cannot be disentangled statistically. For example, when the Current Population Survey underwent a major overhaul, including a new questionnaire and a switch from paper to computer administration, a large experiment ($n = 12{,}000$ in each of 18 months) was mounted that compared the old questionnaire on paper with the new questionnaire on CAPI (Cohany et al., 1994). Unfortunately, it was impossible to distinguish the impact of the mode switch from the impact of the new questions; the two variables were completely confounded. What might well have been the definitive study of the effects of computerization in a face-to-face survey—and on how to word labor force questions—is instead rarely cited.

Such confounds are a feature of many questionnaire design experiments. In the case of the CPS overhaul, it was obviously essential to compare the old questionnaire on paper with the new one on CAPI to measure the overall effect of the changeover. The new questionnaire was deemed too complex to administer on paper and there was no funding to program a CAPI version of the old questionnaire, so the other two experimental groups needed to tease apart the effects of the wording changes from the effect of the switch in mode couldn't be included in the design. Similar arguments are likely to crop up whenever a new questionnaire is a candidate to replace an existing one—it may seem too expensive or too complicated to separate out the impact of any specific design change, and it is likely to seem much more important to determine the overall effect of changing from the old questionnaire to the new. In addition, some combinations of treatments may be seen as impractical (e.g., administering the complex new CPS questionnaire on paper). Still, the gains in scientific knowledge may be considerable when experiments avoid confounds among key variables. One problem is that researchers often think that equal sample sizes are needed to assure that the independent variables aren't confounded; as we shall see in Section 11.2.2, that isn't actually a requirement. Even a "small" comparison group (say, 500 cases per month) who got the old CPS questionnaire via CAPI would have greatly increased the scientific yield from that experiment.

11.2 DESIGN ISSUES FOR QUESTIONNAIRE DESIGN EXPERIMENTS

So one reason that questionnaire design remains an art is the reliance on nonrigorous procedures for testing questionnaires; another is experiments that compare one version of a questionnaire with another version, without allowing us to determine which features of the two versions produced the differences between them. This part of the chapter argues for experimental designs that might yield more definitive conclusions about why some questions produce more reliable and more accurate answers than others. It examines four issues:

- The relative advantages and disadvantages of the laboratory and field as sites for questionnaire design experiments
- The pros and cons of factorial designs versus designs that compare packages of variables
- Different options for randomizing respondents to treatments
- Statistical issues complicating the analysis of results of questionnaire design experiments

11.2.1 Lab versus Field Settings

One issue that arises in the design of questionnaire experiments is where to conduct them. Traditionally, psychological experiments have been conducted in laboratory settings, usually with convenience samples of college students. Periodically, critics of this combination of artificial settings and unrepresentative samples have spurred efforts to move psychological research into field settings (Gergen, 1973; Sears, 1986). In the case of surveys, though, most experiments comparing different versions of particular questions or questionnaires have been done in field settings, often as pilot tests for surveys (e.g., Neter and Waksberg, 1964) or as part of ongoing surveys (e.g., most of the classic experiments reported by Schuman and Presser, 1981, were done as part of the Detroit Area Study, the General Social Survey, or the University of Michigan's monthly Survey of Consumer Attitudes). There are pluses and minuses to each venue, and it's worth reviewing these.

Laboratory experiments offer two major advantages: greater control over the independent variables (i.e., the variables manipulated by design, such as the question wording or mode of interview) and greater control over extraneous variation (cf. Aronson et al., 1998, pp. 125–129). Greater control over the independent variable simply means that it is easier to ensure that everyone gets the intended treatment in a standardized way in a laboratory. In the case of questionnaire design experiments, it is easier to assure that the interviewers actually deliver the different versions of the questions as worded to all the respondents in the lab than it is in the field. Greater control over extraneous variation means that the data collection setting is more standardized and more insulated from outside distractions, interruptions, and so on. Anyone who has ever observed interviewers trying to implement experiments in the field in the face of respondent resistance will immediately recognize these advantages of the laboratory.

Another potential advantage of the lab setting is that it is much easier to collect detailed process data there. For example, reaction times are relatively easy to collect in a laboratory, and reaction time data have proven invaluable in understanding how respondents answer certain types of survey questions (see, e.g., Conrad et al., 1998). Couper (1999) describes the collection of rich usability data by observing and videotaping interviewers as they try to administer computer-assisted questions in a laboratory setting. The think-aloud data and responses to probes that are the hallmark of cognitive interviewing are also generally easier to collect in a lab than in a field setting (Forsyth and Lessler, 1991).

Still, several developments may have closed the gap between the lab and the field as settings for questionnaire experiments. First, when experiments are done from a centralized CATI facility, departures from script seem to be relatively rare (Presser and Zhao, 1992). Central supervision of the interviews seems to be effective at assuring uniform administration of the questions. Even greater levels of standardization are possible with mail questionnaires, ACASI (where the questions are recorded ahead of time and administered by computer), and ACASI's telephone cousin, interactive voice response (IVR). Technological advances have also made it easier to collect reaction time data and other process measures in the field or over the telephone (see Bassili and Fletcher, 1991; Bassili and Scott, 1996).

Both the lab and field are susceptible to their own selection bias problems, but generally these are much worse for the lab than for the field setting. It's difficult to assemble a nationally representative sample in a single site (although it's been done; see McCombs and Reynolds, 1999), and lab studies almost always use volunteer samples of dubious representativeness. On the other hand, in field settings, differential nonresponse (experimental mortality in the terminology of Campbell and Stanley, 1963) can jeopardize the comparability of the groups receiving the different questionnaires/items; differential nonresponse is often a particularly thorny issue when the different treatments involve different methods of data collection as well as different questions. This problem can be reduced if random assignment to treatment takes place *after* respondents agree to do the survey; that way, the different experimental groups have similar nonresponse biases and remain comparable. Last-second assignment to treatment is often practical when the computer application that administers the questions also carries out the randomization to experimental condition.

But the decisive consideration is often cost. An experiment done under realistic survey conditions in the field is likely to be many times more expensive than a similar experiment in a laboratory. The cost difference combined with the greater power possible with lab experiments may drive investigators to use the lab even if they might prefer the field in principle. The lab is often more powerful than the field because of the greater control in the lab and because it is easier to implement more powerful experimental designs (e.g., within-subjects designs) in a lab setting. Despite a very small sample size ($n = 40$), the lab study on ACASI by O'Reilly and his colleagues still managed to find some significant differences across modes of administration. It's hard to believe that a field experiment with the same sample size would have produced significant results. At the same time, a lab setting may give an exaggerated picture of the likely impact of the variable of interest in the field, where people are likely to be paying less attention and are less likely to notice and respond to small changes in wording or other subtle differences in survey procedures.

The lab setting may yield unrealistic results for another reason: Subjects in a lab experiment almost always realize that they are in an experiment, and this can affect their responses (cf. Aronson et al., 1998, pp. 114–116 on participant biases). For example, knowing that they're in a study, respondents may attempt to comply with the perceived experimental demands, or they may attempt to fulfill the experimenter's expectations.

Experiments comparing versions of questions are done with a mix of purposes. At the one extreme, the main point of the experiment may be calibrating the effect of switching questionnaires in an ongoing survey; the CPS experiment falls into this category. In the United States, there isn't a more important time series than the monthly unemployment statistics produced by the CPS. It was, therefore, absolutely crucial to assess the impact of the new questionnaire (and new mode of data collection). At the other extreme, some studies are done to examine issues that are not tied directly to any specific questionnaire. For example, several studies have examined how respondents answer behavioral frequency questions (questions about how often the respondent has done something—visited a doctor, used an ATM, made a long-distance phone call—during a reference period). These studies have investigated the general issue of how respondents answer such questions (see Belli et al., 2000; Burton and Blair, 1991; Conrad et al., 1998; Menon, 1993; Menon and Yorkston, 2000), typically examining multiple items; they are less interested in the effects on any one item than on the overall pattern across questions. Clearly, to the extent that an experiment is attempting to forecast the impact of a questionnaire redesign on an ongoing study, a realistic field experiment with a probability sample of respondents drawn from the same population (with the same sample design) as the ongoing survey is absolutely necessary. To the extent that the goal of the experiment is to address some more basic issue that is not tied to any particular survey, a field setting and a decent sample are usually nice but not entirely necessary. Table 11.1 summarizes this discussion of the merits of the lab and field as venues for questionnaire design experiments.

Table 11.1 Advantages and Disadvantages of Lab and Field Settings

	Laboratory	Field
Advantages	Greater control (over background noise and implementation of treatments) Reduced costs per case	Greater realism More representative samples More projectible to ongoing surveys
Disadvantages	Respondents' awareness of participating in an experiment (heightened participant biases)	Greater distractions

11.2.2 Factorial Designs versus Designs Comparing Packages of Variables

Two types of questionnaire experiment are common in practice. One type attempts to compare several complete questionnaires (or several complete packages of data collection procedures) to find the best one to apply later in the main survey. Such studies are often done in the planning stages prior to some major data collection effort. A good example of this type is an experiment done by Dillman and his colleagues (Dillman et al., 1993) that compared five different versions of the short form for the decennial census. The experimental versions of the short form differed in several ways, including the size and format of the questionnaire, the number of items included, whether a letter from the director of the Census Bureau was printed on the questionnaire or separately, and whether the form asked for the social security number for each member of the household. An additional version reproduced the main features of the form actually used in the 1990 census. The design made no effort to assess the impact of any specific difference across the five versions (with the exception of the request for social security numbers); instead, it compared five specific questionnaires, each of which might have been a reasonable candidate for use in the 2000 census.

At the other extreme are experiments that vary several independent variables, or *factors*, and include all possible combinations of the treatments that define each factor. For example, the ACASI experiment by Tourangeau and Smith (1996) actually looked at the effect of three variables—the format and context of the key items as well as the mode of data collection—on the reporting of sexual partners. Besides comparing three data collection modes (ACASI, computer-assisted self-interviewing without the audio, and CAPI), Tourangeau and Smith examined three response formats and also varied whether the sex partner items followed a series of attitude questions that expressed permissive or restrictive views about sex. In total, then, there were 18 cells representing all possible combinations of these three variables. A *fully crossed factorial design* is one that includes all possible combinations of the independent variables.

Factorial designs have several clear advantages for basic hypothesis-testing research, which explains their great popularity in psychological and biomedical research. First, they allow the researcher to investigate the effects of several variables in a single experiment, in effect letting the researcher conduct several experiments at once. Tourangeau and Smith could have conducted three separate studies that examined, in turn, the effects of mode, question format, and context on the reporting of sex partners. Including all three variables in a single study was obviously the more efficient strategy. If the independent variables are orthogonal to each other (i.e., if they are uncorrelated with each other by design), there is no loss in the precision of the estimates of the effects of one variable despite the presence of the other variables (see Winer, 1971, Chap. 5, and footnote 2 below).

Another important advantage to the factorial design is that it allows researchers to determine the effects of specific combinations of treatments, or *interaction effects*. For example, Tourangeau and Smith (1996) found that the effect of using ACASI to administer the sex partner items depended on whether the prior series of attitude items expressed permissive or restrictive views about sex; the mode of administration mattered less when the previous questions expressed permissive views. These sorts of interactions between variables can only be detected in factorial designs.

The final advantage of the factorial design is that it allows researchers to tease apart effects that might otherwise be confounded. Had the CPS experiment crossed the mode of data collection with the version of the questionnaire, the main effect of each variable could have been distinguished (and their interaction could have been examined). A related point is that factorial designs promote careful analysis of the various components of complex treatments. Consider, for example, the practice on the National Survey of Family Growth (NSFG) of asking some sensitive questions a second time under ACASI after the main CAPI interview has been completed. The ACASI questions net additional reports of abortions (and other sensitive behaviors) and are thought to improve the overall accuracy of the NSFG data (Fu et al., 1998). But there are several possible explanations of this finding—it could reflect respondents' increased willingness to make embarrassing admissions under ACASI, it could reflect gains simply from asking the questions a second time (with resulting improvements in memory), or it could reflect respondents' perceptions that these particular questions must be especially important since they are being asked twice. Factorial experiments are ideally suited to distinguishing the various possibilities, thereby deepening our understanding why things work the way they do in surveys.

The big drawback to factorial experiments is the combinatorial explosion that can result. A design with just three independent variables, each with three levels or treatments, implies 27 experimental cells. Sometimes, such designs are just too complicated to be practical. Computer-assisted interviewing has, however, greatly reduced the logistical difficulties involved in randomizing respondents to treatment and implementing complex designs in the field; factorial designs are likely to become more common in questionnaire design research as a result. In addition, it isn't always absolutely necessary to include all possible combinations of the independent variables. *Fractional designs* include a portion of the treatment combinations (e.g., a half or a quarter of the cells in a design with 2^k cells); typically, higher-order interactions are confounded with the main effects of the independent variables, but if these interactions can be assumed to be negligible, the analysis yields unbiased estimates of the treatment effects (for details, see Kirk, 1968, Chap. 10).

Two misconceptions may discourage researchers from using factorial designs. The first is that such designs require equal numbers of respondents in every cell. This can be a difficult requirement to meet in practice, given nonresponse and the other vicissitudes of field settings. But although equal cell sizes simplify the analysis and ensure that the independent variables are unconfounded, they are not an absolute requirement. The experimental factors will still be unconfounded as long as cell sizes are proportional. (Table 11.2 illustrates the idea of proportional cell sizes in a 2×3 factorial design; the key point is that every level of one independent variable has the same mix of levels of the other independent

Table 11.2 Equal, Proportionate, and Disproportionate Cell Sizes[a]

Equal Cell Sizes			Proportionate Cell Sizes			Disproportionate Cell Sizes		
	A1	A2		A1	A2		A1	A2
B1	50	50	B1	50	25	B1	50	50
B2	50	50	B2	90	45	B2	60	40
B3	50	50	B3	60	30	B3	70	30

[a]The design on the left has equal cell sizes in all six cells. The one in the middle has proportionate cells sizes, with level 2 of factor A getting twice as many cases as level 1. The one on the right has unequal and disproportionate cell sizes; as a result, the impact of A1 and that of B3 are partially confounded.

variables.) Even if proportionality cannot be maintained exactly, minor departures don't greatly reduce the power of the design.[2]

The second misconception is that the power of the experiment depends on the cell size (so that adding more factors necessarily reduces power). In fact, the power of tests for main effects depends on the number of cases *per level* of the factor. More specifically, the power for detecting a main effect of the treatment variable depends on the noncentrality parameter, λ (see Winer, 1971, p. 334):

$$\lambda = \frac{n_a \sigma_a^2}{\sigma_\varepsilon^2} \tag{11.1}$$

in which n_a is the number of cases in per level of factor A, σ_a^2 is the variation in the means across the different levels of that factor, and σ_ε^2 is the error variance (generally, the variation within experimental groups). Error variance (reflecting heterogeneity across respondents or variability in the implementation of the treatments) reduces power; larger effects (reflected in larger variance across the means) and larger sample sizes per treatment increase power. The key implication of equation (11.1) is that including additional factors in a design doesn't reduce power, even though it makes the cell sizes smaller.

Like the decision to conduct a study in the lab or the field, the decision to test packages of variables or to conduct a fully crossed factorial design may ultimately depend on the purpose of the study. If the point of the experiment is to calibrate the impact of replacing an old questionnaire with a new one, the entire package of changes that define the new questionnaire needs to be compared

[2]The issue of confounded factors in an experiment is essentially the same as the issue of collinearity in multiple regression; in both cases, the increase in the variance of estimates of the effects depends on the magnitude of the correlations among the independent variables. The variance of the regression coefficient for a given independent variable is inflated by $1/(1 - r^2)$, where r^2 is the squared multiple correlation between that variable and the other independent variables. The correlations between factors in experiments are rarely large, so the loss of power from failure to maintain proportional cell sizes is usually small. For example, a correlation of 0.20 between the experimental variables in a two-factor design would increase the variance by about 4%

with the old version. It may be too hard to try to separate out the effects of each specific change by conducting a full factorial experiment. On the other hand, if the aim is to test hypotheses about the impact of specific variables, a factorial experiment is the right tool for the job. In my opinion, there are currently too many studies that compare packages of variables and too few that use factorial designs; the relative rarity of analytical experiments is one reason we don't know more about how to design questionnaires.

11.2.3 Completely Randomized Designs versus Randomized Block Designs

Another fundamental design issue is how to assign the sampling units to experimental groups. There are two basic models for doing this. In one model—the *completely randomized design*—each unit is assigned at random to get exactly one of the treatments. In the other model—the *randomized block design*—each unit (or "block") gets all of the treatments. (There are also intermediate designs in which the units get a subset of the treatments; we touch on these later.) When the unit is an individual respondent, a completely randomized design is often called a *between-subject design* (since the comparisons between the different experimental groups are also comparisons between different subjects or respondents). Similarly, when each respondent gets all the treatments, the design is called *within-subjects*. Mixed designs are possible as well, in which units are assigned to one level of at least one of the factors and to all levels of the others.

In many survey experiments, the "block" is not an individual respondent but a cluster of them. For example, the experiment by Dillman et al. (1993) randomly assigned each of five adjacent housing units to get one of the five versions of the short form. This is a randomized block design in which the "block" consisted of five nearby housing units. Each cluster constituted a complete replication of the experiment. Similarly, in their ACASI experiment, Tourangeau and Smith (1996) first selected 32 city blocks as their first-stage units and then sampled individual housing units on each sample block, assigning the units on a block to different cells of the design. In their experiment, the blocking was by clusters of housing units. Other studies have used interviewers as the blocking factor, with each interviewer administering all the treatments. Blocking is like matching, only it is done prior to randomization. For instance, in the study by Dillman and his colleagues, the five experimental groups were exactly matched in terms of geographical distribution. In such designs, sampling error (or some other source of nuisance variation) is completely removed from the estimate of the treatment effect.

This is the major advantage of randomized block designs—they provide exact control of the variation associated with the blocking factor and can yield big gains in power. In addition, randomized block designs are often less expensive than completely randomized designs, since each block contributes an observation to every cell of the experimental design. The savings can be especially dramatic with within-subject designs, since every respondent contributes multiple observations rather than just one.

The big drawback to such randomized block designs is cross-treatment contamination effects. For example, if the blocking factor is interviewers (i.e., each interviewer administers every version of the questions), interviewer expectations or preferences may affect the results. Interviewers rarely like longer questions, so if the experiment compares two versions of a question and the versions vary markedly in length, blocking by interviewers may be a problem; wittingly or unwittingly, the interviewers may sabotage the results for the long version of the question. If the blocking factor is the area segment, the worry is that neighbors will compare notes, perhaps affecting their responses. (This doesn't seem a plausible threat to the conclusions reached by Dillman and his colleagues; it's hard to imagine neighbors in most U.S. neighborhoods getting together to mull over their census forms.) Cross-treatment contamination is especially likely in within-subject designs (imagine giving respondents three or four versions of the same question) or when blocking is done within households (imagine offering members of the same household different incentives for taking part in a survey).

Concern about such contamination effects may lead researchers to assign clusters of cases to the same experimental condition. For example, the design may assign all the cases from an area segment to the same treatment. The drawback to this strategy is the loss of efficiency; just as clustering reduces the efficiency of a sample, it reduces the efficiency of the estimates from an experiment. In a randomized block design, the correlation between observations from the same cluster increases precision; in a completely randomized design (where respondents are clustered and whole clusters are assigned to an experimental condition), it has the opposite effect.

Table 11.3 summarizes our discussion of the advantages and disadvantages of these two common types of design. (See Kirk, 1968, for a similar discussion.)

Experimental design texts (e.g., Kirk, 1968; Winer, 1971) distinguish two other classes of designs—Latin square and balanced incomplete block designs—in which the units get more than one of the treatments but not all of them. In Latin square designs, only a subset of the possible treatment combinations is included in the design and any individual respondent receives only a subset of these. Latin square designs are often used when respondents get all the substantive treatments and the researcher wants to control for possible position effects.

Table 11.3 Advantages and Disadvantages of Completely Randomized and Randomized Block Designs

	Completely Randomized	Randomized Block
Advantages	Reduced concern about "participant biases"	Multiple observations obtained from each block or respondent (reducing sample size and cost)
	Reduced danger of order and similar cross-treatment contamination effects	Exact matching on blocking factor (reducing variance)
Disadvantages	Inefficient; only one observation obtained from each respondent	Each block must accommodate every cell, limiting number of cells

Table 11.4 Illustrative Greco-Latin Square[a]

Group	First	Second	Third
1	ACASI–Income	CASI–Drugs	SAQ–Gender
2	CASI–Gender	SAQ–Income	ACASI–Drugs
3	SAQ–Drugs	ACASI–Gender	CASI–Income

Source: After O'Reilly et al. (1994).

[a]Respondents in the first group get income questions administered under audio computer-assisted self-interviewing (ACASI), followed by questions on illicit drug use administered under CASI (ACASI without the audio), and then questions about sexual behavior administered on a paper self-administered questionnaire (SAQ). Only nine of the 27 possible mode–topic–position combinations are represented in the design. For example, the combination of ACASI and income questions never appears as the second or third section of the interview.

Table 11.4 displays a Greco-Latin square (which controls for the position of two substantive variables), patterned after the design used by O'Reilly and his colleagues in their study of ACASI. They asked questions about three sensitive topics (income, sexual behavior, and drug use) under three modes of data collection. Any one respondent got all three topics and all three modes, but only three of the nine mode–topic combinations. The design controls for any effects of the position in which the respondent received each topic and mode (i.e., first, second, or third). The entire design includes all nine combinations of mode and topic, but only nine of the 27 possible mode–topic–position combinations. Still, the design is balanced in the sense that each topic and each mode appear equally often in each position. For example, a third of the respondents answered questions under ACASI first, another third got questions via CASI first, and the final third got questions on a paper questionnaire first. (Chromy et al., 2004, describe another example of a Latin square design used in a survey setting; this early methodological study was done by Gertrude Cox.)

In a *balanced incomplete block* (BIB) *design*, each unit (typically, each respondent) gets a subset of the treatments, with (roughly) equal numbers of respondents getting each subset to assure balance across the entire sample. The General Social Survey has used a design like this to reduce the length of the GSS interview (Smith, 1988a). All of the respondents get a core set of items; in addition, they get two of three possible modules. Roughly a third of the sample gets each of the three possible pairs of modules.

11.2.4 Statistical Issues for Experiments

This final section on questionnaire experiments will touch briefly on several additional design issues. One theme that has run through the discussion of the design options is the competing goals that motivate questionnaire design studies. One set of goals is quite practical—to select the best questionnaire for a survey or to gauge the impact of shifting from an old questionnaire to a new one. The other set of goals is more theoretical—to identify general characteristics or design principles that lead to better survey questions (questions that yield more reliable and more accurate answers). Depending on the mix of purposes for any particular experiment, the analysts may appropriately place more emphasis on the practical significance of the findings than on their statistical significance. In some cases, this means ignoring small but reliable differences between groups and giving more weight to other considerations (e.g., cost, ease of administration, acceptability to the client) in choosing between versions of a question or an instrument. In other cases, though, emphasizing practical over statistical significance entails playing up small differences that we might otherwise ignore. For example, in the Dillman et al. (1993) study, a 1 or 2% difference in the response rates by version of the short form represented a difference of 1 or 2 *million* households that would require follow-up in the decennial census. In this case, even a small difference was worth detecting and acting on.

In most social science applications, the 0.05 level is conventionally applied to separate the significant from the nonsignificant; this convention means that only about one significant result in 20 is a false positive (a type I error). The convention isn't quite so clear for power, but in many settings, the goal seems to be power of 0.80. Because power is the probability of detecting a real difference of a given magnitude, this implies that the false negative rate (the probability of a type II error) is generally around 0.20 (or higher). Together, these conventions make sense only if a type I error is viewed as a lot more serious than a type II error (in fact, four times more serious). This differential concern about the different types of inferential error isn't always reasonable. When the goal is to decide between two different questionnaires, ignoring a real difference may be just as serious a mistake as seeing a difference that isn't really there.

My guess is that few survey experiments actually have power as high as 0.80 to detect moderate-sized differences. There are several reasons for the low power (see also Cook and Campbell, 1979, pp. 42–44):

- The sample sizes are often small (as in the early studies examining ACASI).
- The treatments in questionnaire design experiments often involve fairly subtle differences.
- Different interviewers may carry out the treatments in different ways (and the same interviewer may vary the treatment across different respondents).
- Field settings tend to be "noisy," with substantial variation across respondents, geographic areas, and so on, as well as ample opportunities for distractions.
- The outcome measures (typically, the responses to the item, but also perhaps rates of unit or item nonresponse) may be unreliable or subject to other sources of variability.

In addition, the sample design features that increase the variance of survey estimates (notably the clustering and weighting of observations) generally have the same impact on the estimates from survey experiments as well. These considerations suggest that it may be worthwhile in some cases to alter the balance between the two types of inferential errors by adopting alpha levels of 0.10 (or some other value) rather than the conventional 0.05 level. Increasing the probability of type I errors by changing the alpha levels necessarily reduces the probability of type II errors and is often the easiest way to increase power.

Unfortunately, pretest experiments with low power force researchers back on their own judgments, forcing them to rely on nonstatistical criteria to figure out what the findings mean. When power is low, what started out as a quantitative endeavor may end up as an expensive qualitative one.

11.3 CONCLUSIONS

The thrust of this chapter is that the only way to put questionnaire design on a fully scientific footing is to adopt proven scientific methods and to apply them as rigorously as possible. Too many pretests consist solely of subjective evaluations of the questions, and the few quantitative pretest studies that are done often consist of low-powered experiments that compare the answers from two complete questionnaires. Even when such studies find significant differences, it may be hard to tell what the findings mean because the studies don't use factorial designs and the necessary comparison groups aren't included; in addition, pretest experiments rarely collect validation or reliability data. As a result, deciding which version of the questions yielded the more accurate answers often depends less on science than on intuition. Calling low-powered or poorly designed studies exploratory doesn't really improve their information yield.

None of this is intended as an attack on the field or its practitioners. All the pretesting tools that are currently in use definitely have their roles to play. At the early stages of research, there's really no alternative to exploratory methods that produce hypotheses rather than conclusions, especially if they prompt follow-up work to test the hypotheses. Those of us who've done cognitive interviews appreciate the rich qualitative data they yield; there's no substitute for seeing how a questionnaire actually works with real respondents. And purely observational studies are sometimes an absolute necessity; no one wants to field a multimillion survey only to see it crash and burn. A small-scale observational pretest is often the best way to prevent such disasters. A corollary is that experiments aren't the solution for every problem. They are best suited to situations in which researchers seek to test general hypotheses that apply to broad classes of items; they are also extremely useful when there are several plausible ways to conduct an important study such as the decennial census or the Current Population Study. But they are generally too cumbersome and too costly when the task is to test the 50 or 100 items that make up a specific questionnaire.

Still, the design of any pretest study needs to take into account both the short-term goals of developing questionnaires for particular surveys and the long-term goal of discovering credible principles of questionnaire design. In my own view, we've often stinted on the long-term goal. We need to strike a better balance, one that gives greater priority to validating principles for writing questions in survey pretests.

CHAPTER 12

Modeling Measurement Error to Identify Flawed Questions

Paul Biemer

Research Triangle Institute

12.1 INTRODUCTION

In this chapter we consider several statistical methods for identifying flawed questions in a survey questionnaire. The primary goal of these methods is to identify survey questions that elicit responses that contain high levels of measurement error, resulting in poor reliability, substantial bias, or both. We also show how the same methods can be used to further explore the sources of the errors in order to gain insights as to their root causes.

The methods we consider assume that multiple measurements (or remeasurements) of the questions to be evaluated are available for a random sample of units. The remeasurements may be provided by (1) one or more reinterview surveys in which a random subsample of the survey sample is recontacted and interviewed again about the characteristics measured in the original survey, (2) a single interview where the repeat measures are embedded in the survey questionnaire, (3) multiple measurements provided by some combination of surveys and administrative records sources or biological measurements, and (4) multiple measurements collected by a panel survey that refer to different time points, such as monthly employment status in a continuing survey of employment and income. Although the techniques we examine can be applied to any categorical data, our methods assume that the final measurements are measured on a nominal scale.

We begin by reviewing briefly some of the statistical methods used for questionnaire evaluation and a general approach for investigating questionnaire problems through statistical analysis. Following this discussion, we present a general

Methods for Testing and Evaluating Survey Questionnaires, Edited by Stanley Presser, Jennifer M. Rothgeb, Mick P. Couper, Judith T. Lessler, Elizabeth Martin, Jean Martin, and Eleanor Singer
ISBN 0-471-45841-4

method for estimating classification error in questions when two remeasurements are available and illustrate its use with data from the U.S. Current Population Survey (CPS) Reinterview Program. Then we present two methods for evaluating questions when three measurements are available: one that assumes the three measurements all refer to the same point in time and one that allows them to reference different points in time. These methods are illustrated using data from the U.S. National Household Survey on Drug Abuse and the CPS. Methods for remeasurements obtained from panel surveys are then discussed briefly. In the last section we discuss the strengths and weaknesses of the methods and their implications for questionnaire evaluation.

12.2 REVIEW OF STATISTICAL METHODS FOR QUESTIONNAIRE EVALUATION

Remeasurement methods include reinterviews, record checks, replicate questions within the same questionnaire, and panel survey measurements. The methods will be referred to collectively as *remeasurement methods*. Models for analyzing remeasurement data were presented by Hansen et al. (1964), who proposed a simple model for interpreting the gross differences in an interview–reinterview table. An important contribution of their work was the development of the index of inconsistency, denoted by I, as a measure of proportion of the variance in the responses to a survey question that is attributable to measurement variance. Hess et al. (1999) show the equivalence of $1 - I$ and κ, where κ is the well-known Cohen's kappa measure of reliability or chance-adjusted agreement (Cohen, 1960). Hansen et al. (1964) proposed an estimator of I based on replicated or test–retest reinterviews (i.e., reinterviews that are designed to replicate the error distribution of the original survey interviews). These methods for estimating response reliability have become common practice in survey research.

Other authors, such as Mote and Anderson (1965) and Tenenbein (1970), proposed models for estimating the false positive and false negative probabilities from remeasurement studies. The false positive probability is defined for dichotomous measurements as the probability that a person who is a true negative (i.e., should respond "no" to a question) is classified by the question as a positive (i.e., responds "yes"). A false negative probability is the probability that a true positive is classified by the question as a negative. Direct estimation of these error probabilities requires knowledge of the true classification for the people in the sample. These *gold standard measurements* can also be used to estimate the bias in a survey estimate. However, as we will see later, indirect estimation of these quantities using latent class modeling approaches do not require gold standard measurements. U.S. Bureau of the Census (1985), Biemer and Stokes (1991), and others have shown the relationship between the misclassification probabilities, reliability, and other components of the mean-squared error of a survey estimator.

Concurrent with early research on measurement error modeling in the survey literature, major developments were being made independently in the psychometric literature. Lazarfeld (1950) and Lazarfeld and Henry (1968) provided a probability modeling framework for analyzing categorical data with classification error. Many authors have since built on their ideas and have extended the structure to log-linear models with latent variables, including Clogg (1977), Goodman (1974), and Haberman (1979). Emerging from this body of work is the notion that the true classification of a person according to some question criteria is unobservable or "latent" and that a person's response to a survey question is an "indicator" of the true, latent classification. That is, individual responses may be highly correlated with the true values but are potentially contaminated by measurement errors. Lazarfeld (1950) showed that when two or more indicators of a latent variable are available, it is possible to model the relationship between the unobserved and the observed in order to estimate the distribution of the unobserved variable.

A latent class model (LCM) describes the relationship between the latent variable, the indicator variables, and possibly other explanatory variables that may be related to the latent variable. The model can then be used to define the probability, or likelihood, of observing various combinations of the indicator and explanatory variables in a cross-classification table. The parameters of the likelihood function are related directly to error in the indicator variables. By identifying the parameters that maximize the likelihood of the observed table—a process known as *maximum likelihood estimation*—the parameters of the model can be estimated.

An advantage of the LCMs over the classical methods of measurement error estimation is that the usual assumptions made in classical measurement error analysis can be relaxed. For example, it is not necessary to assume that one indicator is a gold standard measurement in order to estimate measurement bias in another indicator. In latent class analysis (LCA), all indicators may be assumed to be fallible measurements of the latent variable. Instead, the LCMs make other assumptions relating the errors in the indicators to the values of the classes of latent variable. To identify the best model for a set of data, the analyst posits a number of alternative models for the observed cross-classification table and compares the ability of the model to reproduce the table observed using conventional chi-square goodness-of-fit criteria. The best LCM is typically one that is plausible, parsimonious, and fits the data well. The best fitting model is then used to generate estimates of the classification error probabilities for the measurement processes under investigation.

A disadvantage of LCA is that sometimes rather strong assumptions must be made to achieve an identifiable model (i.e., a model that produces unique estimates for the model parameters). However, this can also be a disadvantage for classical methods, which also make assumptions that are questionable and quite difficult or impossible to verify in practice. LCMs may be complex, involving dozens of parameters, and may make many assumptions about the model parameters which are difficult or impossible to verify.

For evaluating survey questions, both classical and latent class modeling approaches can be quite useful despite the fact that they may be only rough approximations of reality. In the end, the usefulness of a modeling and estimation approach should be judged on how well it achieves its intended purpose. For questionnaire evaluation, the primary purpose of measurement error analysis is to identify questions that are flawed and which elicit unreliable or biasing data from respondents. Thus, if an estimation model or analysis method allows us to achieve this goal more successfully, it has shown its worth as an analytical tool.

It is important to note that whereas a modeling approach may be quite appropriate for question evaluation and improvement purposes, it may be quite inappropriate for other uses: for example, postsurvey adjustment for measurement bias, estimation of mean-squared error components for quality reports, optimization of survey design, and the like. Indeed, it may be quite risky to apply the methods for those purposes without having substantial evidence of the validity of the model assumptions.

As an example, an analysis method may produce estimates of false positive and false negative probabilities that are quite useful for identifying questions that are subject to high levels of misclassification. However, the same estimates may be quite inappropriate for adjusting the survey estimates for these misclassifications for the purpose of producing "unbiased" estimates of the population parameter and then reporting the results as unbiased estimates. This occurs, for example, when the estimates of classification error are themselves biased and using them to correct the survey estimates would eliminate one bias while adding another. It is important to distinguish between the uses of measurement error modeling in discussing the strengths and weaknesses of the methods considered in this chapter.

In that light, the methods described below are intended to be one component in a general strategy of questionnaire improvement. Such a strategy may involve the following four steps:

1. Analyze data to identify potentially flawed questions (i.e., questions with poor reliability or high levels of classification error).
2. Conduct further data analysis to elucidate the probable sources of error for the questions identified in step 1.
3. Verify the sources of the problem through the collection of additional data from field investigations, cognitive laboratory studies, field experiments, debriefing interviews, and so on.
4. Decide on strategies for eliminating the problem and implement these strategies.

The methods described in this chapter are concerned primarily with steps 1 and 2 of this process, and these steps will be illustrated in a number of real-world examples. As we shall see, LCMs provide a particularly useful structure for implementing step 2 of this process due to its generality and ability to explore increasingly complex relationships between both the observed and latent variables.

12.3 METHODS FOR TWO MEASUREMENTS

We first consider a model for representing classification error in a measurement process for the simple case of two dichotomous measurements. Extensions of this simple model to three or more dichotomous or polychotomous measurements will subsequently be considered. First, we introduce a general notation for discussing the models. The sampling structure for all the models is simple random sampling (SRS) from a very large population. Extensions of this methodology to complex sampling has been considered by Clogg and Eliason (1985) and Patterson et al. (2002). Some of those results are illustrated in the examples.

12.3.1 Illustration of the Basic Ideas

In a study reported by Biemer (1997, 2001), a sample of 2908 respondents were recontacted several weeks after an initial interview and reasked a subset of questions from the original survey. The results for one question, "Does anyone smoke inside the home?" is shown in Table 12.1. The cells counts in the table have been labeled a, b, c, and d for later reference. In this study, the reinterview survey was designed to replicate the original interview independently so that the measurements from the two surveys can be assumed to be parallel. By parallel measurements we mean measurements that have the same probability of false positive and false negative errors and whose errors are independent. Provided that these assumptions hold, the computation of I or κ is appropriate.

The formula for the index of inconsistency I is

$$I = \frac{g}{p_1 q_2 + p_2 q_1} \tag{12.1}$$

where $g = (b + c)/n$, referred to as the *gross difference rate* or disagreement rate; $n = a + b + c + d$ is the total reinterview sample size; $p_1 = (a + c)/n$ is the proportion answering "yes" in the original interview; $p_2 = (a + b)/n$ is the proportion answering "yes" in the reinterview; and $q_t = 1 - p_t$, for $t = 1, 2$, denotes the proportion answering "no" for interview and reinterview, respectively. It can be shown that the expected value of I is the proportion of the total variance of a measurement that is error variance. Thus, $\kappa = 1 - I$ is an estimate of the

Table 12.1 Data from NHIS RDD Reinterview Survey

Does Anyone Smoke Inside the Home?		
Reinterview Response	Interview Response: Yes	Interview Response: No
Yes	$a = 616$	$b = 90$
No	$c = 38$	$d = 2164$

reliability of a measurement defined as the proportion of total variance that is due to the true value (or *true score*) variance.

The index of inconsistency for the data in Table 12.1 is $I = 12.3\%$, which corresponds to a reliability of $\kappa = 87.7\%$, which is considered to be high. The *net difference rate* for the table, denoted by NDR, is $p_1 - p_2$, or -1.8%. The NDR is an estimate of the bias in the original interview proportion when the reinterview survey response can be considered as a gold standard or highly accurate response. However, since the reinterview survey was designed as a replication of the original survey, this interpretation of NDR is not appropriate. Instead, the NDR may be used as a check on the validity of the assumption that the reinterview is a replication of the original interview response, since in that case, NDR should not be significantly different from zero. Under simple random sampling, the variance of the NDR is $g(1 - g)/n$ (see, e.g., Hansen et al., 1964). This yields a standard error of the estimate of 0.38%. Since the NDR is highly statistically significant, the assumption of parallel interview and reinterview measurements is rejected for the data in Table 12.1, and thus the estimate of κ is not strictly valid. Nevertheless, the estimate may still be a useful indicator of reliability.

Now consider the application of LCA to these data. We begin by postulating a model of the measurement process. Since the interview–reinterview table contains only four cells, the LCM must necessarily be simple since only three parameters can be estimated. Let π denote the proportion of the population whose true response to the question is "yes"; thus, p_1 and p_2 are estimates of π that are subject to both sampling error and measurement error. Let π_{11} denote the probability that a randomly selected respondent responds "yes" in both the interview and reinterview, π_{10} denotes the probability that a respondent responds "yes" in the interview and "no" in the reinterview, π_{01} denotes a "no–yes" response, and π_{00} denotes a "no–no" response.

Further, let θ and ϕ denote the probabilities of a false negative and a false positive response, respectively. Then, assuming that the interview and reinterview responses are parallel measurements, we can write the probabilities of the four possible response patterns in terms of the error parameters as follows:

$$
\begin{aligned}
\pi_{11} &= \pi(1-\theta)^2 + (1-\pi)\phi^2 \\
\pi_{10} &= \pi\theta(1-\theta) + (1-\pi)\phi(1-\phi) \\
\pi_{01} &= \pi\theta(1-\theta) + (1-\pi)\phi(1-\phi) \\
\pi_{00} &= \pi\theta^2 + (1-\pi)(1-\phi)^2
\end{aligned}
\tag{12.2}
$$

For example, the probability that a person responds with a "yes–yes" response pattern (π_{11}) is equal to the sum of two terms. The first term is the probability that the person is a true "yes" (π) and responds correctly on both the interview and reinterview. The latter probability is the probability of no false negative error on both occasions, or $(1 - \theta)^2$, and thus the first term is the joint probability $\pi(1 - \theta)^2$. The second term is the probability that the person is a true "no" and responds erroneously on both occasions. This joint probability is $(1 - \pi)\phi^2$.

The probabilities associated with the three other response patterns are derived analogously.

Under simple random sampling, the joint distribution of the 2×2 interview–reinterview table is a multinomial with cell probabilities, π_{ts}, where $t,s \in \{0, 1\}$ given in (12.2). Thus, the probability of observing a table with cell counts a, b, c, d (Table 12.1) is

$$P(a, b, c, d) = \frac{n!}{a!b!c!d!} \pi_{11}^{a} \pi_{10}^{b} \pi_{01}^{c} \pi_{00}^{d} \tag{12.3}$$

where the cell probabilities, π_{ts}, for t, $s = 1, 2$ are given in (12.2) and a, b, c, d are given in Table 12.1.

The next step in the estimation process is to use maximum likelihood estimation techniques to determine the unique values of the parameters π, θ, and ϕ that maximize the function in (12.3). Although there appear to be sufficient degrees of freedom with three independent cells to estimate three parameters, it can be shown that the likelihood in (12.3) is a multinomial density with only two degrees of freedom. Since the number of parameters in the model exceeds the number of degrees of freedom for the table, the model is said to be *unidentifiable*, meaning that no unique solution exists.

One method for addressing unidentifiability is to impose further constraints on the parameters to reduce their number. For example, we may consider the constraint $\theta = \phi$ (i.e., the false positive and false negative error probabilities are equal) or that one or the other is zero (i.e., either $\theta = 0$ or $\phi = 0$). For the present example, it may be plausible to assume that $\phi = 0$ (i.e., that respondents almost never erroneously report that someone smokes inside the home. The likelihood function for this model is still given by (12.3) after setting $\phi = 0$ in (12.2).

The estimates of π and θ for this model may be obtained by a grid-search technique in which all the parameters π and θ are systematically varied between 0 and 1 in steps of, say, 0.001. The combination of values that corresponds to the largest value of (12.3) yields the maximum likelihood estimates (MLEs). Using this method, the MLEs are $\hat{\pi} = 0.258$ and $\hat{\theta} = 0.094$. Thus, according to the model, the respondents have a 9.4% chance of erroneously responding "no" to the smoking question and, by assumption, a zero percent chance of erroneously responding "yes." The percent of the population who are true "yes"es is 25.8%, which suggests a bias in the original survey response of $p_1 - \hat{\pi} = 22.5 - 25.8 = -3.3\%$.

With two parameters and two degrees of freedom, there are zero degrees of freedom to test the fit of the model using the usual chi-square goodness-of-fit criteria, X^2. Therefore, the adequacy of the model cannot be assessed formally. However, examination of the observed and estimated cell counts suggests that the fit is poor. For example, under the model, the cell counts b and c should be approximately equal; however, in the data observed, c is almost 2.5 times b.

This illustration shows that obtaining models that are useful for questionnaire evaluation when only two measurements are available can be challenging. In the next section, we extend the ideas of LCA for two measurements and consider models that may have much more practical utility than those considered here.

12.3.2 General Model for Two Measurements

In this section we extend the notation developed in the illustration in Section 12.3.1 to polychotomous response variables. Let S denote a simple random sample of size n from a large population, and let i denote a particular unit in S. Consider the response to a particular question on the questionnaire, such as the smoking question in the previous example. Let X_i denote the true value of the characteristic for unit i, and let A_i and B_i denote two measurements (or indicators) of the characteristic for this unit; for example, A_i may denote an interview response and B_i the corresponding reinterview response. Alternatively, A_i and B_i may denote two measurements of X_i from the same interview, or A_i may be a survey measurement and B_i may be a value obtained from a record check study. In what follows, we drop the subscript i to simplify the notation.

Let the variables A, B, and X take the value 1 for a positive response and 0 for a negative response. Let π_Z denote the probability of Z, for some event Z. For example, $\pi_{A=1}$ denotes $\text{P}(A = 1)$. Thus, $\pi_{A=a}$ or π_a denotes $\text{P}(A = a)$, $\pi_{x\,a}$ denotes $\text{P}(X = x,\ A = a)$, $\pi_{a|x}$ denotes $\text{P}(A = a|X = x)$, and so on, with analogous definitions for $\pi_{x\,b}$, $\pi_{b|x}$, and so on.

Of particular interest here are the classification error probabilities for the indicator variables (i.e., $\pi_{a|x}$ and $\pi_{b|x}$ for $a,b \neq x$). For example, $\pi_{a=1|x=0}$ is the false positive probability for A, and $\pi_{a=0|x=1}$ is the false negative probability. Similarly, $\pi_{b=1|x=0}$ and $\pi_{b=0|x=1}$ are the false positive and false negative probabilities, respectively, for B.

Note that the parallel assumption made in test–retest reinterview surveys is represented by the assumptions (1) $\pi_{a|x} = \pi_{b|x}$ and (2) $\pi_{a\,b|x} = \pi_{a|x}\pi_{b|x}$, where $\pi_{a\,b|x}$ is the joint probability, $\text{P}(A = a,\ B = b|X = x)$. The latter assumption is often called the *local independence assumption* (McCutcheon, 1987). One design feature that attempts to fulfill this assumption is the timing of the reinterview. When the second measurement is designed to be a gold standard or infallible measurement, we assume that $\pi_{b=1|x=0} = \pi_{b=0|x=1} = 0$ (i.e., there is no error in the second measurement and $B \equiv X$). In some cases, the remeasurement or reinterview observation satisfies neither the parallel nor the gold standard remeasurement assumptions. In this case, there may be no restrictions on $\pi_{a|x}$ and $\pi_{b|x}$ other than local independence.

Note that there are four cells in an A $\times$ B cross-classification table, while the general likelihood contains six model parameters, including the parameter for the overall mean. Since the number of parameters to be estimated exceeds the number of cells in the A $\times$ B table, the model is not identifiable and the parameters cannot be estimated. However, this problem can be overcome if restrictions on the probabilities can be introduced to reduce the number of parameters to be estimated.

For example, in the case of the gold standard reinterview, we can set $\pi_{b=1|x=0} = \pi_{b=0|x=1} = 0$ and reduce the number of parameters to three. This permits the estimation of π_x, the true proportion of positives in the population, and $\pi_{a=1|x=0}$ and $\pi_{a=0|x=1}$, the false positives and false negative probabilities for A. As we saw in Section 12.3.1, for a test–retest reinterview, an identifiable model can be obtained by setting $\pi_{a|x} = \pi_{b|x}$ whenever $a = b$ and placing further restrictions on $\pi_{a|x}$. However, as demonstrated in Section 12.3.1, such a model has very limited utility for evaluating survey questions.

A more useful method for obtaining an identifiable model with only two measurements was developed by Hui and Walter (1980). This method introduces a grouping variable G and makes restrictions on the parameters across the levels of G that are both theoretically plausible and more likely to be satisfied by the survey design or the structure of the population. As an example, for two groups, denoted by $g = 1,2$, the method requires a variable G such that the prevalence of the characteristic X in each group is unrestricted (i.e., $\pi_{g=1} \neq \pi_{g=2}$), while the false positive and false negative error in the groups are restricted to be equal across the groups. That is, the two groups have the identical probabilities of false negative and false positive errors.

With these restrictions on the error probabilities, the number of parameters to be estimated is also eight [i.e., the false positive and false negative probabilities in each group, the prevalence of true characteristic X in each group, the proportion of the population in group 1 (or group 2), and the overall mean]. Since the GAB cross-classification table has eight cells, the model is identifiable and all eight parameters can be estimated. Since there are no degrees of freedom left after fitting all the parameters, the fit of the model cannot be assessed.

12.3.3 Application to the CPS

To illustrate use of the Hui–Walter model, we use data from an interview–reinterview study conducted for the U.S. Current Population Survey (CPS). The CPS is a household sample survey conducted monthly by the U.S. Bureau of the Census to provide estimates of employment, unemployment, and other characteristics of the general U.S. labor force population. Since the early 1950s, the Census Bureau has conducted the CPS Reinterview Program by drawing a small subsample (less than 5%) of the CPS respondents and reasking some of the questions asked in the original interview, particularly the labor force questions. Forsman and Schreiner (1991) provide a detailed description of the CPS Reinterview Program.

Hui–Walter Model for Trichotomous Variables Using the notation introduced above, let X denote the true labor force classification for some time point, with $X = 1$ denoting employed (EMP), $X = 2$ denoting unemployed (UNEMP), and $X = 3$ denoting not in the labor force (NLF), and let A and B be defined analogously for the interview and reinterview classifications, respectively. Following Sinclair and Gastwirth (1996), who first applied the Hui–Walter model to the CPS, we use gender for the grouping variable and let $G = 1$ for males and $G = 2$ for females. Table 12.2 displays the interview–reinterview data from the 1996 CPS reinterview.

Table 12.2 Reinterview Results for 1996 CPS Reinterview

Males				Females			
Reinterview Response	Interview Response			Reinterview Response	Interview Response		
	EMP	UNEMP	NLF		EMP	UNEMP	NLF
EMP	237	14	29	EMP	2087	6	60
UNEMP	10	90	27	UNEMP	10	75	41
NLF	75	18	974	NLF	87	33	1639

Recall that in the case of dichotomous indicator variables, the GAB table consists of eight cells, which was exactly the number of unique parameters specified by the Hui–Walter model, and thus the model was fully saturated. For the trichotomous CPS classification, the GAB table consists of 18 cells and again the model is saturated. These 18 parameters are described next.

Consistent with the dichotomous Hui–Walter model, we assume that the prevalence of each labor force category differs for males and females; however, the response probabilities are the same for the two groups. Thus, the parameters to be estimated include the proportions in each labor force category for males and females, $\pi_{x|g}$ (a total of four parameters since $\sum_x \pi_{x|g} = 1$), the response probabilities, $\pi_{a|x}$ (a total of six parameters since $\sum_a \pi_{a|x} = 1$), the response probabilities, $\pi_{b|x}$ (an additional six parameters), the proportion π_g (one parameter since $\sum_g \pi_g = 1$), and the overall mean.

To fit the Hui–Walter model, we use a maximum likelihood estimation method that employs the EM algorithm. Several packages for fitting such models are available commercially as well as in freeware. We used the LEM software (Vermunt, 1997), a general package for fitting a wide range of latent variable models, including LCMs.

To identify the most parsimonious model, we first fit the Hui–Walter model with the additional assumption that the error probabilities for interview and reinterview are identical, represented by the constraint $\pi_{a|x} = \pi_{b|x}$ for $a = b$. Recall that this is the parallel assumption for two independent measurements. Although this constraint is not required for identifiability, it frees up six degrees of freedom for testing model fit.

Using the likelihood ratio criterion for testing model fit, the usual convention for fitting log-linear models is to reject the model if the likelihood ratio statistic, denoted by L^2, exceeds the 95th percentile of a chi-square distribution with six degrees of freedom; otherwise, the model is deemed acceptable. For this model, $L^2 = 31.1$, which corresponds to a p value of less than 0.001, and thus the model is rejected. This suggests that the parallel assumption does not hold for the 1996 CPS reinterview; thus, we will fit the Hui–Walter model without this constraint.

The saturated Hui–Walter model was fit to the data in Table 12.2, and the estimates of the misclassification probabilities from this model appear in Table 12.3. The rows of the table correspond to a person's true status and the columns, the

Table 12.3 Hui–Walter Model Estimates of Response Probabilities[a]

	Observed Status		
True Status	EMP	UNEMP	NLF
EMP	99.6	0.4	0.0
	(0.1)	(0.1)	(n/a)
UNEMP	4.6	67.6	27.9
	(15.2)	(11.1)	(5.3)
NLF	2.6	0.0	97.4
	(1.5)	(n/a)	(1.1)

[a]Standard errors in parentheses.

observed status. For example, the diagonal entries are estimates of the probabilities of a correct response [i.e., $P(A = s|X = s)$ for status s]. These estimates are quite high for EMP and NLF, but only 67.6% (SE = 11.1%) for UNEMP. This suggests that a substantial number (32.4%) of unemployed persons are misclassified in the CPS. The table also suggests that a large proportion (27.9% with SE = 5.3%) of true UNEMP are misclassified as NLF.

Interpretation of the Results It is important to note that the estimates in Table 12.3 depend completely on the assumed classification error model. If the model is misspecified, the estimates will be biased, possibly substantially so. However, there are several features of the CPS labor force analysis that lend validity to the results in Table 12.3. Historical data on the reliability of the CPS data suggest that the classifications of respondents as EMP and NLF are highly reliable and the UNEMP is highly unreliable (see, e.g., Poterba and Summers, 1995). The data in Table 12.3 are consistent with these prior findings and suggest validity in the relative magnitudes of the error probabilities. The correspondence with the current estimates with external estimates of the same parameters from other studies is evidence of the external validity of the estimates.

The model results are also plausible in that knowledge of the labor force classification system would lead one to expect that UNEMP would be the least accurate classification. For example, determining whether a person is employed is not complicated in most cases or is the determination of NLF. The former requires determining whether a person is working or has an income, and for the latter, whether the person is retired, not working by choice, or not able to work: fairly unambiguous concepts. However, the concept of unemployment is often more difficult, involving vague criteria such as whether a nonworking person is looking for work or laid-off with the expectation of being recalled. Thus, the results in Table 12.3 appear to be consistent with theoretical expectations regarding the accuracy of the employment classifications, which is evidence of the theoretical validity of the estimates. Biemer and Bushery (2001) show that the Hui–Walter estimates from the CPS reinterview agree quite well with estimates obtained from traditional methods. Their results support the validity of the LCA estimates of CPS unemployment classification error.

12.4 METHODS FOR THREE MEASUREMENTS

As we have seen, with only two remeasurements, the ability of an analyst to test the model for specification error and to explore alternative error models is quite limited. In addition, an analyst's ability to investigate the root causes of any potential problems identified with further analysis is also limited. However, with three remeasurements, the number of degrees of freedom are increased, often substantially so, which opens many more options for analysis.

12.4.1 Three Measurements at One Point in Time

Extension of the methods discussed in Section 12.3 to three locally independent measurements is straightforward. Let A, B, and C denote the three measurements of the latent variable, X. As for the Hui–Walter model, the three measurements of X need not be parallel (i.e., the measurements can have distinct error distributions). However, unlike the case of two measurements, the likelihood associated with the three-measurement model is identifiable without the introduction of grouping variables. Estimation proceeds as before using maximum likelihood estimation.

Three measurements of the same survey variable that reference the same point in time are difficult to obtain in practice, particularly by using reinterviews methods. Reinterview methods risk problems such as respondent burden and resistance, the response conditioning effects by prior contacts, and high costs associated with repeat contacts with the respondent. Another method for obtaining three measurements of the same variable is to embed the measurements in a single survey instrument and collect all three in one interview. Since the measurements are collected during one interview, a respondent could simply repeat the first response for the other two responses without regard to their accuracy. In this situation, the possibility that the errors in the measurements are correlated (i.e., locally dependent errors) must be considered in the analysis. Altering the wording of the replicate items may help to conceal item redundancy from the respondent and thus avoid respondent resistance to the burden of answering the same questions repeatedly. This may reduce the risk of correlated errors due to memory effects, but it introduces the complexity of nonparallel measurements in the model.

Since local independence models with three measurements are saturated models, models that introduce additional terms for local dependence are not identifiable unless further restrictions are placed on the model (Hagenaars, 1988). For example, by imposing the restriction that the classification error distributions for

A, B, and C are identical, two degrees of freedom are saved (in the case of dichotomous measurements), which can be used to estimate the two additional parameters introduced by relaxing the independence assumption for two of the three indicators (e.g., for $\pi_{b|ax}$). However, this equal-error-probability restriction is not plausible and is likely to be violated if the methods (i.e., question wordings) for obtaining A, B, and C vary within the questionnaire.

As we described for the Hui–Walter methods, another technique for increasing the model degrees of freedom is to introduce a grouping variable G having L levels. Now, the number of cells of the GABC table is L times the number of cells in the ABC table. Equating some model parameters across the L groups to free-up enough degrees of freedom for estimating the correlated error parameters often results in more plausible assumptions for the model than are possible without the grouping variable.

12.4.2 Illustration of Three Embedded Measurements

Biemer and Wiesen (2002) consider the case of three measurements obtained in a single interview in an application to the National Household Survey on Drug Abuse (NHSDA). The NHSDA is a multistage household survey designed to measure the U.S. population's current and previous drug use activities. Before 1999, the NHSDA was primarily a self-administered interview using a paper-and-pencil questionnaire. A number of drug use questions are repeated in the questionnaire since research has shown that some respondents who indicate that they never used the drug, when asked directly will later answer an indirect question about the drug in a way that implies use of the drug. The multiple measurements of drug use can therefore be used to improve the accuracy of drug use prevalence estimates. This redundancy in the questionnaire provided the basis for constructing three remeasurements of past-year marijuana use for an LCA evaluation of each question.

Biemer and Wiesen define three indicators of past-year marijuana use, referred to as A, B, and C, in terms of the questions asked at various points during the interview. Indicator A is the response to the recency of use (or recency) question, which asks about the length of time since marijuana or hashish was last used. Indicator B is the response to the frequency of use (or frequency) question, which asks how frequently, if ever, the respondent has used marijuana or hashish in the past year. Indicator C is a composite of a number of questions on the drug answer sheet. An affirmative response to any one of these is coded as "yes" for C and otherwise C is coded as "no." Their research focused primarily on estimating the false positive and false negative probabilities separately for A, B, and C in order to determine the accuracy of each method. Three years of the NHSDA—1994, 1995, and 1996—where analyzed.

The degree of inconsistency among the three measures in apparent from the results in Table 12.4. The rates in this table are the unweighted disagreement rates for the comparisons listed in the rows of the table. The table suggests fairly substantial inconsistencies among the measurements, which vary by year of the survey. For example, the disagreement rate for the comparison A versus B is varies

Table 12.4 Observed Inconsistencies in the Three Indicators of Marijuana Use

	1994		1995		1996	
Indicator	n	Percent	n	Percent	n	Percent
A vs. B	241	1.35	263	1.48	293	1.61
A vs. C	854	4.80	380	2.14	452	2.48
B vs. C	883	4.96	409	2.31	491	2.69
A vs. B vs. C	989	5.55	526	2.96	618	3.39

between 1.35 and 1.61%, whereas the rates for C versus A and C versus B are considerably higher, particularly for 1994, where the disagreement rate approaches 5%. Since C is a composite of a number of questions, one plausible hypothesis for high rate of inconsistency with C is that C is the more accurate indicator (i.e., a gold standard), and disagreement with C is an indication of classification error in the other two measurements. In support of this argument, estimates of past-year marijuana use are highest for C, which could suggest greater accuracy since marijuana use tends to be underreported in the NHSDA (see, e.g., Mieczkowski, 1991; Turner et al., 1992a). As we will see, this hypothesis can be explored using a LCA of the three measurements A, B, and C.

Modeling the Error in Marijuana for Past-Year Use The models Biemer and Wiesen considered were limited to simple extensions of the basic latent class model for three measurements incorporating multiple grouping variables defined by age, race, and gender. The simplest model considered contained 54 parameters. That model allows the prevalence of past year marijuana use, π_x, to vary by age, race, and gender; however, the error probabilities $\pi_{a|x}$, $\pi_{b|x}$, and $\pi_{c|x}$ are constant across these grouping variables. The most complex model they considered contained 92 parameters and allowed error probabilities to vary by the three grouping variables. Since A, B, and C were collected in the same interview, the possibility of locally dependent errors was also considered in the analysis.

The best model identified in their analysis was a locally independent model containing 72 parameters that incorporated simple two-way interaction terms between each grouping variable and each indicator variable. The reader is referred to the original paper for the full specification and interpretation of this model.

Total Population-Level Estimates of Classification Error Estimates of the classification error rates for all three indicators of past-year drug use were derived from the best model. Table 12.5 shows the estimated classification error rates (expressed as percentages) for the total population for all three years, including a revised 1994 dataset, denoted by 1994′. This dataset is identical to the 1994 dataset for indicators A and B, but differs importantly for indicator C, as described below. Standard errors, which are provided in parentheses, assume simple random sampling and do not take into account the unequal probability cluster design of the NHSDA. Consequently, they may be understated.

Table 12.5 Comparison of Estimated Percent Classification Error by Indicator[a]

True Classification	Indicator of Past Year Use	1994	1994′	1995	1996
Yes (X = 1)	Recency = No (A = 2)	7.29 (0.75)	6.93 (0.72)	8.96 (0.80)	8.60 (0.79)
	Direct = No (B = 2)	1.17 (0.31)	1.18 (0.31)	0.90 (0.28)	1.39 (0.34)
	Composite = No (C = 2)	6.60 (0.70)	7.18 (0.72)	5.99 (0.67)	7.59 (0.74)
No (X = 2)	Recency = Yes (A = 1)	0.03 (0.02)	0.03 (0.02)	0.01 (0.01)	0.08 (0.02)
	Direct = Yes (B = 1)	0.73 (0.07)	0.76 (0.07)	0.78 (0.07)	0.84 (0.07)
	Composite = Yes (C = 1)	4.07 (0.15)	1.23 (0.09)	1.17 (0.08)	1.36 (0.09)

[a] Standard errors in parentheses.

A number of key points can be made from these results:

1. The false positive rates for all three indicators are very small across all three years except for indicator C in 1994, where it is 4.07%, more than four times that of the other two measurements.
2. The false negative error rates vary from 6.93% (1994′) to 8.96% (1995) for the recency indicator (A), from 5.99% (1995) to 7.59% (1996) for the composite indicator (C), and only 0.90% (1995) to 1.39% (1996) for the frequency indicator (B).
3. Across all four datasets the same general results hold: substantial false negative rates for A and C; low false negative rates for B; and low false positive rates for A, B, and C, except as noted in point 2.

The large false positive rate for C in 1994 suggest that the high inconsistency rate between C and the other two indicators is due to classification error in C and not classification error in the other two indicators, as hypothesized earlier. To investigate further, the questions comprising C for 1994 were compared with those in 1995 and 1996 to identify changes in the questionnaire that might explain this finding. This analysis revealed that prior to 1995, one of the questions used for constructing C seemed quite complicated and potentially confusing to many respondents. However, after 1995, this question was eliminated. Therefore, a plausible hypothesis for the high false positive rate for C in 1994 is presence of the complex question referred to in Biemer and Wiesen as question 7.

To test this hypothesis, a new indicator was created, denoted by C', by deleting question 7 from indicator C. The new indicator replaced C in a revised data set denoted by 1994′ and the model selection process was repeated for these data. The error parameter estimates for the best-fitting model to the 1994′ data are shown in Table 12.4 under the 1994′ column.

Note that the false positive rate for C using the revised 1994 data set dropped to 1.23% from 4.07%. Thus, the hypothesis that question 7 being eliminated is the cause of the high false positive rate for C in 1994 is supported. To verify the LCA result, the authors checked the consistency of C' with A and B for 1994 and found it to be similar to that for C versus A and B in 1995 and 1996.

In addition to the large false positive rate finding in point 1, the small false negative rate for B noted in point 2 was also quite unexpected. Why should the false negative rate for B be so much smaller than for A? The authors looked for an explanation in the statement of the survey questions for A and B. Indicator A is based on the question "How long has it been since you last used marijuana or hashish?" while indicator B asks "On how many days in the past 12 months did you use marijuana or hashish?" For A, respondents who use the drug on only a few days must admit to "using marijuana," which would classify them in a group ("marijuana users") they may think is inappropriate, since they used the drug so infrequently. However, for B, respondents can report frequency of use and, thus, some respondents who deny using marijuana in the past year for the recency question (A) may admit to using the drug on 1 or 2 days on the frequency question (B).

Thus, one may hypothesize that respondents who responded falsely to the recency question but answered the frequency question honestly are the infrequent users. To test this hypothesis, responses for the frequency question were cross-classified by the A classification. The hypothesis would be supported if a disproportionate number of respondents who were classified as "No use in the past 12 months" by indicator A and "Yes, use in the past 12 months" by indicator B are light users who responded "1 to 2 days" to the frequency question.

The results of this analysis are reported in Table 12.6 and are consistent with this theory. Among respondents answering "No past year use" for A and "Past year use" for B, 58.62% (weighted) answered in the "1 to 2 days" category for the frequency question. Compare this with only 15.66% in the "1 to 2 days" category among persons consistently classified as users by both indicators.

This analysis demonstrates the utility of LCA for identifying questionnaire problems. In addition, the causal analysis investigating findings 1 and 2 above provides evidence of the validity of the latent class estimates.

12.4.3 Models for Three Measurements at Different Time Points

Latent class models can also be applied when measurements are made at and refer to different time points, as in the case of a panel survey. The panel survey measurements alone can be used to obtain an identifiable model without the requirement of reinterviews or other remeasurements. The models resemble the LCMs described previously except that new parameters must be introduced into

Table 12.6 Distribution of Reported Days of Use in Past Year by Recency of Use in Past Year, 1994–1996 NHSDAs[a]

Number of Days Used Marijuana in Past Year from the Frequency Question	Percent Reporting No Past Year Use on the Recency Question	Percent Reporting Past Year Use on the Recency Question
More than 300 days	5.84	10.00
201 to 300 days	0.96	5.54
101 to 200 days	.93	9.06
51 to 100 days	1.45	10.01
25 to 50 days	2.96	10.50
12 to 24 days	4.76	11.95
6 to 11 days	6.06	11.76
3 to 5 days	18.41	15.51
1 to 2 days	58.62	15.66
Total	100.00	100.00

[a]Based on 53,715 responses.

the models to allow the true value associated with a person to vary across time points. Such models, referred to in the literature as *Markov latent class models* (MLCMs), were first proposed by Wiggins (1973).

The MLCM can be extended in number of ways. Like the models considered in Section 12.4, grouping variables can be included in the models to improve model fit and identifiability or for inferential purposes. Further, models with four waves or more can be modeled simultaneously, which, as mentioned previously, can allow less restrictive models to be identified. Finally, the models can be extended to general polychotomous latent and indicator variables.

One common application of MLCM is for modeling the classification error in labor force panel data. For example, Van de Pol and Langeheine (1997) applied these models to the Netherlands Labor Market Survey, Vermunt (1996) to the SIPP labor force series, and Biemer and Bushery (2001) to the CPS. The latter paper evaluated a number of the assumptions of MLCM for the CPS, including the Markov assumption, and provided evidence of the empirical, theoretical, and external validity of the MLCM estimates of CPS classification error. The authors then proceeded to evaluate the accuracy of the labor force for the revised CPS questionnaire that was introduced in 1994 and compared it with the accuracy of the original questionnaire that had been in use prior to 1994.

In their analysis, Biemer and Bushery found an anomaly in the data that had not been detected in previous research on the CPS redesign. Using MLCMs, they estimated that accuracy for unemployment classification (UNEMP) dropped by about 6.5 points with the introduction of the new design in 1994 (see Table 12.7). The authors speculated that this decline in classification accuracy might indicate a problem with the revised unemployment questions. That is, the revised unemployment questions may be subject to greater measurement error, which manifests

Table 12.7 Comparison of CPS Labor Force Response Probabilities for the Original and Revised Questionnaires

True Class	Observed Class	Original (1992–1993)	Revised (1994–1995)	Original—Revised Diff.	Original—Revised SE
EMP	EMP	98.68	98.84	−0.15	0.40
	UNEMP	0.42	0.39	0.03	0.40
	NLF	0.90	0.78	0.13	0.16
UNEMP	EMP	8.23	10.57	−2.34[a]	0.45
	UNEMP	79.06	73.50	5.56[a]	0.54
	NLF	12.71	15.93	−3.32[a]	0.26
NLF	EMP	2.14	1.99	0.15	0.36
	UNEMP	1.43	1.56	−0.13	0.33
	NLF	96.43	96.45	−0.02	0.18

[a]Significant at $\alpha = 0.001$.

itself in the MLCM analysis as less accuracy in the classification of unemployed persons.

Subsequently, Biemer (in press) applied MLCM to explore the anomaly further. His analysis considered the error associated with the two primary subclassifications of unemployed: persons who are unemployed and on layoff (LAYOFF) and persons who are unemployed and looking for work (LOOKING). The analysis found that the primary cause of the anomaly is a reduction in classification accuracy if persons who are on layoff. Again using MLCA, he found that two questions on the revised questionnaire appear to be responsible. First, his analysis found considerable error associated with the revised global question "*Last week*, did you do *any* work (either) for pay (or profit)?" More that half of the error in the revised LAYOFF classification is contributed by this question. In addition, there appears to be considerable classification error in determining whether people reporting some type of layoff have a date or indication of a date to return to work. This question contributes between 30 and 40% of the LAYOFF classification error. The combination of these two questions appears to explain the reduction in accuracy in UNEMP in the revised questionnaire.

12.5 CONCLUSIONS

Statistical analysis of survey data can identify questions that contribute substantial amounts of measurement error to the survey estimates. Poorly worded or executed questions can contribute to both the bias and variance of survey estimates. One method that has been used traditionally for assessing measurement variance and item reliability is the test–retest reinterview. This method assumes that the interview and reinterview measurements are parallel (i.e., their measurement errors are independent and identically distributed). Reliability is typically estimated by Cohen's κ or, equivalently, $1 - I$, where I is the index of inconsistency. If the parallel assumption holds, the expected value of κ is approximately equal to reliability defined as the ratio of true score variance to total variance.

For estimating measurement bias, one method that is traditionally used is to obtain measurements that are essentially free of measurement error—called *gold standard measurements*. Such measurements may obtain from reconciled reinterviews (see Forsman and Schreiner, 1991), record check studies, biological tests, or any other process where the assumption of error-free measurement is tenable. Measurement bias can then be estimated by the net difference rate (i.e., the difference between the means of the survey observations and gold standard measurements).

One disadvantage of both of these traditional methods is that the assumptions underlying them often do not hold in practical survey situations. Reinterview measurements may not be parallel due to the conditioning effects of the original interview. For example, respondents may remember their previous responses and simply repeat them, inducing correlated error between the measurements. Further, having been asked the question before, respondents may implement a very different response process in the reinterview, violating the assumption of identically distributed errors. [See, e.g., O'Muircheartaigh (1991) for evidence of this for the CPS.]

For estimating bias, gold standard measurements are difficult, if not impossible, to obtain for most survey items. A number of articles show that reconciled reinterview data can be as erroneous as the original measurements they are intended to evaluate (see, e.g., Biemer and Forsman, 1992; Biemer et al., 2001; Sinclair and Gastwirth, 1996) and that administrative records data are often inaccurate and difficult to use (Jay et al., 1994; Marquis, 1978) as a result of differences in time reference periods and operational definitions, as well as errors in the records themselves.

Because of their restrictive assumptions, these traditional methods require special remeasurement studies, such as reinterview surveys or record check studies. Since these data are quite costly to obtain (Biemer, 1988), sample sizes for the special studies are usually quite small and may therefore be inadequate to explore the causes of the inconsistencies or biases in the data. Subsetting the data by geographic area or by characteristics of the interviewers or respondents is often not fruitful, due to the high sampling variances associated with the estimates.

LCA has an advantage over traditional methods in that it can be used when the assumptions associated with traditional analysis fail or when the remeasurements were collected by methods that were not intended to satisfy the traditional assumptions. LCA can be viewed as a generalization of traditional modeling approaches in the sense that when the traditional assumptions hold, LCA and traditional analysis produce essentially the same results. As an example, Guggenmoos-Holzmann and Vonk (1998) develop a general LCM for estimating reliability with two measurements. They show that under certain parameter constraints, the model estimate of reliability is equal to κ. However, when these restrictions are relaxed, a more general measure of reliability is obtained.

For estimating bias when gold standard measurements are available, LCA provides a check on the validity of the gold standard assumption. For example, the latent class model in Section 12.3.2 can be fit with and without the constraint $\pi_{B|X} = 0$ (i.e., no classification error for the second measurement). If there is no improvement in model fit when the constraint is removed, the gold standard assumption is supported and traditional analysis is appropriate. However, if there is a significant improvement in model fit, the hypothesis that the second measurement is error free must be rejected. Then the data can be analyzed using LCA without the constraint.

The use of LCA for questionnaire evaluations should not preclude the use of traditional methods or other methods of analysis. However, when the assumptions associated with traditional analysis do not hold, LCA may be the only method for assessing error in the original measurements. For example, we have shown in Sections 12.3 and 12.4 that even if the second measurement was not intended to be either a replicate or a gold standard measurement, LCA methods could be used to estimate reliability and bias.

Moreover, estimates of the response probabilities themselves, which drive both reliability and bias, provide useful insights as to the nature of the measurement error. For example, a high false negative probability and near-zero false positive probability may suggest a social desirability bias (i.e., a tendency for respondents to deny engaging in behaviors that are socially unacceptable). Conversely, low false negative and high false positive probabilities may suggest a tendency to overreport desirable behaviors. Response probabilities may also suggest the presence of an interviewer bias. For example, Biemer et al. (2001), in an LCA of census enumeration error, found evidence that census enumerators tend to err on the side of counting persons as census day residents in cases where the information regarding residency is ambiguous. This contributed to the finding that the U.S. Census 2000 overstated the total population size by 3 to 4 million persons, due to the counting of nonresidents (U.S. Bureau of the Census, 2001a).

LCA methods are particularly useful for exploring the error in panel survey measurements. Using MLCA, it is possible to estimate reliability, bias, and response probabilities using only the panel data, without the need for special remeasurement studies. Biemer and Bushery (2001) obtained estimates of reliability using the MLCA response probability estimates and the formulation of R in terms of these probabilities given in Biemer and Stokes (1991). They found very close agreement for the MLCA estimates and traditional estimates based on test–retest reinterviews. However, the standard errors of the MLCA estimates were considerably lower since unlike traditional methods, their MLC analysis was not restricted to a small subsample of cases selected for reinterview.

As shown in Section 12.4, MLCA has been applied successfully to the NHSDA and CPS for estimating measurement bias. MLCA can also be used as an exploratory tool to generate hypotheses regarding the nature of measurement error in a data collection process as a form of *data mining* (see, e.g., McLachlan and

Peel, 2000). These hypotheses can then be further investigated and tested using the other methods of questionnaire evaluation as cognitive methods or observation studies. As an example, Biemer and Tucker (2001) applied MLCA to the U.S. Consumer Expenditure Survey to examine the correlates of underreporting for a number of consumer items, such as vehicle expenses, kitchen supplies, clothing, and furniture. Their MLCA results suggest very high levels of underreporting error for some items. Their analysis provided evidence of greater reporting accuracy for longer interviews incorporating consistent use of records and lower accuracy for shorter interviews and less record use.

An important disadvantage of LCA methods is the greater reliance on an assumed model than in traditional methods. Because of the flexibility of LCA approaches, a wide range of models can be fit, some having little or no theoretical foundation other than that the model seems plausible, appears to fit the data well, and the estimates of measurement error seem reasonable.

By contrast, for traditional methods, the design of the data collection for the remeasurements can be controlled (with varying degrees of success) so that the measurement errors theoretically satisfy the traditional model assumptions. As an example, for test–retest reinterview, the data collection can be designed specifically to satisfy the parallel measurement assumptions. The reinterview may be timed so that respondents will be unlikely to recall their original interview responses, a design feature that addresses the local independence assumption. To satisfy the identically distributed errors assumption, the reinterview design may further specify that reinterviewers, the reinterview questionnaire, and the general survey conditions should be as close as possible to that of the original interview. Similarly, for gold standard approaches, the remeasurement process may be designed to obtain the best measurement possible using a highly accurate but costly approach. Thus, by designing remeasurement processes to satisfy the assumptions of traditional approaches, these methods usually have a much stronger theoretical foundation and are considered to be more credible than methods, such as LCA, which may not have a theoretical foundation that is tied to data collection.

Further, it is usually not possible to determine whether the estimates of classification error obtained from latent class methods are valid. Biemer and Bushery (2001) provides an example where the validity of latent class estimates could be established to some extent by external validation; however, situations where the data to support validity analysis are available are extremely rare. A consequence of this inattention to model validity is that many statisticians have little or no confidence in LCA or other latent variable methods and have considered them for survey evaluation purposes. However, the LCA assumptions are quite similar to those made for finite mixture modeling and missing data imputation modeling, and these methods have received much wider acceptance among statisticians.

Nevertheless, for purposes of identifying flawed questions, the use of LCA should be much less controversial since model validity is of secondary importance. The primary issue for questionnaire evaluation work is whether or not a statistical method is successful at identifying questions that are truly flawed. For example, in the NSHDA application (see Section 12.4.2), the assumptions of the embedded measurements model may not be well satisfied for the NHSDA population. Nevertheless, the method successfully identified two problems with the NHSDA questionnaire that would have been difficult and/or costly to detect using other methods.

In the CPS application, a problem in the measurement of persons on layoff was discovered using MLCA that could contribute importantly to the accuracy of unemployment measurement in the United States. Determining whether the error estimated by MLCA is real will require further testing, perhaps using cognitive interviews to investigate the comprehension, recall, or social desirability issues with these questions. If these investigations uncover problems in questions, the utility of the MLCA approach for identifying real problems is supported even though the validity of the MLCA modeling assumptions may never be known.

Finally, experience has shown that it is unwise to rely on a single method for evaluating survey questions. A more prudent approach is to use multiple analysis methods for identifying flawed questions, since methods will differ in their sensitivity and specificity to various types of problems. We advocate the use of traditional and latent class statistical approaches whenever possible. Agreement of the results from multiple methods should engender confidence that the findings of the analysis are valid. Disagreement among the results may lead to further investigation of the underlying assumptions of all the methods. In this way, much more knowledge can be discovered about the underlying causes of the errors than if only one method were used.

CHAPTER 13

Item Response Theory Modeling for Questionnaire Evaluation

Bryce B. Reeve and Louise C. Mâsse
National Cancer Institute

13.1 INTRODUCTION

Each year, new questionnaires are developed or revised from previous measures in the hope of obtaining instruments that are reliable, valid within a study population, sensitive to change in a person's status, and that provide interpretable scores that accurately characterize a respondent's standing on a measured construct. This increasing need for psychometrically sound measures calls for better analytical tools beyond those provided by traditional measurement theory [or classical test theory (CTT)] methods. Applications of item response theory (IRT) modeling have increased considerably in educational, psychological, and health outcomes measurement because it (1) provides more in-depth analysis of items included in questionnaires, (2) facilitates the development of more efficient questionnaires by reducing the number of items to be included in a scale, and (3) allows instrument developers to better handle complex measurement problems such as linking test scores from various instruments or developing computerized adaptive tests. Most existing questionnaires are based on CTT principles and do not take advantage of IRT methodologies.

In this chapter we provide a basic introduction to IRT modeling, including a discussion of the common IRT models used in research, underlying assumptions of these models, and differences between CTT and IRT modeling. The introduction is followed by a demonstration of the information that can be gained by using IRT to evaluate the psychometric properties of a questionnaire. The SF-36

Methods for Testing and Evaluating Survey Questionnaires, Edited by Stanley Presser, Jennifer M. Rothgeb, Mick P. Couper, Judith T. Lessler, Elizabeth Martin, Jean Martin, and Eleanor Singer
ISBN 0-471-45841-4

Mental Health summary scale (Medical Outcomes Trust, 1991) was selected as the model for applying and comparing CTT and IRT methodologies.

13.2 CLASSICAL TEST THEORY

Currently, most psychometric evaluations of instruments have been grounded in CTT. CTT provides the methodology to derive the reliability coefficient, assess individual item properties, compute scale scores, and estimate measurement errors. To compute the reliability coefficient and measurement errors, CTT relies on the basic tenets that are represented by the following equation:

$$x = t + e \tag{13.1}$$

Equation (13.1) assumes that a person's observed score (x) on an individual assessment is comprised of two components: a true score (t) and the measurement errors (e). A person's true score equals the average score that one would achieve by taking an indefinite number of parallel assessments. The inability of a scale to measure a person's true score precisely is represented by the measurement errors. The measurement errors are assumed to be distributed normally and uncorrelated with the true score.

Traditionally, the reliability coefficient is used to determine how well a set of items (comprising a scale) measures the underlying construct (Lord and Novick, 1968; Nunnally and Bernstein, 1994). In CTT, the reliability coefficient assesses the strength of the relationship between the observed score and the true score. The stronger the relationship, the better the observed score reflects the true score, and the smaller the measurement errors (Suen, 1990). The most commonly used measure of scale reliability is Cronbach's α, which is a measure of internal consistency (Chronbach, 1951).

The standard error of measurement (SEM) provides another means to assess how precisely a test measures a respondent's true score. The SEM describes an expected observed score fluctuation due to error in the measurement tool (Embretson and Reise, 2000). In other words, over repeated assessments, the SEM is the standard deviation of a person's error distribution around the person's true score (the average score over repeated assessments). The SEM is linked to the reliability coefficient in that a decrease in the SEM will result in an increase in the reliability coefficient.

Finally, to evaluate the functioning of individual items, CTT typically measures the relationship between an item score and the total scale score, referred to as the *corrected item–total correlation* (Nunnally and Bernstein, 1994). CTT assumes that each item in a scale measures the same underlying construct; therefore, the item is assumed to be correlated with the total score, which is the sum of the responses on the scale. The corrected item–total correlation is the correlation between the item and the total score minus the item evaluated. This index typically has been used to assess which items discriminate and which items should be included in the scale.

13.3 ITEM RESPONSE THEORY

IRT makes the same assumptions about the observed scores and true scores. To derive the indices used to assess item and scale functioning (i.e., reliability and SEM) and to estimate scores, however, IRT does not rely on first and second moments (means, variances, and covariances) that are sensitive to the sample characteristics. Instead, IRT uses higher-order moments (i.e., threshold and slope parameters explained below), which are much more stable and are less affected by the characteristics of the particular people who responded to the items. For example, a CTT evaluation of a scale measuring mental health in a depressed population versus a normal population may result in different conclusions about the scale, because the means and variances of the items will be affected by the distribution of the sample. Under IRT, item and scale properties are unaffected by the choice of respondents used to evaluate the scale, which results in invariant psychometric properties.

13.3.1 IRT Model Basics

IRT models the relationship between a person's response to an item and the underlying latent variable (referred to as *construct*, *trait*, *domain*, or *ability*) being measured by the questionnaire. The underlying latent variable, expressed mathematically by the Greek letter theta (θ), may be any measurable construct, such as mental health, depression, fatigue, or physical functioning. A person's standing (or level) on this continuous construct is assumed to be the only (unidimensional) factor that accounts for a person's response to each item in a scale. For example, a person with high levels of depression will have a high probability for indicating that they "felt downhearted and blue" "most of the time." A person with little to no depression will probably respond "little of the time" or "none of the time."

IRT models the response to each item in a scale by assigning each item a set of properties that describe its performance in the scale. The relationship between a person's response to an item and the latent variable is expressed by the IRT model item characteristic curves [ICCs, also referred to as category response curves (CRCs) or item trace lines]. Figure 13.1 presents the IRT ICCs for the following item of the SF-36 Mental Health scale: "As a result of any emotional problems, have you accomplished less than you would like?" Mental health (θ), the latent variable measured by the 14-item scale, is represented along the horizontal (x) axis in Figure 13.1. People vary in their level of mental health: from people with very poor mental health, located on the left side of the continuum, to people with good mental health, located on the right side of the axis. Numbers on the θ-axis are expressed in standardized units, and for the illustrations in this chapter, the mean mental health level of the study population is set at zero and the standard deviation is set to 1. Thus, a person located at $\theta = -2.0$ indicates a mental health score that is two standard deviations below the population mean. The vertical (y) axis in Figure 13.1 indicates the probability, bounded between 0 and 1, that a person will select one of the item's response categories. Thus, the

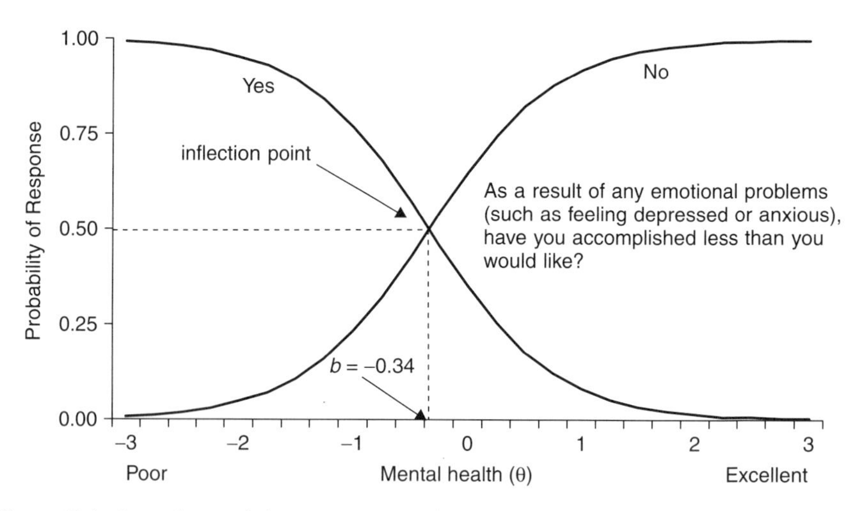

Figure 13.1 Item characteristic curves representing the probability of a "yes" or "no" response to the item "As a result of any emotional problems (such as feeling depressed or anxious), have you accomplished less than you would like?" conditional on a person's level of mental health. The threshold ($b = -0.34$) indicates the level of mental health (θ) needed for a person to have a 50% probability for responding "yes" or "no.".

two ICCs in Figure 13.1 indicate that the probability of responding "yes" or "no" to the item that asks "Have you accomplished less than you would like?" depends on the respondent's level of mental health. Those whose mental health functioning is poor have a high probability of selecting "yes," whereas those whose mental health functioning is high are more likely to select "no" for this item.

The ICCs in Figure 13.1 are represented by logistic curves that model the probability P that

$$\mathrm{P}(X_i = 1|\theta, a_i, b_i) = \frac{1}{1 + e^{a_i(\theta - b_i)}} \tag{13.2}$$

a person will respond "no" (for the monotonically increasing curve) to this item are a function of a respondent's level on the mental health construct (θ), the relationship (a) of the item to the measured construct, and the severity or threshold (b) of the item in the scale. In IRT, a and b are referred to as *item parameters*. When an item has just two response categories, the equation for the monotonically decreasing curve is just a linear transformation of 1 minus the expression on the right side of equation (13.2).

The item threshold or severity level (b) is the point on the latent scale θ where a person has a 50% chance of responding "no" to the item. In Figure 13.1 the item's threshold value is $b = -0.34$, which indicates that people with mental health levels near the population mean have a 50% chance of indicating "no" or "yes" to the question posed by the item. Note that the threshold parameter varies for each item, and it is possible to compare threshold parameters across items to determine items that probably will be endorsed by those with low or high mental health functioning. For example, items with low threshold parameters will have a higher probability of being endorsed by those with low mental health functioning, and vice versa.

The discrimination or slope parameter a in equation (13.2) describes the strength of an item's ability to differentiate among people at different levels along the trait continuum. An item optimally discriminates among respondents who are near the item's threshold b. The discrimination parameter typically ranges between 0.5 and 2.5 in value. In Figure 13.1 the slope at the inflection point (i.e., the point at which the slope changes from continuously increasing to continuously decreasing) is $a = 1.82$. The larger the a parameter, the steeper the ICC is at the inflection point. In turn, steeper slopes indicate that the ICC increases relatively rapidly, such that small changes on the latent variable (e.g., small changes in mental health status) lead to large changes in item-endorsement probabilities (Reise, in press). The a parameter also may be interpreted as describing the relationship between the item and the trait being measured by the scale and is directly related, under the assumption of a normal θ distribution, to the biserial item-test correlation (Van der Linden and Hambleton, 1997).[1] Items with larger slope parameters indicate stronger relationships with the latent construct and contribute more to determining a person's score ($\hat{\theta}$). For example, in the mental health scale, the item that asked, "Did you feel tired?" had a higher discrimination parameter ($a = 2.19$) and item–total correlation ($r = 0.68$) than the item that asked, "Have you been a very nervous person?" ($a = 1.05$, $r = 0.48$, respectively). Thus, the slope of the ICC for the "tired" item will be steeper than the slope of the ICC for the "nervous" item.

IRT also can easily model the responses to items that have more than two response options. The IRT CRCs (also referred to earlier as ICCs for dichotomous response options) for the item that asked, "Did you have a lot of energy?" are presented in Figure 13.2. This item has six response categories, so the IRT model chosen for this data analysis (i.e., Samejima's graded response model, 1969) estimates one discrimination parameter to indicate the relationship of the item with the scale and five threshold parameters (number of response options minus 1) to determine where response curves intersect along the θ scale. In Figure 13.2, people with very poor mental health (i.e., $\theta < -1.48$ = first threshold parameter) have a high probability of answering that they have a lot of energy "none of the time." The IRT model predicts that a person at $\theta = -1.25$ (indicated by a vertical dashed line in Figure 13.2) has a 34% probability of endorsing "none of the time," 48% probability of saying "little of the time," 17% probability of endorsing "some of the time," 2% chance for saying "a good bit of the time," and less than a 1% probability of answering "most of the time" or "all the time." Moving right along the θ-axis, people with better mental health will endorse response categories that are associated with better health.

[1]The *biserial correlation* is an estimate of the correlation between a dichotomous variable and a continuous variable (such as the scale score), in which the threshold of the dichotomous variable is thought to be a point along a normal distribution.

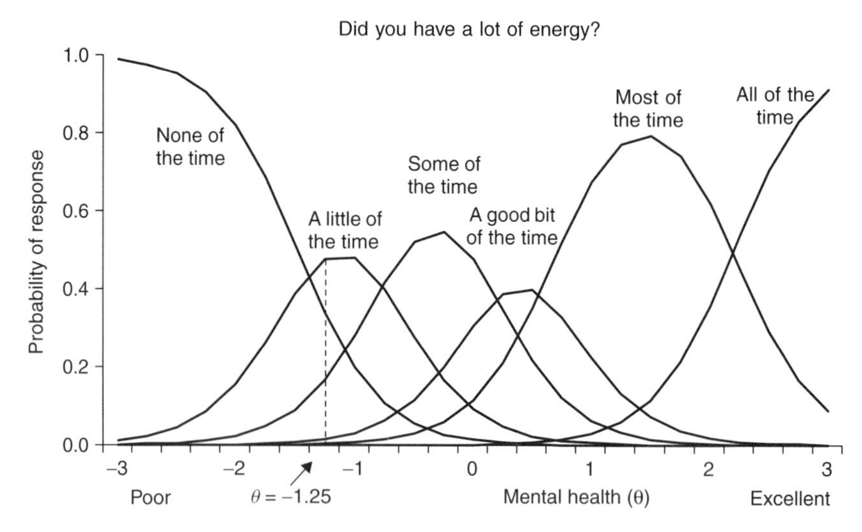

Figure 13.2 Category response curves representing the probability for selecting one of the six response options for the item "Did you have a lot of energy?" conditional on a person's mental health level.

Another important feature of IRT models is the information function, an index that indicates the range over θ (the construct being measured) for which an item or scale is most useful for discriminating among people. In other words, the information function characterizes the precision of measurement (i.e., reliability) for measuring persons at different levels of the underlying latent construct, with higher information denoting more precision. Graphs of the information function place the respondent's trait level (θ) on the horizontal x-axis and information magnitude on the vertical y-axis. Figure 13.3 presents the item information functions that are associated with the two items presented, in Figures 13.1 and 13.2. The shape of the item information function is determined by the item parameters (i.e., a and b parameters). The higher the item's discrimination (a parameter), the more peaked the information function will be; thus, higher discrimination parameters provide more information about individuals whose trait levels (θ) lie near the item's threshold value. The item's threshold parameter(s) (b parameter) determines where the item information function is located (Flannery et al., 1995). In Figure 13.3 the range of the information function (or curve) for the "energy" item is more broad than that for the "accomplished less" item because more response categories (six versus two, respectively) were used, and thus the item can cover a broader range along the mental health continuum. Also, the item information function for "energy" is more peaked than that for the "accomplished less" item, which indicates that the "energy" item contributes more to the measurement of mental health.

Given "local independence" (one of the key IRT model assumptions, discussed later), the individual item information functions can be summed across all the

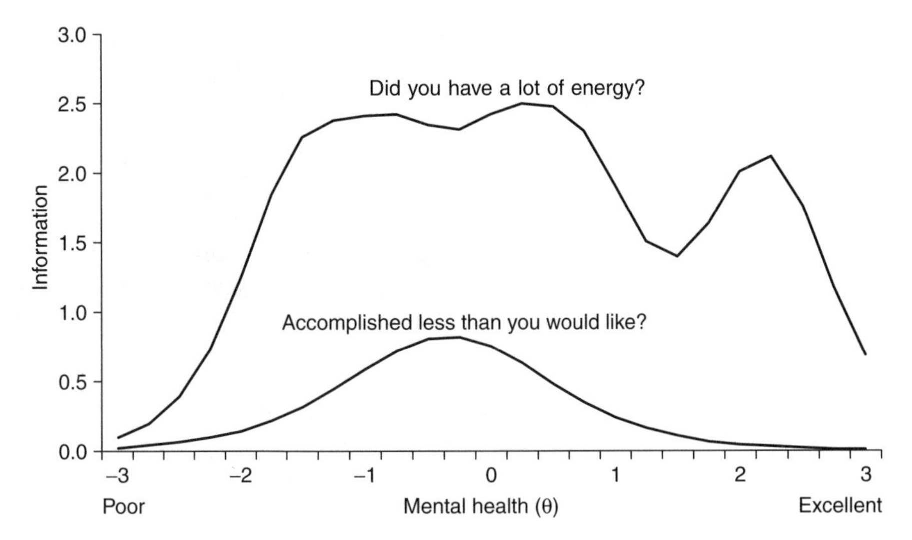

Figure 13.3 Item information functions for the two items "Did you have a lot of energy?" and "As a result of any emotional problems (such as feeling depressed or anxious), have you accomplished less than you would like?".

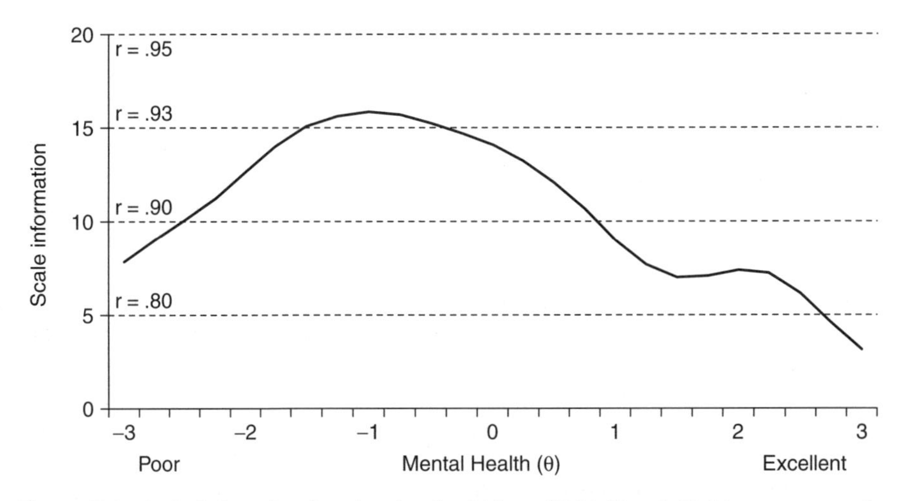

Figure 13.4 Scale information function for the 14-item SF-36 Mental Health summary scale. The horizontal dashed lines indicate the level of reliability (r) associated with different levels of information.

items in the scale to form the scale information function (Lord, 1980). The scale information function for the 14 items making up the SF-36 Mental Health summary scale is presented in Figure 13.4. Along with the information magnitude indicated along the vertical axis in the graph, the associated reliability (r) is provided. Overall, the scale is reliable for measuring a person's mental health across the trait continuum (the curve stays above reliability $r = 0.80$). The function is peaked on the lower end of the scale, which indicates that it measures people with poorer mental health with greater precision. The reliability of the scale decreases for people with excellent mental health.

The information curve serves as a useful tool for instrument developers by allowing them to tailor their instrument to provide high information (i.e., reliability) for measuring their study population. If a developer wants high precision to measure a person at any level of mental health, the information function in Figure 13.4 suggests adding more questions to the scale (or more response options to existing questions) that differentiate among people with good to excellent mental health. Adding appropriate questions or response options will increase the information curve and reflect better measurement precision. Similarly, the developer may wish to develop a diagnostic instrument that determines whether a person needs clinical help. In this case, the information curve should be high in the lower end of the mental health continuum and low in the upper end of the continuum (i.e., it is not important to get an accurate measure of a person who is mentally healthy).

The SEM for each score of the underlying construct can be calculated easily by taking the inverse of the square root of the information function. For example, the SEM for a person at $\theta = 1.5$ (i.e., someone 1.5 standard deviations above the mean population mental health score) is approximately 0.38 (information = 7, reliability = 0.86).

13.3.2 Common IRT Models

IRT models come in many varieties and can handle unidimensional as well as multidimensional data, binary and polytomous response data, and ordered as well as unordered response data (van der Linden and Hambleton, 1997). The IRT models that are used most frequently are presented in Table 13.1. Two of these models can be used with dichotomous items (e.g., yes/no or agree/disagree); whereas the remainder are polytomous models that can be used with items having more than two response categories. All of the models in Table 13.1 are unidimensional models (i.e., designed to measure a single construct). Multidimensional IRT models exist (Wilson, in press) as well as nonparametric IRT models (Ramsay, 1997) but these are not presented in this chapter. For a more complete discussion of these models, see Embretson and Reise (2000), Thissen and Wainer (2001), and van der Linden and Hambleton (1997).

A family of models known as *Rasch* (1960/1980) *models* is identified with an asterisk in Table 13.1. The key difference between the Rasch and non-Rasch models is that the non-Rasch models (also called *two-parameter models*) estimate a discrimination (slope) parameter for each item, suggesting that items are differentially weighted with regard to the underlying construct, whereas Rasch models constrain the discrimination ability to be equal across all items. When data fit a Rasch model, several measurement advantages are gained. The Rasch models' property of *specific objectivity* allows comparison of any two items' threshold

Table 13.1 Commonly Used Item Response Theory (IRT) Models

IRT Model	Item Response Format	Model Characteristics
Rasch model[a]/one-parameter logistic model	Dichotomous	Discrimination power equal across all items. Threshold varies across items.
Two-parameter logistic model	Dichotomous	Discrimination and threshold parameters vary across items.
Graded model	Polytomous	Ordered responses. Discrimination varies across items.
Nominal model	Polytomous	No prespecified item response order. Discrimination varies across items.
Partial credit model[a]	Polytomous	Discrimination power constrained to be equal across items.
Rating scale model[a]	Polytomous	Discrimination equal across items. Item threshold steps equal across items.
Generalized partial credit model	Polytomous	Generalization of the partial credit model that allows discrimination to vary across items.

[a]Models belonging to the family of Rasch models.

parameters independent of the group of subjects being surveyed, and any two persons' trait levels (i.e., scores) can be compared without information on the particular subset of items being administered (Mellenbergh, 1994). The advantages of the Rasch models (i.e., specific objectivity property and ability to handle small sample sizes) (Bond and Fox, 2001), however, are limited by the constraint that the data fit the model, which rarely is observed in latent variable modeling.

13.3.3 IRT Model Assumptions

The parametric, unidimensional IRT models described above make three key assumptions about the data: (1) unidimensionality, (2) local independence, and (3) that the IRT model fits the data. It is important that these assumptions be evaluated before IRT model results are interpreted. Although no real data ever meet the assumptions perfectly, IRT models are robust to minor violations of the assumptions (Cooke and Michie, 1997; Drasgow and Parsons, 1983; Duncan, 1984; Reckase, 1979; Reise, in press).

The unidimensionality assumption posits that the set of items measure a single continuous latent construct (θ) ranging from $-\infty$ to $+\infty$. In other words, a person's level on this single construct gives rise to a person's response to the items in a scale. This assumption does not preclude that the set of items may have a number of minor dimensions (subscales) but does assume that one dominant dimension explains the underlying structure. Scale unidimensionality can be evaluated by performing an item-level factor analysis (Panter et al., 1997) that is designed to evaluate the factor structure that underlies the observed covariation among item responses. If multidimensionality exists, the investigator may want to consider dividing the scale into subscales (based on both theory and the factor analysis) or using multidimensional IRT models.

The assumption of local independence means that the only systematic relationship among the items is explained by the conditional relationship with the latent construct (θ). In other words, if the trait level is held constant, there should be no association among the item responses (Thissen and Steinberg, 1988). Violation of this assumption may result in parameter estimates that differ from what they would be if the data were locally independent; thus, selecting items for scale construction based on these estimates may lead to erroneous decisions (Chen and Thissen, 1997; Wainer, 1996). No commercial software exists for assessing local independence, but researchers can look for redundant items with high inter-item correlations that measure constructs tangential to the domain of interest. The impact of local dependence can be measured by observing how the item parameters and person scores change when redundant items are dropped.

Finally, IRT assesses model fit at both the item and person level. The goal of such evaluations is to determine whether the item and person parameters estimated with IRT can be used to reproduce the observed item responses (Reise, in press). Graphical and empirical approaches can serve to evaluate item fit (Hambleton et al., 2000; Kingston and Dorans, 1985; McKinley and Mills, 1985; Orlando and Thissen, 2000; Rogers and Hattie, 1987) and person fit (Meijer and Sijtsma, 1995; Reise and Waller, 1993; Zickar and Drasgow, 1996). Currently, no standard set of fit indexes is recognized as useful across all research settings. Often, a number of different fit indices are reported from each IRT (or Rasch) software program. Since IRT models are probabilistic models, most fit indices measure deviations between the predicted and observed response-pattern frequencies. For example, the IRT software program MULTILOG (Thissen, 1991) provides for each item the observed and predicted proportion of responses to each category. A likelihood-ratio statistic can be computed to determine significant deviations between the model and observed data. In addition, there are many types of residual analyses for model fit that are referenced in the citations above that can be used to further evaluate model fit.

13.4 COMPARING IRT WITH CTT

CTT views measurement as the determination of the quantity of an attribute that is present in a person. A person's score is a linear combination (i.e., sum) of responses to a set of items (that are sampled from a universe of items measuring a common trait). Thus, the number of items endorsed by a respondent, but not the "type" of questions posed by these items (in terms of difficulty and discrimination ability), determines a respondent's score. On the other hand, IRT is based on the pattern of item responses and uses item properties to estimate a person's

Table 13.2 Contrasting features of Classical Test Theory (CTT) and Item Response Theory (IRT)

	CTT	IRT
Measurement error/precision	Measures of precision fixed for all scores	Precision measures vary across scores
Scale length	Longer scales increase reliability	Shorter, targeted scales can be equally reliable
Invariance of item parameters	Scale properties are sample dependent	Scale properties invariant to sample
Mixed-item format	Mixed-item format leads to unbalanced impact on total test scores	Easily handles mixed-item formats.
Invariance of scale scores	Comparing respondents requires parallel scales	Different scales can be placed on a common metric

score (Nunnally and Bernstein, 1994). Two people could have the same CTT score but different IRT scores because they endorsed different items. Works by Embretson and Reise (2000) and Reise (1999) identify several advantages of the IRT approach. These advantages are summarized in Table 13.2 and described briefly below.

13.4.1 Measurement Error and Reliability

In CTT, both the SEM and reliability coefficient (such as Cronbach's α, internal consistency) are fixed for all scale scores. In other words, CTT models assume that measurement error is the same for all possible scores. In IRT, measures of precision are estimated separately for each score or response pattern, thus controlling for the characteristics of the items in the scale. Measurement precision typically is best (low SEM, high reliability and information) in the middle of the scale-score continuum; and precision typically is least at the tails of the continuum, where items do not discriminate well among respondents. Refer to the IRT scale information curve in Figure 13.4 for an example.

13.4.2 Scale Length

In CTT, scale reliability is a function of the number of items in the scale. Higher reliability requires longer scales. Often, redundant or similar items are included in such instruments. In IRT, shorter and equally reliable scales can be developed with appropriate item placement. Redundant items are discouraged and may violate the assumption of local independence of the IRT model.

13.4.3 Invariance of Item Parameters

CTT scale statistics, such as the reliability coefficient, item–total correlation, SEM, and threshold parameter, are sample dependent. This means that these indices vary across samples, especially for nonrepresentative samples. In IRT, item parameters are assumed to be sample invariant within a linear transformation. This means that item properties remain the same, no matter what subset of respondents from a population is used to estimate them. However, if the questionnaire is administered to a new population that was not included in the initial IRT calibration of item properties, one will need to evaluate the invariance property using the test of differential item functioning (discussed at the end of this chapter).

13.4.4 Mixed-Item Format

Questionnaires often contain mixed-item formats, including dichotomous, polytomous, and open-ended responses. In CTT, mixed-item formats have an unbalanced impact on the total score. Items are unequally weighted, which allows items with a high number of response options to drive the survey score. Methods to correct for mixed-item formats are limited because CTT's statistics are sample dependent. IRT has models that simultaneously can model mixed-item response formats on a common measurement scale (Embretson and Reise, 2000).

13.4.5 Invariance of Scale Scores

Often, there is a need to compare people who have responded to broadly similar but not identical questionnaires (e.g., two different instruments that measure depression). CTT requires that instruments have a parallel format (e.g., equal means, variances, and covariances) to equate scores. This is virtually impossible to accomplish, given the wide variation that exists in current surveys. In CTT, the equating procedure is influenced by any differences that may exist between two surveys (e.g., number of response options and number of items). IRT models control for differences in item properties across surveys. Using a set of anchor items, IRT can place new items or items with different formats on a similar metric to link respondent scores. Once IRT item parameters have been estimated with an IRT model, investigators may calculate comparable scores on a given construct (such as depression) for respondents from that population who did not answer the same items, without having to perform intermediate equating steps (Orlando et al., 2000).

13.5 IRT APPLICATION

In this section we demonstrate the usefulness of IRT methodologies by evaluating the SF-36 Mental Health summary scale administered to a sample of cancer survivors. IRT and CTT analyses were performed to demonstrate the usefulness of IRT in evaluating the psychometric properties of a scale.

13.5.1 Sample Selection

Responses were collected from 888 breast cancer survivors approximately 24 months after breast cancer diagnosis. The women participated in the Health, Eating, Activity, and Lifestyle (HEAL) study, which was designed to evaluate the roles of diet, physical activity, body composition, and hormones in the prognosis and quality of life of women diagnosed with breast cancer. The HEAL study is a collaborative multicenter cohort study of breast cancer patients residing in Los Angeles County, California (University of Southern California School of Medicine), New Mexico (University of New Mexico), and western Washington (Fred Hutchinson Cancer Research Center). Women with newly diagnosed first primary insitu or AJCC stages I, II, or IIIa (AJCC: American Joint Committee on Cancer staging system) invasive breast cancer were identified via the National Cancer Institute's Surveillance, Epidemiology, and End Results (Surveillence, Epidemiology, and End Results Program, 2001) registry rapid case-ascertainment systems.

13.5.2 Instrument Properties

To illustrate the CTT and IRT approaches to evaluating instrument properties, analyses focused on responses to the Mental Health summary scale of the SF-36 (Medical Outcomes Trust, 1991). The SF-36 is a short-form health survey that yields an eight-scale profile of scores as well as summary measures of physical and mental health functioning. It is a generic measure, as opposed to one that targets a specific age, disease, or treatment group (Ware, 2000). The SF-36 Mental Health summary scale is comprised of 14 items that load on four subscales (vitality, social function, role–emotional, and mental health). The four subscales combine to provide an overall measure of a person's mental health. Table 13.3 provides a list of the 14 items included in the mental health scale categorized by the SF-36 subscale to which they belong. Table 13.3 also provides the response categories associated with each item. The response categories and data are organized so that higher scores represent better mental health. In the survey, the order of presentation differed from that shown in Table 13.3.

13.5.3 Data Analysis

CTT was used to conduct an item analysis that consisted of computing the item mean score, the item–total correlation, and Cronbach's α. The item mean score served to assess the item's threshold. The item–total correlation was used to evaluate item discrimination. Any item that had an item–total correlation of less than 0.30 (Nunnally and Bernstein, 1994) was considered to discriminate poorly. Test internal consistency (reliability) was assessed with Cronbach's α, with a value of 0.70 and 0.90 considered adequate for group- and individual-level measurement, respectively (Nunnally and Bernstein, 1994). These analyses were computed using SAS (version 8) PROC CORR.

Table 13.3 Items and Response Categories of the 14-Item SF-36 Mental Health Summary Scale Administered in the HEAL Study

Vitality Subscale

Survey Instructions: The next questions are about how you feel and how things have been with you during the *past 4 weeks*. Please give one answer that comes closest to the way you have been feeling.

	Item	1	2	3	4	5	6
240 244	Did you feel full of pep? Did you have a lot of energy?	None of the time	Little of the time	Some of the time	Good bit of the time	Most of the time	All the time
246 248	Did you feel worn out? Did you feel tired?	All the time	Most of the time	Good bit of the time	Some of the time	Little of the time	None of the time

Social Functioning Subscale

	Item	1	2	3	4	5
237	During the *past 4 weeks*, to what extent have your physical health or emotional problems interfered with your normal social activities with family, friends, neighbors, or groups?	Extremely	Quite a bit	Moderately	Slightly	Not at all
253	During the *past 4 weeks*, how much of the time have your physical health or emotional problems interfered with your social activities (e.g., visiting relatives, friends, etc.)?	All of the time	Most of the time	Some of the time	A little of the time	None of the time

Role–Emotional Subscale

Survey Instructions: During the *past 4 weeks*, have you had any of the following problems with your work or other regular daily activities as a result of any emotional problems (such as feeling depressed or anxious)?

	Item	1	2
234	Have you cut down the amount of time you spent on work or other activities?		
235	Have you accomplished less than you would like?	Yes	No
236	You didn't do work or other activities as carefully as usual?		

Mental Health Subscale

Survey Instructions: The next questions are about how you feel and how things have been with you during the *past 4 weeks*. Please give one answer that comes closet to the way you have been feeling.

	Item	1	2	3	4	5	6
241	Have you been a very nervous person?	All the time	Most of the time	Good bit of the time	Some of the time	Little of the time	None of the time
242	Have you felt so "down in the dumps" that nothing could cheer you up?	All the time	Most of the time	Good bit of the time	Some of the time	Little of the time	None of the time
243	Have you felt calm and peaceful?	None of the time	Little of the time	Some of the time	Good bit of the time	Most of the time	All the time
245	Have you felt downhearted and blue?	All the time	Most of the time	Good bit of the time	Some of the time	Little of the time	None of the time
247	Have you been a happy person?	None of the time	Little of the time	Some of the time	Good bit of the time	Most of the time	All the time

Samejima's (1969) IRT graded response model was chosen because of its ability to model ordinal response data and provide interpretable results. The unidimensional IRT graded response model estimates a discrimination (a) parameter and threshold (b) parameters (number of response options minus one) for each item in the scale. Item parameter estimation was carried out using the MULTILOG software (Thissen, 1991). CRC's information functions and SEM curves were created in Excel 2000. Item analyses of the SF-36 Mental Health summary scale were carried out by inspecting the item parameter estimates, the CRCs, and the item information function. Scale analyses were carried out by inspecting the scale information function and SEM curve.

Confirmatory factor analysis and principal component analysis were used to verify the unidimensionality assumptions of the IRT model. To test the factorial validity of the SF-36 Mental Health summary scale, the Mplus software (Muthén and Muthén, 1999) was chosen for its ability to model categorical response data (i.e., using polychoric correlations and weighted least squares estimation). As suggested by Hu and Bentler (1999) and Yu and Muthén (2001), several indices were used to determine the overall fit of the factorial structure.

13.6 RESULTS

13.6.1 CTT Results

Item Analysis Results of the CTT item analysis are summarized in Table 13.4. The item mean scores show that most of the sample had high scores for social function, role–emotional, and mental health items and scored near the center of the score range for the four items in the vitality subscale. This indicates that the sample, on average, had high mental health functioning. These high scores also suggest that a ceiling effect occurred in the score distribution. All item–total correlations were greater than 0.30, which indicates that all the items in the summary scale were discriminating. The first item, "nervous," had the lowest discrimination (i.e., correlation with the total score), and the second item, "energy," had the highest discrimination.

Scale Analysis The internal consistency, Cronbach's α, for the SF-36 Mental Health summary scale was high (0.90) overall, which is sufficient for individual-level measurement and well above the suggested cutoff of 0.70 for group-level measurement. The column entitled "alpha if item deleted" indicates that the overall reliability of the scale does not change much when an item is removed. Finally, the SEM of the SF-36 Mental Health scale was estimated to be 0.31.

13.6.2 IRT Results

IRT Assumptions The confirmatory factor analysis replicated Ware's (2000) results, which showed that the *a priori* factor structure of the SF-36 Mental Health summary scale was cross-validated in our sample of breast cancer women

Table 13.4 Classical Test Theory Item Statistics and Principal Component Analysis Factor Loîadings[a]

Subscale	Items	Mean (Std. Dev.)	Item–Total Correlation	Alpha if Item Deleted	PCA[b] Factor Loading
Vitality	Pep	3.50 (1.32)	0.71	0.89	0.75
	Energy	3.41 (1.35)	0.73	0.89	0.77
	Worn out	4.29 (1.24)	0.67	0.89	0.72
	Tired	3.92 (1.20)	0.68	0.89	0.72
Social functioning	Social—extent	4.25 (1.03)	0.64	0.89	0.71
	Social—time	4.18 (1.04)	0.65	0.89	0.71
Role emotional	Cut down time	1.80 (0.40)	0.53	0.90	0.63
	Accomplished less	1.60 (0.49)	0.55	0.90	0.64
	Not careful	1.76 (0.43)	0.50	0.90	0.59
Mental health	Nervous	4.97 (1.13)	0.48	0.90	0.55
	Down in dumps	5.40 (1.00)	0.61	0.90	0.68
	Peaceful	3.93 (1.25)	0.63	0.89	0.68
	Blue	4.97 (1.10)	0.67	0.89	0.73
	Happy	4.48 (1.10)	0.59	0.90	0.65

[a]Cronbach's coefficient $\alpha = 0.90$.
[b]PCA, principal component analysis.

(results not shown but available upon request). This suggests that the SF-36 Mental Health items have four subscales and one higher-order factor that represents a dominant factor. To further confirm the unidimensionality assumption, a principal components analysis was conducted to determine how the 14 items loaded on a common mental health scale. The loadings from the one-factor principal component analysis solution are reported in Table 13.4 (see the last column). The high loadings indicate that each item contributes to measuring a single dominant dimension. This single factor explained 47% of the total variance. Note that a second factor explained only 10% of the total variance, which further indicates that the scale has only one dominant dimension.

The assumption of local independence among the items in the SF-36 Mental Health summary scale was assessed by reviewing item content and by examining the inter-item correlations. Two item pairs (first pair: "Did you feel full of pep?"/"Did you have a lot of energy?"; second pair: "Did you feel worn out?"/"Did you feel tired?") had similar content and high inter-item correlations (0.84 and 0.72, respectively). Note that these item pairs belong to the vitality subscale. Including locally dependent item pairs in the estimation process raises the concern that highly redundant items in a dataset may distort IRT parameter estimates for other items in the scale to the extent that the content of locally independent items is unrelated to the overall construct being measured. To examine the impact of these possibly locally independent item pairs, the first item in each pair was dropped and the graded response model parameters were reestimated to examine changes in the discrimination and threshold parameters. Dropping these two items changed the magnitude (but not the order) of the discrimination parameters. The values of threshold parameters did not change enough to affect the interpretation of the results significantly (results not shown). To compare the IRT results with CTT results, we chose to present the 14-item IRT solution rather then the 12-item IRT solution. Note that, in practice, it is recommended that local independence be controlled for when estimating accurate scores on the SF-36 Mental Health summary scale. This would require dropping the two items noted above. To improve the SF-36 Mental Health summary scale, these items should either be dropped or rewritten.

The IRT graded response model was chosen for its flexibility in modeling ordinal response data because the model allows both discrimination and threshold parameters to vary from item to item in the scale. The fit of the IRT model to each item in the scale was evaluated by comparing the observed proportion of sample responses for each response category to the proportions in each category predicted by the IRT model. For all items, the graded response model showed an excellent fit. For example, the item with the largest deviation between observed and expected proportions is: "Have you felt downhearted and blue?" The observed proportions of responses for the six categories are 0.01, 0.03, 0.04, 0.19, 0.34, and 0.38 (e.g., 38% of the sample said "none of the time"), and the expected proportions of the responses predicted from the IRT model were 0.01, 0.03, 0.05, 0.20, 0.33, and 0.38. The chi-square goodness of fit for this item was $\chi^2(5) = 2.5$, $p = 0.78$, indicating a good fit (i.e., residuals are not significant). Similarly, the chi-square p-values for the other items ranged from 0.78 to 1.00, also indicating good fit. Although the focus on model fit generally is performed on an item-by-item basis, the likelihood-ratio statistic was computed to provide an omnibus test for the fit of the IRT model to the data observed. The likelihood-ratio statistics $[G^2(55) = 10.96,\ p = 0.99]$ showed that the IRT model fit the data well.

The standard error associated with each item parameter estimate can provide another way of evaluating model fit. Table 13.5 presents the item parameters and the standard error associated with these parameters. The magnitude of the standard error is related to the number of people that are available for estimating the associated item parameters. Because this is a relatively healthy sample (as noted in the CTT results), it is harder to estimate items that measure poor mental health functioning. The b_1 threshold parameter indicates where the first and second response category curves cross along the mental health (θ) continuum. The magnitude of b_1 can serve to identify those items that discriminate among those who have poor mental health functioning. A lower b_1 parameter indicates that the item attempts to discriminate among those who have poor mental health functioning. As indicated above, items that target poor mental health functioning (i.e., lowest b_1 parameter) will have higher standard errors because the sample is relatively healthy. Note that the "nervous" and "happy" items have the lowest b_1 parameters and the highest standard errors. The closer the b parameters are to the population mean ($\theta = 0$), the lower the standard errors will be. Similarly, the

Table 13.5 Item Response Theory Item Parameter Estimates (a = *Discrimination* and b_1 to b_5 = *Threshold* Parameters) and Standard Error (SE)[a]

Subscale	Items	a (SE)	b_1 (SE)	b_2 (SE)	b_3 (SE)	b_4 (SE)	b_5 (SE)
Vitality	Pep	2.67	−1.64	−0.89	0.06	0.63	2.31
		(0.15)	(0.09)	(0.06)	(0.05)	(0.05)	(0.15)
	Energy	2.90	−1.48	−0.74	0.11	0.70	2.20
		(0.17)	(0.07)	(0.05)	(0.05)	(0.05)	(0.13)
	Worn out	2.04	−2.62	−1.58	−1.01	0.04	1.35
		(0.13)	(0.18)	(0.11)	(0.08)	(0.05)	(0.09)
	Tired	2.19	−2.29	−1.35	−0.65	0.49	2.03
		(0.13)	(0.14)	(0.08)	(0.06)	(0.06)	(0.12)
Social functioning	Social—extent	2.11	−2.80	−1.79	−1.03	−0.22	
		(0.16)	(0.22)	(0.12)	(0.07)	(0.06)	
	Social—time	2.04	−2.85	−1.90	−0.85	−0.09	
		(0.15)	(0.23)	(0.13)	(0.07)	(0.06)	
Role emotional	Cut down time	1.96	−1.08				
		(0.22)	(0.09)				
	Accomplished less	1.82	−0.34				
		(0.18)	(0.07)				
	Not careful	1.65	−0.98				
		(0.18)	(0.10)				
Mental health	Nervous	1.05	−4.89	−3.31	−2.53	−1.03	0.48
		(0.10)	(0.59)	(0.33)	(0.24)	(0.13)	(0.11)
	Down in dumps	1.71	−3.43	−2.87	−2.46	−1.41	−0.49
		(0.14)	(0.33)	(0.24)	(0.19)	(0.11)	(0.07)
	Peaceful	1.72	−2.56	−1.49	−0.52	0.24	2.51
		(0.12)	(0.20)	(0.11)	(0.07)	(0.07)	(0.18)
	Blue	1.73	−3.31	−2.51	−1.91	−0.78	0.43
		(0.13)	(0.30)	(0.19)	(0.13)	(0.08)	(0.07)
	Happy	1.62	−3.55	−2.33	−1.17	−0.44	1.69
		(0.12)	(0.33)	(0.17)	(0.09)	(0.07)	(0.13)

[a] The number of threshold parameters per item is the number of response categories minus one. Items in the Vitality subscale have six response categories per item; thus, each item will have five IRT thresholds. Items in the role emotional subscale have two response categories (yes, no); thus there is one threshold parameter estimate for each item.

magnitude of the standard errors for the discrimination parameters is a function of where, along the mental health continuum, the item discriminates (better if near the population mean) and the magnitude of the discrimination parameter (i.e., steepness of the slope at the inflection point). Steeper curves increase the standard error because more data (people) are required within a narrower range of θ to estimate the slope parameter. (Readers should consult the previously cited references for other statistical methods for assessing model fit.)

Item Analysis In this section the results of the IRT item analysis are presented and compared to the CTT item analysis. The discrimination (a) and threshold (b) parameters derived from Samejima's (1969) IRT graded response model are presented in Table 13.5. The discrimination parameter is an indicator of the relationship that exists between an item and the total score, so it is not surprising to observe a similar pattern between these values and the item–total correlations from CTT (presented in Table 13.4). Similar to CTT, the vitality and social function subscale items were found to have high discrimination values, and the mental health subscale items had the lowest discrimination values (especially the "nervous" item). In this case, the IRT discrimination parameters yielded information similar to the CTT discrimination parameters (i.e., item–total correlations in Table 13.4).

To further interpret the item parameters presented in Table 13.5, it is useful to look at CRCs for some of the items. The CRCs for the item "Did you have a lot of energy?" are shown in Figure 13.2. The response categories were found to function well because each response category maps to a different segment along the mental health (θ) continuum. For example, a person with a mental health score of -1.0 is likely to say that he or she has a lot of energy "a little of the time," and a person one standard deviation above the mean ($\theta = 1$) is likely to say that he or she has a lot of energy "most of the time."

In contrast, Figure 13.5 presents the CRCs for the item "Did you feel worn out?" and shows that one response category does not function well. The response category "a good bit of the time" is overshadowed by the neighboring categories, "most of the time" and "some of the time." Therefore, there is no area along the mental health continuum where the "a good bit of the time" category carries the highest probability of being selected. Thus, the IRT model suggests that this

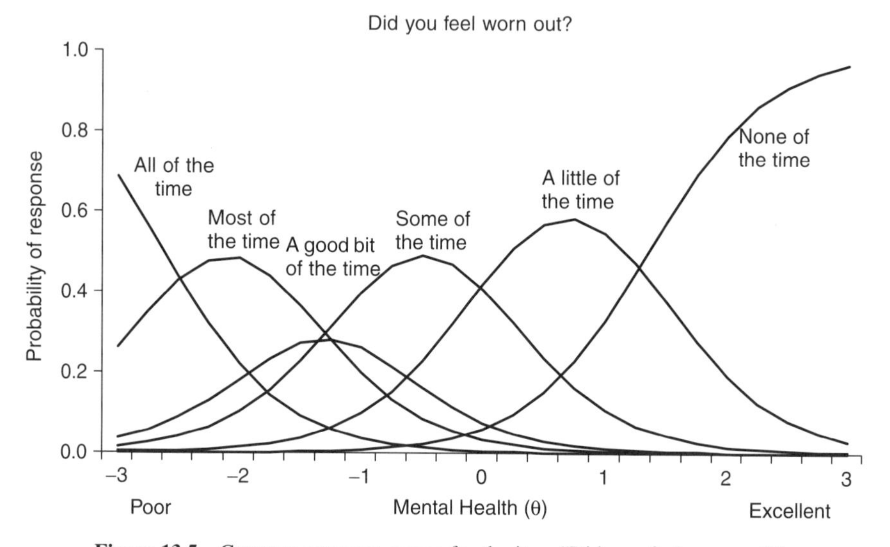

Figure 13.5 Category response curves for the item "Did you feel worn out?".

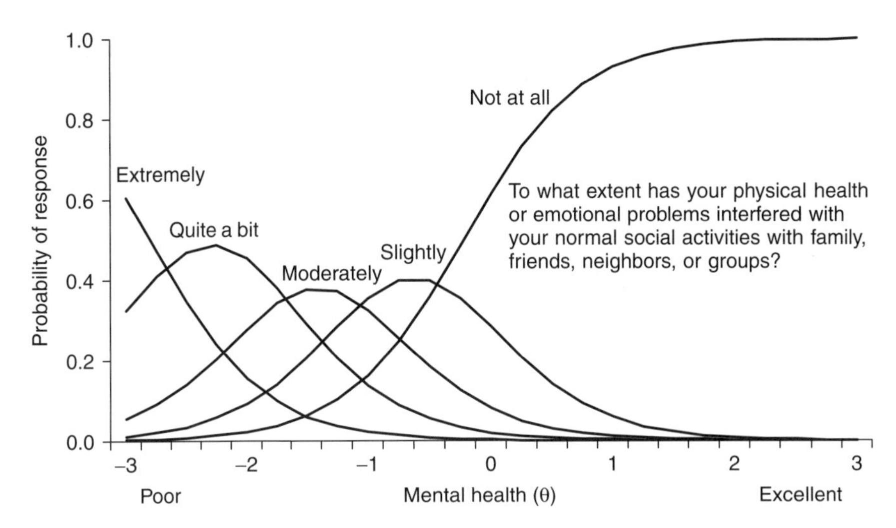

Figure 13.6 Category response curves for the item "To what extent has your physical health or emotional problems interfered with your normal social activities with family, friends, neighbors, or groups?".

response option does not function well for this item. From a content perspective, this may suggest that it is difficult for people to understand how much "a good bit of the time" differs from neighboring categories "most of the time" and "some of the time." In fact, this response option receives little endorsement for most of the items in the SF-36 Mental Health summary scale. This information might be used by test developers to revise the scale to include only five response categories, leaving out the "a good bit of the time" response option.[2]

The CRC in Figure 13.6 shows how a shift in the threshold parameters represents a change in the item content as it relates to the level of endorsement for the item, which is a function of the content of the question. This scale measures mental health, and the threshold represents the level of mental health in the study population that is measured by the item "To what extent has your physical health or emotional problems interfered with your normal social activities with family, friends, neighbors, or groups?" Most respondents with average (starting at $\theta = -0.22$) or good mental health answered "not at all" to this item. The four other response options, "extremely," "quite a bit," "moderately," and "slightly," discriminate among people with below-average mental health. Therefore, this item is ideal for differentiating among those who have low mental health functioning.

Finally, the CRCs for the "Have you felt so down in the dumps that nothing could cheer you up?" item are shown in Figure 13.7. The item content indicates that this item was designed to discriminate between people who have some mental

[2]In the SF-36 (version II), the "good bit of the time" response option is dropped, leaving only five response options in the scale.

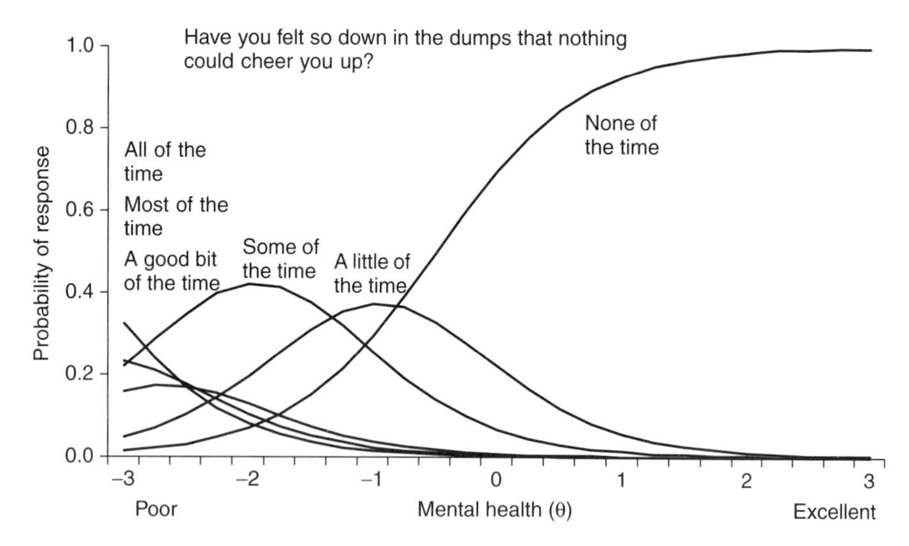

Figure 13.7 Category response curves for the item "Have you felt so down in the dumps that nothing could cheer you up?".

health problems and those with severe mental health problems. Indeed, most mentally healthy people said that they were so down in the dumps that nothing could cheer them up "none of the time." A response to any of the other categories was an indication of poorer mental health. Most people with low mental health, however, selected the "some of the time" or "a little of the time" options. The response categories for this item function more like a three-point response scale than a six-point response scale.

Evaluation of the CRCs provides detailed information about the functioning of the response categories that was not available with CTT. IRT provides another useful tool for evaluating item functioning that is not available with CTT item analysis, examination of the item information curves. The information functions for each of the 14 mental health items are presented in Figure 13.8. Of particular interest is the amount of information (i.e., precision in measurement) provided along the continuum of scale (θ) scores. Some items provide information that is useful in measuring people with poor mental health and other items provide information that applies across a wider range of the mental health continuum. Items such as "energy" and "pep," from the vitality subscale, provided a high level of information across the mental health continuum (similar to the CRCs, where each response option mapped to a different part of the θ scale). Items from the role emotional subscale ("cut down," "not careful," "accomplished less") had only two response categories. Thus, they provided information for a more narrow area of mental health. The range of the information curves for the role emotional items can be broadened by adding more response options that are meaningful to the respondents.

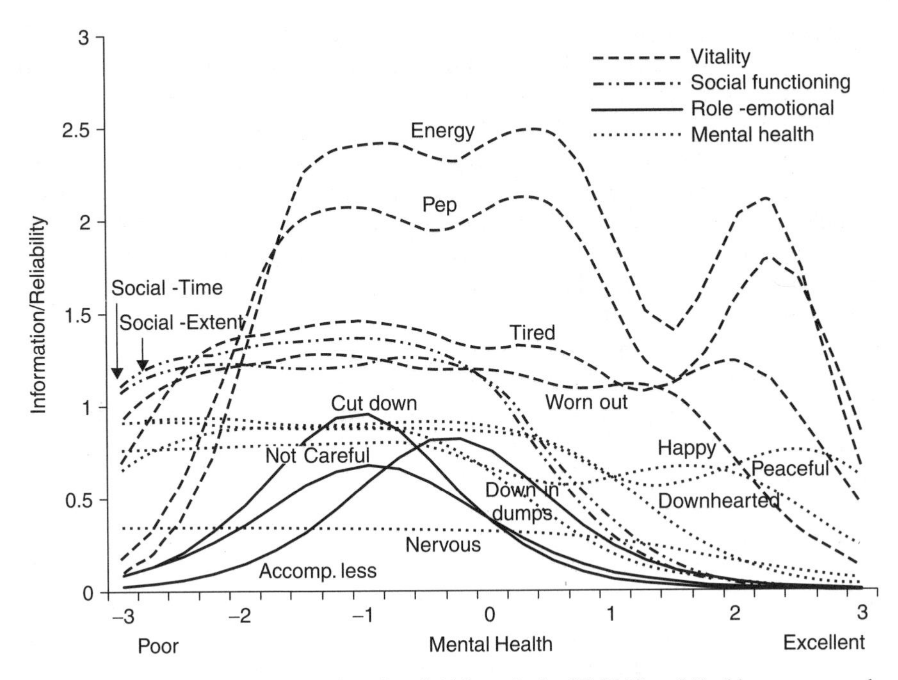

Figure 13.8 Item information functions for all 14 items in the SF-36 Mental Health summary scale.

Instrument developers can use information curves to help them revise an instrument to create shorter scales or to develop diagnostic scales. To shorten the SF-36 Mental Health summary scale, the developers would pick the most informative items of the four subscales to bring a balance of content to the measure of overall mental health. In a shorter test, the "nervous" item may be eliminated because it provides the least amount of information. In addition, redundant items may be eliminated to shorten the scale. From Figure 13.8, the "energy" and "pep" items provide similar information over the same range of θ. Thus, they appear to be redundant. Of the two, the "energy" item provides slightly more information and may be easier to understand than the "pep" item (cognitive testing methods might verify this; see Chapters 2 to 5, this volume). Alternatively, if the goal is to create a diagnostic scale that focuses on discriminating among those who have low to poor mental health, items with higher information curves in the lower part of the θ continuum (e.g., the social functioning items or the "down in the dumps" item) would be included.

Scale Analysis IRT scale analysis consists of evaluating scale information/reliability and the SEM. Figure 13.4 presents the scale information curve and the reliability (r) associated with different magnitudes of information. For example, information in the range of 5 corresponds to a reliability estimate of 0.80. CTT and IRT provide different conceptualizations of reliability. In CTT, a fixed estimate of reliability (usually, Cronbach's α) is associated with all score values.

Cronbach's α was 0.90, which implies that the 14-item SF-36 Mental Health summary scale is adequate for measuring a person's mental health status no matter where the person falls along the continuum. It is more likely, however, that the reliability of an instrument varies depending on who is being measured. This is reflected in the IRT information curves. The information curve in Figure 13.4 shows that reliability varies across the mental health continuum. Reliability is very high ($r > 0.90$) for low to middle levels of mental health (i.e., $-2.5 < \theta < 1$), whereas it is less precise, although still adequate, when measuring people with good to excellent mental health.

In Figure 13.9, the IRT SEM curve indicates the amount of error associated with measuring a person's score along the mental health continuum. Similar to what was observed in the information function, the summary scale produces fewer errors when measuring people with low to moderate levels of mental health, but the errors increase with higher levels of mental health. For moderate to low mental health scores, the SEM is approximately 0.30; whereas for higher scores it increases exponentially to a maximum measurement error of 0.55. In CTT, the SEM for all score levels was 0.31, which differs from the IRT estimate.

Together, the IRT scale information and SEM curves provide test developers a tool for evaluating the performance of their instrument. The curves describe the levels of mental health that can be measured ideally by the SF-36 Mental Health summary scale. A developer who prefers to shorten the scale could remove the information provided by the deleted item and compare the new scale information curve with the original curve to determine how the reliability (across θ) of the instrument changed. These curves also can inform developers about how to improve their instrument. If the developer wants to get an instrument with

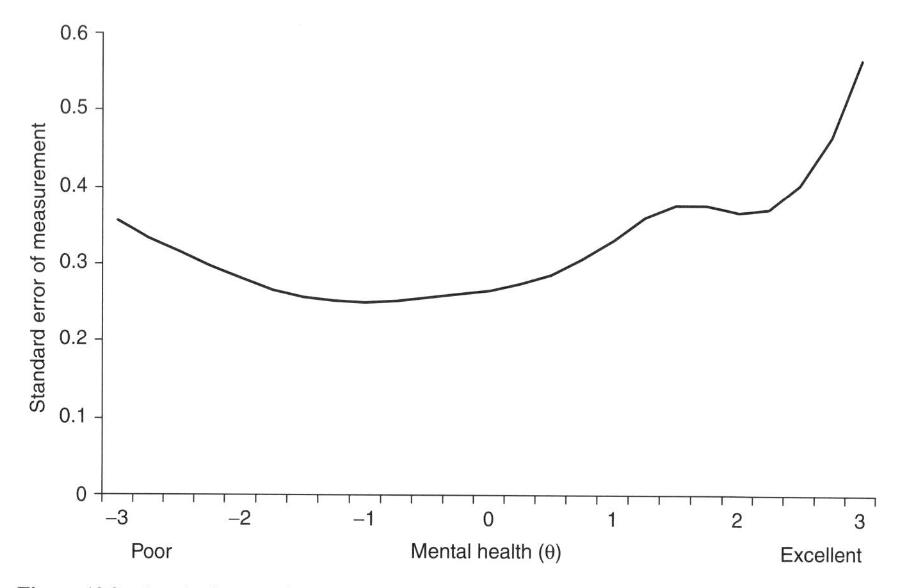

Figure 13.9 Standard error of measurement function for the SF-36 Mental Health summary scale.

reliability that is equally high across the mental health continuum, the information and SEM curves suggest adding more items that can discriminate better among people with good to excellent mental health.

13.7 CONCLUSIONS

We have shown what was learned about the SF-36 Mental Health items and scale by conducting a CTT and IRT analysis. Applying IRT models does not mean that researchers should abandon the CTT methods they traditionally have used to evaluate the psychometric properties of questionnaires. Rather, IRT methods should complement CTT methods, as they make an excellent addition to the psychometrician's toolbox for developing and revising questionnaires. IRT models allow a developer to choose the most informative items in a questionnaire, identify redundancy within item sets, determine the appropriate number of response options per item, and order response categories for developing the scoring system. Also, IRT models allow developers to determine (in terms of reliability/information) whether the properties of the items are appropriate for a particular study population. Although CTT and IRT provide useful statistical information about the items and the scale, it is important to work with an expert familiar with the content area and study population to guide the analyses and interpretation of the results. A researcher skilled in interpreting CTT statistics can identify many of the problems found with IRT. However, the ability of IRT models to describe how a set of items functions across the mental health continuum cannot be done with CTT. Thus, IRT is the only statistical procedure that assesses content representation and determines the optimal number of response categories for individual items.

CTT and IRT are methods to determine the structural properties and internal consistency of a questionnaire (Benson, 1998). A comprehensive assessment of a questionnaire also requires that substantive and external validity be demonstrated. Benson (1998) defines the substantive stage as demonstrating how the construct is defined theoretically and empirically, and external validity as assessing how the construct measured by the questionnaire relates to other constructs (i.e., determining predictive and discriminant validity). In this chapter we have focused on one aspect of construct validity, but a comprehensive evaluation requires a fuller evaluation.

In this chapter, both confirmatory factor analysis and principal component analysis were used to verify the unidimensionality assumption of the IRT model. Although it was not discussed in the results section, scale unidimensionality is an important assumption of CTT, because the item and scale analysis must be conducted on a dominant dimension. Often in CTT analyses, data that are in fact multidimensional will result in low item–total score correlations and decreased reliability. Cronbach's α does not test the dimensionality of a questionnaire. It is possible to have a multidimensional scale with high internal consistency (Cronbach's α). Although high internal consistency has been interpreted to indicate that the scale measures one construct (i.e., one dimension), this is not necessarily the case. A factor analysis is needed to evaluate this assumption. As this chapter focuses on comparing CTT and IRT item and scale analyses, we did not emphasize the role of confirmatory factor analysis in evaluating the psychometric properties of a test. Confirmatory factor analysis, however, does play an integral role in evaluating the psychometric properties of questionnaires.

This chapter has focused on the role of IRT models in evaluating questionnaires, but there are many other applications of IRT in questionnaire development. These applications take advantage of a key feature of IRT models: "invariance." If IRT model assumptions are met, item parameters are invariant with respect to the sample of respondents, and respondent scores are invariant with respect to the set of items used in the scale. After the IRT item parameters are estimated (i.e., calibrated), researchers can choose the most salient items to target a person's level of function with the smallest number of items. This method results in different groups receiving different sets of items; however, items calibrated by the best-fitting IRT model lead to results that are comparable on a similar metric. The IRT principle of invariance is the foundation that researchers use to develop computerized adaptive tests (CATs; such as the Graduate Record Examination and Scholastic Aptitude Test).

CATs combine the powerful features of IRT with improvements in computer technology to yield tailored instruments that estimate a person's level on a construct (e.g., mental health) with fewer items. To accomplish this, CATs use a large pool of items that have been calibrated by IRT, called an *item bank*. Based on a person's response to an initial item, the CAT chooses the item bank's next most informative item (based on the item information curves) to administer to the respondent. After each response, the computer selects and administers from the item bank the next most informative item until a reliable score is obtained. The benefits of CAT technology include decreased burden because fewer items can be administered, reduced "floor and ceiling" effects (often seen in fixed-length paper-and-pencil questionnaires), instant scoring, and widespread availability in many platforms (e.g., Internet, handheld devices, telephones) (Hays et al., 2000; Wainer et al., 2000; Ware et al., 2000).

Also, made possible by the invariance property is the application of IRT modeling for identifying item bias or differential item functioning (DIF). DIF occurs when one group responds differently to an item than another group, even after controlling for differences in sampling distribution on the measured construct. For example, in a depression questionnaire, Azocar et al. (2001) found that a Latino population will endorse a question like "I feel like crying" in greater proportions than will an Anglo population, because Latinos believe that crying is a socially acceptable behavior. This DIF item will result in Latinos receiving a higher depression score on average than Anglos. Scales containing such items have reduced validity for between-group comparisons because their scores are indicative of a variety of attributes other than those intended to be measured (Thissen et al., 1993). DIF analysis has been used to detect equivalence in item content across racial, gender, cultural, and treatment groups; and between two administration modes (e.g., telephone versus self-administered) and two translated versions (Azocar et al., 2001; Fleishman et al., 2002; Morales et al., 2000; Orlando and Marshall, 2002; Panter and Reeve, 2002; Teresi, 2001).

If IRT provides so many useful tools for evaluating and developing instruments, why is its use no more widespread? Several obstacles limit the use of IRT. First, most researchers have been trained in CTT statistics, are comfortable interpreting these statistics, and can generate easily familiar summary statistics such as Cronbach's α or item–total correlations with readily available software. In contrast, IRT modeling requires an advanced knowledge of measurement theory to understand the mathematical complexities of the models, to determine whether the assumptions of the IRT models are met, and to choose the model from within the large family of IRT models that best fits the data and the measurement task at hand. In addition, the supporting software and literature are not well adapted for researchers outside the field of educational measurement. Another obstacle is that the algorithms in the IRT parameter-estimation process require large sample sizes (size varies depending on the model) to provide stable estimates. Finally, IRT was developed to measure latent variables that require multiple item measurement. Such a complex methodology is not necessary when measuring readily quantifiable phenomena, such as the number of cigarettes a person smokes in a week, where a single item can capture the information.

Despite the conceptual and computational challenges, the many potential practical applications of IRT modeling should not be ignored. Knowledge of IRT is spreading as more and more classes are being taught within the academic disciplines of psychology, education, and public health, and at seminars and conferences throughout the world. Along with this, more books and tutorials are being written on the subject, and more user-friendly software is being developed. Research that applies IRT models appears more frequently in the literature, and many concluding comments are directed toward the benefits and limitations of using the IRT methodology in various fields. For all of these reasons, a better understanding of the models and applications of IRT will emerge, and this will result in instruments that are shorter, more reliable, and better targeted toward the population of interest.

CHAPTER 14

Development and Improvement of Questionnaires Using Predictions of Reliability and Validity

Willem E. Saris, William van der Veld, and Irmtraud Gallhofer
University of Amsterdam

14.1 INTRODUCTION

The development of a survey item demands many choices about the structure of the items and the data collection procedure. Some of these choices follow directly from the aim of the study, such as those about the domain of the survey item(s) (e.g., church attendance, neighborhood) and the nature of the responses (e.g., evaluations, norms) (Gallhofer and Saris, 2000; Saris and Gallhofer, 1998). But there are also many choices that are not fixed, including the formulation of the questions, selection of response scales, use of additional components such as an introduction, motivating statements, position in the questionnaire, and mode of data collection. The latter choices can have a considerable influence on the quality of survey items. Therefore, it is important to assess the quality of a survey item prior to its use.

Many methods have been developed to evaluate survey items before they are used in the final survey. These include conventional pretests followed by interviewer debriefings, respondent debriefings (Belson, 1981), interaction or behavior coding (Dijkstra and van der Zouwen, 1982), cognitive interviews, expert panels (Presser and Blair, 1994), coding schemes (Forsyth et al., 1992; van der Zouwen, 2000), and computer programs (Graesser et al., 2000a,b). An overview of many

Methods for Testing and Evaluating Survey Questionnaires, Edited by Stanley Presser, Jennifer M. Rothgeb, Mick P. Couper, Judith T. Lessler, Elizabeth Martin, Jean Martin, and Eleanor Singer
ISBN 0-471-45841-4

of these methods was presented by Sudman et al. (1996), and recent advances are presented elsewhere in this volume.

All these approaches are directed at detecting response problems. The hypothesis is that problems in the formulation of the survey item will reduce the quality of the responses of the respondents. However, the standard criteria for data quality, such as validity, reliability, and method effects, are not evaluated directly.

Campbell and Fiske (1959) suggested that validity, reliability, and method effects can only be directly evaluated if more than one method is used to measure the same traits. Their design is called the *multitrait multimethod* (MTMM) *design*, which is widely used in psychology and psychometrics (see Wothke, 1996), has also attracted considerable attention in marketing research (Bagozzi and Yi, 1991). In survey research, the MTMM approach has been elaborated and applied by Andrews (1984). Subsequently, Andrews's approach has been used for many different topics and question forms in several languages: English (Andrews, 1984), German (Költringer, 1995), and Dutch (Scherpenzeel and Saris, 1997). Andrews (1984) also suggested using meta-analysis of the available MTMM studies to determine the effect of different choices made in the design of survey questions on the reliability, validity, and method effects. Following his suggestion, Saris and Gallhofer (2003) conducted a meta-analysis of 87 MTMM studies and summarized the effects that different question characteristics have on reliability and validity. In this chapter we describe how these results can be used both to predict the quality of survey items before they are used in practice and to improve question formulations when the quality of the original formulation is insufficient.

The structure of the chapter is as follows. We first introduce the MTMM approach and the results of the meta-analysis. We then show how these results can be used to predict the quality of survey items before data are collected. We conclude by discussing the possibilities and limitations of this approach and its future development.

14.2 MTMM STUDIES

Normally, all variables are measured using a single method. Thus, one cannot see how much of the variance of the variables is random measurement error and how much is systematic method variance. Campbell and Fiske (1959) suggested that one needs to use multiple methods for multiple traits in order to detect these error components. The standard MTMM approach uses at least three traits that are measured by at least three different methods, leading to nine different observed variables. In this way a 9×9 correlation matrix is obtained. To illustrate the procedure, Table 14.1 presents a brief summary of a MTMM experiment conducted in the pilot study for the first round of the European Social Survey (2002), in which three traits and three methods were used.

The actual domain of the questions (national politics/economy) remains the same across methods. Also the concept measured (satisfaction) remains the same. Only the way in which the respondents are asked to express their feelings varies.

Table 14.1 MTMM Study in the ESS Pilot Study (2002)

For the three traits, the following three questions were employed:

- "On the whole, how satisfied are you with the present state of the economy in Britain?"
- "Now think about the national government. How satisfied are you with the way it is doing its job?"
- "And on the whole, how satisfied are you with the way democracy works in Britain?"

In this experiment the following response scales were used to generate the three different methods:

Method 1: 1, very satisfied; 2, fairly satisfied; 3, fairly dissatisfied; 4, very dissatisfied
Method 2: Very dissatisfied 0 1 2 3 4 5 6 7 8 9 10 very satisfied
Method 3: 1, not at all satisfied; 2, satisfied; 3, rather satisfied; 4, very satisfied

The first and third methods use a 4-point scale, while the second method uses an 11-point scale. This also means that the second method provides a midpoint on the scale, while the other two do not. Furthermore, the first and second methods use a bipolar scale, while the third method uses a unipolar scale. In addition, the response categories change direction in the first compared with the second and third methods. The first and third methods have completely labeled categories, while in the second method only the endpoints of the scales are labeled.

There are other aspects that remained the same, although they could have been different. For example, direct questions have been used in this study. It is, however, very common in survey research to specify a general question such as "How satisfied are you with the following aspects of society?" followed by provision of stimuli such as "the present economic situation," "the national government," and "the way the democracy functions." Furthermore, all three questions are unbalanced, asking "how satisfied" people are without mentioning the possibility of dissatisfaction. They have no explicit "don't know" option; all three have no introduction, no subordinate clauses, and so the survey items are all relatively short. See Saris and Gallhofer (2003) for a discussion of other item characteristics. Identical characteristics of the three questions cannot generate differences, but those aspects that differ could generate differences in the responses. Many studies have looked at the differences in response distributions (see Schuman and Presser, 1981), but in MTMM studies, the correlations between the variables are inspected and used to derive conclusions about the reliability, validity, and method effects of the questions. Table 14.2 presents the correlations among the nine measures for a sample of 481 respondents from the British population. These results indicate very clearly the need for further investigation of the quality of the various measures, since the correlations among the three questions Q1 to Q3

Table 14.2 Correlations Among the Nine Variables of the MTMM Experiment with Respect to Satisfaction with Political Outcomes

	Method 1			Method 2			Method 3		
	Q1	Q2	Q3	Q1	Q2	Q3	Q1	Q2	Q3
Method 1									
Q1	1.00								
Q2	0.481	1.00							
Q3	0.373	0.552	1.00						
Method 2									
Q1	−0.626	−0.422	−0.410	1.00					
Q2	−0.429	−0.663	−0.532	0.642	1.00				
Q3	−0.453	−0.495	−0.669	0.612	0.693	1.00			
Method 3									
Q1	−0.502	−0.347	−0.332	0.584	0.436	0.438	1.00		
Q2	−0.370	−0.608	−0.399	0.429	0.653	0.466	0.556	1.00	
Q3	−0.336	−0.406	−0.566	0.406	0.471	0.638	0.514	0.558	1.00

are very different for the various methods. In the first method the correlations vary between 0.373 and 0.552, for the second method between 0.612 and 0.693, and for the third method between 0.514 and 0.558. This raises questions such as: How can such differences be explained? What are the correct values? What is the best method?

14.2.1 Data Quality Evaluated by Reliability, Validity, and Method Effects

Given that the same people answered all the questions, the likely explanation for the difference between these correlations is measurement error.[1] It is assumed that each method has its own random errors and systematic errors, also known as *method effects*. Formally, this can be expressed as follows:

$$Y_{ij} = r_{ij}T_{ij} + e_{ij} \qquad \text{for } i = 1 \text{ to } 3 \text{ and } j = 1 \text{ to } 3 \qquad (14.1)$$

$$T_{ij} = v_{ij}F_i + m_{ij}M_j \qquad \text{for } i = 1 \text{ to } 3 \text{ and } j = 1 \text{ to } 3 \qquad (14.2)$$

where Y_{ij} is the measured variable (trait i measured by method j); T_{ij} the stable component of the response Y_{ij}, also called "true score"; F_i the trait factor of interest; M_j the method factor, whose variance represents systematic method effects common for all traits measured with method M_j but varying across individuals; and e_{ij} the random measurement error term for Y_{ij}, with zero mean and uncorrelated with other error terms, with method factors or with trait factors. The

[1]It is also possible that the differences are partially due to the fact that the questions are repeated and that respondents think about these questions between the two observations. Discussion of this point is presented in another paper (Saris et al., 2002).

r_{ij} coefficients standardized can be interpreted as reliability coefficients (square root of test–retest reliability). The v_{ij} coefficients standardized are validity coefficients (with v_{ij}^2 representing the validity of the measure). When standardized, the m_{ij} coefficients represent method effects. Since $m_{ij}^2 = 1 - v_{ij}^2$, the method effect is equal to the invalidity due to the method used. Accordingly, *reliability* is defined as the strength of the relationship between the observed response (Y_{ij}) and the true score (T_{ij}), and the *validity* is defined as the strength of the relationship between the variable of interest (F_i) and the true score (T_{ij}). The systematic *method effect* is the strength of the relationship between the method factor (M_j) and the true score (T_{ij}). There are many more definitions of reliability, validity and method effects, but we will show that the definitions specified above are highly satisfactory.

Figure 14.1 presents the same model for two traits measured using the same method, assuming that the measurement errors are independent of each other, the true scores, and the trait variables. Using path analysis, it can be shown that the correlation between the observed variables $r(Y_{1j}, Y_{2j})$ is equal to the correlation due to the variables we want to measure, F_1 and F_2, reduced by measurement error plus the correlation due to the method effects:

$$r(Y_{1j}, Y_{2j}) = r_{1j}v_{1j}r(F_1, F_2)v_{2j}r_{2j} + r_{1j}m_{1j}m_{2j}r_{2j} \quad (14.3)$$

Note that r_{ij} and v_{ij}, which are always smaller than 1, will reduce the correlation (see the first term), while the method effects, if they are not zero, will increase the correlation (see the second term). This result suggests that it is possible that the low correlations for methods 1 and 3 in Table 14.1 are due to a lower reliability than for method 2. However, it is also possible that method 2 correlations are higher because of higher systematic method effects.

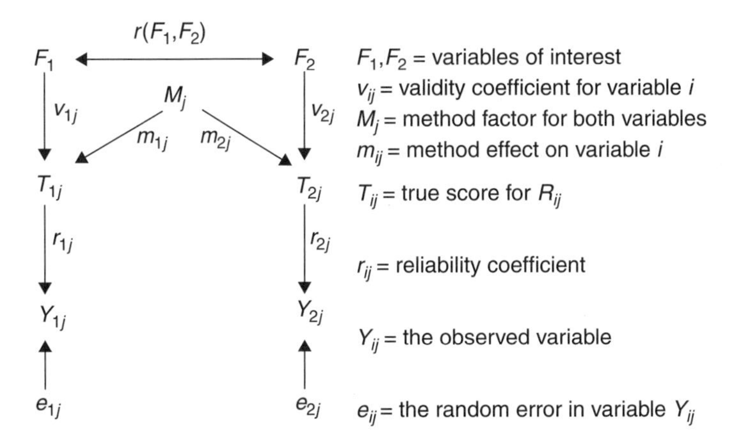

Figure 14.1 Measurement model for two traits measured with the same method.

14.2.2 Estimation of Reliability and Validity

Estimation of the reliability coefficients and validity coefficients is an important issue if the differences in these coefficients can explain the large differences we have shown above in correlations obtained using different methods. If the standard design is used, equation (14.2) represents the basic equation of the MTMM model, and this equation has the structure of a factor loadings matrix, which is presented in Table 14.3. The specification of the structure of this matrix of loadings is commonly accepted, but some authors specify that the method effects (m_{ik}) should be the same for each k. This can be done by specifying

$$m_{ik} = 1 \quad \text{for all } i, k \qquad \text{while the variance } (M_k) \text{ has to be estimated} \qquad (14.4)$$

This approach was used by Scherpenzeel and Saris (1997). Many researchers do not introduce this restriction. There is also a debate currently being waged about the specification of the correlations between the various factors. Some authors leave all correlations free but mention many problems (Kenny and Kashy, 1992; Marsh and Bailey, 1991). Andrews (1984) and Saris (1990) suggested that the traits be allowed to correlate but should be uncorrelated with the method factors and the method factors should be uncorrelated with each other. Zero correlations between factors and the error terms are commonly accepted. Using the model specification of Saris and Andrews (1991) combined with assumption (14.4), scarcely any problems arise in the analysis, as has been shown by Corten et al. (2002).

Using the specification of the model mentioned above and the normal theory ML estimators, the results presented in Table 14.4 are obtained for the standardized factor loadings. These results indicate that the second method of questioning has higher reliability coefficients than the other methods and that the method

Table 14.3 Factor Loading Matrix of the MTMM Model

	F_1	F_2	F_3	M_1	M_2	M_3
T_{11}	v_{11}			m_{11}		
T_{21}		v_{21}		m_{21}		
T_{31}			v_{31}	m_{31}		
T_{12}	v_{12}				m_{12}	
T_{22}		v_{22}			m_{22}	
T_{32}			v_{32}		m_{32}	
T_{13}	v_{13}					m_{13}
T_{23}		v_{23}				m_{23}
T_{33}			v_{33}			m_{33}

Table 14.4 Standardized Estimates of the MTMM Model Specified for the ESS Data of Table 14.1

	Validity Coefficients			Method Effects			
	F_1	F_2	F_3	M_1	M_2	M_3	Reliability Coefficients
T_{11}	0.94			0.35			0.78
T_{21}		0.94		0.33			0.84
T_{32}			0.94	0.34			0.81
T_{12}	0.92				0.40		0.91
T_{22}		0.92			0.40		0.94
T_{32}			0.92		0.40		0.93
T_{13}	0.86					0.51	0.82
T_{23}		0.88				0.48	0.87
T_{33}			0.87			0.49	0.84

effects for this method are intermediate between the other two. If one knows that the correlation between the first two traits was estimated to be 0.69, using equation (14.3), it is easily verified that the measurement quality indicators can produce such different observed correlations as 0.48 for the first method, 0.64 for the second method, and 0.56 for the third method. This means that the differences in correlations obtained are completely explained by differences in data quality between the various measurement procedures. This illustrates quite clearly how important reliability and validity, as defined here, are for social science research. In principle, the MTMM studies can provide estimates of these quality indicators, as we have shown.

Although this approach looks very promising for evaluation of the quality of survey questions, in practice it would mean that in any survey, all questions to be evaluated must be supplemented by two more questions concerning the same trait. Besides, one needs a rather long interview to avoid memory effects. Van Meurs and Saris (1990) suggested that at least 20 minutes should be observed between repeated observations. These requirements would become rather expensive if one has to evaluate a large number of questions. Andrews therefore suggested as early as 1984 that one should employ meta-analysis over the available MTMM studies to summarize the results obtained so far and to use these results to predict the quality of questions before they are asked. At the time of this suggestion the number of studies was limited, but by now the number of studies is rather large and it therefore makes sense to use the meta-analysis of these experiments to predict the quality of survey questions.

14.2.3 Meta-analysis

The experiment presented above is typical of the MTMM experiments that have been conducted over the last 30 years. Similar studies have been conducted by Andrews (1984) and Rodgers et al. (1992) in the United States. Költringer (1995) has conducted a similar study for German questionnaires, while Scherpenzeel and

Saris (1997) in the Netherlands and Billiet and Waege in Belgium have conducted similar studies of Dutch questionnaires. In total, we found 87 MTMM studies containing 1067 survey items. All these studies were based on at least regional samples of the general population. In the United States, the Detroit area was used; in Austria and the Netherlands, national samples were used; and in Belgium, random samples of the Flemish-speaking part of the population were taken. The topics in the various experiments were highly diverse. In general, the MTMM experiments were integrated into normal survey research, where three or more questions of the survey are used for further experimentation. This approach guarantees that the questions are of the type commonly used in survey research. The same is true for the variation in the choices made in the design of survey items. The experiments were designed in such a way that the most commonly used methods (choices) can be evaluated (for more details, see Saris and Gallhofer, 2003).

In an effort to integrate these results, all the MTMM experiments carried out in the three language areas mentioned above were reanalyzed, and the survey items were coded on the characteristics mentioned in Table 14.5. As Scherpenzeel (1995) noted, coding with a common coding schema is necessary to analyze the data of the various countries simultaneously. Therefore, all survey items were coded in the same way.[2] The data from the various studies were pooled and an analysis was conducted over all available survey items, adding a "language" variable in the analysis to take into account the effect of the differences between languages.[3]

Normally, Multiple Classification Analysis (MCA) is used (Andrews, 1984; Költringer, 1995; Scherpenzeel, 1995) in the meta-analysis, but this is not possible with many variables. Therefore, dummy variable regression was used. The following equation presents the approach used:

$$C = a + b_{11}D_{11} + b_{21}D_{21} + \cdots + b_{12}D_{12} + b_{22}D_{22} + \cdots + b_3 N\text{cat} + \cdots + e \qquad (14.5)$$

In this equation, C represents the score on a quality criterion, either the reliability or validity coefficient. The variables D_{ij} represent the dummy variables for the jth nominal variable. All dummy variables have the value zero unless the specific characteristic applies for a question. For all dummy variables, one category is used as the reference category. This category received the value zero on all dummy variables of that set. Continuous variables, such as the number of categories (Ncat), have not been categorized except when it was necessary to take into account nonlinear relationships. The intercept is the reliability or validity of the instruments if all variables have a score of zero. Table 14.5 shows the results of the meta-analysis over the 1067 available survey items.

[2]The codebook can be obtained on request from the authors by e-mailing *Wsaris@planet.nl*.

[3]The analysis shows that the effect of the language is additive, which means that it affects only the absolute level of the quality indicators. If this is true for all languages, this would mean that all comparisons between different choices will hold true for other languages as well. Only the absolute level of the various quality criteria will be incorrect.

Table 14.5 Results of the Meta-analysis: The Effects on Reliability and Validity (×1000)

Variables	Number of Measures	Effect on Reliability			Effect on Validity		
		Effect	SE	Sign.	Effect	SE	Sign.
Domain							
National politics (0–1)	137	52.8	12.3	0.000	44.7	10.9	0.000
International politics (0–1)	64	29.4	18.1	0.104	57.8	15.9	0.000
Health (0–1)	82	16.9	13.9	0.225	21.6	12.0	0.073
Living condition/background (0–1)	223	21.4	8.7	0.014	4.6	7.4	0.541
Life in general (0–1)	50	−76.8	12.6	0.000	−15.9	10.8	0.139
Other subjective variables (0–1)	235	−66.9	14.2	0.000	−1.0	12.4	0.935
Work (0–1)	96	12.8	12.0	0.287	28.2	10.4	0.007
Others	136	0.0	—	—	0.0	—	—
Concepts							
Evaluative belief (0–1)	96	6.1	14.0	0.669	13.8	12.3	0.260
Feeling (0–1)	110	−4.2	10.9	0.704	−7.5	9.4	0.427
Importance (0–1)	96	35.9	15.6	0.021	18.6	13.6	0.171
Future expectations (0–1)	39	2.6	24.0	0.913	−9.0	20.6	0.662
Factual information (0–1)	27	−126.2	21.8	0.000	−150.5	19.2	0.000
Complex concepts	165	−72.3	17.4	0.000	−47.2	15.2	0.002
Other concepts	578	0.0	—	—	0.0	—	—
Associated characteristics							
Social desirability: No/a bit/much (0–2)	1023	2.3	6.2	0.709	8.0	5.3	0.137
Centrality: Very central till not central (1–5)	1023	−17.2	5.2	0.001	−8.9	4.4	0.046

(continued)

Table 14.5 (*continued*)

Variables	Number of Measures	Effect on Reliability			Effect on Validity		
		Effect	SE	Sign.	Effect	SE	Sign.
Time reference:							
Past (0–1)	106	43.9	15.0	0.004	−1.6	12.9	0.901
Future (0–1)	83	−13.3	16.1	0.409	−10.1	13.8	0.465
Present (0–1)	940	0.0	—	—	0.0	—	—
Formulation of questions: basic choice							
Agree/disagree (0–1)	167	4.0	10.9	0.713	41.6	9.5	0.000
Other types	212	0.0	—	—	0.0	—	—
Use of statements or stimulus (0–1)	317	−23.0	12.4	0.065	−12.1	11.1	0.275
Use of gradation (0–1)	809	79.6	14.1	0.000	−22.8	12.4	0.066
Formulation of questions: other choices							
Absolute-comparative (0–1)	98	12.7	16.3	0.436	−8.4	14.5	0.564
Unbalanced (0–1)	411	−3.2	11.2	0.772	−22.3	9.7	0.022
Stimulance (0–1)	92	−11.1	13.3	0.406	−11.7	11.5	0.308
Subjective opinion (0–1)	86	−5.9	19.9	0.767	−34.3	17.2	0.047
Knowledge given (1–4)	358	−12.7	8.8	0.145	−6.3	7.5	0.401
Opinion given (0–1)	101	0.6	14.5	0.964	−10.3	13.1	0.429
Response scale: basic choice							
Yes/no (0–1)	3	−22.2	19.5	0.254	−1.9	17.1	0.911
Frequencies	23	120.8	24.8	0.000	−95.9	21.5	0.000
Magnitudes	169	116.2	20.8	0.000	−115.5	18.3	0.000
Lines	201	118.1	20.9	0.000	−32.7	18.2	0.073
More steps	26	48.7	27.3	0.075	24.5	23.5	0.297
Categories	630	0.0	—	—	0.0	—	—

Response scale: other choices							
Labels: no/some/all (1–3)	1023	33.0	10.0	0.001	−4.5	8.8	0.605
Kind of labels: short, sentence (0–1)	35	−47.5	16.0	0.003	−9.1	13.7	0.506
Don't know: present, registered, not present (1–3)	1023	−6.7	4.8	0.165	−1.9	4.1	0.647
Neutral: present, registered, not present (1–3)	1023	12.6	4.6	0.007	8.4	4.0	0.038
Range: Theoretical range and scale unipolar theoretical range and scale bipolar; theoretical range bipolar but scale unipolar (1–3)	1023	−15.1	9.6	0.116	9.2	8.5	0.277
Correspondence: high–low (1–3)	1023	−16.8	7.5	0.025	1.1	6.5	0.867
Asymmetric labels (0–1)	195	25.5	11.8	0.031	22.3	10.4	0.033
First answer category: negative, positive (1–2)	358	−7.5	8.7	0.387	14.7	7.6	0.052
Number of fixed reference points (0–3)	1023	14.7	4.3	0.001	21.4	3.7	0.000
Number of categories (0–11)	1023	13.5	2.1	0.000	−1.9	1.8	0.298
Number of frequencies (0–5000)	1023	−0.068	0.009	0.000	−0.065	0.008	0.000
Survey item specification: basic choices							
Direct question present (0–1)	841	27.2	15.2	0.074	11.5	13.1	0.379
Question–instruction present (0–1)	103	−43.7	15.4	0.005	−4.2	13.3	0.753
No question or instruction	79	0.0	—	—	0.0	—	—
Respondent's instruction (0–1)	492	−12.7	7.3	0.083	−14.9	6.2	0.017
Interviewer's instruction (0–1)	119	−0.068	10.5	0.995	5.7	9.0	0.524
Extra motivation or information or definitions (0–3) > 0	304	7.1	6.7	0.296	−0.3	5.7	0.959
Introduction (0–1)	515	5.7	12.1	0.637	−10.5	10.3	0.312

(continued)

Table 14.5 (*continued*)

Variables	Number of Measures	Effect on Reliability			Effect on Validity		
		Effect	SE	Sign.	Effect	SE	Sign.
Survey item specification: other choices							
Complexity of the introduction							
Number of interrogative clauses (0–n)	62	−44.6	16.3	0.006	−21.3	14.1	0.132
Number of subordinate clauses > 0	129	29.3	9.8	0.003	7.6	8.6	0.377
Number of words per sentence > 0	510	−1.3	0.867	0.134	1.4	0.75	0.063
Mean of words per sentence > 0	510	0.064	1.1	0.954	−0.373	0.9	0.699
Complexity of the question							
Number of interrogative clauses (0–n)	192	12.7	9.8	0.199	−8.3	8.6	0.335
Number of subordinate clauses (0–n)	746	13.6	6.8	0.048	−17.7	5.9	0.003
Number of words (1–51)	1023	0.809	0.749	0.280	−1.3	0.644	0.041
Mean of words per sentence (1–47)	1023	−2.2	0.926	0.014	1.1	0.807	0.161
Number of syllables per word (1–4)	1023	−32.5	9.6	0.001	−10.4	8.2	0.207
Number of abstract nouns on the total number of nouns (0–1)	1023	2.9	27.7	0.917	−13.9	23.7	0.558
Mode of data collection: other choices							
Computer assisted (0–1)	626	−3.8	12.6	0.760	−38.3	10.7	0.000
Interviewer administered (0–1)	344	−50.8	22.9	0.027	−104.1	19.5	0.000
Oral (0–1)	219	10.4	12.2	0.397	25.3	10.3	0.014

Position in questionnaire							
In battery (0–1)	225	−10.3	12.3	0.403	28.9	10.7	0.007
Position of question	1023	0.304	0.064	0.000			
Position 25 (1–25)	396				1.5	0.402	0.000
Position 100 (26–100)	458				0.420	0.137	0.002
Position 200 (101–200)	129				0.267	0.062	0.000
Position 300 (> 200)	12				0.098	0.100	0.333
Language used in questionnaire							
Dutch (0–1)	731	−20.3	22.8	0.373	−76.0	19.8	0.000
English (0–1)	174	−72.0	26.6	0.007	−2.9	22.9	0.899
German (0–1)	118	0.0	—	—	0.0	—	—
Sample characteristics							
Percentage of low educated (3–54)	993	−0.911	0.596	0.127	1.1	0.511	0.027
Percentage of high age (1–49)	1023	−0.410	0.560	0.464	−0.753	0.488	0.123
Percentage of males (39–72)	1023	−0.030	0.690	0.966	0.405	0.596	0.497
MTMM design							
Design: one or more time points (0–1)	713	4.36	16.3	0.790	−36.9	14.3	0.010
Distance between repeated methods (1–250)	1023	−0.169	0.094	0.072	−0.249	0.081	0.002
Number of traits (1–10)	1023	−0.370	2.0	0.855	−1.7	1.7	0.320
Number of methods (1–4)	1023	0.959	2.6	0.715	−2.3	2.2	0.314
Intercept		825.2	69.5	0.000	1039.4	60.4	0.000
Explained variance (adjusted)		0.47			0.61		
Correction for single-item distance		−42.3			−62.25		
Starting point for single item		782.9			977.15		

Table 14.5 indicates the effects on the quality criteria validity and reliability of the various choices that can be made in the design of a survey question.[4] Each coefficient indicates the effect of an increase of one point in each characteristic indicated, while all other characteristics remain the same. Many of the characteristics are nominal variables. For example, the domain variable is such a nominal variable with several categories. In such cases, the nominal variable is transformed into several dummy variables. If a question concerns national politics, the question belongs to the first domain category ($D_{11} = 1$ for this category, while for all other domain variables, $D_{i1} = 0$) and the effect on the reliability and validity[5] will be positive, 0.052 and 0.044, respectively, as can be seen in the table. If a question concerns life in general, the fifth category applies ($D_{51} = 1$) and the effects are negative, −0.077 and −0.016, respectively. From these results it also follows that questions concerning national politics have a reliability coefficient of 0.052 + 0.077, 0.129 higher than the questions about life in general. Questions concerning consumption, leisure, family, personal relations, and race were coded as zero on all domain variables, and this set of questions can be seen as the reference category. For these questions the effect on reliability and validity is zero. This procedure has been applied to all characteristics with more than two nominal categories such as "concepts" and "time reference."

Other characteristics with at least an ordinal scale are treated as metric. For example, "centrality" is coded in five categories, from "very central" to "not central at all." In this case an increase of one point gives an effect of −0.0172 for the reliability, and the difference between a very central or salient item and an item that is not at all central is $5 \times -0.0172 = -0.0875$.

There are also real numeric characteristics, such as the "number of interrogative sentences" and "the number of words." In that case, the effect refers to an increase of 1 unit (i.e., one word or one interrogative sentence). A special case in this category is the variable "position" because it turns out that while the effect of position on reliability is linear, it is nonlinear for validity. To describe the latter relationship, the position variable is categorized and the effects should be determined within each category.

Another exception is the "number of categories in the scale." For this variable we have specified an interaction term because the effect was different for categorical questions and frequency measures. So, depending on whether the question is categorical or a frequency question, a different variable has to be used to estimate the effect on the reliability and the validity.

Space does not permit us to discuss the table completely here. More results can be found in Saris and Gallhofer (2003). In the remainder of this chapter we illustrate how the table can be used to predict the quality of survey questions before data collection.[6]

[4]The effects on the method effects have not been indicated because they can be derived from the validity coefficients.

[5]The effects in Table 14.5 were multiplied by 1000.

[6]The description of the categories used in the coding of the characteristics of questions is very brief and incomplete in Table 14.5. If one would like to code questions to make quality predictions, one

14.3 SURVEY QUALITY PREDICTION

Suppose that one would like to evaluate the quality of proposed survey items using the information summarized for this purpose in Table 14.5. To do this, one must first code all the items on the variables in the classification system and then use the coefficients in the prediction table to calculate a predicted total score for reliability and validity. This requires considerable effort. It would therefore be desirable to have a computer program that could evaluate all questions of a questionnaire automatically on a number of characteristics. The designer of the survey could then, on the basis of this information, determine which items require further study to improve the quality of the data collected.

14.3.1 Two Survey Quality Prediction Programs

A prototype of such a program for the Dutch language questionnaires, called the *survey quality predictor* (SQP), has been developed by van der Veld et al. (2000). The program contains the following functions: (1) reading and automatic coding of the survey items in a questionnaire, (2) prediction of item quality on a number of criteria, (3) information about the reasons for the quality level, and (4) suggestions for improvement of the item.

It would be rather complex and time consuming to develop such programs for many languages. The automatic coding of the questions is particularly complex and requires different rules for different languages. Therefore, a nonautomatic version of SQP has also been developed where the coding is done by the researcher. The researcher answers questions about the characteristics of the survey item and the program uses this information to estimate the reliability and validity and the total quality of a question on the basis of the results of Table 14.5. We illustrate both procedures below.

14.3.2 Automatic SQP

Virtually all questionnaires are now written with text processors and are computer readable. In principle, this makes it possible to analyze the text automatically in order to classify the different survey items with respect to characteristics that affect the quality of the data collected. An automatic coding procedure has been developed by van der Veld et al. (2000) based on the characteristics studied by Scherpenzeel and Saris (1997).[7] This coding procedure is implemented in the SQP prototype. In this program the survey items are converted into files that can be

needs the code book developed for this purpose. This code book can be obtained from the authors (*Wsaris@planet.nl*).

[7]It is interesting to note that the automatic procedures for coding were more valid than human coders because the human coders are often inconsistent in their judgments, whereas the program always applied the same and correct rule. This was detected by comparing the codes of two human coders with the codes of the program. This result has been found for Dutch questionnaires (Saris and Gallhofer, 1998) and German questionnaires (Gallhofer and Saris, 2000).

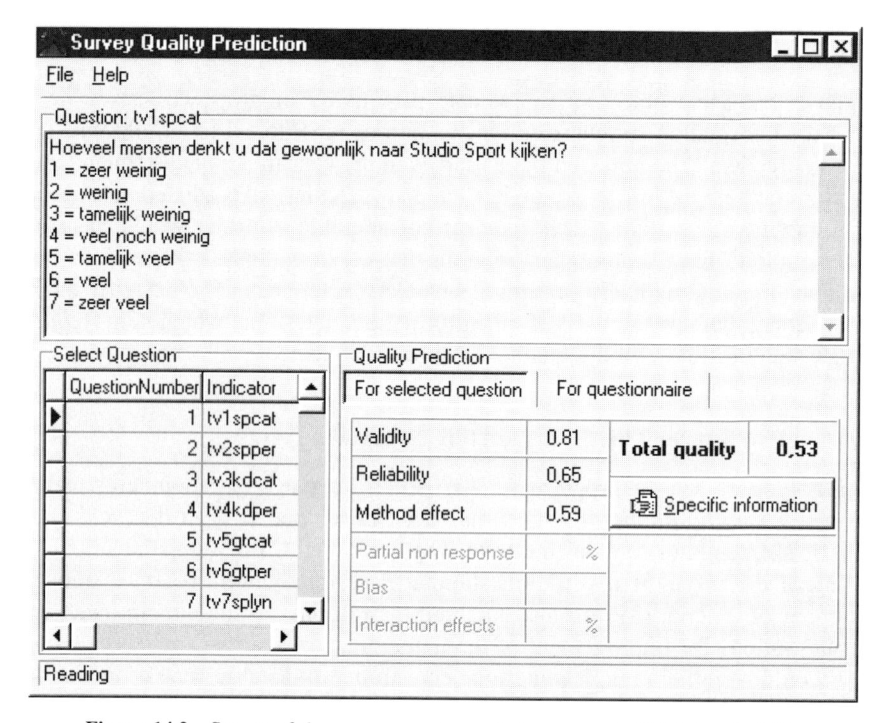

Figure 14.2 Screen of the survey quality prediction using MTMM data in SQP.

read by the program. For those characteristics that cannot be coded automatically, questions are presented to the user, and the user is asked to provide the codes. All these codes are used in the next steps for prediction.

This prototype of a program for the prediction of reliability, validity, method effect, and total quality demonstrates the feasibility of our approach in that it facilitates the required coding and calculations. Without such an aid, carrying out this procedure could be so time consuming that researchers would avoid using it. The first screen of the predictor is shown in Figure 14.2. This screen shows that for the survey item named "tvsp1cat" (reading: "Hoeveel mensen denkt u dat gewoonlijk naar Studio Sport kijken?", that is, "How many people, do you think, usually watch *Studio Sport* [on TV]?"), the reliability is 0.65, the validity is 0.81, the method effect is 0.59, and the score for the total quality is 0.53. These results have been calculated on the basis of the automatic coding of the survey item and the linear prediction equation (14.5).[8]

On the basis of the information provided for each survey item, a user of the program can decide whether or not the survey item satisfies the required quality level regarding the various criteria. If not, one can also obtain information about what might be the cause(s) of the problem. Given that information is available

[8]These calculations were done taking into account that the (method effect)^2 = 1 − (validity coefficient)^2, and the total quality coefficient = (reliability coefficient) · (validity coefficient).

about the effect of the various item characteristics (or choices) the program can also indicate what the contribution is of each of the choices to the quality predictors and can also suggest improvement by changes in these choices.

14.3.3 Semiautomatic SQP

In the semiautomatic procedure, the researcher must first answer a series of questions about the choices he or she has made in developing the survey item.[9] Subsequently, the program estimates the reliability, validity, and total quality of the question based on these codes and the quality predictions shown in Table 14.5. This approach has been applied to one of the questions proposed for the European Social Survey discussed in Table 14.1. The question used in this example reads as follows:

And on the whole, how satisfied are you with the way democracy works in Britain? Are you ...

1 very satisfied
2 fairly satisfied
3 fairly dissatisfied
or 4 very dissatisfied?
8 don't know

Table 14.6 presents the results of the coding of this question on the characteristics mentioned in Table 14.5 and the estimation of the total quality of the question. In the first column the names of the characteristics (variable) of the question are given that were also used in Table 14.5, while the next column (code) gives the score on that characteristic. The last two columns give the effects of the choices on the reliability and validity of a question. For a detailed discussion of the definition of the characteristics and scores of the code book, we refer to Saris and Gallhofer (2003).

The coding of the question and the addition of the effects of the different characteristics of the question leads to an estimate of the reliability coefficient of 0.760 and of 0.893 for the validity coefficient. It is interesting to compare this result with the results obtained in the experiment summarized in Table 14.4. There, the reliability coefficient was estimated to be 0.81 and the validity coefficient, 0.94. These estimates are slightly higher than the predictions in Table 14.6, but one should realize that the coefficients in these experiments are overestimated by 0.042 with respect to reliability and by 0.062 with respect to validity, because of the presence of repeated observations in one survey.[10] Correcting for

[9]Note that the choices made by the researcher become characteristics of the survey items and become predictor variables of quality after they have been coded.

[10]It may be observed that the correlations between variables in a questionnaire increase when the questions are repeated. This suggests that people realize better the relation between these questions. This could lead to higher estimates of the reliability and validity. This is one of the reasons that a part for design effects is included in Table 14.5. The effects in that part indicate to what extent the reliability and validity are too high because of design effects. Therefore, the correction factors are given in the table for single questions. These correction factors have been applied here.

Table 14.6 Prediction of Reliability and Validity Coefficients (× 1000) for Satisfaction with Democracy

Coefficient	Code	Reliability	Validity
Intercepts		782.9	977.15
Domain	National politics	52.8	44,7
Concept	Feeling	−4.2	−7.5
Associated characteristics			
Social desirability	No	0	0
Centrality	Medium	−51.6	−26.7
Time reference	Present	0	0
Formulation of question			
Basic	Other type	0	0
Use of stimuli/statements	No	0	0
Use of gradation	Yes	79.6	−22.8
Other choices			
Absolute statement	No	0	0
Unbalanced	Yes	−3.2	−22.3
Stimulance	No	0	0
Subjective opinion	No	0	0
Knowledge given	No	−12.7	−6.3
Opinion given	No	0	0
Response scale: basic choice			
	Categories	0	0
Response scale: other choices			
Labels	Yes	99	−13.5
Kind of labels	Short	0	0
Don't know	Present	−6.7	−1.9
Neutral	Not present	12.6	8.4
Range	Theoretical range and scale bipolar	−30.6	18.4
Correspondence	Low	−50.4	3.3
Asymmetric labels	No	0	0
First category	Positive	−15	−29.4
Fixed reference points	No	0	0
Number of categories	4	54	−7.6
Number of frequencies	No	0	0
Present in survey item			
Question	Present	27.2	11.5
Question–instruction	Not present	0	0
No question or instruction	No	0	0
Respondent's instruction	Not present	0	0
Interviewer's instruction	Present	−0.07	5.7
Extra motivation	Not present	0	0
Introduction	Not present	0	0

Table 14.6 (*continued*)

Coefficient	Code	Reliability	Validity
Other choices: complexity of the introduction			
Number of interrogative sentences	0	0	0
Number of subordinate clauses	0	0	0
Number of words	0	0	0
Number of words per sentence	0	0	0
Other choices: complexity of the question			
Number of interrogative sentences	1	12.7	8.3
Number of subordinate clauses	0	0	0
Number of words	27	21.8	−35.3
Mean of words per sentence	27 in 1	−59.4	29.7
Number of syllables per word	47 in 27	−56.6	−18.1
Number of abstract nouns on the total number of nouns	3 from 3	2.9	−13.9
Mode of data collection			
Computer assisted	No	0	0
Interviewer administered	Yes	−50.8	−104.1
Oral	Yes	10.4	25.3
Position in the questionnaire			
In battery	No	0	0
Position	55	16.7	23.1
Language			
Dutch	No	0	0
English	Yes	−72	−2.9
German	No	0	0
Sample characteristics			
Percentage of low educated			
Percentage of high age			
Percentage of males			
Total result		759.33	903.33

this degree of overestimation, the values would be expected to be around 0.768 for reliability and 0.878 for validity. The corrected results of Table 14.4 are very close to the predicted values reported in Table 14.6, but this prediction is based on the existing knowledge presented in Table 14.5 without collecting any new data. The agreement between the two approaches will not always be so close. There will be cases where the differences are larger because the predictions are not perfect. As shown in Table 14.5, the explained variance of 50% for reliability and 60% for validity indicate that the predictions will be quite good, but one should realize that the coding of the questions certainly contains errors, and the same is true for the estimates of reliability and validity.

14.3.4 How to Use these Predictions

In general, we prefer observed variables with reliability and validity coefficients as close as possible to 1.0. The product of these two coefficients gives an indication of the total quality of the measure, and the square of the product gives the variance of the observed variable explained by the variable to be measured. In the specific case discussed above, the explained variance is $(0.67)^2 = 0.46$. This result is rather low but close to the mean value in survey research suggested by Alwin and Krosnick (1991). Thus, 46% of the variance in the observed variable comes from the variable of interest. The rest (54%) is error; the systematic error variance is 12% $((0.76)^2 \times [1 - (0.893)^2])$ and the random error variance is 42% $[1 - (0.76)^2]$. This shows that the random error is considerable and the quality of the measure rather low.

Looking at equation (14.3) once more, one can see that the systematic effect of the method (second term) increases the correlations between variables. Let us see what we can say about this effect for the satisfaction questions of Table 14.1. All questions are measured in the same way. So let us assume that these questions have exactly the same reliability and validity coefficient as the question predicted in Table 14.6. In that case we could expect an extra correlation between these measures of $(0.76)(0.46)(0.46)(0.76) = 0.12$. This effect on the correlations is due to the fact that the same method has been used for all three questions. Fortunately, in this case, the effect of the equality of the method on the correlations is rather small. However, that need not be the case.

In this case, the lack of quality of the two measures is a more serious problem. The total quality is equal to the product of the reliability and validity coefficient for each measure. The total quality is 0.67. Assuming that the quality is the same for each measure, equation (14.3) suggests that this (lack of) quality reduces the correlation between the opinions by a factor of $(0.67)^2$, or 0.46, which means that the correlation observed will be 46% of the correlation between the variables of interest. In this case, this reduction is rather large, an effect due to lack of quality of the measurement procedure.

The primary reason for the lack of quality in this case is random measurement error, so there are good reasons to try to improve at least the reliability of the measures and to keep the validity approximately the same or even to improve that quality criterion as well. There are two possible ways of improving a survey item: (1) by changing the characteristics with the most negative effect on the quality criteria, and (2) by changing the characteristics that can lead to an improvement of the quality criteria. There are, however, limitations to both approaches. For example, one cannot change the domain or the concept, and even the mode of data collection is often determined by the budget and not by quality considerations.

In this specific case we see the following possibilities for improvements:

1. One could decide to use a category scale with categories 0 to 10, producing a scale with 11 instead of four categories.
2. One could also decide to use fixed reference points, for example, by labeling the endpoints as "completely satisfied" and "completely dissatisfied" and the middle category as "neither satisfied nor dissatisfied."

The corrections specified above lead to the following reformulation of the question:

And on the whole, how satisfied are you with the way democracy works in Britain? Choose a number on a scale from 0 and 10, where 0 means completely dissatisfied, 10 means completely satisfied, and 5 means neither satisfied nor dissatisfied.

Note that one cannot say that the reliability increases by 7×13.5 because of the increase in the number of categories, and by 3×14.7 by the introduction of three fixed reference points. This is not the case because other characteristics of the question also change, such as the number of words, the direction of the ordering of the categories, and the instruction to the respondents. But with one of the two programs mentioned above, the new item could be evaluated again. These estimations and improvements are then made on the basis of the complete knowledge we have about the effect of such choices on the quality of survey questions with respect to reliability and validity.

14.4 CONCLUSIONS

The purpose of the SQP programs is to provide the user a convenient way to make use of the existing knowledge of the effects of survey item characteristics on data quality. This information is not summarized in tables, as is customary in the literature, but in programs for quality prediction. This information is also not separately provided for each item characteristic (choice option), as is mostly the case, but for many variables at the same time. The SQP program thus provides users with the prediction of the quality of survey items and suggestions for the improvement of the items. The user has the possibility of using this information in the design of his or her own questionnaires at a point when there is still time to improve the questionnaire. In the same way, the nonresponse and bias could in principle be predicted on the basis of the study of Molenaar (1986). More quality indicators can probably be given in the future, but they require further research.

We think that such an expert system presents real advantages for survey researchers, since knowledge concerning the quality of survey items is by now so dispersed over the methodological literature that making use of it is very time consuming. A program like SQP can bring this information together and use it to predict the quality of the survey items that the user is designing for his or her research.

14.4.1 Limitations of the Approach

This analysis presents an intermediate result. So far, 87 studies have been reanalyzed, with a total of 1067 survey items. This may seem a large number, but it is not enough to evaluate all variables in detail.

Important limitations are:

- Only the main categories of the domain variable have been taken into account.
- Questions concerning consumption, leisure, family, and immigrants could not be included in the analysis.
- The concepts of norms, rights, and policies have been paid too little attention.
- Open-ended questions and WH questions have not yet been studied.
- Mail and telephone interviews were not sufficiently available to be analyzed separately.
- There is an overrepresentation of questions formulated in the Dutch language.
- Only a limited number of interactions and nonlinearities could be introduced.

The database will be extended in the future with survey items that are now underrepresented.

Nevertheless, taking these limitations into account, it is remarkable that the analysis has shown that the choices discussed earlier can explain almost 50% of the variance in reliability and 60% of the variance in validity for 1067 survey items. In this context it is also relevant to refer to the standard errors of the regression coefficients, which are relatively small. This indicates that the multicollinearity between the variables used in the regression as independent variables is relatively small. If one takes into account that all estimates of the quality criteria contain errors, while in the coding of the survey item characteristics errors are certainly also made, it is remarkable that such a highly explained variance can be obtained.

This does not mean that we are satisfied with this result. Certainly, further research is needed, as we have indicated above. But we also think that it is the best summary we can make at the moment of the effects of the choices made in designing a questionnaire on reliability and validity. We therefore recommend the use of these results for the time being, until the new results have been incorporated in the meta-analysis and a new Table 14.5 can be produced.

14.4.2 Future Development

For the future, we expect first of all to collect new MTMM experiments, which will be integrated into the existing database and will lead to new versions of the SQP programs. If there are more datasets available, the model used for estimation of the effects will also change, because more interaction effects can be built into equation (14.5). We are aware of the fact that more conditional relationships would be required to make more precise predictions, but that requires more data than we have at present. Whether the new data will change the predictions very much is an interesting question for empirical investigation.

We also plan to study the prediction of other criteria. We think first of adding "item nonresponse" and "interaction problems." For the first topic there are already enough data available to study how much the survey item characteristics determine item nonresponse. If a table such as Table 14.5 were generated for item nonresponse, the prediction of this criterion could be integrated into the programs. For interaction analysis, the same approach could be used, but for that criterion there are not yet enough data available.

ACKNOWLEDGMENTS

We appreciate the helpful comments of Judith Lessler on earlier versions of this chapter. Comments on this draft should be sent to Willem Saris: *Wsaris@planet.nl*.

CHAPTER 15

Testing Paper Self-Administered Questionnaires: Cognitive Interview and Field Test Comparisons

Don A. Dillman
Washington State University

Cleo D. Redline
National Science Foundation

15.1 PURPOSE

Historically, cognitive interviewing has been used predominately to identify wording problems in questions. However, self-administered questionnaires are composed of much more than questions. Typically, they include mailing packages with envelopes and letters. And they are administered through the visual channel, which means that more than just the words of a question convey meaning to respondents (Redline and Dillman, 2002). Although discussions of cognitive interviewing methods often mention their application to self-administered questionnaires (e.g., DeMaio and Rothgeb, 1996; Forsyth and Lessler, 1991), little practical information exists in the literature on how such procedures might need to be expanded to elicit information about these additional features.

Despite cognitive interviewing's widespread adoption, it remains unclear to what extent the information obtained from cognitive interviews, which are typically done in very small numbers and without the representativeness associated

Methods for Testing and Evaluating Survey Questionnaires, Edited by Stanley Presser, Jennifer M. Rothgeb, Mick P. Couper, Judith T. Lessler, Elizabeth Martin, Jean Martin, and Eleanor Singer
ISBN 0-471-45841-4

with sample surveys, can or should be relied on to make reliable decisions about changing even the wording of questions, let alone survey procedures. In one published comparison, Gordon and Schecter (1997) found that results from the field strongly supported predictions from cognitive interviews when it came to changes in question wording.

This chapter has two objectives. The first is to discuss briefly the ways in which the objectives and procedures for cognitively testing self-administered questionnaires need to be expanded from those that apply to interviews. The second objective is to report three case studies in which cognitive interviews and field experiments were conducted simultaneously to evaluate three self-administered questionnaire design issues: response rates, improvement of reduction of branching error rates, and decreasing the amount of item nonresponse. Our emphasis in these comparisons is to ascertain the relative value of information from the cognitive interviews and experimental data for improving self-administered mail questionnaires.

15.2 THEORETICAL BACKGROUND

15.2.1 Design of Self-Administered Questionnaires

The questions in self-administered questionnaires require that information expressed in as many as four languages—verbal, numerical, symbolic, and graphical—be processed and understood to ascertain the intent of questions (Redline and Dillman, 2002). In addition to the verbal language relied on to convey meaning, numerical language or numbers might be used to communicate what information needs to be processed and the order for doing that. Furthermore, symbolic language, which consists of symbols such as arrows, check boxes, and parentheses, provides information on navigation and interpretation of questionnaire information, apart from the information provided by the previous two languages. Finally, graphical language (i.e., attributes of printed information such as the brightness and color, shape, and location of information) conveys how information is to be grouped and interpreted.

One of the challenges for designing self-administered questionnaires is to utilize some or all of these languages in concert to achieve optimal question meaning. Thus, it is important that cognitive interviews for self-administered questionnaires be done in a way that will provide an understanding of how each of these languages affects respondent behavior, jointly, and sometimes individually.

A much-utilized model for analyzing response behavior is that developed by Tourangeau (1984), whereby that action is divided into comprehension, recall, judgment, and reporting. In self-administered questionnaires, this model needs to be expanded to include perception (i.e., whether or not respondents perceive the existence of the question and actually read it) (Redline and Dillman, 2002). This conclusion is based on Matlin (1994), who has described the complex mental processes, such as top-down and bottom-up processing, necessary for achieving pattern recognition prior to comprehension.

Numerical, symbolic, and graphical languages may be particularly important in determining not only whether the verbal language is perceived, but also the order in which it is perceived, which in turn may affect the way it is comprehended. Eye-movement analysis has been used in other fields, most notably reading and scene perception, to study cognitive processing (e.g., Rayner, 1992). In the survey field, preliminary eye-movement research with branching instructions provides evidence that the graphical layout of information can affect whether the branching instruction is perceived, the order in which it is perceived, and the result this has on respondents' branching behavior (Redline and Lankford, 2001). It cannot be assumed that respondents are perceiving information on the questionnaire, nor that they are perceiving information in the order that it was intended.

For self-administration, questions need to be packaged into a physical structure (i.e., the questionnaire), which in turn is placed in an envelope to which a cover letter is added to communicate the intent of the questionnaire. It is important to learn from cognitive interviews first, if respondents perceive all of this information, and second, how they are interpreting what they perceive. How they perceive and interpret information may provide additional insight into how the information is affecting their motivation to complete or not complete the entire questionnaire; thus, the envelopes, cover letters, and the questionnaire's construction may become the focus of cognitive interviews, too, quite apart from questions about the impacts of questionnaire wording.

In self-administered questionnaires, any information that the sponsor believes should be available to the respondent must be provided in writing. Thus, cognitive interviews for self-administered questionnaires may need to ascertain whether and/or how respondents utilize and interpret such information. The result is that cognitive interviews may need to focus on features and issues not previously considered much. It is also apparent that with so many features of self-administered questionnaires being subject to design considerations, the specific objectives for sets of cognitive interviews may differ substantially from one another.

Concurrent think-aloud and retrospective techniques for conducting cognitive interviews, our primary focus in this chapter, are only two of many cognitive laboratory methods that might be used to evaluate the content of questionnaires. Forsyth and Lessler (1991) have identified four general types of methods (expert evaluation, expanded interviews, targeted methods, and group methods) and 17 specific methods for understanding the thought processes that respondents use to interpret and answer survey questions. Except for "interactional behavior coding" that is used to assess the interaction between respondents and interviewers, all of the 17 techniques seem applicable to evaluations of question wording in self-administered questionnaires.

However, because of the visual components of the questionnaire, the range of concerns to be addressed using these methods may be expanded. For example "expert analysis" can be expanded to address issues of whether a respondent is not likely to return the questionnaire and the reasons, as well as issues surrounding question wording. Indeed, all three of the case study issues reported here (i.e., response rate impacts of mailing packages, whether respondents follow branching instructions correctly, and the effects of number and graphic changes) could be addressed in this way.

It remains to be seen how well some of these methods do in evaluating such issues. For example, one of the group methods that Forsyth and Lessler report, focus groups, was used to gain insight into whether a mandatory message should be used on the envelope containing the 2000 Census. Participants in all locations strongly recommended not using this message (Dillman et al., 1996a), yet when placed into a national experiment this technique improved responses rate from 8 to 10 percentage points, more than any other technique tested for use in the Census. Additional research needs to be done on whether such motivational issues can be evaluated effectively by these methods.

15.2.2 Think-Aloud versus Retrospective Interviews

One of the strengths of a think-aloud cognitive interview protocol is that respondents mention what is going through their minds as it occurs to them. It has been argued that if they were to wait until the interview is completed and then be asked to recall what they were thinking when they answered one of the early questions, they may not remember or remember correctly (Ericsson and Simon, 1984; Forsyth and Lessler, 1991). However, a think-aloud protocol that requires respondents to read questions and say out loud everything going through their minds results in an interaction, regardless of how slight, between the respondent and the cognitive interviewer. It may encourage respondents to read words and phrases that they may have passed over more quickly or perhaps skipped altogether, if left to read silently at their own pace. Furthermore, having to read out loud and respond to interviewer encouragement and probes at the same time would seem to divide the respondent's attention between the questionnaire and the interviewer. To the extent that this happens, it is questionable whether this procedure emulates what actually occurs when a respondent completes a questionnaire by himself or herself at home. In studies where the major aim is to evaluate question wording, regardless of whether respondents would have read those words if left to their own devices, the emphasis on reading aloud and responding to the interviewer do not appear problematic. However, if an additional concern is to learn whether respondents naturally read information, an alternative to the think-aloud may be desirable.

Retrospective methods have been developed to counteract these possible features of the think-aloud (Dillman, 2000a; Dillman et al., 1996b). Under the retrospective method, respondents are asked to complete a questionnaire as if they are in their home or office alone, and to ignore the presence of the cognitive interviewer. The interviewer observes the answering process while noting errors (e.g., missed branching instructions), hesitancies, facial changes, and other possible indicators of problems. After the questionnaire is completed, the interviewer then asks questions to ascertain why those events might have occurred. In either the think-aloud or retrospective situations, additional debriefing questions, or probes, may also be asked.

Forsyth and Lessler note that think-aloud techniques may change a respondent's behavior and that "... retrospective techniques permit researchers to use relatively unobtrusive methods to observe a survey interview" (1991, pp. 399–400). Various procedures may be used to minimize possible effects of the interviewer's presence while observing the respondent behavior. One procedure used by the Washington State University (WSU) Social and Economic Sciences Research Center (SESRC) is to employ closed-circuit television that projects images of the form (from an overhead camera) and the respondent onto a television monitor that can be observed from another room. A second procedure used there is an interviewing room that is large enough to allow the interviewer to move away from the respondent to another table from where observations of the respondent's face, hands, and form can still be made and notes taken without distracting the respondent.

Based on Gerber et al. (1997), the concurrent interview may lead respondents to pay more attention to the questionnaire's content than they would in a self-administered setting. A separate comparison of 28 concurrent think-aloud interviews with 27 retrospective interviews evaluated whether respondents navigated differently in the two methods (Redline et al., 1999). A difference was noted on only one of 12 navigational indicators coded for the interviews. However, the difference occurred on the only indicator that was dependent entirely on graphical language; think-aloud respondents were more likely to make a navigational error. Thus, it was reasoned that the interactional aspects of the interview might have made it more difficult for respondents to focus on the graphical language than was the case for retrospective respondents. Further studies of this issue need to be done. Meantime, it seems appropriate to conduct cognitive interviews for self-administered questionnaires using both think-aloud and retrospective methods.

Use of retrospective methods requires the development of somewhat different protocols than do think-aloud interviews (Dillman, 2000a). It is necessary to emphasize to the respondent that they should fill out the questionnaire just like they would do it at home, and to formulate debriefing questions or probes. In addition, while the respondent is answering the questionnaire, it is necessary for the interviewer to formulate debriefing questions, based on observations being made, to ask after the response task is complete. This is in contrast to the think-aloud, where it is necessary to train respondents on how to talk out loud while responding to the questionnaire and to develop common probes and encouragement phrases that may need to be provided to the respondent during the interview.

15.2.3 Comparing Results from Cognitive Interviews and Field Experiments

Although cognitive interviews and field experiments may address the same general question (e.g., are respondents more likely to respond to one questionnaire mail-out package than another), the specific questions that can be addressed are often quite different. In field experiments, statistical inferences at stated probability levels can be used to reach clear-cut decisions on whether a particular design feature will influence response rates or data quality. A shortcoming of such experiments, however, is that they often provide little insight into why the difference occurs. Even when design variations across experimental treatments are limited to one discernible characteristic (e.g., color of the questionnaire), an experiment typically reveals nothing about why the characteristic has an effect.

In contrast, cognitive interviews may provide insight into why respondents, for example, like or dislike certain mail formats and are affected by them. The open-ended nature of cognitive interviews allows respondents to express ideas that the research investigator has not yet asked, (e.g., "This color is really ugly") or might never have intended to ask about (e.g., "I'm not sure where I am supposed to go next"). Thus, possible barriers to completing a questionnaire accurately and deciding to return it may be identified. However, the intensive efforts required for recruiting, getting commitment from, and meeting with respondents means that relatively few interviews get completed. If the population to be surveyed is fairly heterogeneous and the issues to be investigated are many, it seems unlikely that respondents will represent well the range of issues likely to be encountered by the eventual survey population. Sometimes, investigators specify characteristics of volunteers (e.g., three different age categories, four different race and ethnicity groups, two different housing situations) in order to get insights from a range of people. The result is to draw conclusions on the basis of only one or two people to whom a particular question or set of categories applies.

Cognitive interviews exhibit richness as a source of hypotheses about whether differences in outcomes are likely to occur as a result of using certain questions, questionnaires, and delivery/retrieval procedures. They also serve as a source of hypotheses as to why such differences may occur. In contrast, field studies have the richness of providing generalizable evidence as to whether differences in outcome actually occur and the magnitude of those differences. In the three case studies examined here, the simultaneous conduct of cognitive interviews and field experiments allowed us to address three questions: Did the outcomes suggested by the cognitive interviews actually happen in the field studies? Second, did the cognitive interviews suggest possible explanations for patterns observed in the field experiments? Third, how were subsequent study and questionnaire design decisions affected by these two types of data?

15.3 CASE STUDY 1: EVALUATION OF MARKETING VERSUS OFFICIAL GOVERNMENT APPEALS FOR DECENNIAL U.S. CENSUS FORMS

15.3.1 Test Objectives

In 1994, three potential short forms for use in the 2000 Census were evaluated through cognitive interviews and field tests. One of the forms (green-folded) had evolved from previous census tests (Dillman et al., 1993). Two additional forms (gold vertical and yellow-folded) were developed by a private contractor who used bright colors and graphics to give them a marketing appeal. The forms and mailing packages differed from each other in multiple ways [see Dillman (2000a, pp. 306–310), Dillman et al. (1996b), and Pullman (2000) for detailed descriptions, including pictures].

Development of the marketing appeal forms was motivated by the belief that making them contrast with the predominant red and blues then prevalent in mass mailings would encourage response. Design of the plainer green-folded form, which evolved from previous census tests, was motivated by evidence that association with government would produce higher response rates (Heberlein and Baumgartner, 1978).

15.3.2 Research Methods

Fifty-five cognitive interviews were conducted during the spring of 1995: 35 at WSU and 20 at the U.S. Bureau of Census in Suitland, Maryland. A shared protocol was used in the two locations where all the interviews were done. Half of the respondents were assigned randomly to think-aloud interviews, and half were assigned to retrospective interviews. During each of the interviews, which lasted less than an hour, the respondent completed one of the short forms and was then asked to look at and comment on each of the other forms. A standard set of debriefing questions was asked in each of the interviews [see Dillman et al. (1996b) for additional details]. Results from the cognitive interviews were presented in a detailed report, completed prior to the results from the field experiment being known (Dillman et al., 1996b). The information presented here about conclusions from the cognitive interviews is drawn from that report. The same forms were used for three of 12 panels in a much larger experiment that tested multiple forms and question formats for possible use in the 2000 Census. Each of the three experimental forms was sent to a national probability sample of 2400 households. The results reported here are drawn from an analysis by Leslie (1996).

15.3.3 Cognitive Interview Results

Conductors of the cognitive interviews reported considerable tabular data (Dillman et al., 1996b). Those data suggested that respondents were most likely to recommend use of the green form and least likely to recommend use of the gold vertical form. Respondents also indicated that the envelope containing the green form looked more like it was from the government and less like junk mail than those containing the marketing appeal forms. They were also more likely to have seen the mandatory message on that envelope and indicated a greater likelihood that they would respond because of the message.

Based on these data, the report concluded: "Any marketing strategy that is used must be designed carefully not to undermine the authority and official look of the census form." It went on to state that the "bright loud color on the envelope, the use of colored buttons containing the slogan or the mandatory message, and the graphical representation of the Capitol reminded many respondents of junk mail, and these features appear to undermine the authority, seriousness, and urgency of the mailing package for the gold and yellow forms" (Dillman et al., 1996b, p. 56).

The report also concluded that the mandatory message appeared to be the most powerful visual determinant of respondents' reactions and noted the difficulty of seeing that message on the gold and yellow envelopes and concluded with a recommendation that future mailing packages should prominently display a message like that used on the green form's (white) envelope. The report stopped short of stating that use of the gold or yellow forms was likely to lower response rates. That issue could not be addressed in a meaningful way.

15.3.4 Field Experiment Results

In the national field test it was found that response to the official government green form was 75.6%, while response rates to the yellow and gold forms were, respectively, 9.4 and 4.9 percentage points lower (Leslie, 1997). In addition, it was found that several other forms similar to the green form, except for greater length (12 pages versus 4 pages), performed better than the marketing appeal forms. Leslie offered several possible reasons for the lower response rates from the marketing appeal forms, including negative effects from colors, slogans, graphics, a prominent due date, and the word "TEST" on the envelopes. Because there were so many differences between forms, a definitive conclusion could not be provided: "Additional testing should be conducted to attempt to isolate the specific design feature(s) responsible for producing this effect" (Leslie, 1997, p. 339).

15.3.5 Outcome

The results of these two studies were quite complementary. Although the potential problems identified in the cognitive interviews (e.g., color of envelope, poor visibility of the mandatory message, and perception of junk mail) had pointed to some potentially serious problems, no conclusion could be reached about the quantitative impact on response rates. The field experiment, on the other hand, while documenting a negative effect on response, provided minimal insight into possible reasons.

15.4 CASE STUDY 2: INFLUENCE OF ALTERNATIVE BRANCHING INSTRUCTION DESIGNS ON NAVIGATIONAL PERFORMANCE IN CENSUSLIKE AND CENSUS QUESTIONNAIRES

15.4.1 Test Objectives

It has been theorized that modifying the visual languages of branching instructions will influence whether or not respondents see those instructions and respond to them as intended (Jenkins and Dillman, 1997). A test of these ideas was proposed for the Census 2000 long form in which 19 of 91 items asked of each person in the household included the feature that respondents who chose certain answer choices were instructed to branch ahead, while other respondents were supposed to continue with a different question.

Two formats were proposed as possible alternatives to the existing Census Bureau practice of placing branching instructions several spaces to the right of the answer choice in a font identical to that used for the answer choice. Under the existing format, the check box was placed to the left of the first word in the answer choice. A person's vision is sharp only within two degrees (Kahneman, 1973), which is equivalent to about 8 to 10 characters of 12-point type on a questionnaire. Thus, it was hypothesized that respondents might not see the branching instruction because it was outside the normal foveal view of respondents when they were in the process of marking the check box.

The first alternative, which we labeled the *prevention method*, included a warning message prior to the beginning of each question that contained a branching instruction, placement of the check box to the right of the answer categories, and location of the branching instruction immediately next to the box as shown in Figure 15.1. The second alternative, which we labeled the *detection method*, relied upon arrows to direct respondents to their appropriate next question, use of an enlarged and bolder font for expressing the branching instructions, and words [e.g., "(If No)"] placed at the beginning of the next listed question to clarify who should answer that question. Each manipulation of the branching instructions involved several visual language concepts, including changes in words and symbols, as well as graphics (i.e., location, brightness, contrast, and size of font). To prepare for a possible test in the 2000 Census, an initial test questionnaire was developed that could be asked of students, in which the three formats were designed for cognitive testing and evaluated simultaneously in a classroom experiment.

15.4.2 Research Methods

A four-page questionnaire was developed that had the look and feel of the Census questionnaire but consisted of general questions about lifestyles and activities, all of which could be answered by young adults. The questionnaires contained a total of 52 questions, 24 of which contained branching instructions. A variety of types of branching instructions were developed (e.g., every response option skips, location at the bottom of the page, skips after closed-end items as well as open-ends). A total of 1266 students enrolled in 34 classes at WSU were randomly assigned to complete one of the three questionnaire versions in the fall and winter of 1998. The questionnaires were completed at the beginning or end of a regularly scheduled class period (Redline and Dillman, 2002; Redline et al., 2001).

Simultaneously, 48 cognitive interviews were conducted to learn whether and why people perceived, comprehended, and correctly or incorrectly followed branching instructions in the three forms. Half of these interviews was completed

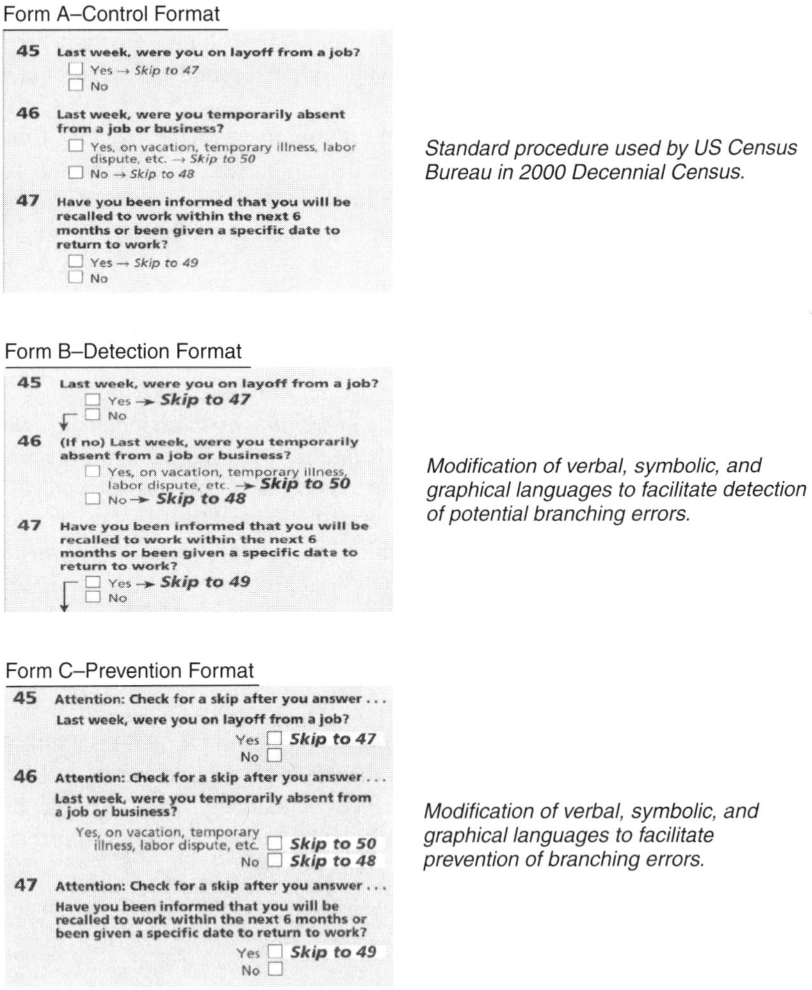

Form A–Control Format

45 **Last week, were you on layoff from a job?**
☐ Yes → *Skip to 47*
☐ No

46 **Last week, were you temporarily absent from a job or business?**
☐ Yes, on vacation, temporary illness, labor dispute, etc. → *Skip to 50*
☐ No → *Skip to 48*

47 **Have you been informed that you will be recalled to work within the next 6 months or been given a specific date to return to work?**
☐ Yes → *Skip to 49*
☐ No

Standard procedure used by US Census Bureau in 2000 Decennial Census.

Form B–Detection Format

45 **Last week, were you on layoff from a job?**
☐ Yes → ***Skip to 47***
☐ No

46 **(If no) Last week, were you temporarily absent from a job or business?**
☐ Yes, on vacation, temporary illness, labor dispute, etc. → ***Skip to 50***
☐ No → ***Skip to 48***

47 **Have you been informed that you will be recalled to work within the next 6 months or been given a specific date to return to work?**
☐ Yes → ***Skip to 49***
☐ No

Modification of verbal, symbolic, and graphical languages to facilitate detection of potential branching errors.

Form C–Prevention Format

45 **Attention: Check for a skip after you answer . . .**
Last week, were you on layoff from a job?
Yes ☐ ***Skip to 47***
No ☐

46 **Attention: Check for a skip after you answer . . .**
Last week, were you temporarily absent from a job or business?
Yes, on vacation, temporary illness, labor dispute, etc. ☐ ***Skip to 50***
No ☐ ***Skip to 48***

47 **Attention: Check for a skip after you answer . . .**
Have you been informed that you will be recalled to work within the next 6 months or been given a specific date to return to work?
Yes ☐ ***Skip to 49***
No ☐

Modification of verbal, symbolic, and graphical languages to facilitate prevention of branching errors.

Figure 15.1 Branching instruction formats evaluated in Case Study 2.

by staff at WSU (Dillman et al., 1999). The other half was completed by staff at the Census Bureau in Suitland, Maryland. Cognitive think-alouds and retrospective methods with probes were assigned randomly to half of the videotaped interviews conducted at each location. In each case respondents were randomly assigned to complete one of the questionnaires. In the case of the think-aloud, they were asked to read aloud anything they would normally read to themselves. They were then debriefed on items where branching errors were made. Specifically, they were asked to articulate as best they could why each branching error was made and, to the extent possible, to identify the visual elements that might have led to the making of each error. They were then informed that the same questionnaire could be developed in different ways and asked to complete several branching questions on the alternative forms. Finally, they were asked some comparative questions about the three forms.

15.4.3 Cognitive Interview Results

Many of the cognitive interview respondents made branching errors. In addition, substantial differences existed by question in the making of branching errors. The question commission error rates (i.e., not skipping ahead when directed to do so) averaged 12% for questions on the control form or standard census format versus 3.4% for questions on the detection form and 4.1% for the prevention form. In addition, it was found that wide variations existed in the error rates for particular questions, ranging from 0 to 45% on the control form, 0 to 35% on the detection form, and 0 to 18% on the prevention form. In contrast, the omission error rates (i.e., the percent of time that respondents skipped ahead when not directed to do so) were about the same for all forms (control, 1%; arrow, 1.1%; and right box, 0.7%).

Most of the conclusions from the cognitive interviews were drawn from respondents' behavior or performance in the cognitive interviews and not from their verbal reports. While we were able to determine features of the branching instructions that appeared to be in need of improvement, it seemed impossible to choose between the alternative new designs (prevention versus detection methods), based on respondents' verbal reports.

15.4.4 Classroom Experiment Results

Error rates for questions on the 1266 questionnaires completed for the classroom experiment paralleled those from the cognitive interviews (Redline and Dillman, 2002). The control form had a much higher rate of commission errors for questionnaire items than did either of the other forms (20.7% versus 7.6% for the detection method and 9.0% for the prevention method). The item omission error rates were 3.7% for the detection method, 3.3% for the prevention method, and 1.6% for the control form. In addition, the percent of commission errors by question ranged from 0 to 51.6%, and for omission errors 0 to 33.9%, with the higher rates involving most of the same questions that exhibited high error rates in the cognitive interviews.

15.4.5 Outcome

This case study differs significantly from the first one with regard to being able to assess more than respondents' verbal reports. In this study, branching error rates could be measured in both the cognitive interviews and field experiments. In the first case study the dependent variable of interest (response rates) could not be measured in the cognitive interviews. Although the number of cognitive interviews was relatively small (48) in this second case study, the pattern observed in the error rates of the cognitive interviews was generally similar to that observed in the field experiment. This similarity provided confidence for using respondent explanations for why they had missed making a certain prescribed skip from the cognitive interview to make changes for the proposed Census 2000 experiment.

We began this study with a fairly clear theoretical conception of how and why visual languages would influence respondents. One of the direct outcomes of the cognitive interviews was to change some of those theoretical conceptions. For example, we concluded that some respondents deliberately ignored skip instructions because they were interested in seeing what question they were being asked to skip. This led in later formulations of the theory to posit that visual language manipulations could probably reduce branching errors but were unlikely to eliminate them completely. In addition, a final step of the cognitive interview protocol, in which all three forms were taken away from respondents and they were asked which side of the answer categories the boxes were on, made it apparent that placing boxes on the left versus the right was visually transparent to most respondents.

The fact that university students, who already have more education than the average U.S. resident, made a substantial number of errors suggested that conduct of the proposed national experiment in Census 2000 was warranted. The branching instructions were revised slightly, based on insights derived from the cognitive interviews. In the Census 2000 experiment an average commission error rate of 20.8% occurred for questions on the control form, compared to 14.7% for the prevention form and 13.5% for the detection form. Omission error rates were 5.0%, 7.0%, and 4.0%, respectively (Redline et al., 2003). The results of this national test were quite consistent with those first observed in the cognitive interviews and the classroom experiment, and provided convincing experimental evidence that navigational performance could be improved through visual redesign.

15.5 CASE STUDY 3: INFLUENCE OF GRAPHICAL AND NUMERICAL LANGUAGES ON ITEM NONRESPONSE TO THE FIRST QUESTION ON GALLUP Q-12[1] QUESTIONNAIRES

15.5.1 Test Objectives

The Gallup Q-12 is a set of questions developed by the Gallup Organization for measuring satisfaction of employees in a way that can be related to organizational performance. It consists of 13 items asked of organization employees in order to obtain an understanding of their opinions about working for their employer. The first question is a satisfaction question. It is followed by 12 agree/disagree questions (see Figure 15.2). It was noted in repeated administrations of this questionnaire to employees of many different organizations that the item nonresponse to the initial question, which asked for a rating of employee workplace satisfaction, was substantially higher than the item nonresponse to the agree/disagree

Form A: All items modified for purposes of test

SECTION 1

	Extremely Dissatisfied 1	2	3	4	Extremely Satisfied 5	Don't Know
0. On a five-point scale where 5 is extremely satisfied and 1 is extremely dissatisfied, how satisfied are you with Washing State University as a place to go to school?	☐	☐	☐	☐	☐	☐

On a five-point scale where 5 is strongly agree and 1 is strongly disagree, please indcate your level of agreement with each of the following items.	Strongly Disagree 1	2	3	4	Strongly Agree 5	Don't Know
1. As a student at this university, I know what my instructors expect of me	☐	☐	☐	☐	☐	☐
2. I have all the tools and materials I need in order to complete my class work	☐	☐	☐	☐	☐	☐
3. In classes, I have the opportunity to do what I do best every day	☐	☐	☐	☐	☐	☐
4. In the last seven days, I have received recognition or praise for doing good school work	☐	☐	☐	☐	☐	☐
5. I have instructors or classmates that seem to care about me as a person	☐	☐	☐	☐	☐	☐
6. There is someone at the university who encourages me to learn and grow	☐	☐	☐	☐	☐	☐
7. In classes my opinions seem to count	☐	☐	☐	☐	☐	☐
8. The mission of this university to provide students with good educations makes me feel my class work is important	☐	☐	☐	☐	☐	☐
9. My fellow students are committed to doing quality work	☐	☐	☐	☐	☐	☐
10. I have a best friend in one or more of my classes	☐	☐	☐	☐	☐	☐
11. In the last six months, someone has talked to me about my academic progress	☐	☐	☐	☐	☐	☐
12. This last year, I have had opportunities in school to learn and grow	☐	☐	☐	☐	☐	☐

Figure 15.2 Form A with all items used for test of numerical and graphical language effects on a Gallup Q-12 questionnaire modified for student test purposes only. (Copyright © 1992–1999, 2002, The Gallup Organization, Princeton, NJ. All rights reserved.)

items that followed. In 53 data collections by Gallup from a variety of companies, the average nonresponse rate to the initial satisfaction question was 10.5% versus 1.6% to the first agree/disagree item. In only one instance was the item nonresponse to the latter question larger than that for the satisfaction question. The range of nonresponse was 0.7 to 23.5% for the satisfaction question versus 0 to 12.6% for the first agree/disagree item that followed (S. Caldwell, personal communication).

Sixteen different formats had been used for asking these questions in systematic but nonexperimental efforts to find ways of lowering the item nonresponse to the initial satisfaction question. Certain characteristics of the various questions appeared to be related to the item nonresponse, as shown in Table 15.1. A graphical separation was created by several means, including drawing a line around the first item and answer spaces and drawing another line that enclosed the 12

Table 15.1 Relationship Between Specific Design Features and Item Nonresponse on 53 Gallup Q-12 Surveys in 16 Specific Formats Administered by the Gallup Organization

		Mean Item Nonresponse	
Design Feature	n^a (Data Collections)	Initial Satisfaction Question	First Agree/ Disagree Item
Graphical separation between satisfaction and agree/ disagree items (e.g., separate boxes and/or larger spacing)	11	19.7%	0.8%
Satisfaction item not numbered; first agree/disagree item numbered 1	13	16.3%	0.9%
Satisfaction item numbered 0, first agree/disagree item numbered 1	19	10.4%	1.3%
Satisfaction item numbered 1	21	7.0%	2.2%

[a]Number of surveys adds to more than 53 because some surveys had more than one of these design features.

items below it (see Dillman et al., 2000, for examples). It can also be seen that not identifying the initial question as 1, either by giving it no number at all or labeling it zero, is associated with a higher item nonresponse than labeling it as 1. The unconventional start with either no number or a zero for the satisfaction/dissatisfaction item was desired by the survey sponsors in part because the Q-12 is a copyrighted scale developed by the Gallup Organization, and for cultural reasons it was deemed desirable to avoid the items being labeled as Q-13. It was also deemed unacceptable to change the order of questions, placing the satisfaction item later, because of data that showed the correlations between the 12 agree/disagree items, and the satisfaction item would be affected significantly.

Cognitive interviews and a field experiment were conducted in an effort to learn the extent to which graphical separation and the zero start, as well as unknown factors, might be contributing to the high item nonresponse. An overarching objective of these experiments, which affected the design significantly, was to find ways for lowering item nonresponse to the first question.

15.5.2 Research Methods

An experiment was designed in which two manipulations of the Gallup Q-12, which was modified slightly for assessing student rather than employee perceptions, were made in a 2×2 design. A 0 versus 1 start and graphical separation versus no separation were cross-classified to form four treatment groups. The

Form B

SECTION 1

	Extremely Dissatisfied 1	2	3	4	Extremely Satisfied 5	Don't Know
1. On a five-point scale where 5 is extremely satisfied and 1 is extremely dissatisfied, how satisfied are you with Washington State University as a place to go to school?........	☐	☐	☐	☐	☐	☐

On a five-point scale where 5 is strongly agree and 1 is strongly disagree, please indcate your level of agreement with each of the following items.	Strongly Disagree 1	2	3	4	Strongly Agree 5	Don't Know
2. As a student at this university, I know what my instructors expect of me..........	☐	☐	☐	☐	☐	☐

Form C

SECTION 1

	Extremely Dissatisfied 1	2	3	4	Extremely Satisfied 5	Don't Know
0. On a five-point scale where 5 is extremely satisfied and 1 is extremely dissatisfied, how satisfied are you with Washington State University as a place to go to school?........	☐	☐	☐	☐	☐	☐
On a five-point scale where 5 is strongly agree and 1 is strongly disagree, please indcate your level of agreement with each of the following items.	Strongly Disagree 1	2	3	4	Strongly Agree 5	Don't Know
1. As a student at this university, I know what my instructors expect of me..........	☐	☐	☐	☐	☐	☐

Form D

SECTION 1

	Extremely Dissatisfied 1	2	3	4	Extremely Satisfied 5	Don't Know
1. On a five-point scale where 5 is extremely satisfied and 1 is extremely dissatisfied, how satisfied are you with Washington State University as a place to go to school?........	☐	☐	☐	☐	☐	☐
On a five-point scale where 5 is strongly agree and 1 is strongly disagree, please indcate your level of agreement with each of the following items.	Strongly Disagree 1	2	3	4	Strongly Agree 5	Don't Know
2. As a student at this university, I know what my instructors expect of me..........	☐	☐	☐	☐	☐	☐

Figure 15.3 Form B–D formats used for test of numerical and graphical language effects on a Gallup Q-12 modified for assessment of student attitudes. (Copyright © 1992–1999, 2002, The Gallup Organization, Princeton, NJ. All rights reserved.)

experiment was included on the first page of a four-page questionnaire. The experimental variations are shown in Figures 15.2 and 15.3.

Inasmuch as the overarching objective was to identify ways of lowering item nonresponse to the first question and it was recognized that multiple graphical design features were likely to affect response rates, we decided to construct the questionnaire using knowledge of other graphical design factors that were likely to lower item nonresponse. In particular, a colored background was used to highlight white boxes where respondents were expected to check their answers. This decision was based on graphical design theory, and concepts that suggest this construction make the check boxes more visible and send signals of consistency and regularity to respondents who might lower overall questionnaire item nonresponse, as well as for the first question on page 1 (Jenkins and Dillman, 1997). Multiple nonexperimental cognitive tests of designs using white boxes against colored background fields on test forms for the 2000 Census and the Gallup Organization suggested that respondents could more easily see the check box. Each of the four treatment questionnaires was sent to a random sample of 450 undergraduate students (1800 total) in residence on the Pullman campus of Washington State University (WSU) in early April 2002. Three contacts were made: the questionnaire plus $2, a thank-you postcard, and a replacement questionnaire to nonrespondents. An overall completion rate of 57.9% ($n = 1042$) was achieved.

A total of 22 cognitive interviews were conducted with undergraduates not in the sample, each of whom had no previous knowledge of the questionnaire. They were recruited through contacts of SESRC staff, and interviewers and paid $15 after completing the interview. Both think-aloud and retrospective procedures were used to conduct the interviews. Each respondent was asked to complete one of the four questionnaires (randomly assigned) and was then debriefed using a semistructured questionnaire, details of which are provided in Sawyer and Dillman (2002).

It was known that the behavior of interest (item nonresponse) was not likely to occur in many, if any, of the cognitive interviews. The incidence rate of item nonresponse to the first question was expected to be no higher than 6 to 8%, and because of the attention given to aspects of graphical design, was expected to be somewhat lower. We were also concerned that the cognitive interview environment might lead to particularly conscientious behavior at the beginning of the interview, of answering the first question, whereas at home that might not be the case. Two design steps were taken as a result of this concern.

First, the introduction to the interview stressed to the respondent that they should answer the questionnaire exactly as they would at home. The following information was given to respondents as part of the introduction: "If you are likely to read every question or every word at home, you should do that here. But, if you are not likely to do that, then you should not do that here. If you are someone who is likely to answer every question if you were doing this at home, then you should answer every question here. But, if you aren't likely to answer certain questions at home, then please do the same thing here. In other words, you should answer this questionnaire just like you were at home and I was not here. That's what will help us the most."

Second, an extensive debriefing was undertaken consisting of open-ended as well as closed-ended questions. Several of the earlier questions asked about general impressions of the questionnaire and whether the respondent might have responded differently to any of the questions than they would have at home. The last part of this semistructured interview included five questions that invited all respondents to comment on whether any features of questionnaire construction might in their opinion lead to item nonresponse by other respondents.

This series of items was designed to make respondents aware of each of the differences and then ascertain in an open-ended fashion why they might or might not expect differences. The protocol used in this case study was based on the premise that although respondents could exhibit the behavior of interest to us (not providing an answer to the first question), that behavior was likely to occur infrequently. It included a second, projective (to other respondents) level of information gathering that we hoped would provide insights into why respondents might be inclined to skip answering the first questions. Detailed results of the analysis of these data are provided elsewhere (Sawyer and Dillman, 2002).

15.5.3 Cognitive Interview Results

Only 1 of the 22 respondents skipped answering the first question. It happened on Form B, which had a graphical separation and started with a 1. When probed as to why, she responded, “Actually, I didn’t read it. I just skipped it. I didn’t see this question.... Maybe it’s because it’s two areas. I think the question start here [points to question 2]. I didn’t pay attention to the first one because usually they’re put together, right? But here it’s kind of separated.” These comments suggest that the graphic presentation of the information affected her perception of the information and consequently, the navigational path she followed through the questionnaire. It also provides an illustration of questionnaire information that cannot be comprehended because it has not been perceived to exist.

In response to the series of projective questions about why other respondents might skip the first question and the reasons, six of the respondents mentioned the number zero, three mentioned the graphical separation, and six indicated that the first item looked like an example. Several comments combined these ideas to explain why people might not answer the first question. For example, “Maybe because it looks like it’s a sample question. You know how sometimes, like on the SATs, they have a separate box and it shows that this is how you should mark your answer. And, well, if it’s a separate box and it says zero, your eyes will just skip right over it because that’s what you do when there’s a sample question like that. Because it’s a separate box.”

In response to other questions, the vast majority of respondents (16) thought respondents were more likely to complete the first question if it started with a 1 (versus 0), and the remaining six thought that there would be no difference. When the 16 respondents were asked which, from among all four forms, was most likely to have the first question answered, 10 respondents identified Form D (1 start with no separation), four chose Form B (1 start with separation), one chose Form A (0 start with separation), and none chose Form C (0 start with no separation). The other respondent thought that there would be no difference. Thus, the trend is toward Form D, which started with 1 and had no graphical separation, to perform best.

Additional findings, however, suggested that item nonresponse might result from respondents perceiving the question correctly and understanding it, but *judging* the information it elicited as sensitive, that is, as possibly having harmful consequences. Six of the 22 respondents suggested that some people might skip answering the first question because of not wanting to express dissatisfaction with WSU as a place to go to school.

15.5.4 Field Experiment Results

The field results showed that a total of only 17 of 1042 respondents, or 1.7%, did not answer the initial satisfaction item compared to only one person (0.1%) who failed to answer the first agree/disagree item. As expected, this percentage was lower than the rates previously observed by Gallup, in part, we believe, because of the use of the colored background with white spaces to highlight answer spaces. However, the distribution of results across treatment groups was in the expected direction. Nine (or 3.5%) of those who did respond to Form A (0 start plus graphical separation) did not answer the satisfaction item, compared to none where there was no separation and a 1 start (Form D). Each of the remaining forms had four respondents who did not respond to that first question.

15.5.5 Outcome

Results from the cognitive interviews and field test were quite consistent. In neither case did more than a few respondents skip answering the first question (1.7% in the field test and 4.5% or one person in the cognitive interviews). Indirectly, the field experiment suggests that using the other graphical manipulations that were employed for this test (e.g., the use of white check boxes in a blue background) may have kept the item nonresponse at fairly low levels even when 0 and/or graphical separation was employed. However, the cognitive interviews suggest that respondents draw information from a zero start (numerical language) and graphical separation (graphical language) that do not encourage its completion. The cognitive tests also resulted in identifying dissatisfaction with school (and by implication, employment) that might cause item nonresponse independent of the use of visual languages.

15.6 CONCLUSIONS

Conducting cognitive interviews on self-administered questionnaires involves many of the same issues as those specified by Forsyth and Lesser (1991) more than a decade ago, with respect to understanding the meaning of questions. However, additional issues that pertain specifically to the use of self-administered questionnaires, such as how entire mailing packages influence respondent behavior and the use of nonverbal languages (numbers, symbols, and graphics) also need to be addressed. In this chapter we have proposed that such issues can be addressed effectively through the use of cognitive interviews and that the results of such interviews provide important data for decision making that cannot be provided by traditional field experiments, and vice versa.

We have reported three case studies in which cognitive interviews and field experiments were conducted simultaneously to evaluate the performance of test questionnaires on three different issues (response rates, branching errors, and item nonresponse to a particular question). In each case, the conclusions drawn from the cognitive interviews were informative of why certain field results were obtained.

For all three case studies the cognitive interviews provided important information that could not be obtained from the field experiments. In Case Study 1, multiple differences existed among the three test instruments so that the field experiment did not allow conclusions to be reached as to why the different response rates occurred. The cognitive interviews suggested a number of possibilities, large and small, that provided design guidance for subsequent redesign decisions. In Case Study 2, the cognitive interviews provided information about how and why individual respondents made specific branching errors with specific questions. This information, none of which was available from the field experiment, was used to refine the proposed branching instructions in more than a dozen ways for the national test in Census 2000. In Case Study 3, the cognitive interviews provided limited evidence that the visual and graphical languages affected how respondents thought about questionnaires. They also identified another issue—the social desirability of not providing a response to the satisfaction question—as a possible additional reason for the item nonresponse that was unanticipated by those involved in conducting the study.

When practical design decisions are at stake, as they were in each of these case studies, the procedure of individually manipulating every possible variation in combination with all other variations is not possible, as illustrated here in particular by Case Studies 1 and 2. In these cases cognitive interviews provided information that helped scientists select which issues to explore in more detail. Rather than depending on theory alone to pick the most interesting hypotheses for further pursuit, it was possible to use additional observations of respondent behavior to help select which proposed questionnaire features seemed most worthwhile to pursue.

The consistency between the cognitive interviews and field experiments observed here may lead some to conclude that the field experiments were unnecessary for making redesign decisions. Such a conclusion is inappropriate. As the conductors of the case study interviews we were far more impressed with our inability to draw logically definitive conclusions from them than we were with the eventual compatibility and mutual support of the interview observations with findings from the field experiments. The cognitive interviews were quite inadequate as convincing evidence of the likely quantitative effects of the treatment manipulations. However, as a potential explanation for quantitative findings from the experiments, and sources of ideas for redesign of survey questionnaires and procedures, they were essential. Used in tandem cognitive interviews and field experiments are a powerful means of testing and refining procedures for the conduct of self-administered surveys.

CHAPTER 16

Methods for Testing and Evaluating Computer-Assisted Questionnaires

John Tarnai and Danna L. Moore
Washington State University

16.1 INTRODUCTION

The increasing sophistication of computer-assisted-interviewing (CAI) systems has made the survey questionnaire designer's task more complex. CAI has increased our ability to handle complex branching; to use fills, calculations, and rostering; and to randomize sequences of questions and response categories. A particularly difficult task in light of this complexity is adequate testing and evaluation of CAI questionnaires, especially since this is an easily neglected task. That this is a problem for survey researchers has been documented in the survey research literature (Kinsey and Jewell, 1998; Tarnai et al., 1998). Over half of survey research centers nationwide report that they have written procedures for testing and debugging CATI questionnaires, and yet over 65% of them also report having had to recontact survey respondents at least once because of errors found in a CATI survey. This is a significant issue because the possibility of errors increases as the complexity of the questionnaire increases (McCabe and Watson, 1994), and such errors have implications for both data quality and costs. In this chapter we summarize the literature on testing and evaluating CAI questionnaires. We describe current practices of survey research organizations, and we present a taxonomy of CAI errors and some approaches to testing for them. We also report some results of a comparison of two approaches to testing CAI instruments.

Methods for Testing and Evaluating Survey Questionnaires, Edited by Stanley Presser, Jennifer M. Rothgeb, Mick P. Couper, Judith T. Lessler, Elizabeth Martin, Jean Martin, and Eleanor Singer
ISBN 0-471-45841-4

The purpose of CAI testing is to confirm that a questionnaire is working as specified and that a CAI system will react robustly to unexpected events. By specification we mean a description of the questions, responses, branching and routing instructions, and all other aspects of a questionnaire needed for proper implementation. For most CAI questionnaires, a paper-and-pencil questionnaire is an insufficient specification. House and Nicholls (1988) present a checklist of CAI issues that must be specified in advance before they are translated into CAI format. Even if they are specified completely, however, CAI specifications may themselves contain errors.

How well a CAI instrument is implemented depends first on the accuracy and thoroughness of its specification. In general, the more thoroughly an instrument is specified, the more likely it is that the programming will take place as specified. However, this is easier said than done. Additionally, there may well be errors in the specification itself, so in addition to testing for specification errors, we may also be testing for variances from client expectations. A categorization of software errors (Perry, 1995) has application to CAI testing. Some errors represent things that are "wrong" in that the specifications have been implemented incorrectly. Some errors represent specified requirements that are "missing." And sometimes, a CAI instrument may include something that was not specified, and is "extra."

To test a CAI questionnaire completely and be absolutely certain that there are no errors is virtually impossible, particularly since the resources of testing (time and personnel) are finite. Instead, the goal is to design a testing process that efficiently and effectively catches as many errors as possible within the time and budget constraints of the particular project. As Berry and O'Rourke (1988) have noted, making changes to a CAI questionnaire is so easy to do that there is always the temptation of trying to make the instrument too perfect, which risks introducing unwanted errors. We agree with the view that the goal of the testing process should be to reduce risks to an acceptable level (Perry, 1995).

Computer-assisted telephone interviewing (CATI), computer-assisted personal interviewing (CAPI), and other forms of CAI offer an exponential expansion of opportunities to increase complexity over what is possible with a paper questionnaire. Groves (2002) has suggested that with "modern CAI we have lost all constraints on the complexity of measurements we can introduce." Branching and skipping around the questionnaire is practically limitless. Calculations based on a person's previous answers and fills created from previous answers can be incorporated into any part of a CAI questionnaire. Sample information can be imported into and used in various parts of a questionnaire. The order of questions and response options can be randomized. All of these capabilities represent real opportunities to improve questionnaires and to tailor them to the specific needs of different segments of a survey population. However, our testing procedures have not kept pace with these developments. The increased complexity of CAI instruments means significant increases in the amount of testing that must be done to ensure an accurately fielded survey.

16.2 RESULTS OF A SURVEY ON CAI TESTING

To identify the methods that survey centers currently use to test CAI questionnaires, we designed an informal e-mail survey of people who had attended the International Field Directors and Technology Conference in past years. This conference is attended by staff from academic, government, and nonprofit survey organizations. A total of 410 invitations to the survey were sent out, and 129 people responded. These 129 respondents represented 57 of the 98 organizations that attended the conference from 2000 to 2002. Several organizations contacted us to indicate that only one person would be responding for their organization. For those organizations with more than one respondent, we selected one respondent based on their level of involvement in CAI testing (very involved) and whether they had provided a written description of their organization's CAI testing procedure. The remainder of the discussion focuses on these 57 respondents or organizations.

About 83% of respondents indicated that their organization has a process for testing CAI questionnaires. However, only about a third of these reported that they had a written process for testing CAI questionnaires that testers followed in a standardized way. This suggests that for most survey organizations, CAI testing is a rather ad hoc process that is not performed in a standardized formal way that is reliable or can be replicated by testers. More than half of these organizations consider CAI testing before production interviewing to be a "somewhat" or "great" problem for their organization, and almost half say that errors in CAI questionnaires have been somewhat or very costly for their organizations. Most organizations reported using six of seven CAI testing methods, including having professionals and designers conduct tests, conducting mock interviews, pretesting with actual respondents, question-by-question testing by interviewers, scenario testing by study directors, and scenario testing by interviewers. Only CAI data simulation was not a commonly used method to test CAI questionnaires.

We asked respondents to tell us what they did not like about their current CAI software with respect to testing. Many of the same types of responses came up repeatedly, and they suggest that current CAI software and testing systems are inadequate for what survey researchers have to accomplish. The main complaints made by many respondents included the following:

- Testing is labor intensive and detecting errors takes too much time.
- Changing specifications often creates additional errors.
- There is no way to know for certain if all errors have been found and corrected.
- Sample administration features are often difficult to test.
- There is no way to create a flowchart to evaluate the structure of an interview.
- No simulation or auto test option is available for testing a CAI questionnaire.
- Inadequate CAI testing time is allowed in the survey schedule.

- Questionnaire complexity precludes testing of all branching.
- CAI/CAPI testing and usability in actual field environments are inadequate.
- Maintaining a good log of problems tested is difficult.

Many respondents complained that current approaches to CAI testing are too manual and unstructured, and they wished that there was a way to automate the testing process. Particularly frustrating to many respondents is the need to retest CAI questionnaires whenever specification changes are made.

16.3 TESTING CAI QUESTIONNAIRES

16.3.1 Current Approaches

To identify current approaches to testing CAI instruments, we reviewed the survey literature for references to CAI testing. We also reviewed the software testing literature to assess whether this had any application to CAI testing. Researchers at several survey organizations talked with us about their testing processes, and some shared their written procedures. We also obtained information about CAI testing procedures from an e-mail survey of government and nonprofit survey organizations. Some of the procedures used are illustrated below.

Beaule et al. (1997) describe a process for testing CAI instruments that has been used at the University of Michigan's Survey Research Center. Many of the elements of this process are used by other organizations and are also described in the survey research literature on testing. The process may involve identification of a testing coordinator for each survey project. This person is responsible for forming a testing team, developing testing goals and procedures, enforcing a testing schedule, and providing a single point of contact. The testing team may be comprised of project managers, programmers, and interviewers. Two types of tests are generally done by different people, and they often identify different types of errors. Initial testing is usually done by the CAI programmers and their staff to ensure that the instrument meets basic project specifications and is functional. Then the instrument is turned over to a project team for testing, including project staff, research assistants, survey managers, and interviewers and supervisors. This team reviews survey specifications, identifies and tests critical paths, and reviews the test data.

Research Triangle Institute has a similar CAI testing process involving a formal testing plan and methods, identifying a testing team, determining what methods to use to identify and correct CAI errors, and developing a testing schedule. Testing methods generally involve a question-by-question process of checking all wording and response paths, followed by testing of routing, consistency, and range checks, and other instrument features. Programmers and questionnaire authors are best at checking for specification errors, and interviewing staff seem best for usability and functionality testing (Kinsey and Jewell, 1998).

The Social and Economic Sciences Research Center (SESRC) at Washington State University has a semiformal process for testing CAI questionnaires. Although there are written procedures describing how CAI testing should take place, this is more of a guide than a procedures manual. The primary responsibility for CAI testing resides with the project manager and the CAI programmer. Sometimes this is one and the same person, especially for smaller surveys. For larger surveys, others, such as telephone supervisors and telephone interviewers, are included in the testing process. The CAI testing process generally involves attempts to "break" the questionnaire within a sequence of activities designed to find errors. No single method of testing is able to catch all the errors, so we have adopted a system that involves multiple methods. This generally begins with a functionality test to ensure that a CAI instrument loads properly onto an interviewer's PC, and that it is operational and can be navigated. This is generally followed by question-by-question checks of the testing elements described in Table 16.1, either separately or in combination. This may then be followed by testing of specific respondent scenarios. At some point there may also be unstructured testing done by interviewers, or more structured pretesting with actual respondents. The data resulting from this testing is then reviewed by project staff and compared with what is expected from the survey specifications as a final check for errors.

16.3.2 Taxonomy of CAI Errors

There are two main reasons to test CAI questionnaires: (1) testing to evaluate the functionality of the instrument, and (2) testing to evaluate how well the CAI instrument agrees with the questionnaire specifications. The purpose of testing for functionality is to ensure that the CAI software is able to implement the questionnaire from beginning to end, without the system ending abnormally or encountering any other performance problems. This type of testing may be a hit-or-miss proposition, since a questionnaire may appear to be working normally at one time, but may end up "crashing" at a different time for a variety of reasons. Testing for agreement with questionnaire specifications requires that the specifications be published and shared with survey staff. In general, CAI instruments require more detailed specifications than paper-and-pencil instruments to enable the conversion to CAI format (Berry and O'Rourke, 1988; House and Nicholls, 1988). We expect that the more clearly and completely a survey questionnaire is specified, the easier it is for programming staff to convert it correctly to CAI, and consequently, the fewer errors that are found in testing. There is the issue of what to do if the specifications are wrong, which may become apparent during testing. It is also common for specifications to be changed and consequently for CAI questionnaires to be modified, resulting in additional testing.

Table 16.1 describes some common specifications for CAI questionnaires and possible errors that can occur in programming and methods for detecting and correcting these types of errors. We make no presumption that all possible CAI errors are included in Table 16.1; instead, this is meant to be representative of probable

Table 16.1 Taxonomy of CAI Questionnaire Errors and Methods for Detecting Them

CAI/Questionnaire Feature	Possible Errors	Detection Methods
Screen appearance	Poor visual design, inconsistent formatting	Q-by-Q testing
Preloaded sample data and sample administration	Incorrect data formats, incorrect data order, appearance	Q-by-Q testing Testing by task Pretesting
Question wording	Inaccurately worded questions, missed words, spelling errors	Q-by-Q testing Testing by task Pretesting
Response ranges and formats	Formats don't match specs, missing response options, inappropriate formats	Q-by-Q testing Testing by task Data testing
Missing values	Refusal, don't know, not applicable; other response options not used consistently or not at all	Q-by-Q Testing Testing by task Data testing
Skip patterns—unconditional, conditional, missing values	Not all response options branch correctly, skips to wrong question	Testing by task Data testing Scenario testing Simulation
Calculations and fills	Division by zero, missing values, incorrect formulas, insufficient space reserved for fill variables	Q-by-Q testing Testing by task Pretesting
Randomization	Biased processes	Testing by task Data testing Simulation
Function keys	Not accessible, inaccurately worded	Testing by task
Rosters	Incorrect branching, insufficient calls to a roster	Q-by-Q testing Testing by task Scenario testing Pretesting
Attempt tracking	Insufficient variables to track call attempts, inappropriate call slots	Testing by task Pretesting
Screening questions	Inaccuracies in determining eligibility	Q-by-Q testing Scenario testing Data testing
Termination questions	Insufficient termination codes	Q-by-Q testing
System issues	Abnormal terminations, corrupt output files	Scenario testing Testing by task

types of errors and methods that are typically used to detect them. Commonly used approaches to detecting and correcting CAI errors are described below.

The length and complexity of the questionnaire affect which procedures are used to test it. The shorter the questionnaire, the less involved these procedures may be, and the longer and more complex the questionnaire, the more rigorous the testing procedures have to be to locate errors. Typical CAI testing might involve interviewers, CAI programmers, project managers, and survey clients. Additionally, some CAI systems have the ability to simulate survey responses, adding another procedure that may be used to test CAI questionnaires.

A key issue in testing CAI questionnaires is how the results of the testing should be reported and communicated with others. One common method is simply for each tester to make notes on a paper copy of the questionnaire being tested, and for all testers to turn these in to the CAI programmer or survey manager. Although this works for smaller questionnaires, it quickly becomes unmanageable for larger and more complex questionnaires that have multiple authors and testers. A newer approach is to have testers enter questionnaire problems or errors into a database, which can then be sorted by question number, tester, or other characteristics (Sparks et al., 1998). Newman and Stegehuis (2002) describe such a database system that automatically enters errors into a problem-tracking system. As problems are encountered, testers press a function key to access a pop-up window that captures CAI data automatically, and allows the tester to enter a description of the problem.

16.3.3 Main Testing Approaches

There are six main approaches to testing for CAI errors: (1) question-by-question (Q-by-Q) testing, (2) testing by task, (3) scenario testing, (4) data testing, (5) pretesting with survey respondents, and (6) simulation. In actual practice, combinations of these approaches may be used in testing specific CAI questionnaires, because each has advantages and disadvantages.

Question-by-Question Testing In this approach each question or screen in the survey is tested thoroughly before moving on to the next question. The testing includes checking overall appearance, question wording, response options, branching questions, missing values, fills, computations, and other aspects of the question. For surveys that may have hundreds to thousands of questions, this method is very time consuming and may not be possible at all. However, it can also be very thorough. It requires great attention to detail, and it is easy to miss the larger issues of how questions and responses relate to one another.

Testing by Task Another method is to divide up different testing tasks to individual testers. For example, one tester may focus only on checking the wording of questions and response categories; a second tester may check only the branching into and out of every question; a third tester may check only the valid response ranges and missing values for every question. This method allows a single tester to focus on one or more features of a questionnaire. However, it also lacks the cohesiveness that a single tester can bring to the task. A high level of teamwork and coordination must exist among the various testers for this method to identify all CAI errors accurately. When using this method, a testing coordinator is essential to keep track of all problems and revisions.

Scenario Testing For some complex questionnaires, it is helpful to construct various scenarios of responses and then enter these into a CAI questionnaire and observe the outcome. This method is useful for example to test whether the questionnaire accurately routes people who respond a certain way to the appropriate questions that ask for more detail. This method involves the testers answering survey questions and entering the data into the program following a set of predetermined criteria. It is essentially a matter of role-playing many of the possible scenarios within the context of the questionnaire and checking to make sure that the resultant interview is logical and makes sense. This method has the advantage of viewing the questionnaire from a broader point of view. However, the method is inherently flawed in that a comprehensive list of all possible scenarios is rarely achievable. For these reasons, it should never be used as the sole method for testing a new CAI questionnaire. It works best in conjunction with one of the other methods or for "updated" versions of a CAI questionnaire tested previously using the Q-by-Q method. Compiling a suitable list of scenarios can be a daunting task for complex surveys.

Data Testing In addition to the approaches described above, it is essential to examine a preliminary data output from the CAI program to make sure that the output conforms to expectations and to the survey specifications. This is often accomplished by interviewers or other staff, who enter responses from pretests or practice interviews. This can be a very revealing type of testing, since CAI output provide the raw data for survey analysis, and examining the data can readily identify branching errors and response range problems. Data testing and examining survey output is one of the few ways to ensure that randomization of questions, response categories, or respondents is working as intended. However, data testing by itself cannot ensure that all possible combinations of survey responses have been checked. There are simply too many possible paths through most CAI questionnaires and too many opportunities for error for any of the common testing methods to be able to identify them all.

Pretesting with Survey Respondents Pretesting was the third most frequently used technique reported by respondents from our informal survey. This approach has the advantage of including elements of all the other approaches. It can also serve as a final test of a CAI questionnaire, after other CAI test procedures have been used. However, we would not recommend pretesting as the only method of testing CAI questionnaires, since some types of errors are less likely to be identified this way: for example, specific respondent scenarios, randomization, and all possible skip patterns.

Simulating Survey Data About one-fourth of respondents in our survey reported that they use their CAI software "sometimes or always" to conduct computer simulations. Such software can produce random responses for hundreds of respondents and output a survey dataset for review by survey staff. Several current CAI software packages make this type of simulation testing possible and relatively easy to do. These systems can generate simulated data that can then be examined for branching and other types of logical errors. Most commonly, the CAI software simply generates random responses for a predetermined number of interviews.

16.3.4 Automated Approaches to CAI Testing

Because CAI testing is so time consuming and laborious, there have been attempts to automate some of the testing tasks. Simulating data is a relatively new feature of CAI systems, and there is little information about this approach. However, about 40% of organizations in our survey say that they have used simulation to test CAI questionnaires. In our experience, some of the more serious CAI errors can only be found by examining CAI output and/or through analysis of the data on a large number of interviews. An example is an experiment that is testing three versions of a question where equal thirds of the survey respondents are allocated to each of three branching sequences. Although normal CAI testing procedures may suggest that people are being routed randomly to the three branches, any systematic errors in the randomization may not be detected until after data on many cases are available. Simulated data testing is particularly cost-effective for detecting these types of errors.

Some survey organizations conducting large-scale ongoing surveys are in the beginning stages of developing techniques for automating at least some of the CAI testing functions. Levinsohn and Rodriguez (2001), for example, describe a prototype system, *RoboCAI*, that allows researchers to create *test scripts* or questionnaire scenarios that are then implemented on a Blaise programming system to test a CAI questionnaire. The entire process is monitored to detect errors, and the data output from this testing are also examined to identify errors. However, currently the system does not test screen text, answer text, or question paths and calculations not included in the test scripts or scenarios. For very large and complex questionnaires, this system has proven to be very useful, since it automates the process of testing scenarios and producing test datasets that can be examined for errors. Despite its usefulness, it is still only one aspect of the entire CAI testing process, which includes traditional testing approaches described above.

Newman and Stegehuis (2002) describe a process for automated testing of complex Blaise instruments. This process involves using a commercial testing tool that provides ways of recording and playing back mouse movements and keystrokes. These "scripts" can then be compared with baseline scripts to identify errors. The authors note that there is a substantial learning curve in understanding how to use the testing tools, and thus automated testing is most appropriate for very complex and lengthy surveys that might overwhelm the more traditional testing approaches.

Pierzchala (2002) notes that there are many unanswered questions with this type of automated testing. For instance, what is the cost of building test scripts? How does one update these scripts when the specifications change, as they often do? At what point does one implement automated playback of scripts? Implementing this type of automated testing will be a challenge because the procedures and techniques are not well known and the tools are not yet readily available to most survey researchers.

Automating the process of questionnaire development documentation and specification also holds promise for improving the testing of CAI instruments. Schnell and Kreuter (2002) describe a software system that permits documentation of all versions of a questionnaire and all versions of questions. The software reduces the burden of tracking questionnaire changes that are made as result of CAI testing. Bethlehem and Hundepool (2002) describe similar software that they have called a *tool for the analysis and documentation of electronic questionnaires* (TADEQ). This software is intended to ease the process of creating a readable version of an electronic questionnaire, in a way that facilitates analyzing the questionnaire and identifying possible errors. Both of these recent approaches represent significant advances in the ability to produce error-free documentation that should help to reduce the CAI testing burden.

16.3.5 What Can Be Learned from Software Testing

Many of the issues confronted by CAI instrument testing can also be found in the software and personal computer testing literature (Beizer, 1990; Perry, 1995) and the usability testing literature (Barnum, 2002; see also Chapter 17, this volume). A number of testing techniques have been borrowed from the software testing field, and although there may be differences in terminology, many of the ideas and techniques apply to both CAI and software testing. The software industry has produced a variety of ways to conduct testing that may have application to CAI testing (see, e.g., Watkins, 2001).

A major difference between software testing and CAI testing is that the former is concerned with testing the performance of a single application under a variety of conditions, while CAI testing is concerned with testing every implementation of an instrument on a single software platform that is presumed to be error free. This means that some testing techniques are more important for CAI testing than for software testing.

A problem that arises in both software and CAI testing, especially when many testers or programmers are involved, is controlling the version of the CAI instrument that is being managed. Since testing is an iterative process, it is easy for programmers to make a change that may then be overwritten by other programmers working on different parts of the instrument. Newman and Stegehuis (2002) describe methods to control CAI source code files using commercial software that restricts access to master copies of questionnaire files. They argue in favor of the principle that only one master copy of each source file exists; otherwise, it becomes too easy for changes made by one programmer to be copied over changes made by another programmer. The purpose of version control is to ensure that only one master copy of the instrument exists and that all programmers are working with the same instrument. Version control may be less of a problem for small surveys involving only a single programmer, and it provides an additional level of security, ensuring the reliability of the final instrument.

The software testing industry has also developed ways of streamlining the error-reporting process. Newman and Stegehuis (2002) describe a software application that is fully integrated into the CAI system, so that as errors are encountered by testers, a function key can be used to bring up a problem report screen. This screen automatically records information about the question number, date and time, and instrument version, among other things. The tester then enters comments about the problem, and the report is saved in a database of problem reports. Similar procedures are in place at survey organizations conducting large complex surveys or those involving multiple authors or programmers. Sardenberg and Gloster (2001) describe a similar reporting system. The testing log is a database that keeps a log of all CAI problems encountered during the testing process and the resolution of each problem. The advantages of these automated error-reporting systems cannot be overstated. It makes the tester's job easier when reporting problems; it reduces paper, which can be lost or be unreadable; it provides more information about errors, more immediately and automatically; and the database can be sorted by several variables to help isolate and deal with problems.

When to stop testing, one of the most difficult decisions to make, is based largely on time and resource constraints. After several iterations of testing, making changes to an instrument, and retesting, continued testing may jeopardize survey deadlines. Additionally, tester fatigue and boredom can set in, and the likelihood of finding errors may diminish as well. The software industry has adopted a system in which the final testing of software is done by users of the software. A beta version, not a final release version of the software, is released to users willing to use the software and test it before a publicly released version is attempted. Survey researchers don't have the luxury of fielding beta versions of a CAI questionnaire. By the time we field a survey, it must be as error free as possible given the constraints of time and budget.

16.4 MODEL FOR TESTING CAI APPLICATIONS

In this section we describe an ideal model for testing CAI questionnaires, which combines some of the current practices in survey research for testing CAI instruments with automation techniques borrowed from the software testing literature. How thoroughly this model should be followed depends on the complexity of a survey, time constraints, and the resources available. However, whenever any CAI questionnaire is created and a need for testing is anticipated, a number of questions must be addressed.

- How should the testing be organized?
- Who is responsible for the testing process and for forming the testing team?
- What types of goals do we have for testing?
- What testing plan should be adopted?
- What test scenarios should be used?

The first step in the CAI testing process is to prepare for testing by forming a testing team, developing testing plans, and determining the goals of testing. This process also assumes that the specifications for the questionnaire and the survey output dataset have been identified. It is essential that these be in written form and stated clearly in such a way that the testing staff know what to look for. For most smaller-scale surveys, this may be nothing more than a written version of the questionnaire with programming instructions described on the questionnaire.

A person or a team of people who have responsibility for developing a testing plan and implementing the testing activities are identified and assigned responsibility for managing the testing process. The testing team determines the goals, the timeline for testing, and the specific procedures to use in conducting the testing. A CAI instrument must first be configured successfully and installed on a computer before other types of testing can begin. Any problems that occur at this stage must be fixed, usually by programming staff, and the questionnaire updated.

Implementing the testing techniques that have been decided on is the next step in the process. Some type of Q-by-Q testing will probably be necessary for all CAI instruments, as this ensures that specifications have been met and the questions appear on the screen as desired. Additionally, testing for key scenarios is essential to ensure that the most important paths through the questionnaire are accurate. If possible, this testing should be supplemented by data testing and/or simulation of respondent data. This may catch additional errors and will help to ensure that the output dataset matches specifications.

The testing team is responsible for ensuring that the proper version of the CAI instrument has been released for testing and that testers are trained in how to use the techniques and how to report and log errors. Whether paper is used to record CAI problems or whether a database is used to collect this information depends on the resources and expertise of the organization and the size and complexity of the survey being tested. The more important issue for this step is how well this problem collection system is organized.

After problems have been identified, reviewed, and decisions made about changes to the CAI instrument and to the specifications, if necessary, someone has to make the changes to a master version of the instrument, and the changes must be verified. This may involve a single person, or multiple programmers for more complex surveys. In either case, version control will be an important issue to ensure that corrections are not overwritten inadvertently. The revised questionnaire then goes on to a regression testing phase to check that errors have been corrected and that these corrections have not created other errors. Regression testing may use any of the testing techniques described earlier (Q-by-Q testing, scenario testing, data testing, simulation), focusing on the changes and corrections. The process ends when no further errors are found or when a decision is made to stop testing, either because time for testing has run out or because the errors that are being found are too minor to justify the cost of continued testing. At this point, a version of the questionnaire may be fielded for pretesting. Errors may still be found during pretesting, in which case the questionnaire must be revised, and regression testing may have to occur again.

This is a very generalized model of an ideal process for testing CAI questionnaires that is based on current practices and suggested techniques borrowed from the software testing literature. Actual practices may vary considerably from this model, depending on time and budgetary limitations. Groves (2002), for example, suggests that testing might benefit by being more focused on the specific features of an instrument that provide the best ratio of benefits to cost per error found. As CAI instruments become ever more complex, the possibility of testing every possible path through a questionnaire becomes less likely. Thus, the best model for testing may be one that is designed to catch the most serious errors through a focused effort, and then some percent of all other errors, recognizing that it may be cost prohibitive to identify and correct all possible errors.

16.5 COMPARING Q-BY-Q AND SIMULATION APPROACHES TO CAI TESTING

To examine the effectiveness of different approaches to CAI testing, we designed an experiment to compare a traditional Q-by-Q approach with two simulation approaches. The first data simulation approach is a step-by-step simulation of data in which the computer enters a response to each question in a survey but the interviewer must press a key to continue on to the next question. This approach is most similar to the Q-by-Q approach, in which interviewers view each question on the screen and enter a response to each question in order to continue. The second simulation approach we tested was running the simulation in background mode and generating 500 respondent cases randomly.

We created a short questionnaire of mostly categorical questions about health insurance and programmed a CAI version of the questionnaire. We created both a long version (70 questions) and a short version (35 questions) of the questionnaire by deleting some of the questions from the long version. We then introduced the same 31 errors into both versions of the questionnaire. A group of 30 interviewers were randomly assigned to one of four groups. Half of the interviewers conducted the questionnaire testing using the traditional Q-by-Q testing method. The other half conducted the questionnaire testing using the step-by-step simulation feature of the CAI software. Half of the interviewers received the long version of the questionnaire, and half received the short version. Interviewers were trained for about half an hour in how to conduct Q-by-Q testing of a CAI questionnaire or a step-by-step simulation testing of a questionnaire. Interviewers were instructed to test the questionnaire as they normally would when testing a new version of a questionnaire. Interviewers recorded all errors they found on a paper form, by noting the question number and a description of the error that they found.

16.5.1 Simulating Data for a CAI Questionnaire

Separately, we simulated data from 500 respondents for the short version of the questionnaire, using the simulation feature in our CAI system. The data output from this simulation was then submitted to five professional staff (two CAI programmers, two survey managers, and one survey supervisor), who were instructed to examine the data and to identify all errors, which they recorded on a paper form similar to the one used by the interviewers for testing CAI questionnaires.

The simulation produces a printout of the questionnaire, which shows the questions, the response categories and ranges, and the skip patterns. It also prints the number and percent of responses to every question. The simulation results displayed along with the questionnaire were evaluated by the five professional staff. The evaluation was done by examining the number of responses made to each question, and whether that number was expected given the branching pattern in the questionnaire. For example, if 234 of 500 responses are "no" to question 2, and all "no" responses are supposed to skip to question 5 and "yes" responses somewhere else, we should expect to see a total of 234 responses at question 5. For numerical questions where a range of responses are accepted, say between zero and 180, we should expect to see responses distributed over the acceptable range. This method of evaluating a CAI instrument is very similar to what testers might do in evaluating results from a pretest. The advantage of the simulation is the ability to identify errors before they are made with respondents from a fielded survey.

16.5.2 Results

We compared the three methods of testing CAI questionnaires on the total number of question errors found as well as the rate of error detection by each group. The results are displayed in Table 16.2 for the three testing approaches. We have combined the results for the short and long versions of the questionnaire, since no significant differences were found between them. None of the groups found all of the 31 errors in any of the questionnaires. The best performance was produced by the professional staff using the simulated data output. As a group, the five professional staff found 23 of the 31 errors, or 74% of all possible errors. Individually, their performance was not as good; of 155 possible errors (five testers times 31 errors), only 76 errors or 49% were found by these five testers.

The interviewer groups using the traditional Q-by-Q approach and the step-by-step simulation approach found 18 and 19 errors, respectively, of the 31 possible errors, and there were no significant differences between the two approaches used. Even though there were three times the number of testers using these two approaches, they detected fewer errors (60% and 62%, respectively) then were found by the professional staff. The overall rate of error detection was quite low. Whereas the professional staff overall had a 49% rate of detection, the Q-by-Q testers only had a rate of 24%, and the stimulation testers had a rate of only 18%.

Table 16.2 Error Detection Results from Testing a CAI Instrument

Measure	Professional Staff Simulated Data	Q-by-Q Review by Interviewers	Step-by-Step Simulation Review by Interviewers
Number of testers	5	15	15
Total errors identified	76	112	84
Total errors possible	155	465	465
Total errors detected	23	18	19
Rate of detection	49%	24%	18%
Percent of errors detected	74%	60%	62%

The worst group, which reviewed the short questionnaire and used the step-by-step simulation approach, had only a 15% rate of error detection. The best group, which was the professional staff using the simulation approach, had only a 49% rate of detection.

There are some differences in the types of errors that interviewers found. Those using the Q-by-Q method were more likely than those using the step-by-step method to detect wording/spelling errors and response category errors. Those using the step-by-step method were more likely to catch errors involving branching, out-of-range errors, and the random response category errors. The differences are attributable to how the responses to each question are generated. In the Q-by-Q method, interviewers type in whatever response they decide to make to a question. In the step-by-step simulation mode, the responses are generated randomly by the computer. The random pattern of responses in the step-by-step simulation mode is very likely allowing interviewers to identify some errors that they would normally miss. On the other hand, it appears that the step-by-step simulation mode takes away interviewers' focus on the wording of questions, since fewer of these types of errors are caught by this method.

Overall, professional staff took more time to review questionnaires than did interview staff. For each type of error, professional staff reviewing simulated CAI output was considerably more effective than interviewing staff reviewing simulated output. Interviewing staff were at least twice as effective in detecting each type of error when Q-by-Q screen review was used compared to simulated questionnaire output. Neither professional staff nor interview staff was completely effective at detecting all types of errors. Professional staff was most effective at detecting question fill and branching errors, and least effective at detecting random category errors, wording/spelling errors, and response category errors. Interviewing staff was most effective at detecting question fill errors and wording/spelling errors. In Q-by-Q testing, they were least effective in detecting open-ended numeric response range errors, random responses errors, and branching errors.

The results demonstrate some of the principles of CAI testing that have been presented. These include the principle that different types of people are necessary to test CAI instruments, that they catch different types of errors, and that different testing methods tend to catch different types of errors. We expected to find that both CAI simulation approaches would be superior to the Q-by-Q method, and that the most efficient method (in terms of time and errors found) would be the data simulation method. In fact, the most effective and efficient testing was done by the professional staff using the simulated data. The interviewers, perhaps due to a lack of experience and motivation to find errors, performed significantly worse in finding errors, despite spending almost as much time as the professional staff.

The most effective and efficient approach to testing may be a combination of these approaches, involving both interviewers and professional staff testing different aspects of a CAI instrument. Screen features and question wording may be detected more easily by interviewers. Branching errors and response range errors may be more easily caught by examining simulated data. Combining these two approaches with different staff may be more effective than having all staff test all aspects of a CAI instrument.

16.6 CONCLUSIONS

We have described the basic problem of testing CAI questionnaires and instruments and some of the current approaches that survey research organizations use to accomplish this. Our survey reveals many similarities in how survey organizations approach this task. Basic testing procedures are labor intensive, involving interviewing and programming staff repeatedly reviewing questionnaires either by specific tasks or specific scenarios, and checking the survey output for errors. Some of the largest survey organizations are beginning to explore the possibility of automating at least part of the testing process. Because of the size and complexities of CAI surveys conducted by these organizations, these automated methods hold much promise for containing costs and detecting errors. However, the investment in time and effort required to automate the testing process may not be affordable for most small to medium-sized survey organizations.

Simulating random survey data may offer many survey organizations their best hope of automating the CAI testing process. Current software makes data simulation relatively easy to do and inexpensive. The output provides another valuable view of the questionnaire, one that can reveal some types of errors more readily than can be found by viewing a computer screen. Traditional Q-by-Q approaches, along with pretesting of instruments and examination of preliminary data, will remain useful ways to test CAI questionnaires. However, simulation of survey data offers advantages in identifying some types of survey errors. Ideally, it could lead to a system of testing in which simulation can identify the majority of programming errors (i.e., due to branching, response categories), and Q-by-Q or scenario testing would find the remaining errors that cannot be found by an automated system.

A significant improvement in CAI testing may be achievable if the available CAI systems could automate the process of moving from questionnaire specifications to CAI questionnaire. This would not obviate the need for CAI testing, but it could reduce the amount of testing that has to be done. Another way to reduce the testing burden is to prepare some sections of a questionnaire as stand-alone modules that can be tested independently and reused in CAI questionnaires for other surveys (Pierzchala and Manners, 1998).

ACKNOWLEDGMENTS

We express our appreciation to several people who read and commented on previous drafts, especially Mick Couper and Mark Picrzchala for their critical review and helpful comments, which resulted in a much improved paper. We also want to thank the interviewers and staff of the Social and Economic Sciences Research Center at Washington State University for their assistance in testing CAI instruments and the many ideas they offered for improving CAI testing procedures.

CHAPTER 17

Usability Testing to Evaluate Computer-Assisted Instruments

Sue Ellen Hansen and Mick P. Couper
University of Michigan

17.1 INTRODUCTION

Social science research has employed computer-assisted data collection methods for more than 30 years (Couper and Nicholls, 1998). Over the last decade, advances in technology, reductions in cost, and the resulting growing prevalence of personal computers have led to increased use of computer-assisted interviewing (CAI), both telephone (CATI) and personal (CAPI), as well as Web and e-mail surveys, audio computer-assisted self-interviewing (ACASI), and interactive voice recognition (IVR). Despite the long history and widespread use of CAI, only recently have researchers begun to recognize the effect that computer assistance has on the interviewers and respondents who use CAI instruments. It is not sufficient to focus exclusively on question wording in the development and evaluation of CAI instruments. It is necessary also to design for and evaluate the usability of the instruments.

Usability focuses on how effectively and efficiently the user (interviewer or respondent) makes use of the system (CAI instrument) to achieve the desired goals. Whereas functionality focuses on how well the system works (see, e.g., Chapter 16, this volume), usability focuses on the user rather than the system. In this chapter we review recent research on the impact of CAI on interviewers and respondents, provide a model of the computer-assisted interview, present guidelines for CAI design and key usability evaluation methods (focusing primarily on the usability test), and describe usability test findings from four case studies.

Methods for Testing and Evaluating Survey Questionnaires, Edited by Stanley Presser, Jennifer M. Rothgeb, Mick P. Couper, Judith T. Lessler, Elizabeth Martin, Jean Martin, and Eleanor Singer
ISBN 0-471-45841-4

Much of the research on CAI has focused on attitudes of interviewers toward CAI and on its impact on data quality. In regard to the former, empirical and anecdotal evidence suggests that interviewers have reacted quite positively to CAI (Baker and Wojcik, 1992; Couper and Burt, 1994; Edwards et al., 1993). The most often reported reasons were an increased sense of professionalism (Nicholls and Groves, 1986; Weeks, 1992) and the elimination of postinterview editing (Couper and Burt, 1993). A review of research literature on direct comparisons of data from paper-and-pencil (P&P) surveys converted to CAI has shown data quality to have been largely unaffected by the transition to CAI (Nicholls et al., 1997).

In the last several years there has been increasing recognition of the effects that computers may have on interviewer behavior and performance (Couper and Schlegel, 1998; Couper et al., 1997) and on interviewer and respondent task demands. Research has begun to address issues of usability in the design and evaluation of computer-assisted data collection systems (Bosley et al., 1998; Hansen et al., 1998), and there is a growing body of evidence that seemingly innocuous CAI design decisions (e.g., separation of questions across screens and the numbering of response options) convey expectations to interviewers and respondents about what types of information are required, and thereby influence the responses obtained [for an overview, see Couper (2000)].

Such work has been informed by research on human–computer interaction (HCI), which suggests that design of the computer interface plays a large part in determining the usability of CAI instruments. Designing for usability means designing CAI instruments so that interviewers or respondents can perform their tasks easily and efficiently. Evaluating usability means assessing how interface design, and the ways in which computer assistance constrains or enhances interaction, affect the interviewer's or respondent's performance and the quality of data collected. Usability is achieved through the availability, clarity, and consistency of certain design features and system functions, and through providing effective feedback following interviewer or respondent actions. HCI, which has a firm theoretical grounding in cognitive and social psychology, computer science, communication, ethnography, and other disciplines, focuses on the social and cognitive processes involved in a person's interaction with a computer (Carroll, 1997). Thus, it extends instrument design issues beyond those of the visual presentation of information to the user's communication with the computer itself.

Many principles that govern good design (Couper, 2000; Redline and Dillman, 2002) apply both to paper and to computer-assisted survey instruments. Regardless of method of administration or technology, good design makes it clear to the user of the instrument what he or she must do, including the order in which to ask or respond to questions, when to skip questions, and how to record or enter responses. If the survey is interviewer administered, good design also conveys when to probe for additional information from the respondent, when to refer to response option cards and other interviewer aids, when to verify information, and whether or not to provide clarification or definitions.

However, the introduction of the computer as an interactant in the survey interview raises additional design and evaluation considerations in the development of

CAI instruments. Unlike the paper instrument, the computer is more than a neutral tool. It plays an active part in the survey interview. Consistent with research on the impact of technology on human–machine interaction, such as using copiers (Suchman, 1987), medical consultations (Greatbatch et al., 1993), and emergency call dispatching (Whalen, 1995), our work has shown that interviewers and respondents attend to the information provided by the computer, and the feedback it provides in response to their actions, in determining their own next actions (Couper and Hansen, 2002). Although as yet unexplored, the research of Greatbatch et al. (1993) and Whalen (1995) suggests that in interviewer-administered surveys, respondents also attend to the interaction between the interviewer and computer. This attention may affect the respondent's cognitive processes and contributions to the interaction.

Because the computer displays the instrument one screen or question at a time, segmenting the questionnaire, the computer user has a limited sense of context for any particular screen (Groves et al., 1980; House and Nicholls, 1988). Computer assistance also may divide the user's attention between the immediate task (record a response) and how to perform that task (type the number associated with the response and press [Enter]). Thus, the designers of CAI instruments must anticipate interviewer or respondent needs and provide the appropriate feedback and context information to the user at each screen the computer displays.

These aspects of CAI call for an expansion of traditional models of interaction and cognitive processes in the survey interview and for designing and evaluating questionnaires in terms of the impact of the computer and interface design on interviewer and/or respondent performance.

17.2 MODEL OF THE COMPUTER-ASSISTED INTERVIEW

Traditional models of question-response processes tend to focus primarily on either (1) the effects of interviewer–respondent interaction and the effects of the *interviewer* on respondent performance (Cannell and Kahn, 1968), (2) the impact of the structure of the *survey instrument* and question types on interviewer and respondent task demands and resulting responses (Sudman and Bradburn, 1974), or (3) the cognitive aspects of the question-response process, primarily *respondent* cognition (Tourangeau, 1984). These models are complementary (Schaeffer, 1991a), emphasizing one or more elements of the survey interview: interviewer, instrument, and/or respondent.

Our research on the effect of computer assistance on user (interviewer or respondent) behavior and on data quality has been guided by a model of interviewer–computer–respondent interaction that acknowledges the interaction among the interviewer, the respondent, and the CAI instrument. This model is shown in Figure 17.1. In an interviewer-assisted interview there are two interactions, interviewer–respondent and interviewer–computer, each affected by the other, as conveyed by the dashed line. In this three-way interaction, the interviewer, respondent, and computer (through software and instrument interface

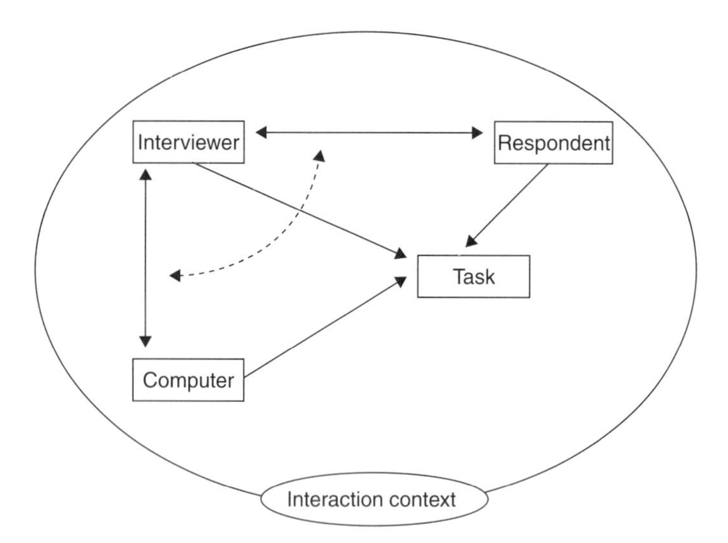

Figure 17.1 Computer–interviewer–respondent interaction.

design) all potentially affect the tasks of the others. In addition, the interaction is affected by the environment or survey setting, including mode of administration. For example, because the contexts or environments of personal and telephone interviews differ, the three-way interaction between the interviewer, respondent, and computer may also differ. An interviewer conducts a face-to-face interview in the presence of the respondent using a laptop computer, which makes the computer salient to the respondent. In contrast, in telephone interviews an interviewer typically conducts the interview using a desktop computer (with larger keyboard and screen display), and the respondent sees neither the interviewer nor the computer. This increases the interviewer's attention to the computer relative to the respondent, while the computer becomes less salient to the respondent.

In exploring the differences in interaction between paper and computer-assisted survey interviews (interviewer administered), we have developed process models for interviewer–respondent interaction in the P&P interview and for computer–interviewer–respondent interaction in the CAI interview. These models are based on an integrated model of social interaction and cognition that assumes that for each type of interaction there is a basic conversational sequence (Cannell and Kahn, 1968; Schaeffer and Maynard, 1996), through which collaborative understanding is achieved. The basic conversational sequence for interviewer (I) and respondent (R) is (1) I: ask question, (2) R: provide answer, and (3) I: receive response. For the three-way interaction among computer (C), interviewer, and respondent, the basic conversational sequence is (1) C: display question, (2) I: ask question, (3) R: provide answer, (4) I: input response, and (5) C: receive response.

In the course of interpreting questions and providing responses, respondent cognitive tasks involved in question interpretation and response formulation could lead to additional conversational turns through which interviewer and respondent collaborate to achieve a joint understanding of question and response. Similarly, interviewer tasks involved in the interpretation of computer displays and the entry of responses may lead to additional interaction with the computer. For example, if the respondent asked for a definition that the interviewer needed to retrieve from the computer's online help system, the sequence could unfold as follows (10 instead of the basic five steps):

1. C: Display question
2. I: Ask question
3. R: Request definition
4. I: Access computer online help
5. C: Display help
6. I: Provide definition
7. I: Exit computer help
8. R: Answer question
9. I: Input response
10. C: Receive response

In cases where the interviewer does not have to ask a question, the sequence would be similar to that of a self-administered questionnaire (1) C: display question, and (2) I: input response, and (3) C: receive response. Such would be the case if the computer displayed only an interviewer instruction ("Press [1] to continue") or a question for the interviewer, neither of which involves interaction with the respondent. Thus, at each screen displayed, the interviewer must interpret what is displayed on the computer screen to determine what actions to take. Elements to identify and interpret include response option show card instructions, questions (including optional and emphasized text), probe instructions, online help indicators, and so on. Screens may also include information based on prior responses (from the current or a previous interview), follow-up questions or conditional probes, or error messages. If the interviewer is to succeed at this stage, clear and consistent design of the screen elements and appropriate context information must make the specific task(s) required of the interviewer readily apparent (see Section 17.3).

Figure 17.2 shows a simplified model of the five-step basic conversational sequence (boxes labeled 1 through 5) and associated tasks of the computer, interviewer, and respondent in a computer-assisted interviewer-administered interview. Figure 17.3 shows the parallel simplified model for a self-administered interview, with conversational steps 1 through 3 and associated tasks. These models provide a framework for developing measures for assessing the usability of a particular CAI instrument.

In each of the models, tasks of the computer user (either interviewer or respondent) appear in gray. The term *display screen*, a computer action, involves either displaying a new question or screen in sequence, or providing feedback on input

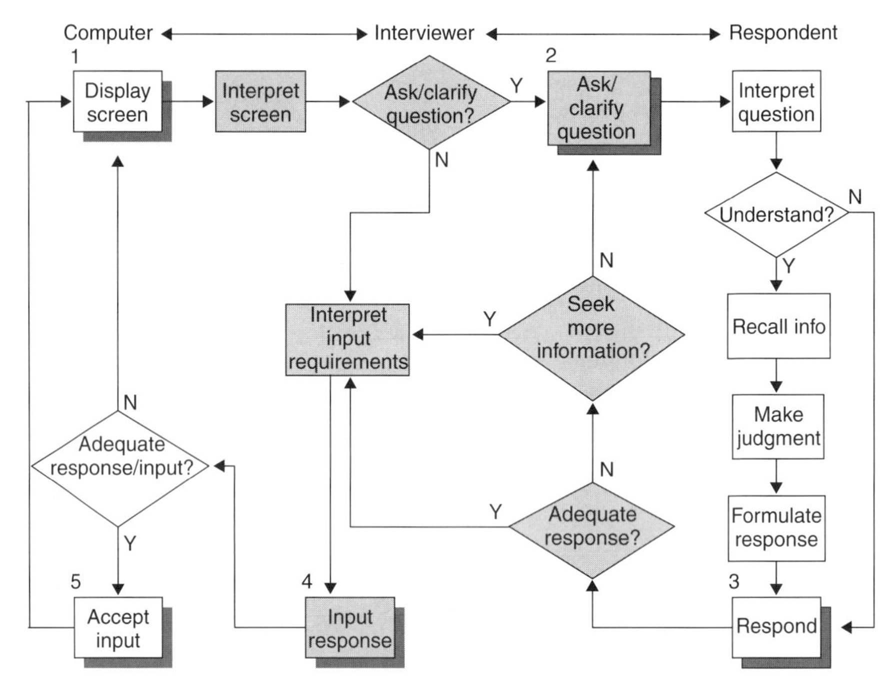

Figure 17.2 Model of interviewer-administered CAI interview (interviewer as computer user).

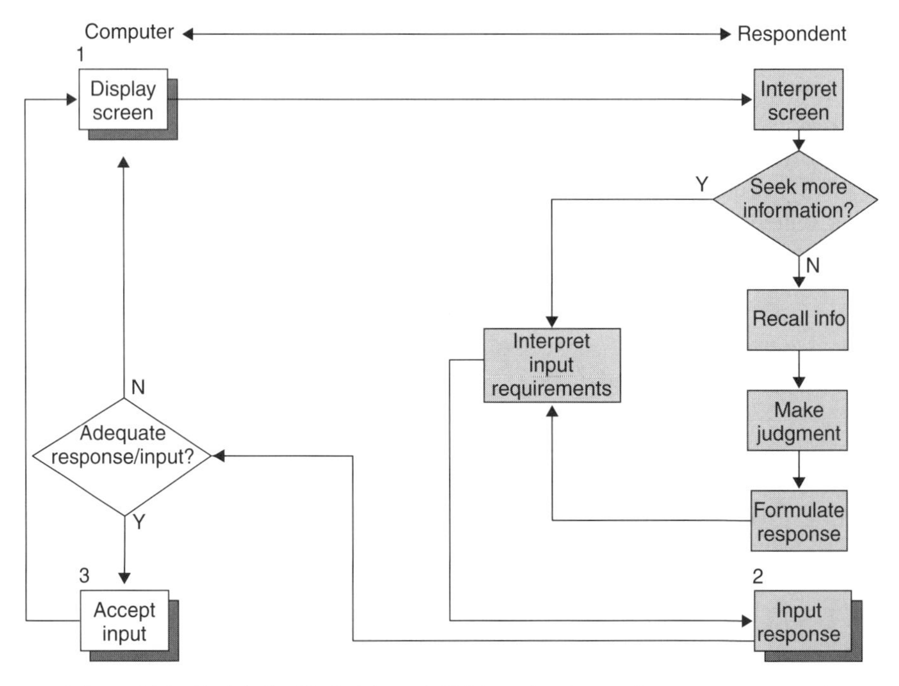

Figure 17.3 Model of self-administered CAI interview (respondent as computer user).

to the previously displayed screen, based on the outcome of the computer's evaluation of whether the response or input was adequate. *Interpret screen*, an interviewer task, involves perceiving and interpreting instructions and other information on the screen that will facilitate the interviewer or respondent performing the appropriate tasks, in the order required. During this sequence, there may be impediments to successful computer–interviewer interaction. There are several indicators that one can use to detect usability problems at this stage in the interaction. These include failure to perform a required task, such as refer to a show card, to read all or part of a question, or to follow probe or data-entry instructions. Other indicators of a potential usability problem are long silences, pauses, or affect- or task-related comments ("What does it want me to do?") before performing the task.

"Seek more information?" during the question–response sequence involves the user returning to the computer for further interpretation of the screen, to request online help, to backup and refer to prior questions and responses, and so on. Indicators of usability problems at this stage are user affect- or task-related comments (e.g., "I can't find the 'help'"), higher than usual item-level counts of use of the CAI backup function, or unsuccessful attempts to access item-level help.

At "Adequate response/input?" the computer determines what to display next (new question, response error message, or input error message). If the input is adequate (i.e., it meets question response and data input requirements), the computer accepts the input and displays the next screen or question in sequence. If not, the computer provides feedback to the user indicating the nature of the inadequacy, and does not display the next screen or question. At this stage of the response process, indicators of potential usability problems related to the input of responses can include item-level counts of error messages or consistency checks, item-level counts of backing up to change answers, user affect- or task-related comments ("Just a moment, while I record your response on the computer," or "The computer won't let me do this"), or higher than overall item times or times between the display of the screen and the entry of a response.

CAI design elements that decrease usability are most likely to arise at the "interpret screen" and "input response" stages of the process, because these are not addressed by traditional questionnaire evaluation techniques, such as cognitive interviewing, which focus on the respondent's interpretation of the question.

Step 1 in the conversational sequences of the models in Figures 17.2 and 17.3 (C: display screen) is critical to the outcome of the succeeding interaction, that is, successful interpretation of the screen and successful entry of a response. Norman (1986) describes seven stages of user activity in relation to the computer, which include perceptual and motor activity as well as cognitive processes: (1) identify the next goal; (2) form an intention in relation to the goal; (3) specify an action in pursuit of the goal; (4) execute the action; (5) perceive the computer's reaction to the action; (6) interpret the computer's reaction; and (7) evaluate the computer's reaction in relation to both goal and intention.

The first four stages apply to any interview, paper or computer-assisted. Both HCI and CAI research (Carroll, 1997; Couper, 2000; Norman, 1986) suggest that CAI screen and instrument design influence interviewer or respondent success at stages 1 and 5 through 7. These stages occur at the beginning of the conversational sequence, when the computer displays or updates the screen, and they are critical to the outcome of the succeeding interaction. This step in the conversational sequence (C: Display screen) is the primary focus of evaluating the usability of survey instruments. The success of the user in response to this stimulus depends on aspects of CAI screen or interface design. Methods of usability evaluation seek to identify problems in user–computer interaction that can be linked to CAI screen design.

17.3 CAI SCREEN DESIGN AND USABILITY EVALUATION METHODS

17.3.1 CAI Screen Design

That design has an impact on the way the interview is conducted is not a particularly new idea. It has long been accepted that the design and layout of paper questionnaires can affect the data collection process. Nonexperimental studies have demonstrated this for both interviewer- and self-administered surveys (Sanchez, 1992; Smith, 1995). House and Nicholls (1988) noted 15 years ago that in CAI interviews the introduction of the computer "modifies the role of both survey interviewer and questionnaire designer. The interviewer becomes more dependent on the work of the designer."

There are many examples of the impact of CAI design on the way the interview is conducted and on the data collected (Couper and Schlegel, 1998; Frazis and Stewart, 1996; see Couper, 2000, for an overview). Informed by HCI research, such research on the impact of CAI on interviewer behavior and performance has led to the development of theory-based guidelines for design of CAI instruments (Couper et al., 2000). Good design respects the natural flow of the user's tasks. It should make the interviewer or respondent more efficient, prevent errors, and reduce training or retraining needs, through making required tasks obvious. For successful delivery of the survey question and recording of the response, upon seeing a screen for the first time, the user's eyes should be drawn immediately to the key features of the screen. Ideally, a relatively untrained interviewer or respondent seeing a particular screen for the first time should be able to distinguish the action items from the information items and know what the task should be. This requires:

- Consistent screen design
- Visual discrimination among the various elements (so that CAI users learn what is where, and know where to look)
- Adherence to normal reading behavior (i.e., start in upper left corner)

- Display of instructions at points appropriate to associated tasks (e.g., show-card instruction precedes question and entry instruction follows it)
- Elimination of clutter and unnecessary information or other display features (e.g., lines and toolbars) that distract users from immediate tasks

The question is the most important feature of the CAI screen, and it should be the most visible element of the screen, immediately identifiable. In addition, different question types and response input formats (e.g., currency, dates, grids, and open text) and other elements of the screen (e.g., response options, online help indicators, section and other context information) should all be formatted consistently. In addition to ease of computer use, these are the aspects of design that usability evaluation methods address.

17.3.2 Usability Evaluation Methods

Usability research and evaluation techniques are widely used for predictive performance modeling and for evaluation of computer software, hardware, or other tools. Two such sets of techniques are particularly useful for evaluation of CAI instruments: usability inspection methods and end-user evaluation methods.

Usability Inspection Methods Usability inspection or expert methods generally involve the evaluation of a system by one or more HCI or usability experts (see, e.g., Nielsen and Mack, 1994). These methods, also referred to as *heuristic evaluations*, are analogous to expert review in the evaluation of survey questionnaires (Presser and Blair, 1994).

Usability inspections typically involve experts being provided with a small set of heuristics or evaluation criteria distilled from a larger number of interface design principles. The criteria generally expand on such key HCI principles as be consistent, provide feedback, minimize user memory load, and so on, and are modified to the explicit system or tool under evaluation. For example, the general heuristic "make it easy to navigate" might be made more explicit for specific use of keyboard, mouse, or stylus. The experts then test the system and note violations of particular heuristics. In general, from three to five experts are recommended for this approach, which could be completed within a few days. Lansdale and Ormerud (1994) note that heuristic evaluation can be highly unreliable but can find many problems relatively cheaply. There are a number of examples in the research literature of the use of formal heuristic evaluation in the survey world (Bosley et al., 1998; Levi and Conrad, 1995; Nicholls and Sedivi, 1998; Sweet et al., 1997).

End-User Evaluation Methods End-user evaluation methods involve the ultimate users of the systems, and can be laboratory- or field-based, and either experimental or nonexperimental. They include collection of field-based performance data, observational methods, direct questioning of users, cognitive walk-throughs, or combinations of these methods. They can involve either scripted activities or natural interactions (free exploration of systems), and can range from more intrusive (in a laboratory with the evaluator interacting with the user) to less obtrusive (automatic collection of performance data without the knowledge of the user). Two of these methods have been particularly useful in our own usability research: analysis of field-based performance data, and laboratory-based usability tests of CAI instruments, both experimental and observational.

Analysis of Field-Based Performance Data One of the benefits of computer-assisted data collection is that the survey process generates a great deal of automated data, and these data can, in turn, be used to evaluate the survey process. Such *paradata* (data that describe the process) include case management information (e.g., response rates, calls per case, and average interview length) and audit trail files created by CAI systems. Some audit trails also include every keystroke the interviewer presses (keystroke files). They can be used to produce rates of use of certain functions (e.g., backing up, using online help, switching languages, and entering interviewer comments), at the survey item, interviewer, or interview level. Couper et al. (1997), Caspar and Couper (1997), and Couper and Schlegel (1998) provide examples using audit trail and keystroke data to evaluate interviewer or respondent performance and to identify usability problems in CAI instruments.

Laboratory-Based Usability Research Laboratory-based usability evaluation techniques, or usability tests, are widely used in HCI research to collect data on behavior and preferences of representative users of systems in controlled settings. They may be either experimental, designed to confirm or refute research hypotheses about design features, or observational, intended to expose design problems. Experimental studies are employed in early stages of development, using design prototypes and small-scale instruments with a larger number of subjects (40 or more), randomly assigned to treatments. Observational studies generally are conducted with a smaller number of subjects (five to 15), sometimes iteratively, on more fully developed instruments. In our work we have conducted both experimental and observational usability studies. In the next section we describe usability testing more fully.

17.4 USABILITY TEST

There are five basic characteristics of a usability test (Dumas and Redish, 1994): (1) the primary goal is to improve the usability of the CAI instrument; (2) participants represent real users, that is, interviewers or respondents; (3) participants do real tasks; (4) testers or monitors observe and collect data on what participants do; and (5) analysis of the data collected leads to recommendations on how to improve design of the instrument. Usability testing requires a test area or laboratory, development of procedures or a test plan, identification or recruitment of appropriate participants, and an experienced test monitor. Based on data gathered during the test, including audit trails, the tester's or monitors' notes, participant debriefing questions, and/or posttest coding of each test, the final product is a report that describes usability problems or issues identified and provides recommendations for changes in the instrument design.

17.4.1 The Laboratory

Usability testing requires a setting in which participants can be observed unobtrusively. The typical cognitive laboratory, in which there is generally a one-way mirror for observation, can easily be adapted for usability testing. Although videotaping or automatic recording of user actions is not essential, it is very useful to have such recordings, especially of each computer screen that is displayed during the CAI interview, with associated audio. The usability laboratory at the University of Michigan can videotape up to three images or views of a usability test interview. The laboratory has two ceiling-mounted cameras in the test room and three monitors in the observation area, one of which displays a scan-converted image of the interviewer's or respondent's computer screen. It is possible to display two additional video images, from the two ceiling cameras: for example, the interviewer's hands on the computer keyboard, and the interviewer as he or she interacts with the respondent. Figure 17.4 shows three videotaped images as displayed on the monitors in the observation area of the laboratory.

The Michigan laboratory was equipped in 1997 at a cost of approximately $20,000, including the additional equipment and software for reviewing and editing videotapes. With advances in technology, it is possible to design a laboratory

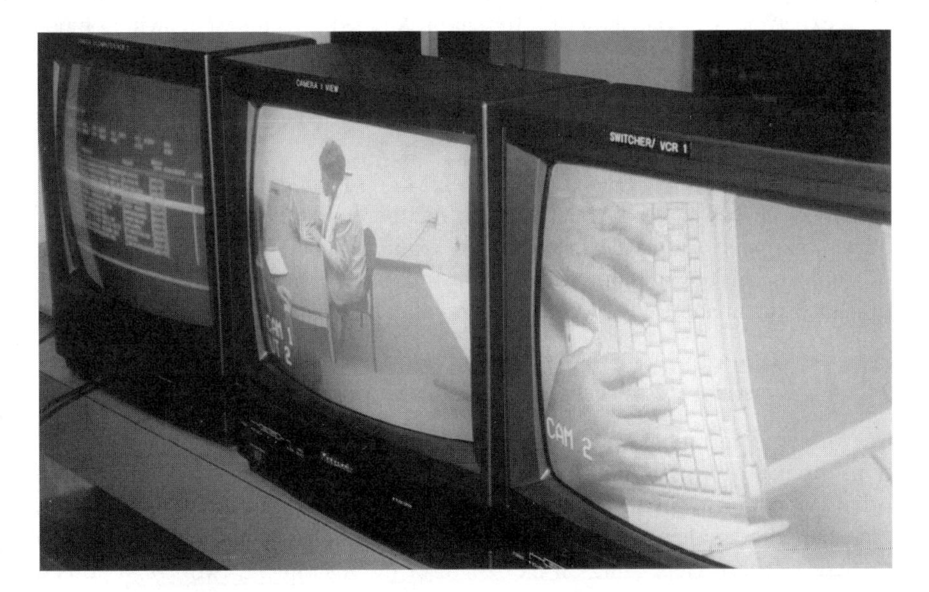

Figure 17.4 Monitors in observation area of the Michigan Laboratory.

today at a significantly lower cost. Through the use of ceiling-mounted digital video recorders and software that creates digital video streams of computer screens with associated audio, it is now possible to record digital images directly to disk. Such digital images are easier to store, review, code, and edit.

17.4.2 Usability Test Plan

For every usability test, it is important to have a clear test plan that "addresses the how, when, where, who, why, and what of your usability test" (Rubin, 1994). Although less formal approaches may be taken, formal advance documentation of a usability test helps focus the test, from initial to final stages. It serves as a tool for communication among test developers, monitors, and analysts, and helps identify required internal and external resources. A typical plan generally includes the following (Rubin, 1994): purpose and test objectives, user profile, method (test design), participant task list, test environment and equipment, test monitor tasks, evaluation measures, and report content and presentation.

17.4.3 Selecting and Recruiting Respondents

The approach to selecting and recruiting usability test subjects will depend in part on whether the test design involves self- or interviewer-administered instruments, the number of subjects required, and whether the test involves experimental research on design prototypes or evaluation of an actual survey instrument that interviewers or respondents will be required to use. If the test is a small usability evaluation of an interviewer-administered instrument involving three to five interviewers, it is best to use actual CAI interviewers. On the other hand, if the test is an experimental evaluation of different prototyped screen designs with a much larger number of subjects, recruiting users from the local community may be preferable, particularly when the pool of on-staff interviewers is small. Recruiting from the local community is also preferable for tests of self-administered instruments.

For some studies we have found it useful to do mock interviews with scripted scenarios. In such cases, CAI interviewers or other research staff can act as respondents, following scripts designed to test explicit features or paths of the CAI instrument. This obviates the need to get Institutional Review Board (IRB) or Office of Management and Budget (OMB) approval, which can be a lengthy process.

17.4.4 Timeline, Resources, and Cost of the Usability Test

The timeline, resources, and cost of usability tests depend on whether observational or experimental, and the number of and types of participants. An evaluation of a CAI instrument with 10 to 12 test interviews need only take a week or so to conduct, with another two to three weeks needed to analyze data collected and to prepare a report. Depending on the length of the instrument, two to five tests a day are feasible. An experimental study with a larger number of respondents may take longer, requiring a larger effort to recruit, screen, and schedule respondents, and to schedule interviewing staff if interviewer-administered.

We have found the cost of conducting an observational usability test of a CAI instrument with 10 to 12 respondents to be between $5000 and $10,000, which covers the cost of supplies, duplicating, and recruiting and paying respondents, as well as non-interviewer staff hours, for preparation, conducting the tests, and posttest review, coding, and reportwriting. It is possible to conduct usability tests at an even lower cost. A usability test that recorded only the image of the computer screen (the primary source of usability information), with audio, would be adequate for most tests. Also, the cost would be less if coding and a formal report were not required. However, the latter help ensure that more objective criteria are used in assessing the severity and extent of problems.

17.5 CASE STUDIES

17.5.1 Observational Usability Test of NHIS 2000 Cancer Supplement

In 1999, the University of Michigan Institute for Social Research conducted a usability evaluation of the cancer supplement to the National Health Interview Survey (NHIS) 2000 CAPI instrument. The supplement had five major sections: Hispanic acculturation, nutrition, physical activity, tobacco, cancer screening, and genetic testing, none of which had been implemented previously. The goal was to evaluate the usability of the cancer supplement, not the core NHIS instrument, which had been evaluated previously for usability.

Procedures Over a one-week period in June and July 1999, four interviewers from the Detroit Regional Office of the U.S. Bureau of the Census conducted videotaped usability interviews using the NHIS 2000 pretest instrument in the laboratory described previously. Each of the four interviewers conducted two to three interviews each, with a total of 11 interviews that averaged 71 minutes (21 minutes for the supplement). Respondents were recruited from the local area through a newspaper advertisement and were paid $30 to participate. At the end of each day's interviews, the interviewers gave their observations and comments on the interviews they had conducted and their reactions to the supplement questions.

We used four sources of information to identify problems in the NHIS 2000 cancer supplement: (1) observations made during the interviews, (2) interviewer debriefings, (3) a review of a printed copy of the supplement, and (4) systematic coding of videotapes of the supplement sections in each interview. These revealed both usability and questionnaire design problems.

Findings The 400 problems identified were distinguished as usability, questionnaire design, training, and combinations of the three. Those classified as usability problems focused on screen design issues: that is, placement of information on the screen, the way screen elements were formatted and made distinct from other elements, and the consistency of design across screens. Questionnaire design problems were those that a traditional pretest would be expected to identify, including interviewer and respondent difficulties resulting from question wording and order. Some problems suggested a need for focused interviewer training that would prepare the interviewer to deal better with issues unique to a survey on health care and medical issues. These were identified as training issues.

In very few cases were attempts made to suggest solutions to the questionnaire design problems identified, which it is felt are best left to substantive experts. However, there were general solutions proposed for the major usability issues identified:

1. *Flashcard (response option card) instructions that were skipped or referred to after reading the question.* Distinguish flashcard instructions from other instructions, and consistently place them above the question text.
2. *Inconsistent response options and formats that led to interviewer confusion and data-entry error.* Make response options and categories consistent through the instrument.
3. *Emphasized text that interviewers did not read consistently.* Use emphasis sparingly; reserve capitalization for instructions; use boldface text for emphasis; use boldface text and capitalization consistently throughout the instrument.
4. *Inconsistent presentation of related items in a question series that led to nonstandardized question reading.* Present multiple items on a screen; display question stems as optional text on second and later items in a series.

Overall, emphasis was placed on maintaining consistency across screens, within the supplement, and with other sections of the instrument.

Many problems found in this usability test were also found in a simultaneous large field pretest of the same instrument (approximately 300 interviews), and many of the problems identified could apply to either P&P or CAI instruments. The clear advantage of the videotaped usability test was that interviews could be viewed repeatedly, and segments used to make clear the impact of problems reported, especially the impact of the computer on the performance of tasks.

One problem in particular revealed how the computer can affect interaction and interviewer performance in the CAI interview. This occurred on a question at which the respondent was asked which language(s) he or she spoke as a child: Spanish, English, or both (Figure 17.5). The respondent answered "Italian," and a several-minute interaction ensued, during which the interviewer tried first to see if she could record such a response. When it was clear that she could not, she negotiated with the respondent until she could enter a response of "both," after he finally told her that by the time he was 3 or 4, he was able to speak some English and some Spanish. In a paper interview, the interviewer probably would have entered a marginal note rather than force field coding of the response. In the CAI interview, the interviewer's options were to enter a code for "don't know" or "refused," or to negotiate a response, any of which would have resulted in loss of the original response.

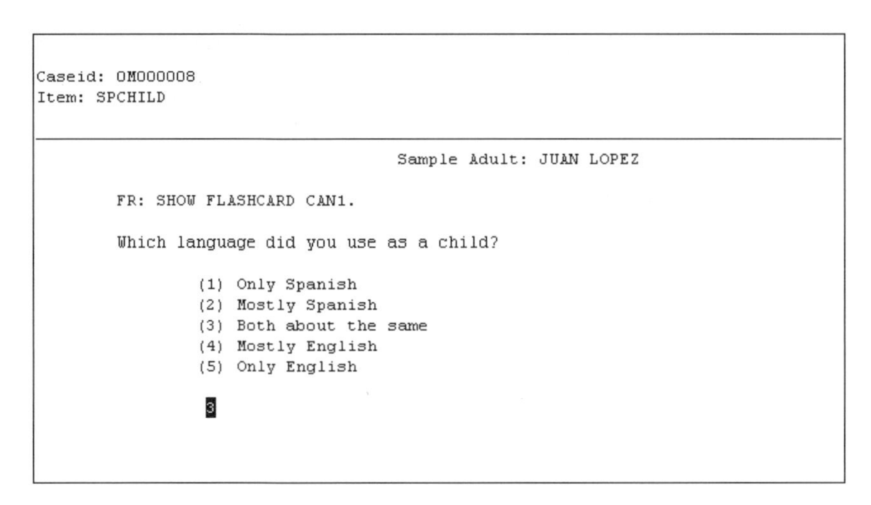
Caseid: OM000008
Item: SPCHILD

Sample Adult: JUAN LOPEZ

FR: SHOW FLASHCARD CAN1.

Which language did you use as a child?

(1) Only Spanish
(2) Mostly Spanish
(3) Both about the same
(4) Mostly English
(5) Only English

3

Figure 17.5 CAI item that prevented entry of response "Italian."

Other questionnaire evaluation techniques (e.g., cognitive interviews or an expert review) also might have revealed this question design problem. Many conventional techniques overlap in the types of questionnaire design problems they reveal. However, other techniques would not necessarily have revealed the aspect of CAI design that affected interviewer performance and data quality: The responses options, as well as the CAI software, made it difficult for the interviewer to continue the interview without first entering an invalid response[1]. CAI software should facilitate recording an "out of frame" response, one that is easily identifiable and recoded at the coding phase of data collection. That it did not in this case is a usability issue.

17.5.2 Experimental Usability Test of the Effect of Design on User Performance

To test the application of basic CAI design principles, we conducted an experiment that compared the effects of two approaches to screen design on user performance (Hansen et al., 2000). The first approach, using a *standard* design (e.g., Figure 17.6), was to display information to the user as it typically is displayed in most CAI instruments that we have encountered: question text and instructions in the same font size, with little (or haphazard) use of color and context information, and with instructions capitalized. The second approach, using an *enhanced* design (e.g., Figure 17.7), was to (1) display question text larger than instructions and other information on the screen; (2) display instructions in a smaller font, blue, and in mixed case; (3) display key context information at

[1]It is also worth noting that although usability evaluation may uncover more human–computer interaction and computer user performance issues than other techniques, other techniques could also reveal usability issues during evaluation of computer-assisted instruments.

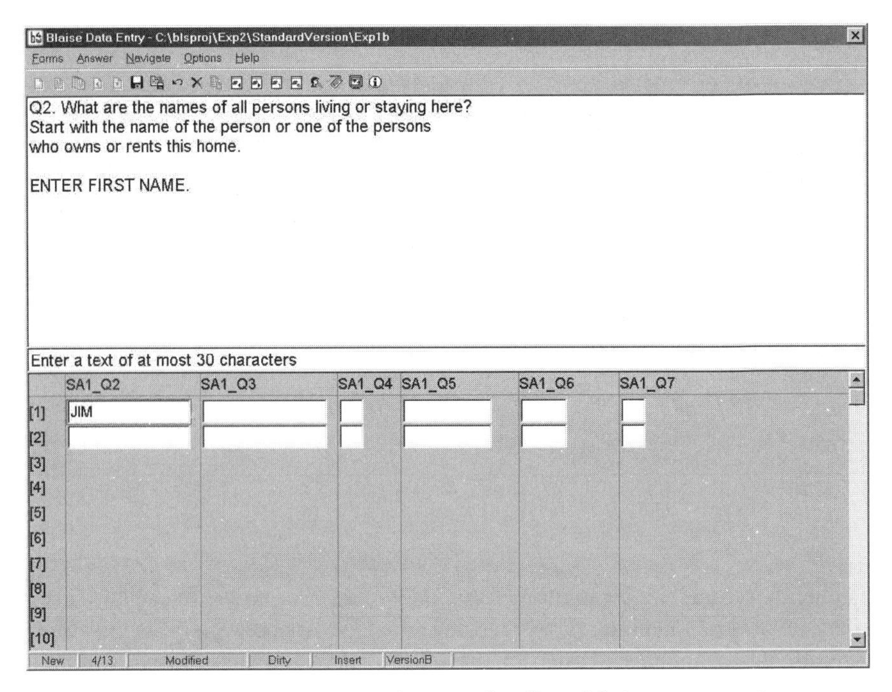

Figure 17.6 Standard format in experiment on the effect of design on user performance.

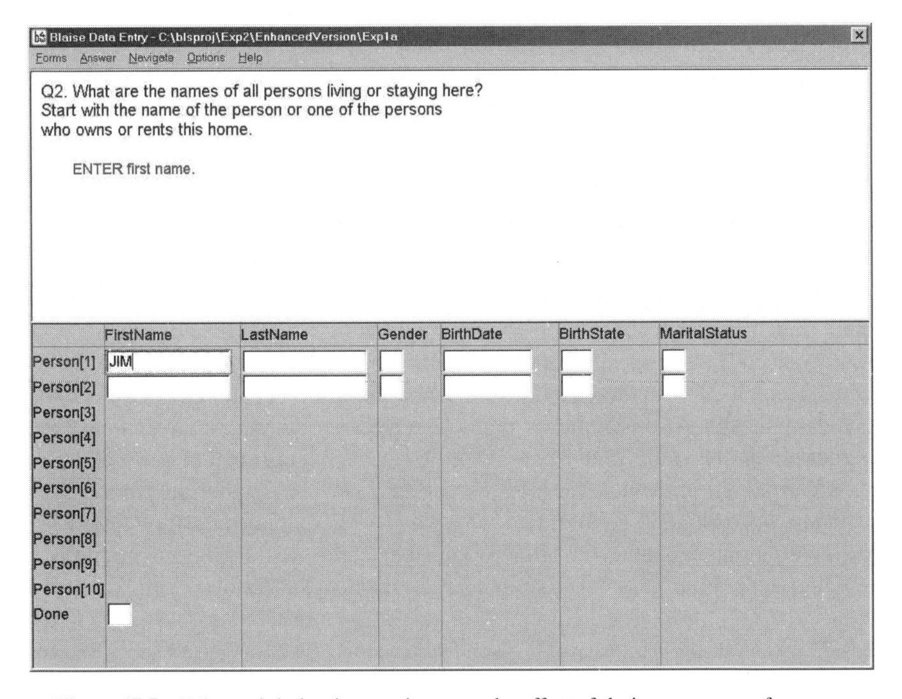

Figure 17.7 Enhanced design in experiment on the effect of design on user performance.

some items; and (4) not display unnecessary information (such as a status line at the bottom of the screen). It was hypothesized that the enhanced design, which distinguished questions from instructions and presented a less cluttered and consistently formatted screen to the user, would increase the speed with which users process information displayed.

Procedures A short instrument based on the Consumer Expenditure Quarterly Interview Survey was developed to test the effect of the enhanced design versus the standard design. The instrument required participants in the experiment to (1) enter a household listing; (2) give counts of rooms, bedrooms, and full and half bathrooms; (3) indicate whether they had specific household appliances; and (4) specify costs of acquiring owned property.

The experiment was conducted in March and April 2000 in the University of Michigan usability laboratory. Forty subjects, recruited from the Ann Arbor area, were paid $20 for their participation. All answers that participants provided were based on four scenarios developed to test the effect of the enhanced screen design on user performance. The scenarios were presented in the form of a series of flip cards alongside the computer, one card corresponding to a specific screen or sequence of screens in the instrument. Each subject completed the four scenarios in order, two in the standard version of the instrument and two in the enhanced version, randomized to complete either the standard or enhanced scenarios first. The use of scenarios and randomizing the order in which versions appeared minimized variation in the type of information entered by subjects and allowed for controlling for instrument order and learning effects.

Following the completion of the four scenarios, subjects filled out a self-administered debriefing questionnaire to elicit their reactions to the alternative versions of the instrument. In addition to the subjective reactions from the debriefing questionnaire, an audit trail with item names, responses, and item time stamps was captured from the CAI system.

The goal of this study was to obtain empirical evidence that the enhanced design improves interviewer performance. However, because the subjects were not professional interviewers, the experiment was not an ideal test of interviewer performance, and reactions to the two versions are not those of trained interviews. Therefore, following the experiment, 15 University of Michigan interviewers were asked to use the first and second scenarios to walk through the standard and enhanced instruments, respectively, and then fill out the same debriefing questionnaire completed by the experiment's subjects. This allowed a comparison of subject and interviewer reactions to the standard and enhanced screen designs.

Findings The results of this experiment suggested that the enhanced screen design makes an instrument easier to use, and increases efficiency in terms of time to complete an item. Major findings were:

- Users generally took less time to answer questions in the enhanced version. The overall completion time for the enhanced version (469 seconds) was about 9% less than that for the standard version (515 seconds).

- There was no significant overall difference between versions in user errors ($p = 0.079$).
- Both subjects ($n = 40$) in the experiment and trained interviewers ($n = 15$) generally preferred the enhanced version of the instrument, and found it easier to use.
- Subjects generally completed sections of the enhanced version faster. Three sections in the enhanced version were completed from 12 to 20 seconds faster, with p values from 0.001 to 0.014, while one section took 4 seconds longer than the standard version ($p = 0.058$).
- The difference between standard and enhanced average completion times tended to be greater if subjects preferred the enhanced version.

In summary, this experiment on screen design found that designing CAI screens for usability generally improves user performance, in terms of time to complete an item, and is preferred by users. While the enhanced version had a higher incidence of errors across items (6% versus 5% in the standard version), the difference was not significant ($p = 0.079$). We concluded that the enhanced design did not affect data quality, but that further research is needed to assess the impact of various aspects of design on data quality.

The time differences we found between versions are likely to be smaller with experienced interviewers, and may well diminish as interviewers gain experience. However, these results are likely to apply to initial learning of a new interviewing system. Given the high cost of interviewer training, this is a benefit even if the effects do not persist over time. Furthermore, as Dillon (1987) found, a well-designed interface results in more effective transfer of knowledge and a deeper understanding of the underlying system. In other words, learning and remembering the system becomes easier over time with a user-centered design.

We did not conduct this experiment with trained interviewers for reasons of cost and time and because we did not want to risk contamination of the pool of interviewers available for future research. This study shows that experiments with noninterviewers can inform screen design for interviewers, which is reinforced in this experiment by generally parallel findings on interviewer and subject preferences for the enhanced over the standard design.

17.5.3 Experimental Usability Test of Mouse versus Keyboard Data Entry

Designers of self-administered instruments face a number of design challenges, since respondents do not have the familiarity with the instrument, computers, or surveys in general that the trained interviewer has. Thus, computer-assisted self-interviewing instruments call for a simpler design that facilitates an untrained respondent's reading of questions and data entry. Sparked by reports from other survey research organizations of inaccurate data when using keyboard entry and poor screen stability using touch-screen entry, in the winter of 1998 the University of Michigan conducted an experimental usability test exploring whether the mouse was a feasible alternative to the keyboard for ACASI data entry.

Procedures Questions were taken from the National Survey of Family Growth (NSFG) and included questions about sex partners, birth control, and other issues of a sensitive nature. The ACASI screen was designed with a series of numbers and icons at the bottom of the screen that were to be used for mouse data entry (Figure 17.8). With the exception of the bar at the bottom of the screen with the icons, the keyboard version of the screen was identical. Forty female respondents were recruited from the local community and paid $30 incentives to participate. Screening procedures selected a higher number of respondents with low education and little or no computer experience. Respondents were asked to complete two versions of the survey, one using the mouse for data entry, the other the keyboard. The order was randomized.

The mouse version required the respondent to click the right arrow after entering a response in order to move to the next question, and the keyboard version required the respondent to press the [Enter] key. This allowed the respondent in either mode to get visual feedback of the response entered prior to advancing to the next question. Scripted instructions and icon/function keyboard templates were provided for each version. As part of the instructions, respondents were shown how to adjust the audio volume on the headset, which allowed respondents to listen to the questions as they read them. The speed and size of the mouse cursor on the laptop were carefully adjusted prior to the test so that it would be comfortable for most respondents to use. Respondents were asked to

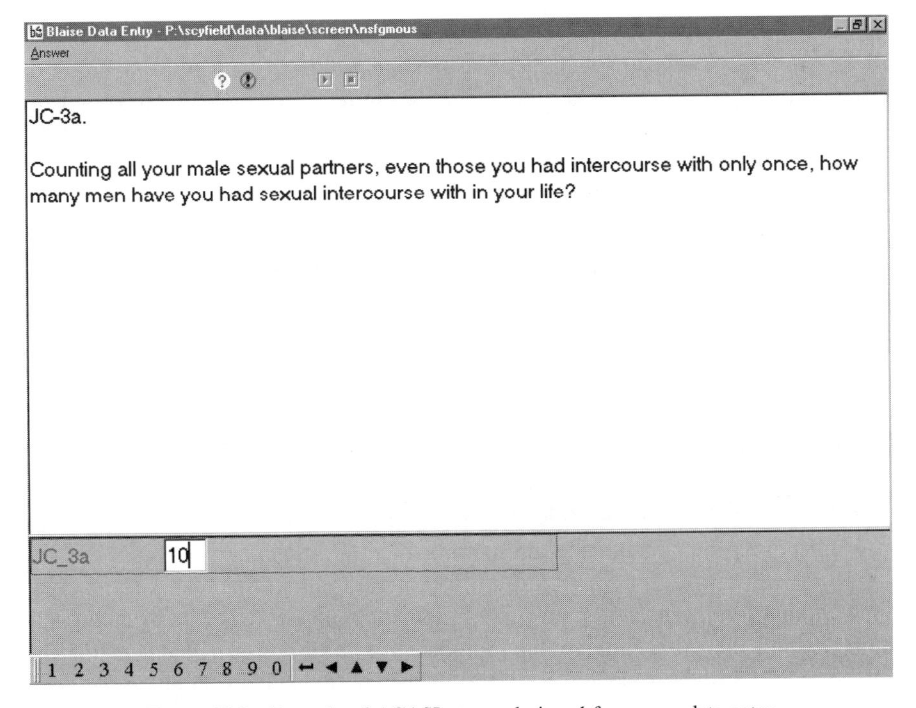

Figure 17.8 Example of ACASI screen designed for mouse data entry.

alert the monitor at the end of the first version so that he or she could provide instructions for the second version and change the template. All respondents were debriefed at the end of the sets of interviews, and the debriefings were recorded.

Findings The results of the test revealed that at least for respondents with little or no computer training, the keyboard was both easier to learn to use and was the preferred mode of data entry. Women in the study who had never used a mouse before took some time to figure out how to hold, move, and use the mouse to select items. For example, some held the mouse sideways with the right hand while clicking with the left index finger. Some held the mouse sideways, with one hand in front of the mouse and one in back of it. Others held the mouse cord while aiming at icons, appearing to use the mouse as a steering device. Some respondents talked to the mouse ("Just get over here!) while positioning the cursor, and others sometimes seemed to instinctively reach for the [Enter] key on the keyboard rather than use the mouse to select a response.

In the debriefing interviews, 65% (26) of the 40 respondents preferred keyboard data entry, and 55% (22) felt that the keyboard was easier to learn to use. This varied by education, with 87.5% with less than a college degree finding the keyboard easier to learn to use. For those that preferred the mouse, major reasons were (1) selecting a response seemed faster, (2) locating a response was easier, (3) using a mouse was fun, (4) it was easier to change answers, and (5) the mouse could be used as a pointing device to follow text on the screen while the audio was playing. Keyboard data entry was faster (5.7 minutes on average) than mouse data entry (9.4 minutes on average).

The tests also revealed a number of design problems with both versions of the instrument. It was thought that the following design changes would improve the usability of the mouse version:

- Increase the size of the icons and space them farther apart to facilitate selection.
- Make the keypad larger and place it higher on the screen.
- Slow the speed of the mouse and make the pointer larger.
- Place the "Advance" key higher on the screen and make it larger.
- Provide a mouse pad.

The following suggestions were made for improving both versions:

- Increase the amount of entry space for open-ended questions.
- Improve the voice used to audio record the questions

Although there was clear preference among test respondents for keyboard data entry, which was also faster, we concluded that with design enhancements the mouse could still become a useful and accurate data-entry option for ACASI. Additional research with improved instruments could test this further.

17.5.4 Observational Usability Test of the NHIS Redesigned 2004 Instrument

In December 2002 the University of Michigan conducted a usability test of portions of the 2004 NHIS instrument, which had been redesigned in Blaise, a Windows-based CAI system. The goal was to identify usability problems in time to make changes in the complete instrument prior to a June 2003 pretest.

Procedures Four U.S. Census Bureau Detroit regional office interviewers conducted 30 interviews. Two of the interviewers conducted 10 interviews each, and the other two conducted five each. A University of Michigan staff person acted as a mock respondent for each interview. She followed 15 scenarios for these interviews, using each twice. Scenarios varied on family size, demographic characteristics, and health conditions of family members. No scenario was repeated with the same interviewer.

The interviews were videotaped, and digitized recordings were captured for 28 of the 30 interviews (two were lost due to technical difficulties). These digitized recordings showed the CAI instrument as the interviewers conducted the interviews, as well as keystrokes and mouse movement. These also included audio recordings of interviewer–respondent interaction. The Michigan staff member who acted as the mock respondent debriefed the interviewers on each of the test days, giving them an opportunity to comment on the new Blaise NHIS instrument. The digitized recordings and debriefings were the primary sources of data for the analysis of usability problems in the new instrument.

Each of the digitized interviews was played back, during which events (interviewer or respondent behaviors) were noted that might indicate usability or design problems. These were classified according to source of problem.

Findings The nearly 600 problems identified fell into five broad categories: (1) computer assistance, (2) CAI software, (3) CAI programming logic, (4) screen design, and (5) question design. As in other usability tests, they suggested solutions that fell into three categories: usability, questionnaire design, and training.

The majority of the computer assistance problems involved interviewers trying to use the mouse pointer on the computer laptop in situations in which it would have been much more efficient to use keyboard commands, for example, getting out of auxiliary windows, such as online help and consistency checks. It was recommended that interviewer training focus on the keyboard navigation and commands in the Windows environment, which was new for the NHIS interviewers.

Most of the usability issues associated with screen design were associated with inconsistent placement and formatting of elements on the screen, for example, show card instructions, data-entry instructions, and mandatory and optional text, and with the formatting of auxiliary windows, such as help screens, lookup tables for response options, consistency checks, and online help. Recommendations were made that would increase consistency and improve usability of these screen elements and windows.

The two most notable usability problems were related to constraints placed on design by the CAI software. In the first, there were at least 50 instances in which interviewers had difficulty at "enter all that apply" questions. Interviewers were accustomed to using [Enter] to separate responses, but the software does not allow this option for separating entries on such items. The problems generally occurred at items critical to the NHIS: lists of household members to which a situation (e.g., physical limitation or source of income) applied, and lists of conditions that caused a household member's limitations. At these questions, interviewers were instructed to separate entries with a comma. However, they often pressed [Enter] after the first entry, moving to the next question. This either led to data-entry error if they had not realized that a second entry had become the response to a question they had not asked, or they had to back up and enter additional responses before proceeding. The software also allows using the spacebar to separate responses, which was recommended. It was also recommended that training focus on data entry for "enter all that apply" questions. It was felt that these were only partial solutions, since all interviewers experienced this problem at least a few times, even if generally handling data entry at these items correctly.

The second usability problem associated with the CAI software occurred on multipart questions, such as "times per period" questions (e.g., number of days, weeks, or years). Difficulties experienced on these questions were exacerbated by the split-screen design in the CAI software, in which the question text area

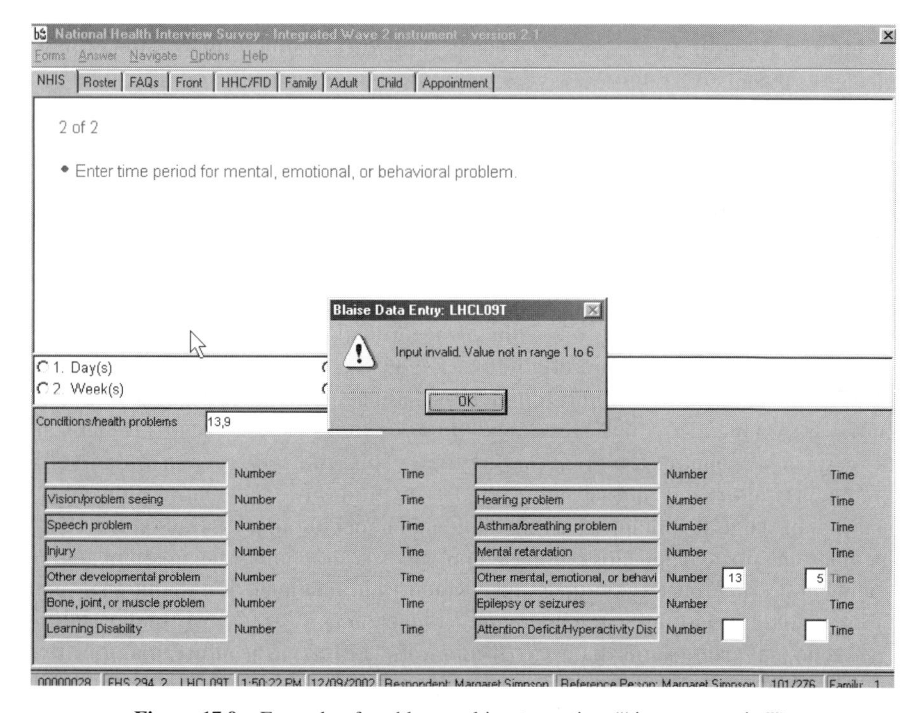

Figure 17.9 Example of problem multipart question ("times per period").

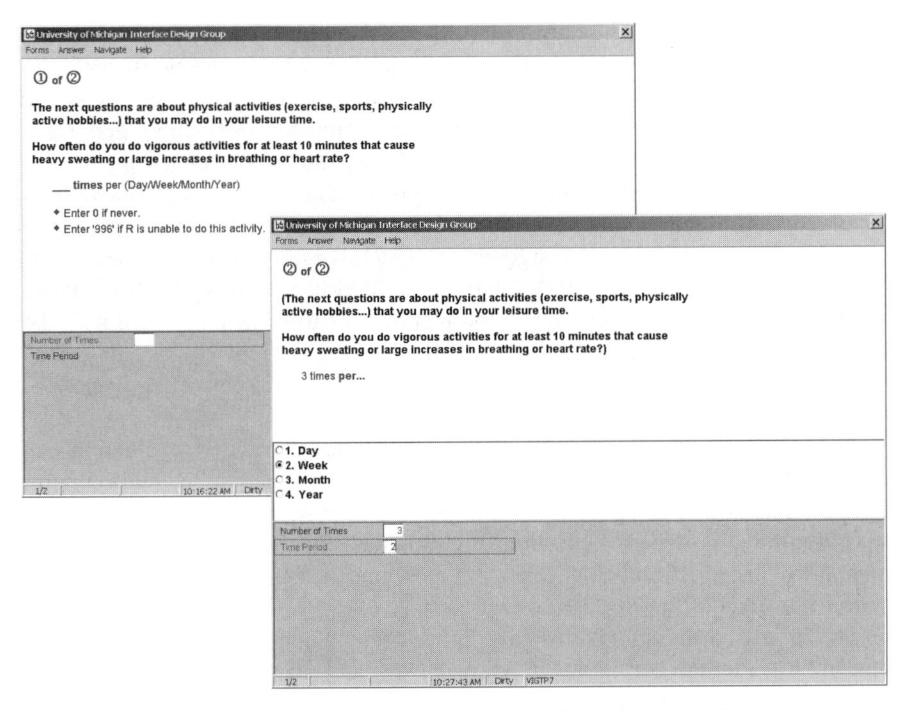

Figure 17.10 Example of design alternative for multipart question.

is separated from the data-entry area of the screen. This design shifts the interviewer's focus from the question text area (and instructions and context it may provide for data entry) to the input area of the screen. An example of the second part of a two-part question in the NHIS is shown in Figure 17.9. In this example, the interviewer was trying to enter the number of years a child had had a "mental, emotional, or behavioral problem," code 13 in the prior question. The input field labels for this two-part question are "Number" and "Time," which led the interviewer to reenter "13" as "Number" and "5" as "Time," rather than "5" and the code for "Years," when the respondent answered "5 years."

As solutions to this problem, it was recommended that (1) interviewer instructions in the question text area distinguish these items as multiple-part questions, (2) question text on the second part of the question include the response to the first part of the question, and (3) labels for input fields be changed to minimize confusion when the interviewer's attention is focused on the input area of the screen. Figure 17.10 shows a prototype of a design alternative for a multipart question that has these features.

17.6 CONCLUSIONS

Usability evaluation complements other questionnaire evaluation techniques, which focus on the crafting of questions and the question-response process:

in particular, respondent cognition. Although the unit of analysis is often the survey question, usability evaluation shifts the focus from questions per se to interviewers and respondents as users of computers, the computer screen on which the question is displayed, the human–computer interaction that takes place, and the efficiency with which interviewers and respondents are able to perform their tasks at each question or screen. There are several techniques for evaluating usability, falling into two basic categories: expert evaluation and end-user evaluation. The former tends to be less costly and can be done very quickly, although it may be less reliable than end-user techniques that involve systematic observation or coding of the behavior of users as they work with the CAI instruments. The latter include usability testing, which we have described in detail, with examples from case studies that demonstrate the effectiveness of this technique.

As with other questionnaire evaluation techniques, usability evaluation can take place throughout the questionnaire development cycle, and should be iterative. Such iterative evaluation could include cognitive walk-throughs and usability tests of initial prototypes of questionnaire modules, expert review of CAI screens, more formal heuristic evaluation of screens, and a usability test of the entire instrument coincident with a formal pretest.

The focus of discussion in this chapter has been on interviewer–computer–respondent interaction during the CAI interview and on measuring usability as questions are asked. However, there are additional effects of usability that may be more difficult to assess but are very important. For example, further research is needed to assess the effect of usability on interviewer training and learning retention.

As we continue to apply technological advances to the development of CAI instruments, through innovative application of the use of graphics and multimedia, and we increase the complexity of survey instruments (e.g., dependent interviewing and increased tailoring of question text), usability becomes more important in the design and evaluation of survey instruments. A broader understanding of design and the visual presentation of information is required, and usability evaluation techniques should be among the standard techniques used to evaluate the design of CAI instruments.

CHAPTER 18

Development and Testing of Web Questionnaires

Reginald P. Baker, Scott Crawford, and Janice Swinehart

Market Strategies, Inc.

18.1 INTRODUCTION

Questionnaire development and testing for traditional paper-and-pencil questionnaires have tended to focus on such elements as question type selection, wording, and order. Graphical design has also been an area of focus where self-administered (principally, mail) questionnaires are used (Dillman and Jenkins, 1997; see also Chapter 15, this volume). The transition to computer-assisted methods has added a new set of concerns including technical correctness (e.g., branching, calculations, text fills) and, more recently, usability (Couper et al., 1997; Chapter 17, this volume).

The advent of Web-based interviewing has further complicated the questionnaire development and testing phases of survey design and implementation. Concerns about Web-based communication styles, unique question types and formats, screen layout and use of color, added question complexity, and technical performance (e.g., speed of page delivery, appearance under different browsers) are major new issues that are mostly peculiar to Web. Add to this the self-administered nature of Web surveys and testing becomes even more critical.

In this chapter we describe what we believe to be the essential elements of a robust process for testing Web questionnaires. We begin by presenting a

Methods for Testing and Evaluating Survey Questionnaires, Edited by Stanley Presser, Jennifer M. Rothgeb, Mick P. Couper, Judith T. Lessler, Elizabeth Martin, Jean Martin, and Eleanor Singer
ISBN 0-471-45841-4

holistic view of Web questionnaires, a view that marries the traditional concept of the survey questionnaire to the technology with which it is bound in the online interview. We then discuss the major goals that a robust testing process must address. Finally, we describe specific approaches to testing and the key components of an effective testing process. In the course of doing so, we rely primarily on our own experience in deploying hundreds of Web surveys. Through a combination of case study and analysis of testing results, we attempt to show the practical utility of various testing strategies and how they are combined to define a robust process that reduces error, contributes to the continued development of Web questionnaire standards, and ultimately provides the framework for future Web questionnaire development.

18.2 DEFINING WHAT WE MEAN BY "WEB QUESTIONNAIRE"

We use the term *Web questionnaire* to delineate a subset of online data collection techniques. We mean specifically a questionnaire that resides on the Internet, executes on a Web server, and that respondents access via a standard Web browser. This is different from an e-mail survey where a questionnaire might be distributed via Internet with the interview taking place either within the e-mail application or after download to the respondent's personal computer (PC). It also differs from a questionnaire that might be downloaded from a Web server and then executed on the respondent's PC, either as a standard executable file or via a proprietary nonstandard browser. These qualify as online or Internet surveys, but they are a small subset of online survey activity.

We further narrow our focus to surveys that are interactive and dynamic, that is, consist of multiple screens with functionality common to CATI and CAPI questionnaires, such as automatic conditional branching, text fills, use of calculated variables, question randomization and rotation, and so on. We distinguish these questionnaires from static HTML scrollable forms. Although common in the early days of Web interviewing, HTML surveys are declining in the face of rapidly emerging, high-functionality interviewing software and the need to deploy increasingly complex questionnaires.

18.2.1 How Web Questionnaires Are Different

For the survey designer, a Web questionnaire has many parts (i.e., questionnaire, interviewing software, hardware platform, etc.), but to the respondent it appears as a single entity, not unlike a Web page, running on his or her PC. Thus, we suggest a holistic definition of the term *Web questionnaire*, one that encompasses all of the elements that influence how a respondent experiences and interacts with the survey instrument.

Arguably the most unique feature of Web questionnaires is their dynamic nature. Traditional paper questionnaires are fixed and static. Respondents may take different paths through the instrument, but the questionnaire itself is always the same. Interviewers introduce some variability, but those are issues of training that generally fall beyond the scope of questionnaire testing. Questionnaires used in various computer-assisted modes (e.g., CATI, CAPI, CSAQ, etc.) are generally more dynamic than paper questionnaires, but they typically are executed in a controlled environment: whether the technical infrastructure of a CATI call center or interviewer laptops. They run in technical environments that are standard and under the complete control of the survey organization. As a consequence, they are consistent and predictable.

By their very nature, Web questionnaires must run in the uncontrolled and largely unpredictable environment of the respondent. PC configuration, Internet connection quality, and browser software all have a significant impact on how the questionnaire is presented and behaves. Ensuring that all respondents experience the survey in a reasonably standard way is a primary goal of Web questionnaire testing. This dynamic nature of Web questionnaires requires that our testing protocols extend beyond the techniques now employed both for paper and other computer-based instruments.

18.2.2 Components of the Web Questionnaire

A holistic conception of the questionnaire that marries the traditional questionnaire to the hardware and software on which it runs is essential to the development of effective testing protocols. One such conception is presented in Figure 18.1, where we depict six components or layers. The components are grouped into two broad categories: those that are mostly specific to individual surveys and those that comprise the general infrastructure, primarily hardware, software, and connectivity.

The *respondent interface* consists of the layout and graphical format used to display the instructions and questions on respondents' screens. It includes the general screen layout, colors, fonts, font effects, graphics, and user interface objects, including navigation buttons.

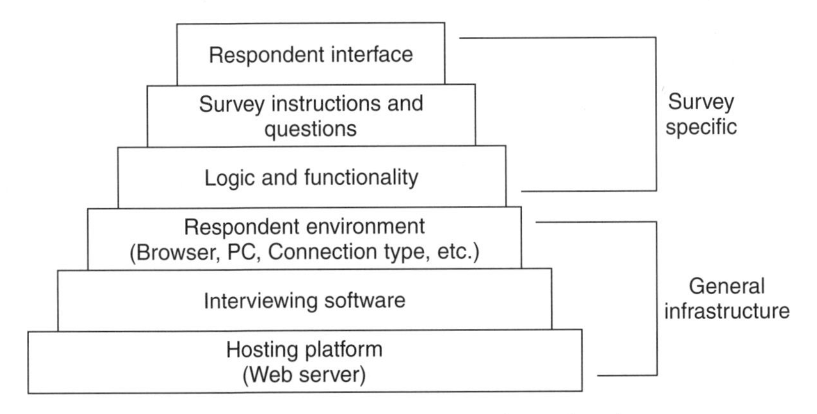

Figure 18.1 Components of the Web questionnaire.

The *survey instructions and questions* component is similar to that of a conventional, self-administered questionnaire. It includes the question-and-answer text, general and question-specific instructions for answering questions and moving through the questionnaire, and any introduction or transition text used.

Logic and functionality refers to the routing of respondents through the questionnaire and application of various question types. The former includes conditional branching, randomization, and other navigational aspects, such as going backward to change previous answers. Functionality includes question validations, such as range checks, enforcement of single and multiple response items, calculations of constant sums, display of items, insertion of text fills, and calculated items.

The *respondent environment* consists of all of those elements of the respondent's hardware and software configuration that affect how he or she interacts with the Internet. Primary concerns typically include browser compatibilities with the interviewing software, presence or need for specialized browser plug-ins, monitor display properties, page download speed, CPU, memory capacity, and operating system.

Interviewing software refers to the software system being used or the computer language in which the questionnaire has been coded. It guides the respondent through the interview, enforces the questionnaire rules (e.g., branching, answer validation, data store), manages interactions with the respondent, and stores the data collected. We are strong believers in the use of a programmable survey software system rather than of general-purpose tools such as HTML, CGI, Java, or ColdFusion. The Web survey software market is now sufficiently developed that there are several systems that have the functionality needed for complex instruments, support fast development of applications, and are more stable in production than the custom applications developed in general-purpose tools.

Finally, the *hosting platform* is the hardware and software infrastructure that defines the hosting server or servers. The primary issue with this component typically is its ability to handle peak loads without significant performance problems, such as slow delivery of pages or rejection of attempts to access the survey. Other issues include security settings, firewalls, and backup and failover systems, among others.

Poor or inconsistent performance in any one of these six components will affect how respondents experience the questionnaire, with almost certain implications for data quality. A poorly designed interface may make questions difficult, if not impossible, to read. Variations in how different browsers display screens may cause respondents to interpret questions differently. Errors in logic or functionality can result in respondents skipping questions or being asked questions that make little sense. Slow page loadings or a clumsy mechanism for recording answers can create frustration and even premature termination. The potential problems lurking in the complexity of a Web questionnaire are indeed many, and the potential for substantial survey error great.

18.3 TESTING APPROACH

We recognize that our experience leads us to advocate a certain approach to testing that may not be universal in its applicability. Although we have considerable experience with a variety of CASIC systems and the long, complex questionnaires often found in government-sponsored surveys, much of our Web-specific experience has been within a commercial market research firm, where questionnaires tend to be somewhat less complex and where there is a strong emphasis on rapid development and deployment of surveys. This need for speed has caused us to design a testing process that has two important characteristics. The first is to test those elements of the questionnaire, common to all surveys, that are off the critical path of any survey (i.e., the general infrastructure of hosting platform, interviewing software, and the respondent environment referred to in Figure 18.1). The second is to develop and rely on standards that if followed will ensure that any one Web questionnaire will perform in a predictable way. There always will be surveys whose designs invite departure from current standards, and the standards themselves will change as more research is done, experience gained, the tools improve, and the Web itself continues to evolve.

18.3.1 Testing in the Survey Life Cycle

While the respondent experiences the Web questionnaire holistically, both the development of the questionnaire and its testing are structured by component. Differences in component life cycles, in the testing tools used, and in the types of staff skills needed all argue for a highly compartmentalized development and testing process. At first blush, this might seem to add complexity to testing that would make the process more difficult to manage and control. In practice, it has the opposite effect. By breaking the questionnaire down into its component pieces, development and testing of individual components can be done on separate schedules with separate staff and, in most cases, off the critical path of any specific survey. Individual components can be developed and tested independently, making it easier to isolate problems and in general speeding up both the development and testing processes for surveys.

Differences in component life cycles offer a major advantage in the testing process. For example, although the survey instructions and questions will generally have a life cycle identical to that of the survey for which they are used, the hosting platform will have a much longer life cycle, one that spans many surveys. The major concern with the hosting platform is typically with its robustness, that is, its ability to perform under the load of a large number of respondents and a variety of surveys fielded simultaneously. Additional concerns might include its load-balancing capabilities and its ability to execute its failover protocol should a hardware component fail during the survey field period. Tests of these and other features should be designed and conducted on schedules that are largely independent of individual surveys so that the questionnaire developer has available a fully tested component that can be added to and used in the specific survey questionnaire with a minimum of additional testing responsibility. Understanding the life cycles of all questionnaire components and scheduling their testing appropriately is a major organizational challenge, but one that falls outside the scope of a specific survey.

18.3.2 Reliance on Standards

A second important element in any Web questionnaire testing protocol is an infrastructure of standards. Standards provide the framework within which all components of a Web questionnaire are designed and developed. A variety of standards are needed to cover everything from the fonts and colors to be used, to the programming techniques for accomplishing specific questionnaire tasks. They are the rules of the game within which all of the participants in the Web questionnaire development process must play. They reduce variability among individual components and ensure that all of the pieces of a survey will fit together into a stable application that behaves in a predictable way. The more variability allowed into the Web questionnaire development process, the more time consuming the testing, and the more uncertain the result.

Important as standards are, they are not cast in stone. Rather, they should evolve over time based on testing outcomes and empirical data. This evolution should be controlled so that it occurs across the entire survey infrastructure and all surveys rather than varying by individual survey at the whim of the designer. Some survey designs will always require that we depart from standards. In these cases, thorough testing of all affected components is essential. Lessons from these tests should then not only be incorporated in the design of the current survey but also be used to modify standards as appropriate. Changes in infrastructure components, such as the interviewing software and hosting platform, also require thorough testing and updating of standards.

18.4 TESTING TOOLS

We now turn to a discussion of specific tests and testing tools that we have come to rely on in an organization that deploys Web surveys of all sizes to respondent populations of many different types on a consistent and ongoing basis. Other organizations may have different approaches, but we believe that the tests and tools described here are the minimum that should be considered.

18.4.1 Standard Questionnaire

Any survey organization that deploys computer-assisted surveys will be well served by developing and maintaining a standard questionnaire that includes all of the question types and questionnaire functionality that the organization uses routinely in its work. The more comprehensive the questionnaire in terms of reflecting the types of questionnaires the organization deploys, the more effective it is as a testing tool. At a minimum, such a questionnaire should include the following:

- At least one use of every question type (e.g., single choice, multichoice, matrix, open end, constant sum) used by the organization's questionnaire designers.
- At least one use of every questionnaire function (e.g., simple and complex skip logic, text fill, new variable creation and use, randomization and rotation of responses and questions, dynamic question creation, interface with other applications) used.
- At least one use of every validation technique (e.g., mandatory responses, range checks, data format checks, respondent authentication) used.
- Screen display specifications (e.g., fonts and colors used, overall screen layout, question formatting) that reflect the organization's standards for screen design.

The primary use of the standard questionnaire is testing of interviewing software. Programming the questionnaire in the interviewing software as specified in the organization's current screen design standards (discussed in Section 18.4.2) will expose gaps in the software's functionality and identify the major programming challenges to be faced in questionnaire development. Functionality gaps can be documented and provided to software developers as candidates for future releases. Programming techniques and workarounds required for various question types can be developed and documented. Standards can be revised to reflect what is possible with the current software. When vendors release new software versions, this test should be repeated, and the questionnaire can also be used to evaluate competing software products for possible adoption.

The programmed standard questionnaire also is an excellent tool for testing how the survey organization's Web questionnaires will perform in different respondent environments. Variations in browser version and settings, connectivity to the Internet, and PC hardware can produce substantially different respondent experiences. Testing the performance of the programmed standard questionnaire in the variety of environments likely to be encountered will help to isolate areas of problematic performance.

Finally, the standard questionnaire can serve as a vehicle for load testing the hosting platform. Survey organizations must have a means of simulating actual survey conditions so that the capacity of the hosting platform can be measured. The standard questionnaire is an excellent tool for doing so.

18.4.2 Conformance to Standards

In Section 18.3.2 we stressed the importance of standards in the survey development and questionnaire testing process. Standards are a pervasive element of the infrastructure that a survey organization needs to create and administer Web questionnaires. It is essential not only that standards exist, but that the testing process include specific review tasks to ensure that they are being followed.

An organization may start by developing a set of initial standards *a priori*, that is, from an appropriate literature or from widely accepted industry standards. Dillman (2000a), for example, has condensed a broad base of survey methods research into a set of design standards for Web questionnaires. The literature on Web usability (Nielsen, 1999) offers a plethora of standards for screen design, use of color, text formats, and so on. Section 508 of the U.S. Rehabilitation Act (Center for IT Accommodation, 1998) specifies standards for Web applications to ensure access by people with disabilities. Software and hardware providers offer a variety of standards for configuration that optimize performance and offer at least some guarantee of consistent behavior across different browsers. IT organizations have standards for structuring applications, and survey organizations have standards for the construction and use of different question types.

Survey designers are wise to leave the issues surrounding technical standards, such as operating systems, programming techniques, program version control, and naming conventions, in the hands of technical specialists. We note here only that it is essential that such standards exist and be followed. The central issue for the Web questionnaire designer is the way in which the questionnaire will appear on the respondent's computer screen, with a special concern for ensuring that it appears and behaves consistently across the variety of platforms likely to be encountered.

Screen design standards for Web questionnaires typically address three major issues:

- Overall screen layout, including screen color and background; definition of the elements to be used (e.g., logos, rules, navigation buttons, contact information); and their placement.
- Text formatting, including font type, color, and treatment to denote specific questionnaire elements, such as question text, answers, selective emphasis, or error messages.
- Question formatting, describing how each of the question types (e.g., closed ends, multiple response, grid questions, open ends) is to be laid out on the screen.

As an example, Table 18.1 shows the screen design standards that we currently use. These work well for us, given the software we use, the populations we survey, and the clients we serve. Other survey organizations will have other needs and their standards will vary accordingly.

Standards and the Interviewing Software The specific standards a survey organization wishes to use must be supportable within the organization's interviewing software. Standards that are difficult or impossible to implement will need to be

Table 18.1 Example of Screen Design Standards

Screen design

Use a white screen, no background color or image.
Place the logo (Market Strategies or client) in the upper left (if used).
Place the contact information (toll-free number and/or e-mail address) in the upper right (if used).
Place a rule at the top to separate the logo and contact information from the survey.
Place a "next screen" button at the lower left. If a "previous screen" button is used, place it as far to the right of the "next screen" button as possible.
Do not use a "stop" or "quit" button on the screen.
Place a rule at the bottom, separating the question-and-answer text from the navigation buttons.

Text

Set the text in black Arial font, 10 point.
Set the question text in bold the answer text in normal type.
Use blue bold typeface for selective emphasis.
Use red typeface for all error messages.
Put instructions in regular italics, enclosed in parentheses, and placed one line below question.

Question presentation

Use one question per screen with some exceptions (grid questions) as long as "next screen" button remains visible on the screen.
Provide an instruction once and whenever a new question type is introduced.
Position fully labeled scales vertically below a question (except when presented in a grid format).
Position non-fully labeled scales horizontally below a question. Use a grid layout if possible to combine questions with identical scales.
Require answers on closed ends but not on open ends. Always provide some form of "opt out" code (e.g., "Rather not say" or "Don't know").
Grid questions:
Have only one reserved code (such as "Don't know" or "Rather not say") if an answer is required.
Never design grids with more than 12 columns.
Space all grid columns containing response options consistently such that none stand out from the others.
Fit all questions on one screen. Column headers should always be visible without requiring scrolling.
Use alternating light gray bars (color code = #CCCCCC) to improve row readability.
Label but do not explain scale points in a question stem.

Other

Do not use a progress indicator.
No question numbers should appear on the screen.
Use standard and clearly written error messages.
Use drop-down boxes only for single-response options with many codes in which the respondent is likely to know the answer (e.g., state, country, or month). Never preload the box with an answer. Always begin the list with an instruction: for example, "Select one" or "Select state."
Do not use list boxes.
Size open-ended text boxes to reflect the data requested and fit on the screen.

revised. Once established, standards will need to be tested again as new versions of the interviewing software are released or replacement candidates are evaluated.

We noted at the outset a strong preference for standard interviewing software systems over the use of general-purpose programming tools. Development and maintenance of standards is easiest with a standard system, and the applications that result behave more predictably.

Standards and the Respondent Environment Variations in respondent environment are the second major challenge in standards development. Browser software, connectivity speed, and PC configuration (CPU, video, operating system, and memory) will vary across the target population and may cause the questionnaire to appear and behave differently for individual respondents.

Web browsers, in particular, show significant variation in screen presentation and behavior across versions and providers. Older versions of Web browsers adhere to older standards for HTML. These older versions lack functionality available in the current standard (HTML 4.1). Thus, a question element that relies on new functionality in the current standard will display unpredictability for a respondent with an older browser. Variations in browser performance also exist across providers. The two main providers, Microsoft and Netscape, are inconsistent in their adherence to published standards for HTML. Neither supports all of the HTML tags in the 4.1 standard, and both support their own "extensions" to that standard that are inconsistent with one another (see, e.g., Musciano and Kennedy, 1998).

The same is true for JavaScript, the most widely used language for creating special functions beyond what is supported by HTML or the interviewing software being used. Older browsers rely on older versions of JavaScript, and it is important that any JavaScript code used in a Web questionnaire be tested with the full range of browsers likely to be encountered in a survey. In the most extreme cases, respondents may actually have disabled JavaScript altogether. Although only a small minority (less than 5%) of respondents are likely to have JavaScript disabled, survey organizations should nonetheless use it sparingly and do whatever is possible to minimize the impact of its failed execution on these respondents.

The respondent's hardware platform and operating system will also affect screen presentation and performance. Variations in screen resolution, number of colors that can be displayed by the video board/monitor combination, speed of the CPU, available memory, cache space, operating system, and, of course, speed of the connection to the Internet will all have an effect on how the screen appears, how quickly pages load, and how well special functions perform.

Standards Testing and Maintenance The goal of the standards development is to arrive at a set of instructions that can be used by questionnaire designers and programmers to build questionnaires that will display and behave consistently across all of the variations expected in the target population. We suggest a four-step process for doing so:

1. Define the most prominent respondent platform, that is, the environment that we might expect to encounter in the vast majority of cases. The main components of the environment typically are the type of PC (IBM-compatible or Mac), the operating system, the CPU speed and memory configuration of the PC, the type of display and its likely settings, browser version, and Internet connection type and speed.
2. Develop and program in the interviewing software a questionnaire that conforms to the organization's standards and includes all question types currently used. If an organization maintains a standard questionnaire as described in Section 18.4.1, it should be used. This questionnaire should be tested continually during development on the most prominent platform, and standards revised as needed to achieve the desired results in this environment.
3. Define and implement the necessary test beds. These tests should vary three main conditions: browser, operating system, and screen resolution. Information about the browser, operating systems, and screen setting variations most likely to be encountered is not difficult to come by. A number of online sources provide estimates of current browser penetration (see, e.g., *http://www.thecounter.com* or *http://www.upsdell.com/BrowserNews/*) and other technical data. Examination of the server logs on the survey organization's hosting platform can show how its current respondent population breaks down by browser, operating system, and screen resolution and help pinpoint where variation from the predominant platform will require further testing. Once the most important alternative configurations are determined, test beds are created with those environments.
4. Run the test questionnaire on all of the test beds specified, comparing the results achieved on each to the baseline of the most prominent platform, that is, the one expected to occur most often in the field. As variations are discovered, they should be corrected to the maximum extent possible via programming changes in the standard questionnaire. At the same time, the applicable standard should be updated to reflect the differences. Important variations to look for include the following:
   - Consistent display of all colors and fonts, especially where differences (such as color shifts) are so extreme that they may affect readability.
   - Screen formatting, including placement of objects and question presentation.
   - Proper functioning of standard questionnaire features (e.g., range checks, fills, randomizations).
   - Functioning of special questionnaire features (such as JavaScript scripts).
   - Whether there are significant delays when going from screen to screen.

Although this is an interactive process, it is typically not time consuming, and a final set of standards evolves relatively quickly.

It is important to note that we do not suggest that these tests be performed for every survey, but rather, that a set of standards be created to be applied to all surveys that the organization performs. Standards should be revisited and tests redone only when something changes, such as a new question type, a new version of the interviewing software, or a new browser version. Standards should evolve in tandem with the other components of the questionnaire. For any individual questionnaire or survey, the key test is whether it conforms to the standards established.

18.4.3 Usability Testing

In this context, *usability* refers to the ease of use of a software application such as a Web questionnaire. Typical ways in which usability is measured include the speed with which a task can be performed, the frequency of errors in performing a task, and user satisfaction with an application's interface. One good online resource for understanding the basics of Web usability can be found at *http://www.usabilityfirst.com*. In the specific case of Web surveys, usability issues revolve around the interface, that is, presentation of the questions and the survey tasks in ways that respondents find easy to understand and execute. We have found two techniques to be especially valuable for usability testing: qualitative interviews and analysis of paradata.

Qualitative Interviews We typically conduct usability tests with small groups or even single individuals. We begin by developing a fully functional Web questionnaire using our current or proposed standards for the interface and then invite a number of people who are typical of the respondents expected to do an in-person, in-depth interview. We first ask the respondent to complete the Web interview while we observe and answer any questions that he or she may have, noting areas of confusion. When the respondent has completed the interview, we ask a few standard questions aimed at uncovering especially difficult questions or sections, or cumbersome tasks. We then go back to the beginning of the questionnaire and proceed through each screen with the respondent noting any ambiguities or difficulties that he or she may have encountered. We typically follow this process with 10 to 12 respondents, or until we feel that the respondents' reactions suggest that we are unlikely to encounter major usability problems in actual field use.

We note here that we conduct this type of usability testing only intermittently, and then typically only when we are considering changes in our interface standard or are about to survey a population that may have some unique needs. For example, we were asked to conduct a survey of users of *http://www.medicare.gov* to assess their overall experiences and reactions to the site. We expected the population would be older, less Web savvy, and perhaps have some disabilities (such as poor eyesight) compared to many of the respondent populations we survey. A series of in-depth interviews with seniors having a variety of Internet experience helped us to fine tune this questionnaire's presentation in ways that would meet the special needs of this particular population (MSInteractive, 2001).

Paradata Analysis A second, quantitative form of usability testing is analysis of paradata, that is, data about the actual Web questionnaire completion process. One of the many advantages of CASIC applications in general is that they can collect information about the actions of their users, whether interviewers or respondents, as they complete surveys. Couper et al. (1997) have demonstrated the utility of paradata for improving our understanding of how interviewers use CAPI applications. Jeavons (2002) and Heerwegh and Loosveldt (2002) have shown similar utility for data about respondent behavior when completing Web questionnaires.

Survey researchers are only beginning to realize the value of paradata for Web questionnaire testing and analysis of respondent behavior. The specific types of information that we can collect are limited only by our imaginations and technical skills. Some measures can be captured within the interview application, while others require working with the hosting platforms server logs where every interaction with a respondent is recorded. At present the most useful information would seem to fall into five categories: characteristics of the respondent's technical environment, respondent reaction time, errors made, navigation behavior, and last question answered. The first of these, information about the respondent's environment, such as browser version and operating system, is discussed elsewhere in this chapter. In this section we focus on the remaining four.

Respondent Reaction Times Most full-function interviewing software can put a time stamp on a question to record when a question is displayed for the respondent and when a response is received at the server. Using these time stamps, researchers can calculate the amount of time it took each respondent to read and answer each question. Analysis of these reaction times can identify those questions that are especially long and difficult for respondents to process and answer. Experimental designs that vary screen designs and question presentation can generate important information about optimal design for respondent ease of use.

Tourangeau et al. (2003) have studied the impact of positional cues in presentation of scales to Web survey respondents. In their experiments, respondents were shown five-point fully labeled scales arranged in highest-to-lowest order and the same scales with the midpoints placed either at the beginning or the end of the list. Analysis of response times showed significant differences among these treatments, suggesting confusion and increased difficulty in responding when responses are presented "out of order."

Error Counts Most interviewing software also provides a way to point out errors to respondents as they move through a questionnaire. Those errors might include keying out-of-range responses, failing to answer a question, or not completing an exercise properly. Capturing these interactions and combining them with more standard data on missing items and use of reserved codes (refusals, don't know responses, etc.) provides another technique for identifying problematic questions, due to either their content or the task required of the respondent.

As of this writing, we are unaware of any research that has analyzed these data in any depth. We note that capturing of such data can be technically challenging, but their value may be worth the effort.

Navigation Behavior A third potentially useful element of paradata is information on respondent navigation, that is, tracking how respondents use whatever navigational freedom the designer gives them to move through the questionnaire. If respondents are allowed to go back to previous questions, or if they are allowed to move forward without being required to answer a question, this can be recorded. One can also collect data on respondents' use of special features in questionnaires, such as links to help files, definitions, and even other Web sites. Analysis of these data can help us to understand how respondents use this type of special functionality and therefore guide use of it in questionnaire design.

For example, Conrad et al. (2003) have looked at respondent use of hyperlinks in surveys to provide on-demand definitions of terms used in survey questions. The research created several experimental conditions by crossing the type of term (technical versus nontechnical), the usefulness of the definition, and the number of mouse clicks required to access the definition. The preliminary results show that a surprisingly low percentage of respondents (less than 20%) actually used the definitions with various patterns around the difficulty of the term, usefulness of the definition, and difficulty of accessing it.

Last Question Answered As Web survey practitioners know all to well, one of the major challenges of conducting surveys by Web is retaining respondents for the full interview. The challenges of length and salience are as real for the Web as for any other mode. Web questionnaires, however, have the additional challenge of presenting the answering task in ways that minimize difficulty and respondent fatigue. One way to identify particularly problematic parts of Web questionnaires is to pay close attention to where in the interview respondents are terminating.

Figure 18.2 portrays graphically the results of a last-question-answered analysis for a survey of information technology (IT) managers sponsored by a major hardware manufacturer. It is a Pareto chart of the number of terminations by

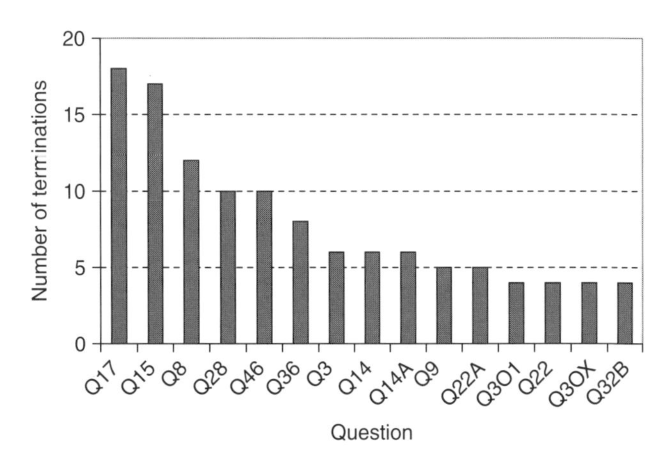

Figure 18.2 Terminations by question.

question for the top 15 termination points. A large number of terminations on a question suggests a problem in question design, in technical performance, or in subject matter. They indicate a problem so severe that respondents quit in frustration.

Examination of the questionnaire revealed that all five of the questions with the highest number of terminations shared a common design. Respondents were asked, for example, to break down their sources of information on new IT products into 13 different categories. The questionnaire required that the total equal 100%, but provided no easy means for the respondent to track his or her running total. The questionnaire would not allow the respondent to go forward until the entered data totaled 100%. To make the task even more difficult, the software required that every box be filled in with some number, even it were zero. Blanks were treated as errors. The difficulty of the task caused respondents to get stuck. They would fill out the screen, click to go to the next screen, and get an error message telling them that their choices did not sum to 100%. Presumably after multiple tries and being unable to move forward, many just quit. Based on this analysis, we redesigned constant sum questions to provide running totals so that respondents can monitor their distributions as they record their answers.

Paradata Analysis and Standards Development Analysis of paradata can teach us a great deal about how respondents interact with different designs. Some paradata are easy to collect and can be generated on every survey. When they are monitored during data collection, they help spot problems or improve designs in midstream. What we learn from these analyses constantly feeds into the standards development process, so that the mistakes of one survey are not replicated in the next.

Collection of other elements of paradata may require varying levels of technical effort and therefore only be appropriate as part of controlled experiments designed to evaluate competing designs. With time and experience, access to some of the paradata that are more difficult to collect may become standardized and thus their incorporation into the ongoing survey process, possible. Regardless, every organization doing Web-based data collections should have a plan for collection and analysis of some forms of paradata on an ongoing basis. Their standards for Web questionnaire design should reflect what analysis of those data teach them.

18.4.4 Testing for Correctness

Correctness tests focus on ensuring that the Web questionnaire meets the specifications of the questionnaire designer. This is generally a case of comparing the Web questionnaire to its paper antecedent or any other questionnaire specification vehicles (e.g., a flowchart) that the organization might use. The questions that these tests must answer include:

- Are all questions and answers present, in the proper order, with no spelling or grammatical errors?

- Have the organization's interface standards been followed?
- Are all navigation buttons working correctly?
- Are all specified fills, whether from sample preloads, from respondent answers, or generated internally by the application program correct and appearing as specified?
- Are all answer types (e.g., numeric or alphanumeric) specified correctly, checked for the specified range, and being saved in the correct format?
- Does the Web questionnaire logic execute as specified?
- Are any special questionnaire functions (e.g., randomizations or rotations) functioning properly?
- Are any specified hyperlinks working, and are the materials they link to (e.g., graphical displays, help files, and support e-mail address) correct?
- Are the expected outputs being written correctly?

Overall, the tests for correctness in Web questionnaires are not materially different from those needed for computerized instruments generally. The main exception typically is the user interface. Most CATI systems, for example, have standard screen formats that all applications use. The software places objects automatically in predefined places on the screen, and the designer is concerned only with issues such as selective emphasis or overall question-and-answer length. Beyond that, the correctness tests for Web questionnaires are essentially the same as those of other CASIC applications. In our case, the process we use for Web is derived directly from a similar CATI process developed and honed over a decade and used on thousands of surveys.

Testing for the correctness of Web questionnaires may involve a number of different actors, each bringing specific expertise to the process and contributing in different ways. The programmer, the questionnaire designer, the quality assurance staff, the survey sponsor, and almost anyone with an Internet connection and a willingness to become involved can provide useful feedback in the various testing processes involved.

Survey researchers have begun to recognize that testing computerized questionnaires has some of the same properties as software testing. The software literature offers at least some general concepts that are potentially useful for testing Web questionnaires. Two especially popular testing techniques in software development are white and black box testing (see, e.g., Beizer, 1990, or Mosley, 1993). Both involve extensive exercising of program code to assess overall stability and correctness. The methods differ in knowledge of the application possessed by the tester. In white box testing the tester has detailed knowledge of how the application is supposed to work and uses that knowledge to find missing functions, incorrect functions, and incorrect outputs. In black box testing the tester has only very limited knowledge of the specifications and contributes to the testing process by testing the application's overall robustness, ensuring that its operations are intuitive, providing real-world input, and creating outputs that can be evaluated by others with greater knowledge of what the application should produce.

Although the extremes of white and black box testing are useful as ideal types, we have found that the most effective testing processes for Web questionnaires covers a range that runs from white box to gray box to black box.

White Box Testing The essential element of white box testing is detailed knowledge of how the application is supposed to work. Three roles in the survey development process have that knowledge: the questionnaire designer, the programmer, and the primary questionnaire tester or quality assurance staff, known in our system as Webcheckers.

The questionnaire designer has the most complete knowledge, at least in the beginning, and that knowledge must be communicated to the other key players in the process before they can test effectively. So we conceive of the testing process in two phases: a review phase to uncover any obvious errors and to familiarize the tester with the design, and a testing phase where the programmed Web questionnaire is tested systematically and compared to the paper document.

Programmers have primary responsibility for ensuring that the programmed questionnaire is correct. Programmers are charged not only with programming the questionnaire as specified but also finding misspecifications in the original questionnaire. Prior to beginning programming, programmers review the questionnaire and provide feedback to the designer that is, in turn, used to develop the final instrument. This review is by inspection and typically uncovers only the most obvious of errors as well as identifying ambiguities that must be resolved before programming. This review phase has seven broad steps:

1. *Review for conformance to specification standards.* We discussed earlier the importance of standards, and one important set of standards dictate how the designer specifies the questionnaire to the programmer. In this step, the programmer ensures that the specification standard has been followed.
2. *Check question numbering.* Review to ensure that all questions are numbered, that they conform to the organizational numbering standard, that they are unique, and that they follow in sequential order. Although we do not typically display question numbers on the screen, they are required to routing instructions, documentation, and analysis.
3. *Read the question text.* Check all question-and-answer text for overall content, grammar, and punctuation. Ensure that transitions and question introductions are placed appropriately. Note any likely difficulties with or departures from the organization's screen formatting standards.
4. *Review instructions.* Check that any instructions (such as scale definitions or multiple versus single response items) are specified correctly.
5. *Check the answer codes.* Ensure that all answer codes are numbered consecutively, in ascending order and without duplicates, match the question text, and are meaningful for numeric answers. Also check that reserved codes (i.e., "Refused," "Don't know," etc.) are present and specified properly.
6. *Check text fills and sample preloads.* Ensure that all references to fills and preloads are valid, that is, are specified elsewhere in the questionnaire.

7. *Check filters and routing instructions.* Be sure that filters are specified properly; that all referenced question numbers, preloads, and calculated items are valid; that the question skipped to exists; and that the logic is programmable.

By the end of this process the programmer has a detailed working knowledge of the questionnaire. He or she provides feedback to the questionnaire designer, who makes the needed changes, obtains the needed approvals, and produces a final questionnaire for programming. This final questionnaire becomes the key working document for all additional testing, whether done by the programmer, by the designer, or by quality assurance staff.

Once the programmer has completed programming, he or she is also responsible for testing and debugging the instrument to ensure that it is working as specified. As part of this test, the programmer ensures that all of the data items specified are being written (outputs) in the expected locations and in the required format. In an ideal world the programmer eventually produces a completely correct questionnaire program that matches both the specifications in the final questionnaire and conforms to all of the survey organization's standards. In reality, this is almost never the case, and extensive additional testing by others is required. In our process, the first test step after programming is done by the Webchecker.

The Webchecker behaves in much the same way as the programmer. He or she first reviews the paper questionnaire and notes any especially difficult or problematic sequences. Once familiar with the questionnaire and its design, the Webchecker systematically tests to the maximum extent possible all the required displays and functions of the questionnaire. Although the primary objective is to ensure that the Web questionnaire is a faithful representation of the paper questionnaire in terms of text and functionality, this is also the first time that anyone sees the questionnaire as respondents will see it. At this stage, the person doing quality assurance plays two roles. One is to ensure that the programmed questionnaire conforms to organizational standards and the questionnaire specifications. The second is to view the questionnaire as a respondent will view it. Put another way, the Webchecker reviews the work of two people upstream in the questionnaire development process: the programmer and the questionnaire designer. Specific tasks are as follows:

1. *Proofread.* Ensure that all question-and-answer text is present, correct, and as specified in the questionnaire.
2. *Question-by-question review.* Read each question carefully, study its intent, and be sure that the possible answers make sense and are all-inclusive. Carefully evaluate instructions for clarity and correctness.
3. *Check continuity.* Note the general flow of the questionnaire to ensure that there is a logical progression with one question leading to the next, and note any abrupt shifts that lack transitional devices.

4. *Check screen displays.* Review each screen to ensure that it conforms to the organization's interface and screen display standards.
5. *Check ranges.* Check to see that any range checks on individual items have been implemented correctly.
6. *Check fills.* Verify that all text fills and sample preloads are displaying correctly.
7. *Check skips.* Check every skip instruction (including filter outs) to ensure that it is correct and look for possible "missing" skips, that is, instances in which a question appears to be in conflict with an earlier question or answer.
8. *Test special functions.* Check any special functions such as rotations, randomizations, constant sum questions, and so on, to ensure that they are working correctly.
9. *Test error trapping.* Ensure that all error messages are displaying properly.

In many cases, the most difficult part of testing is verifying that all conceivable respondent paths through the questionnaire are programmed correctly. When preload items drive the path through the questionnaire, mock sample records with all valid preload values can be aggregated into a database, then used by the Webchecker in the testing process. Where the path is driven by responses rather than sample preloads, a similar process can be used, but it requires temporarily converting skips or functions driven by respondent answers into skips or functions driven by sample preloads. This approach is similar to that used for testing commercial software applications in which a database of test records is created and then read by the application being tested. We note, however, that this has its dangers. Eventually, the questionnaire code must be converted back to take this input from the keyboard or mouse rather than sample preloads, and in the process, errors may be introduced. The preferred method is to provide specifications to the Webchecker that include the inputs to drive skips. The Webchecker then selects records from the database and assumes the role of the respondent, answering the questions as they are presented and checking the fills and routing against the questionnaire specification. Further testing of complex logic can also be done later in the process using autogeneration, described below under "Black Box Testing."

This is, of course, a very simple example. More complex examples for more complex questionnaires are easily imagined, and with greater complexity the time needed to design and construct the test database and accompanying specifications increases. Multiple Webcheckers may also be needed, with each focused on specific parts of the questionnaire.

Once the Webchecker has certified the programmed questionnaire, it is passed to the questionnaire designer for additional testing. The questionnaire designer is the person with the most complete knowledge of how the questionnaire should work. He or she has written (or coordinated the writing of) the questions; specified the routing instructions, branching conditions, text fills, and so on; specified any preloads to be included from the sample file; and specified any special functions needed. The typical weakness of the questionnaire designer is in understanding the outputs, their structure, and their content. Thus, the questionnaire designer may spot not only programming errors but also errors in how the questionnaire was originally specified to programming.

Gray Box Testing The programmer and the Webchecker do the heavy lifting in correctness testing. In our experience one or both of these people uncover the great majority of errors. The questionnaire designer is more likely to uncover design problems of the sort that are not detected until the questionnaire is formed into an application that can be viewed on the Web and as a respondent will view it.

Survey organization staff, even those with few technical skills, also can add value to the testing process. This is especially the case with very complex questionnaires, where there may be modules that are very complex and therefore likely to contain errors. Staff can be assigned specific sections for testing and by virtue of their ability to focus, test much more thoroughly than someone like the Webchecker, who may be charged with testing the entire questionnaire.

Clients also can play a role in testing, especially given the ease with which they can access the questionnaire over the Web. In fact, clients are sometimes the first testers to see the questionnaire from outside the survey organization's firewall or on a slower connection. More than any other group of the testers, they will see the questionnaire as the respondent will see it, and that, coupled with their knowledge of the research goals and the overall questionnaire design, gives them a unique and valuable perspective. Still, in our experience the value of testing at this level is often psychological. It serves to reassure the client about the quality of the survey prior to fielding, and it provides additional peace of mind that the questionnaire has been implemented properly and will be well received by respondents.

Black Box Testing A common criticism of the testing techniques described thus far is that even imperfect knowledge of how the application works limits the effectiveness of the testing process. People in the testing roles described above all have at least some knowledge of how Web questionnaires are supposed to work, so that knowledge limits their behavior in ways that respondents may not imitate. Thorough testing, this argument goes, is best assured by including at least some of the black box approach.

Black box testing is a software system testing approach that consists of exercising an application without knowledge of its internal structure or operations. Its primary goal is to determine whether the system meets its functional requirements. Does the system do what it's supposed to do without failing? Black box testing requires a functioning version of the entire system or application, so it happens relatively late in the development process and at a time where problems are both difficult and time consuming to correct.

One approach to black box testing in survey organizations goes under such colorful names as *key banging* and *chimp testing*. Testers are challenged to "break" the application by answering and behaving unpredictably, deliberately going outside the rules specified in the questionnaire. Although such testing sometimes uncovers instability in application software, we have never found it to be particularly effective for testing overall correctness.

A more effective black box testing method in our estimation is autogeneration of survey responses (Chapter 16, this volume). A computer program assumes the role of respondent and moves through the Web questionnaire just as a respondent would. At each question the program randomly generates a valid response. The data from each "respondent" is saved as if the survey is in the field. The program can be set to simulate as many respondents as the tester feels necessary to adequately exercise all of the programmed questionnaire code. The more complex the questionnaire, the greater the number of respondent iterations advised. Using the data check techniques described in the following section, the resulting database can be used to find incorrect range specifications, missing answer categories, and incorrect skips.

Our CATI system includes such a tool. Its use is a standard part of our protocol for testing CATI questionnaires, and we have found it to be extremely effective. A similar tool for Web questionnaires is in development. We also use this tool for load testing, as described in Section 18.4.6.

Other Software Testing Techniques As survey organizations become more proficient CASIC developers and questionnaire complexity increases, we will probably develop more powerful automated tools for testing. Applications of graph theory might automatically generate testing regimens that ensure that all conceivable paths in a questionnaire are tested (see, e.g., Jorgensen, 1995, or Rosario and Robinson, 2000). Autogeneration applications might be made smarter and augmented to read and execute testing plans developed in this way. Questionnaire flowcharting, long advocated for aiding questionnaire programming but far from widely used, might be another valuable input to the testing process. With the right tools, databases of test records might elevate our testing processes to a new level in terms of their thoroughness and speed. But as compelling as automated solutions may seem, it is difficult to imagine a future in which the role of the Webchecker will disappear. In the end, a questionnaire is a human artifact designed to interact with a human respondent, and "correctness" is as much an issue of judgment as it is technical assessment.

18.4.5 Data Checking

At the end of the day, "it's the data, stupid." Too often, the ultimate proof that a Web questionnaire is performing correctly is the data collected. One of the major advantages of Web surveys is the easy availability of survey data as soon as they are collected. Analysis of these data can uncover major errors such as bad skips or out-of-range values or even minor design problems such as undesirable response patterns on individual questions. The nature of Web surveys makes it relative easy to detect such problems and to correct them quickly.

We believe that a program of ongoing data examination throughout the survey's development and field period is an essential component of any quality assurance program. These checks might be designed to monitor the functioning of particularly complex parts of the questionnaire or general high-level reviews of frequencies and cross-tabulations geared to spot more obvious problems. Data checks are the last line of defense against potentially disastrous errors in Web questionnaires. We advocate three stages of data check: prerelease, slow release, and intermediate.

Prerelease data checks are an artifact of the correctness testing process. Saving data from checking processes such as Webcheck and autogeneration provides databases that can be examined both to ensure that data are being saved in the proper format and that the questionnaire logic and range checks are operating correctly.

Slow-release data checks provide the first opportunity to evaluate the questionnaire using real data from responses. We typically prerelease enough of the survey sample to generate between 25 and 50 completes that can then be examined either individually or via frequencies and cross-tabulations to ensure that the questionnaire is functioning properly.

Finally, intermediate checks may be done throughout the field period, especially in the case of very complex questionnaires. Small samples such as those examined in slow release may not be sufficient to test thoroughly those elements of the questionnaire where a very small percentage of respondents are routed. Checks of larger samples as data collection progresses are often necessary to verify that these sections are working properly.

Unlike some of the other tests described here, it is essential that data checks be performed on every survey. We have found that when data checks are used consistently in combination with the correctness tests described above, serious errors in Web questionnaires can be virtually eliminated.

18.4.6 Load Testing

It is a somewhat peculiar problem of Web surveys that the number of respondents wishing to do an interview at any point in time may exceed the survey organization's capacity for taking them. Particularly when large numbers of respondents are contacted via bulk e-mail, they may come quickly and in such numbers that they exceed server capacity. The primary testing issue is ensuring that the server or servers on which the survey will run can support the expected load, that is, the number of respondents taking the survey simultaneously. Although there are a variety of technical metrics to measure the use of CPU, memory, and bandwidth, the survey designer is concerned primarily with the impact on the respondents' experience. An overloaded hosting platform will affect the respondent in one of three ways: slow response time, server crash or loss of connection during the interview, or inability to connect at the time the respondent wishes to do the survey. Although different in severity, any one of these conditions is a potential source of survey error.

In Section 18.4.4 we describe autogeneration software that simulates a respondent moving through the questionnaire. For load testing, this software must also allow for specifying the total number of respondents to simulate, the questionnaire to be used by each, the frequency with which respondents will come to the server, and the per question response time. The software should track the amount of time it takes to load the survey for each respondent (the setup time), the time from receiving the answer to a question until the next question is displayed (response time), and any special events such as failure to display a question or server halt.

We have developed a tool like this and found it to be extremely helpful for estimating load capacity. In a typical test we might configure the software to launch a new "respondent" every 10 seconds, and once a question is displayed, to wait 5 seconds before submitting an answer. Output from such a test shows us how setup time and response time increase as the number of respondents active on the server increases. We consider 5 seconds for setup and less than 2 seconds per question response time to be ideal. Our experience suggests that setup times longer than 15 seconds and response times longer than 5 seconds are problematic. Using these tolerances we can estimate the maximum number of simultaneous respondents that we can handle effectively. We then try to manage our sample contacting strategies across projects so that we stay within these limitations.

Fortunately, tests of this sort are not necessary for every survey. They typically are performed only when there is a change in the server infrastructure, when there is some major change in another component or standard, or when the load (e.g., number of respondents simultaneously taking surveys) is expected to exceed what the survey organization thinks of as "normal." Especially complex questionnaires with heavier than normal server resource requirements also may signal a need for load testing.

18.5 CONCLUSIONS

In this chapter we have described the principal building blocks of a robust Web questionnaire development and testing process. We have specified the various components of a Web questionnaire and described the techniques we have developed and found most effective for testing each. The challenge to the reader is to put these techniques together into a process or testing protocol that works within his or her organizational setting. Variations in the size and complexity of questionnaires fielded, subject matter, target populations, volume of Web surveys, types of clients being served, and technology used no doubt will produce different protocols across organizations. Nevertheless, the goal remains the same: to field correct questionnaires that impose the lowest possible burden on respondents and ensure to the maximum extent possible that every respondent experiences the questionnaire in the same way.

We have stressed two overarching themes to guide the development of a testing protocol. The first of these is the modularity of Web questionnaires, and therefore the need to test individual components with techniques and in time frames appropriate to the component's use and life cycle. There are components of the Web questionnaire that we refer to as general infrastructure; that is, they are used by all of the organization's Web surveys. These need not be tested within the context of every survey unless a decision has been made to go outside the organization's standards. These components are survey building blocks with known performance properties and reliability. Changes that impact these components, such as a new browser release, a change in interviewing software, or a reconfiguration of the hosting platform, will mandate periodic retesting and validation, but these events should be off the critical path of any specific survey and outside the scope of the questionnaire designer's responsibility.

The second major theme is the need to establish, enforce, and constantly update standards for every component of the Web questionnaire. The more variation that is allowed from survey to survey, the more arduous the testing process and the more unpredictable the outcome. We recognize that the science of Web surveys is still very much in its infancy, and so the standards we advocate today will not be the best standards for tomorrow. Standards inevitably must evolve and change, but in a controlled and deliberate way.

The evolution of CASIC technologies continually presents us with new challenges. Chief among them has been the need to be increasingly precise about how we design, specify, and implement questionnaires. The Web presents us with perhaps the greatest challenge thus far. We will need to become more disciplined, more structured, and exert significantly more control over our questionnaire development and testing processes if we are to meet it successfully.

CHAPTER 19

Evolution and Adaptation of Questionnaire Development, Evaluation, and Testing Methods for Establishment Surveys

Diane K. Willimack
U.S. Bureau of the Census

Lars Lyberg
Statistics Sweden

Jean Martin
Office for National Statistics, United Kingdom

Lilli Japec
Statistics Sweden

Patricia Whitridge
Statistics Canada

19.1 INTRODUCTION

An establishment is usually defined as an economic unit, generally at a single physical location, where business is conducted, or industrial operations are performed (U.S. Federal Committee on Statistical Methodology, 1988). An establishment survey collects information from or about establishments or economic units composed of establishments (Cox and Chinnappa, 1995). The units may be, for example, individual establishments, companies, enterprises, or legal entities, as well as hospitals, schools, institutions, farms, or government agencies.

Methods for Testing and Evaluating Survey Questionnaires, Edited by Stanley Presser, Jennifer M. Rothgeb, Mick P. Couper, Judith T. Lessler, Elizabeth Martin, Jean Martin, and Eleanor Singer
ISBN 0-471-45841-4

All countries conduct establishment surveys, many of which support systems of official statistics and provide short-term economic indicators and data to inform decision-making about education, transportation, health care, and other program areas.

Improving quality in establishment surveys has traditionally focused on coverage error, nonresponse, sample designs, and timely release of survey data. Less attention has been devoted to the development of questionnaires and the treatment of measurement errors. Only a small portion of the extensive literature on questionnaire-related issues addresses establishment survey applications, and activities for developing establishment survey questionnaires are not well documented in the public domain.

The goal of this chapter is to present an overview of questionnaire development, evaluation, and testing methods used in establishment surveys around the world, particularly among government and national statistics institutes. We limit our scope to surveys of businesses, organizations, and institutions, excluding agricultural surveys of farms. We begin in Section 19.2 by describing characteristics of establishments that lead to a survey response process more complex than that for household surveys, resulting in modifications to survey data collection procedures and questionnaires. In Section 19.3 we describe our literature review, which we augmented by conducting a small international study to assess how questionnaire development methods have evolved or were adapted to the establishment survey environment. Our findings are presented in Section 19.4. In Section 19.5 we discuss an alternative questionnaire development culture for establishment surveys and conclude with recommendations.

19.2 COLLECTING DATA FROM ESTABLISHMENTS

19.2.1 Characteristics of Establishment Surveys

Distinct differences between household surveys and establishment surveys are described by Cantor and Phipps (1999), Cox and Chinnappa (1995), Dillman (2000a,b), and Rivière (2002). Highlights of these differences include:

Mandatory Reporting Unlike most household surveys, responses to many establishment surveys are required by law because they collect critical economic data. As a result, the respondent burden placed on businesses is greater than that for households.

Many statistical agencies are required to monitor burden and compliance costs. Nevertheless, survey managers tend to measure survey length in terms of the number of questionnaire pages or the number of questions. As a result, survey managers are often extremely reluctant to add pages or questions, believing that this will add to the burden and the associated compliance costs, even when breaking a complex concept into multiple simple questions may be easier for respondents.

Analysis of Change in Key Variables Data collected to monitor the economy must support estimation of changes in key economic indicators, as well as their levels. This calls for panel or longitudinal sampling designs in which the same units remain in the sample for periodic (e.g., monthly, quarterly, or annual) surveys, contributing to the reporting burden. In addition, measurements must be comparable, so that changes observed in economic variables may be attributed to changes in economic conditions rather than the questionnaire. Thus, questions must remain stable over the life of a panel. Although this permits respondents to institute routines that ease the period-to-period reporting burden (Dillman, 2000b), reporting errors may be perpetuated (Willimack et al., 1999).

Skewed Target Populations Populations of establishments tend to be comprised of a few large units that contribute disproportionately to population totals, and a large number of smaller units that contribute much less. As a result, the largest businesses are usually selected into samples with certainty to ensure that resulting statistics have small variances and reflect adequate coverage. Moreover, common household survey nonresponse adjustment techniques, such as donor imputation, are inappropriate for accounting for nonresponse from large businesses. Their statistical importance is another reason for mandatory reporting. However, the consequence is substantial respondent burden.

Reliance on Data Existing in Business Records As inputs to the national accounts, data collected by establishment surveys measure technical concepts with precise definitions. Survey designers expect these data to be present in records maintained by businesses (Griffiths and Linacre, 1995). To ensure consistent measurement, detailed instructions often accompany establishment survey questionnaires. Yet data requested may not be comparable to the form or the timing of data available in records, contributing to specification error (Dutka and Frankel, 1991). Questionnaires may be tailored to specific industries or sizes of businesses to reduce specification error, which may result in more questionnaire variants than can be tested (e.g., there were more than 500 industry-specific questionnaire versions for the 2002 U.S. Economic Census).

In addition, concepts, definitions, and reference periods often vary across surveys within and between organizations (Fisher et al., 2001a; Goldenberg et al., 2002b), contributing further to specification error. The impact on respondents is that multiple surveys appear to request the same, or similar, data, but in different ways. This, too, contributes to reporting burden.

Prevalence of Self-Administered Data Collection Establishment surveys are predominantly self-administered (Christianson and Tortora, 1995; Nicholls et al., 2000) to encourage respondents' use of records. Respondents are commonly given a choice of mailed paper or electronic forms, fax, touchtone or voice recognition data entry, electronic data interchange, or Web-based reporting (Nicholls et al., 2000). Indeed, the prevalence of multimode surveys has become a hallmark of establishment survey data collection (Dillman, 2000c), helping to maintain response rates. Yet survey organizations require additional resources to maintain, coordinate, and integrate multimode processes (Clayton et al., 2000a; Rosen and

O'Connell, 1997), which may reduce resources available for questionnaire design and testing.

Emphasis on Timeliness Because establishment survey data are used to monitor rapidly changing economic conditions, timely reporting is often emphasized over other aspects of data quality. Hence, considerable resources are typically allocated to postcollection activities, such as editing to correct reporting errors, with callbacks to respondents to clarify suspicious or edit-failing data. Other postsurvey evaluations of data quality are based on item nonresponse rates, imputation rates, or internal or external consistency checks (Biemer and Fecso, 1995).

19.2.2 Response Process in Establishment Surveys

The characteristics of establishment surveys result in a response process profoundly different from that in household surveys. Willimack and Nichols (2001), following Sudman et al. (2000), offer a model of the response process in establishment surveys that incorporates both organizational and cognitive factors:

1. Encoding in memory/record formation
2. Selection and identification of the respondent or respondents
3. Assessment of priorities
4. Comprehension of the data request
5. Retrieval of relevant information from memory and/or existing records
6. Judgment of the adequacy of the response
7. Communication of the response
8. Release of the data

Since establishment surveys rely heavily on data in business records, the first step, *encoding*, includes *record formation.* For recorded data to be available for survey response, knowledge of those records must also be encoded in a person's memory. Data recorded in business records are determined primarily by management needs, regulatory requirements, and accounting standards. Often different data reside in different information systems that, although automated, may not be linked. Thus, information and persons with particular knowledge are likely to be distributed across organizational units, particularly in large businesses (Groves et al., 1997; Tomaskovic-Devey et al., 1994).

This distributed knowledge and associated division of labor have consequences for *respondent selection.* In their cognitive response model for establishment surveys, Edwards and Cantor (1991) consider respondent selection implicit in the encoding step, because the preferred survey respondent is someone with appropriate knowledge of available records. However, other organizational factors influence respondent selection, such as distributed knowledge, division of labor, organizational hierarchies, and the authority versus the capacity of a particular organizational informant to respond to a survey (Tomaskovic-Devey et al., 1994). These factors have consequences for survey response. First, response to a single survey covering multiple topics may require gathering information from multiple people or record systems distributed throughout the business. Second, institutional knowledge of various data sources typically rests with those in authority, who may be removed from direct access to those sources. Instead, these authorities delegate the response task, implying that respondent selection is performed by the business rather than the survey organization.

The third step, *assessment of priorities*, links organizational factors with a respondent's motivation or attentiveness to the response task. Priorities set by the business dictate the relative importance of activities performed by its employees. Willimack et al. (2002) found survey response to have low priority because it represents a non-revenue-producing activity with a tangible cost to the business. In addition, personal motivation is affected by job performance and evaluation criteria, also organizationally determined, as well as professional standards.

These three steps—encoding/record formation, respondent selection, and assessment of priorities—set the context for the four core cognitive steps constituting Tourangeau's survey response model (1984): *comprehension, retrieval, judgment, and communication*. In establishment surveys, the cognitive steps apply not only to respondent interactions with survey questions, but also to their interactions with records and other data sources within the business.

The final step in the model, *releasing the data*, returns the response process to the organizational level, since survey response is among the ways a business interacts with its external environment (Tomaskovic-Devey et al., 1994; Willimack et al., 2002). Reported figures are reviewed and verified against other externally available figures to ensure a consistent picture of the business to the outside world. Since data requested by establishment surveys are often proprietary in nature, concerns for confidentiality and security come into play as well.

The complexity of the response process in establishment surveys has consequences for questionnaire design and the application of testing methods. These include:

- The labor-intensive nature of survey response may be exacerbated by the need for multiple respondents and data sources. As a result, completion of a long or complex questionnaire may take hours, days, or even weeks of elapsed time.
- Data availability is a function of record formation, respondent access to data, and retrieval activities. Yet establishment survey questionnaire design tends to presume respondent knowledge of available data.
- Businesses exercise varying respondent selection strategies (Nichols et al., 1999; O'Brien, 2000a). Consequently, the respondent's role in the business may result in variation in the core cognitive steps, which in turn has implications for data quality (Burns et al., 1993; Burr et al., 2001; Cohen et al., 1999; Gower, 1994).

- The survey questionnaire bears a great burden in mediating between respondent and response due to the predominance of self-administered data collection. Detailed instructions are provided to equalize varying degrees of respondent knowledge and motivation and to convey needed information in a consistent manner to all respondents.

19.2.3 Implications for Establishment Survey Questionnaire Development, Evaluation, and Testing Methods

Four major interrelated themes emerge that we believe have implications for the application of questionnaire development, evaluation, and testing methods in establishment surveys:

- The *nature of the data requested*—measurement of technical concepts with precise definitions—has consequences for many aspects of data collection, including personnel involved in questionnaire design, choice of data collection mode(s), design of questionnaires, and the reporting ability of business respondents.
- *Survey response is labor-intensive* and represents a tangible but nonproductive cost to the business.
- Due to statistical requirements related to target population characteristics and data use, extensive *respondent burden* represents a serious constraint to questionnaire testing, experimentation, and evaluation.
- *Uses of economic data* require specific measurements, affecting the nature of the requested data, and favor timeliness over other attributes of data quality.

We believe that these four factors implicitly or explicitly affect many of the establishment survey testing methods currently in use. We will now describe our methodology for examining their impact.

19.3 RESEARCH METHODOLOGY

19.3.1 Review of the Literature

The goal of our literature review was to learn about the use of questionnaire development and testing methods for establishment surveys. We also looked at related papers on editing and on the development of new technologies for data collection. We searched survey research journals, monographs, conference proceedings, and unpublished papers over the last 10 years, focusing on conference programs since 1998.

In contrast to household questionnaire testing methods, there were almost no articles on establishment survey testing in relevant peer-reviewed survey research journals. Traditional textbooks on questionnaire design (e.g., Fowler, 1995; Mangione, 1995; Sudman and Bradburn, 1982; Tanur, 1992) contain little discussion of practices for designing survey questionnaires for establishments, institutions, or organizations, although Dillman (2000a) devotes a chapter to business survey methods.

The majority of the papers we found were in the monograph or proceedings published from the two International Conferences on Establishment Surveys (ICES) (American Statistical Association, 1993, 2000; Cox et al., 1995). In addition, the monograph from the International Conference on Computer-Assisted Survey Information Collection (Couper et al., 1998) contains several useful chapters, indicating the significant role of establishment surveys in the advent of electronic reporting.

Altogether, including unpublished papers, we assembled more than 130 items deemed relevant for our research. Although the number is more than we expected at the outset, this literature is very sparse in comparison to the household side. In addition, we discovered a clear North American bias; only a few papers originated outside the United States and Canada.

Nearly 40 papers described applications of cognitive methods in establishment surveys. Exploratory site visits and respondent debriefings were the next most frequently reported methods, appearing in nearly 20 papers each. Focus groups, pilot tests, data user input or content reviews, and methodological expert reviews or appraisals were described in approximately 10 to 12 papers each. These studies were not mutually exclusive, since many papers described multiple methods used in an iterative manner, particularly for new surveys or major redesigns of recurring surveys. In addition, almost one-third of the papers described research and development activities associated with electronic data collection, reflecting its importance to establishment surveys.

A number of papers described empirical evaluations of data quality. The literature on editing emphasizes the need for improving survey questions (Granquist, 1995). Fewer than 10 papers described split-sample experiments embedded in pilot studies, and only one paper employed a factorial experimental design (Vehovar and Manfreda, 2000).

19.3.2 International Study

To fill gaps in the literature, we conducted a small study of selected survey organizations around the world to learn more about questionnaire development, evaluation, and testing methods used in establishment surveys. Our purposive sample does not support statistical inferences. Instead, our goal was to identify organizations undertaking significant establishment survey development and testing activities, so we could augment our literature review with detailed descriptions of their applications.

We limited our scope to surveys of nonfarm businesses and institutions. We focused on surveys that collect data describing the business, including, but not limited to, financial variables (e.g., sales, shipments, revenues), employment data, descriptive categorical variables, or in the case of institutional surveys, data collected from person-level records maintained by the institution. Out of scope were surveys that queried individuals within businesses or institutions for information about themselves, such as organizational climate surveys. Otherwise, we covered surveys regardless of their topic, industry, frequency, or mode.

We purposively selected 56 survey and statistical organizations throughout the world. We attempted to represent the breadth of organizations likely to conduct business or institutional surveys by including national statistics institutes (NSIs), federal government statistical agencies, central banks, universities, and private-sector organizations, as well as international organizations such as Eurostat and the United Nations. We sent a brief questionnaire by e-mail to midlevel or senior-level contacts. The questionnaire featured a few questions aimed at identifying the types and purposes of methods being used. We asked about personnel, standards or guidelines, and special studies related to questionnaire development. We also queried methods used for recurring surveys, one-time surveys, new surveys, and/or major redesigns of recurring surveys.

Of the 56 contacts, two were undeliverable and four conducted no surveys. We received 43 completed questionnaires from the remaining 50 organizations and selected 11 for case study follow-up interviews conducted in person or by telephone. Using a protocol organized around the steps in the survey process, we gathered detailed descriptions of questionnaire development practices used for selected surveys, which we summarized and categorized into generally accepted taxonomies of testing methods.

19.4 ACCOMMODATING THE ESTABLISHMENT SETTING IN QUESTIONNAIRE DEVELOPMENT, EVALUATION, AND TESTING

We turn our attention to describing how the application of questionnaire development, evaluation, and testing methods accommodate the establishment survey setting. We cannot judge the prevalence of these activities. Based on our own and our international study participants' experiences, it is our impression that application of many of these methods is not pervasive among survey organizations worldwide. Nevertheless, evidence from the literature and, to some degree, our international study demonstrates how these methods have evolved and adapted to accommodate the circumstances associated with the establishment setting identified in Section 19.2.3. We next consider how each of those four major factors influences the application of various questionnaire development, evaluation, and testing methods.

19.4.1 Nature of Data Requested

Data requested in establishment surveys are based on technical concepts having precise definitions, and they rely on data found in business records. These characteristics have implications for personnel involved in questionnaire development and testing, as well as the methods used.

Consultations with Subject-Area Specialists, Stakeholders, and Data Users Establishment survey personnel tend to be organized by survey or program area rather than along functional lines, so they acquire expertise in the subject area covered by the survey. They are integrally involved in all stages of the survey process—development, collection, callbacks, data review and summary, tabulation, and publication—and draw on experiences at later survey stages to suggest modifications to questionnaires. A drawback of this structure is a lack of cross-fertilization, which could lead to innovation in testing methods and greater consistency in instrument designs across surveys.

Questionnaire design experts typically work outside these dedicated survey areas, although some organizations assign survey design specialists directly to a survey program. We found that when they are involved, questionnaire design experts are usually enlisted as consultants after an instrument has been drafted.

Because of stringent data requirements (e.g., the national accounts) and the technical nature of the data, stakeholders and data users play a substantial role in questionnaire development, progressing well beyond simply providing conceptual underpinnings and survey requirements. Stakeholders may even draft survey questions themselves or collaborate closely with questionnaire design experts (Goldenberg et al., 2002a; Stettler and Willimack, 2001). In addition, international statistical coordinating organizations, such as Eurostat, develop and administer regulations and set requirements for economic measurements that member NSIs must follow, so that surveys of businesses and institutions rarely start with a "clean slate."

Traditionally, input is also sought from data users, including subject area experts, academic researchers, advocacy group members, and from industry experts, such as representatives from trade or professional associations. Not only do data users suggest topics for survey content, they also review the appropriateness of proposed measures of survey concepts (Andrusiak, 1993; Fisher et al., 2001b; Ramirez, 2002; Underwood et al., 2000). Trade association representatives may also be target respondents, and thus represent both perspectives during questionnaire development when they suggest terminology for survey questions and offer advice on the availability of data requested.

A number of papers indicate the prevalent role of consultations with stakeholders, sponsors, data users, industry experts, and internal survey personnel to support questionnaire development (Freedman and Mitchell, 1993; Goddard, 1993; Mueller and Phillips, 2000). Our study participants describe using the following forms of consultation during questionnaire development:

- *Expert/user/advisory groups.* These groups, consisting of representatives from the primary data users for a particular survey and senior survey managers, are consulted regularly during the questionnaire development process (Francoz, 2002). Members prioritize new questions and changes to existing questions and are responsible for final approval of the questionnaire.
- *Iterative rounds of consultations.* Data users generate a list of requested data items that are subsequently ranked by noted experts in the field and prioritized further by major stakeholders. This is followed by contact with trade association representatives to assess data availability, while survey

personnel identify data items that have previously been obtained with high quality. The result is a questionnaire that balances needs for new data with obtainable data.

- *Large-scale content review.* Review of draft questionnaires is sought by mailing sample forms to stakeholders, data users, trade associations, and key respondents. This activity was automated via the Internet for the 2002 U.S. Economic Census.
- *Exploratory focus groups.* Focus groups with survey personnel or data users are used to help clarify concepts to be measured, output tables, and data publication goals, thus aiding in the specification of questions, definitions, and terminology (Oppeneer and Luppes, 1998; Phipps et al., 1995). Focus groups with survey personnel throughout the organization are conducted to ensure that survey questions match survey concepts. Focus groups have also been used to assess data users' needs and prioritize data requests (Andrusiak, 1993; Carlson et al., 1993; Fisher et al., 2001a; Kydoniefs and Stinson, 1999).
- *Observers.* When pretest interviews are conducted by questionnaire design experts, survey personnel or key stakeholders often participate as observers (Fisher et al., 2001b; Ware-Martin, 1999). Using their subject area knowledge, they help assess whether respondents' answers meet the question objectives.

Exploratory/Feasibility Studies and Site Visits Because establishment surveys rely on business or institutional records, investigations into data availability play an integral role in questionnaire development. These typically take the form of exploratory or feasibility studies conducted as company or site visits (Birch et al., 2001; Burns et al., 1993; Francoz, 2002; Freedman and Rutchik, 2002; Gallagher and Schwede, 1997; Mueller and Phillips, 2000; Mueller and Wohlford, 2000; Schwede and Ott, 1995). Nearly all study participants reported their use. The primary goals are to understand the concepts from the respondent's point of view, determine the availability and periodicity of records containing the desired data, and ascertain how well recorded data conceptually match desired data. Other goals include assessing respondent burden and quality and identifying the appropriate respondent.

Sample sizes for exploratory site visits are usually small and purposive, targeting key data providers. Survey personnel conduct meetings with multiple company employees involved in survey reporting. Usually, survey personnel follow a protocol or topic agenda and discuss definitions and the types of data that can be supplied by company reporters. Information gained during these visits helps determine whether survey concepts are measurable, what the specific questions should be, how to organize or structure the survey questions, and to whom the questionnaire should be sent.

Since site visits are costly and telephone calls tend to be laborious for staff and bothersome for respondents, self-administered approaches have been attempted (Kydoniefs, 1993). One study participant reported selecting a small subsample of businesses in a recurring survey, enclosing a supplemental self-administered questionnaire along with the regular survey questionnaire, and asking respondents to assess the availability and quality of data requested by proposed new questions. These are topics that typically would have been covered during company visits, if resources had permitted.

Record-Keeping Studies Conducted as occasional special projects, formal studies of business record-keeping practices tend to be comprehensive, investigating the structures and systems for keeping records, the location of particular categories of data, and who has access to which data (U.S. Bureau of the Census, 1989). More common are studies associated with individual surveys to determine whether specific data exist in records and how well they match requested items (Gower, 1994; Kydoniefs, 1993; Leach, 1999; Ware-Martin, 1999). In addition, several study participants mentioned studying the extent to which electronic records are available among target institutions.

Using Questions from Other Sources Because data requested are often subject to precise definitions and specifications, survey organizations review other surveys and borrow existing questions, definitions, and even formatting, where possible (Goldenberg et al., 2002a; Gower, 1994). Examples from our international study include data harmonization projects in several statistical organizations, where the goal is to develop a single integrated questionnaire to avoid duplication of data requests (Andrusiak, 1993; Crysdale, 2000), and the development of "question libraries" containing standard definitions (Underwood et al., 2000). Use of this practice is not pervasive, even within the same survey organization, where different questions or questioning strategies may be used for items that appear to be the same, but have definitional differences satisfying specific data uses (Fisher et al., 2001a). Regardless, survey questions borrowed from other sources may not receive additional testing in their new context.

Methods for Testing Instructions Given the heavy reliance of establishment surveys on instructions, it is curious that for the most part, testing and evaluation of instructions appear to be incidental to questionnaire testing, rather than its focus. Only a few international study participants mentioned testing the placement of instructions. The literature includes a few cognitive interview studies of respondents' use of instructions and pilot tests of their placement and formatting (Schwede and Gallagher, 1996; Zukerberg and Lee, 1997). Ware-Martin et al. (2000) conducted respondent debriefings to obtain feedback about redesigned questionnaires that directly incorporated instructions that had previously appeared in a separate booklet.

19.4.2 Labor-Intensive Survey Response Process

Certainly, one goal of questionnaire redesign is to simplify questionnaires so that the amount of work required of respondents is reduced. Nonetheless, because the survey response process is labor-intensive, efforts are made to minimize additional workload for respondents when pretesting questionnaires.

Informal Interactions with Respondents Christianson and Tortora's (1995) international survey of statistical organizations found that most pretesting conducted with establishment survey respondents was informal or ad hoc. Our international study suggests that little has changed in the intervening years. All case study participants reported using informal practices for obtaining feedback on new or changed questions or definitions on recurring surveys. Informal activities seem to be characterized by small convenience samples targeting large businesses, with unstructured probing in telephone calls by survey personnel with little or no formal training in the causes or consequences of response error. Yet, their probes request appropriate information about the response process, such as feedback on problematic terminology, concepts, and definitions, the availability of data in records, as well as suggestions for making questions clearer and the response task easier.

Respondent Focus Groups Focus groups with respondents have been conducted for exploratory purposes prior to questionnaire development (Gower, 1994; Kydoniefs, 1993; Palmisano, 1988; Phipps et al., 1995). Once a draft questionnaire is available, content evaluation focus groups examine the appropriateness and clarity of question wording, use and usefulness of instructions, general reactions to the questionnaire and its design, and the compatibility of requested and recorded data (Babyak et al., 2000; Carlson et al., 1993; Eldridge et al., 2000; Gower, 1994; Snijkers and Luppes, 2000). Several researchers indicated difficulties recruiting business respondents for focus groups (Bureau, 1991; Freedman and Rutchik, 2002; Gower and Nargundkar, 1991), and a few of our study participants noted their limitations, because businesses tend to be reluctant to reveal information to competitors. To remedy this, one case study participant conducted focus groups with business respondents via teleconference, thereby maintaining participants' anonymity.

Cognitive Pretesting Methods A number of conference papers describe applications of cognitive testing, particularly during the design of new surveys or when major redesigns are undertaken. Gower (1994) lists traditional cognitive methods such as in-depth interviews, concurrent think-aloud interviews, and paraphrasing, while others also describe use of retrospective debriefings and vignettes to explore the four cognitive steps in the establishment survey response process. Traditional cognitive probes are often augmented to explore estimation processes (judgment), request observation or descriptions of records and their contents (retrieval), or examine use of complicated structures for entering data, such as grids, often found on business survey questionnaires (communication of the response). Because the establishment survey response process is more complex, cognitive interviews often take on aspects of exploratory studies as well, resulting in a cognitive hybrid that explores data availability and respondent roles along with cognitive aspects of the response process (Freedman and Rutchik, 2002; Goldenberg and Stewart, 1999; Sykes, 1997).

Traditional cognitive methods have also been adapted to minimize additional work by respondents. A number of papers (DeMaio and Jenkins, 1991; Gerber and

DeMaio, 1999; Stettler et al., 2001), along with several of our case study participants, describe using hypothetical probes to explore steps involved in retrieving data from records, since it is often impractical for researchers to observe this labor-intensive process directly. These probes identify data sources, discover respondents' knowledge of and access to records, recreate steps likely taken to retrieve data from records or to request information from colleagues, and suggest possible estimation strategies.

Hypothetical scenarios or vignettes have also been used during pretesting with business respondents (Morrison et al., 2002). In some cases, these vignettes were used in a similar manner to household studies—to evaluate judgment processes regarding unclear, multidimensional concepts (Stettler et al., 2001). In addition, a number of studies used vignettes consisting of mock records containing simulated data, allowing researchers to observe respondents' retrieval, judgment, and communication strategies when retrieval from their own records was too time consuming to be completed in advance or during the appointed interview (Anderson et al., 2001; Goldenberg et al., 2002b; Kydoniefs, 1993; Moy and Stinson, 1999; Schechter et al., 1999; Stettler et al., 2000). Both applications allow researchers to evaluate probable response errors, since correct answers associated with the vignettes are known.

Researchers often require flexibility and take greater latitude than may be permitted in traditional cognitive interviews using standardized protocols. This allows researchers to react to the variety of uncontrolled and uncontrollable situations encountered during testing outside the laboratory at the business site. Researchers may apply a funnel approach, in which successively more directive probes are asked to help focus a respondent's thinking.

Traditional cognitive methods, coupled with these various modifications, enable collection of more detailed information about the establishment survey response process (Davidsson, 2002; Dippo et al., 1995; Goldenberg et al., 2002a; Laffey, 2002; Rutchik and Freedman, 2002). However, observation of respondent behaviors remains somewhat artificial, since the response task often is not performed in full view of the researchers (Birch et al., 1998; Schwede and Moyer, 1996). Thus, issues associated with conducting pretest interviews with establishment respondents are not solved entirely by adapting cognitive methods.

Pretesting Logistics Procedures for pretest interviews, including cognitive interviews, have been modified to accommodate the characteristics of establishments. First, pretest interviews are conducted at the business site rather than in a laboratory, so that respondents have access to their records (Babyak et al., 2000; Eldridge et al., 2000). Although this is more costly and time consuming for the survey organization, Moy and Stinson (1999) found that on-site testing provided a richer, more realistic context than conducting comparable interviews in a laboratory. Due to resource constraints, cognitive interviews occasionally have been conducted by telephone (Cox et al., 1989; Stettler and Willimack, 2001). Davis et al. (1995) embedded open-ended cognitive probes in a self-administered questionnaire and compared results with traditional in-person cognitive interviews, finding that the latter provided more complete and detailed information.

Most pretest interviews with business respondents are limited to 60 to 90 minutes in duration, because taking respondents away from their job responsibilities represents a tangible cost to the business. Since the entire questionnaire may be lengthy and require laborious data retrieval from records, testing is often limited to a subset of questions (Jenkins, 1992; Stettler et al., 2000; Ware-Martin, 1999). Thus, researchers must be careful to set the proper context for the target questions.

Sample sizes for pretest interviews vary widely. In the United States, collection of data from 10 or more respondents requires approval from the Office of Management and Budget (OMB), so many studies are conducted with samples of nine or less. (OMB allows some federal agencies a streamlined clearance process for pretesting with larger sample sizes.) Survey organizations in other countries face similar sample size restrictions. Moreover, recruiting business and institutional respondents presents a significant challenge, because they are being asked to take time away from their jobs. Studies cited in this chapter used as few as two to five cases (Babyak et al., 2000; O'Brien et al., 2001) and as many as 60 cases, depending on study goals and resources. Multiple rounds of pretest interviews are preferred, with the questionnaire revised between phases (Goldenberg, 1996; Goldenberg et al., 1997; Schwede and Gallagher, 1996). However, based on comments from some international study participants, we suspect that only one round of interviews with a small number of cases is not uncommon for cognitive studies. An alternative strategy is to use co-workers as substitutes for respondents, particularly during early development (Kydoniefs, 1993) or when resources are limited (Goldenberg et al., 2002b).

According to our case study participants, formal cognitive pretest interviews are usually conducted by questionnaire design experts. Survey personnel trained in cognitive interviewing techniques may also conduct pretest interviews, guided by questionnaire design experts, who debrief survey personnel, summarize research findings, and collaborate on recommendations.

Usability Testing of Electronic Instruments Computerized self-administered questionnaires and Web survey instruments add another layer to the interaction between respondents and the questionnaire—that of the user interface. Although there are many aspects of software graphical user interfaces, testing thus far has tended to focus on navigation and embedded edits. A handful of papers describe the current state of *usability testing*—which is the broad name given to methods for testing the respondent–computer interface—for establishment surveys (Anderson et al., 2001; Burnside and Farrell, 2001; Fox, 2001; Nichols et al., 1998, 2001a,b; Saner and Pressley, 2000; Sperry et al., 1998; Zukerberg et al., 1999). A synthesis of these research activities follows.

Since the timeline for electronic instrument development must include programming, testing may begin with paper prototypes or nonfunctioning screen shots preceding software development. Usability testing is typically conducted with partially or fully functioning electronic prototypes or with existing instruments undergoing redesign.

Usability testing relies heavily on direct observation of business or institutional respondents performing tasks such as entering data, navigating skip patterns, correcting errors, and printing or submitting completed forms. Usability tests are often videotaped to facilitate review of respondent behaviors, which may also be coded and analyzed to identify usability problems.

Like cognitive testing, usability testing with establishment respondents typically takes place on site at the business or institution, although there has been some success with bringing institutional users into the laboratory. Some researchers point out the value of observing the task being completed in a user's natural environment, looking at the atmosphere of the workspace and the ways in which the environment affects the reporting task.

In addition to direct observation, researchers may incorporate other cognitive methods into usability testing, such as concurrent probes, retrospective debriefings, and user or respondent ratings. Vignettes and other hypothetical scenarios have been used to test alternative design options. After interacting with the instrument, users may be asked to complete self-administered debriefing questionnaires or participate in focus groups to provide feedback on their experiences.

19.4.3 Respondent Burden

Reducing respondent burden is an objective of questionnaire testing, and yet the accumulation of burden often constrains testing activities with establishments. In the United States, the expected number of hours associated with survey response must be reported to the OMB. Several European countries are required to measure the amount of time businesses actually spend testing and completing survey questionnaires, and to associate a monetary cost. Establishment survey testing is thus hampered by efforts to minimize additional contacts, resulting in the substitution of methods that avoid respondent contact or take advantage of the experiences of survey personnel who routinely interact with respondents. Since they require additional contacts with business respondents, pilot tests are reserved for exceptional circumstances, such as major new surveys or redesigns, and their goals tend to be broader than questionnaire evaluation. Under the best circumstances, developmental prototypes or other research activities may be embedded within recurring surveys, rather than undertaken separately, to avoid additional respondent contacts.

Expert Reviews and Cognitive Appraisals One way to reduce respondent burden associated with testing is to bypass respondents altogether. To this end, we found prevalent use of expert reviews (also called *desk* or *paper reviews*) of the survey instrument conducted by questionnaire design experts. Although the literature tends to report their use in concert with other methods (DeMaio and Jenkins, 1991; Eldridge et al., 2000; Fisher and Adler, 1999; Ware-Martin et al., 2000), our international study suggests that expert reviews may often be the only evaluation a questionnaire receives. This is particularly true for one-time quick-turnaround surveys (Labillois and March, 2000). Questionnaire design experts appraise the questionnaire, applying generally accepted questionnaire design principles and knowledge based on their own pretesting experiences. The questionnaire expert's task is to identify potential problems that may result in reporting errors or inconsistencies, and to suggest solutions, often in a written report. In some of our case study organizations, questionnaire design specialists collaborate with survey personnel to arrive at mutually acceptable solutions to design problems. Unfortunately, expert review outcomes are rarely documented in the literature, thus diffusion of "good" questionnaire designs for establishment surveys is limited.

We found only occasional use of the systematic cognitive appraisal coding system developed by Lessler and Forsyth (1996) and adapted for establishment survey evaluation by Forsyth et al. (1999) (O'Brien, 2000b). Fisher et al. (2000) used expert appraisal to guide development of a cognitive protocol. O'Brien et al. (2001) compared appraisals with cognitive interviews and found that the cognitive appraisal identified some problems as major that did not turn out to be terribly problematic for respondents in cognitive interviews, suggesting a need for further refinement of the appraisal methodology. One case study participant indicated that the appraisal scheme was useful as a tool for guiding novice staff members or subject area personnel in their reviews of questionnaires.

Electronic Instrument Heuristic Evaluations, Style Guides, and Standards Electronic instruments are also subjected to expert reviews, which are called *heuristic evaluations* because they are typically assessed relative to human–computer interaction principles or heuristics (Nielsen and Mack, 1994). Heuristic reviews have been conducted by individual usability experts, expert panels (Sweet et al., 1997), and members of development teams (Fox, 2001).

The respondent burden of testing electronic instruments can be minimized by adhering to a style guide or standards for the user interface when developing electronic instruments (Burnside, 2000; Harley et al., 2001; Murphy et al., 2001). These rules for navigation, layout, screen design, graphics, error handling, help, user feedback, and other internal electronic mechanisms are based on usability standards and best practices, as well as previous testing and operational experiences with electronic instruments.

Feedback from Survey Personnel Nearly all case study organizations reported relying on feedback from interactions between survey personnel and respondents. Survey personnel act as proxies for respondents to informally evaluate questionnaires. These interactions include "Help Desk" phone inquiries from respondents (McCarthy, 2001) as well as follow-up phone calls to respondents by survey personnel investigating data that failed edits. Since anecdotal feedback may sometimes be overstated, a notable example from our international study involves systematic recording of respondent feedback in an integrated database. Survey personnel typically convey this respondent feedback early in the questionnaire development cycle for recurring surveys. Rowlands et al. (2002) conducted interviews with survey personnel who review data and resolve edit failures to identify potential candidate questions for redesign or evaluation by other methods.

Methods familiar from household surveys have been applied to interviewer-administered establishment surveys. Goldenberg et al. (1997), Oppeneer and

Luppes (1998), and Paben (1998) conducted interviewer debriefings to identify problem questions. Goldenberg et al. (1997) and Mueller and Phillips (2000), together with a few case study participants, noted that having survey personnel observe live interviews helps to identify problems. Behavior coding of interview interactions was rare (Goldenberg et al., 1997). One case study participant pointed out that interviewers for establishment surveys tend to be allowed greater flexibility, due to the technical nature of the requested data, so coding schemes must be modified accordingly. In addition, such research activities are more likely to point out interviewer training issues than needed questionnaire modifications.

Pilots and Pretests The literature indicates a number of formal pilot studies or pretests for new surveys or major redesigns to evaluate data collected by new questionnaires (Birch et al., 1999; Cox et al., 1989; DeMaio and Jenkins, 1991; Fisher and Adler, 1999; Fisher et al., 2001a; Goldenberg and Phillips, 2000; Kydoniefs, 1993; Mueller and Phillips, 2000; Mueller and Wolford, 2000; Phipps, 1990; Phipps et al., 1995; Schwede and Ellis, 1997; Ware-Martin et al., 2000). However, our international study does not permit an assessment of the pervasiveness of this activity. Moreover, neither the literature nor our study participants make a clear distinction in terminology, often using the terms *pilot* and *pretest* interchangeably. While a *pilot* may be designed as a "dress rehearsal" to test all the steps and systems in the data collection process (Kydoniefs, 1993; Mueller and Wohlford, 2000; Phipps et al., 1995), Gower (1994) defines a *pretest* to be more limited in scope and questionnaire oriented. Kydoniefs (1993) and Mueller and Phillips (2000) describe iterative pretests involving multiple panels targeting establishments of different sizes and industries. The pretests evaluated respondents' understanding of the questions and definitions and tested the feasibility of the data collection mode, while investigating data sources, timing relative to survey deadlines, and respondents' use of estimation strategies. Pilots and pretests often take advantage of other evaluation methods, such as interviewer or respondent debriefings.

Sample sizes and designs for pretests and pilot studies vary widely (Goldenberg et al., 2002a). Pretests are usually smaller in scale and may draw on convenience or purposive samples. Pilot studies, on the other hand, usually entail sufficiently large random samples to generate statistical estimates of some variables (Fisher et al., 2001a; Goldenberg et al., 2000; Kydoniefs, 1993; Mueller and Wohlford, 2000; Schwede and Ellis, 1997). One case study participant conducted a pilot test using a stratified random sample of nearly 10,000 businesses, although the target population was limited to businesses with characteristics pertinent to the redesigned questions.

Split-Sample Experiments Split-sample experiments (also known as *split panels* or *split ballots*) allow empirical evaluation of alternative questions, questionnaires, or designs (Gower, 1994). A few study participants also used split-sample experiments to evaluate the positioning of instructions. Several studies incorporated split-sample experiments into pilot studies for question evaluation (DeMaio and Jenkins, 1991; Goldenberg et al., 2000; Phipps, 1990; Schwede and Ellis, 1997). Burt and Schappert (2002) embedded a split-sample experiment within a production survey to empirically evaluate questions considered "marginal" after cognitive testing and respondent debriefings. Other studies conducted experiments to evaluate mode selection (Dodds, 2001; Manfreda et al., 2001) or assess data quality and respondent burden among different modes (Sweet and Ramos, 1995; Vehovar and Manfreda, 2000). Embedding split-sample experiments into pilot studies or production surveys minimizes the added burden of involving business survey respondents in questionnaire design research. We believe that this strategy, particularly embedding experiments in production surveys, should be used more extensively to evaluate alternative questionnaire designs.

Development Strategies for Electronic Instruments Many of the available papers on electronic reporting describe survey organizations' strategies for developing, introducing, and implementing alternative modes for establishment surveys. We observed international variation in adoption of new technologies, including touchtone data entry in the United Kingdom (Kinder and Baird, 2000; Thomas, 2001), mixed-mode collection using CATI after mail in Canada (Parent and Jamieson, 2000), and varying forms of Internet-based data collection in the United States (Clayton et al., 2000b; Gaul, 2001; Ramos et al., 1998; Sedivi et al., 2000), Australia (Burnside and Farrell, 2001), and New Zealand (McBeth et al., 2001).

In general, initial development and testing of new technologies began on a small scale and grew as knowledge and confidence increased. To reduce respondent burden, testing early in the development process might be conducted in the laboratory using the survey organization's own personnel playing the role of respondents (Nichols et al., 2001a; Ramos and Sweet, 1995). This exploratory usability testing narrowed the design options subjected to confirmatory usability testing with establishment reporters (Gaul, 2001).

Development strategies have relied heavily on pretesting with a small number of cases, often fewer than 100 (Clayton and Werking, 1998; Meeks et al., 1998; Nichols and Sedivi, 1998; Ramos and Sweet, 1995; Rosen et al., 1998). Pretest respondents were identified by screening members of the survey sample for their interest and ability (i.e., hardware requirements) to report electronically. Thus pretesting of electronic instruments tends to be embedded within operational survey collection.

Results from iterative testing inform decisions about many design issues and identify other unforeseen problems to be resolved before rolling out the application on a large scale. Nearly all these pilots and pretests obtained respondent feedback in some way, either by embedding evaluation questions within the instrument itself, or by following up with debriefing telephone interviews or self-administered questionnaires. Meeks et al. (1998) convened workshops with potential users and advisory panels to solicit input on instrument design.

Electronic instrument use can also be assessed empirically once the instrument has been fielded. Burr et al. (2001) conducted multivariate analysis examining variables related to survey procedures that may be associated with mode selection. Others have analyzed electronic instrument event logs, which track keystrokes and timing, to identify user problems with the electronic interfaces and functionality (Saner and Pressley, 2000; Sperry et al., 1998).

19.4.4 Uses of Economic Data

The use of establishment data as inputs to systems of national accounts requires that precise definitions be followed. For their use as economic indicators, timeliness takes priority over accuracy, and time series must be preserved. These surveys employ panel designs in which the same businesses or institutions are contacted in successive survey cycles, so changes in question wording and other questionnaire features are discouraged. Therefore, collected data receive greater postcollection scrutiny than on the household side.

Postcollection Empirical Evaluations Postcollection evaluations of data quality are routinely conducted in recurring establishment surveys. Our international study showed that tabulating item nonresponse and imputation rates, along with examination of outliers and edit-failing records, plays a significant role in quality evaluations, although the literature we reviewed contained only a few examples of these activities (Birch et al., 1999; Monsour, 1985; Monsour and Wolter, 1989; Schwede and Ellis, 1997; U.S. Bureau of the Census, 1987; Wolter and Monsour, 1986). Results of these routine quantitative analyses are not systematically fed back into questionnaire design. However, our study participants reported a few notable examples of high item nonresponse or imputation rates that led to removing questions or redefining response categories.

The quality of collected data may also be evaluated through external consistency checks involving comparisons or reconciliation with data from other sources (Biemer and Fecso, 1995; Erikson, 2002). Significant discrepancies may warrant further investigation of survey questions (Mueller and Phillips, 2000).

Respondent Debriefings A number of studies debriefed respondents to evaluate the quality of collected data. Like pretesting, there appears to be a continuum from informal activities to formal evaluations. Formal respondent debriefings, also known as *response analysis surveys* (RASs), are conducted after data collection using structured questionnaires. Survey respondents are recontacted and asked about the response strategies and data sources actually used to provide answers to specific questions. The extent to which reported data meet the definitions is evaluated empirically, along with respondents' use of records and estimation strategies. Although commonly associated with pilots of new or redesigned surveys, RASs may also be used to evaluate data quality in recurring surveys. [For examples, see Goldenberg (1994), Goldenberg and Phillips (2000), Goldenberg et al. (1993), Palmisano (1988), Phipps (1990), and Phipps et al. (1995).]

Sample sizes and designs for respondent debriefings vary. Our case study participants reported sample sizes as small as 20 or as large as several hundred.

Sample designs may be probability-based, or selection may be purposive to focus on respondents who made reporting errors (Phipps, 1990) or to compare early and late respondents (Ware-Martin et al., 2000). Hak and van Sebille (2002) conducted respondent debriefings with timely reporters using retrospective questions to identify reporting errors at different stages in the response process model identified in Section 19.2.3.

Whereas most respondent debriefings reported in the literature are based on telephone or in-person interviews, DeMaio and Jenkins (1991) conducted self-administered respondent debriefings. Similarly, one case study participant described enclosing self-administered questionnaires with new or redesigned survey questionnaires, asking respondents to evaluate the questionnaire in general and note problem questions, difficult terminology, and burdensome record retrieval.

Reinterview Studies and Content Evaluations Another method used for post-collection evaluations is reinterview studies, which may or may not include reconciliation with previously reported figures (Biemer and Fecso, 1995; Paben, 1998). The primary goal of reinterview studies is to estimate bias in summary statistics due to reporting errors; hence, sample sizes and data collection methods must be sufficient for this purpose. Corby (1984, 1987) describes reinterview studies, called *content evaluations*, that examined the components and sources of reported figures, their reliability, and sources of reporting errors. Although some recommendations were made for revised question wording and instructions, the primary goal was to evaluate data quality empirically. Van Nest (1987) piloted a content evaluation study, only to find so much variation in response strategies that a full study was deemed infeasible. Instead, results supported numerous recommendations for changes to data collection questionnaires and instructions. Thus results of such research can also be used to improve questionnaires.

19.5 CONCLUSIONS

19.5.1 Alternative Questionnaire Development Culture in Establishment Surveys

It is clear from our literature review that the amount of publicly available establishment survey questionnaire development literature is much smaller than that for household surveys. Moreover, it is relegated to specialty sources such as conferences or monographs that focus on business surveys. Thus, it is easy to overlook pertinent applications of these methods in establishment surveys. For example, the earliest citation we found describing use of cognitive research methods in establishment surveys was Palmisano (1988), only four years after Tourangeau's landmark piece. Comparison of the proceedings from the two International Conferences on Establishment Surveys, occurring eight years apart, shows that adoption has increased. However, there remains a North American bias in the literature, which probably reflects slower adoption of these methods internationally (see, e.g., the *Proceedings of the Questionnaire Evaluation*

Standards Workshop, U.S. Bureau of the Census, 2001b). In addition, current establishment survey literature focuses on applications of the methods, with only a few papers comparing methods and none addressing their validation.

Much of the questionnaire work so far within establishment surveys appears to have taken a different route from household surveys. Our international study participants, together with a number of recent papers (Davidsson, 2002; Francoz, 2002; Laffey, 2002; Rivière, 2002; Rowlands et al., 2002), identify several characteristics of establishment surveys (such as those discussed in Section 19.2) that affect questionnaire development methods. Thus, our research suggests several ways in which questionnaire development methods have been adapted to accommodate the distinctive characteristics and response process of establishment surveys.

First, the greater complexity of the establishment survey response process calls for different testing methods than are used for household surveys. Activities investigating data availability and respondent selection seem on par with those investigating core cognitive processes. Methods respecting the roles of stakeholders and data users in questionnaire development also seem warranted, given the severe consequences associated with inaccurate statistics.

We recognize that a number of valid considerations associated with establishment survey characteristics call for different approaches to improving questionnaires. For example, maintaining questionnaire stability stems from the need to measure change in economic variables. Concerns for increasing reporting burden on business respondents, who are already surveyed frequently by multiple statistical organizations, also limit testing. However, often the consequence is what appears to be inattention to, and lack of rigor in, questionnaire development and testing.

Because of the various constraints and limitations to questionnaire development and improvement, establishment survey practices tend to emphasize identifying and correcting response errors through editing and imputation. Results of rigorous postcollection evaluations and formal respondent debriefings are used to varying degrees, or may not be used at all, to revise questions and instructions to prevent response errors in the first place. Although feedback from survey personnel about their interactions with respondents provides a less formal evaluation of the collected data, it is more likely to result in questionnaire revisions.

The establishment survey literature has no shortage of descriptions of new surveys or major redesigns of recurring surveys in which multiple methods were applied iteratively and results were scrutinized to ensure data quality and the integrity of ongoing data series. This is not unlike household surveys (e.g., the major redesign in the early 1990s of the Current Population Survey, which supports estimation of U.S. unemployment statistics). However, we found few recurring surveys that routinely conducted research and evaluation activities, in parallel with production data collection, specifically for the purpose of informing questionnaire design decisions, thereby effecting continuous quality improvement in data collection processes and instruments. We believe that such efforts should be commended and emulated more often.

19.5.2 Research Recommendations

Our findings clearly indicate an establishment survey culture that emphasizes postcollection processing (e.g., editing) to correct response errors rather than attention to questionnaire design to prevent them in the first place (Granquist, 1995; Underwood et al., 2000). To reverse this emphasis, we recommend that better advantage be taken of activities that occur already. Feedback loops from postcollection evaluations to questionnaire development need to be strengthened. What do item nonresponse, imputation, and edit-failure rates indicate about the quality of questions? What does analysis of event logs suggest about the efficacy of electronic instrument features? Which survey questions consistently generate comments from respondents indicating confusion, misunderstanding, or problems associated with data availability? Better use could be made of informal evaluations by survey personnel based on their interactions with respondents to clarify data or correct errors. Respondents' comments should be documented more systematically, summarized, and analyzed routinely (using, e.g., content analysis), so that aberrations would not be overstated. By and large, statistical analyses are underutilized as a tool for identifying questions or other design features that warrant follow-up investigations of the causes of response error and possible solutions to the problems. Shifting the traditional establishment survey paradigm from error correction to error prevention through improved instrument design should reduce the amount of postcollection processing and associated costs—both of which can and should be measured.

While Cantor and Phipps (1999) suggest a research agenda to aid questionnaire design, we also note the need for continued improvement of questionnaire development, evaluation, and testing methods, along with their validation and comparison. Commendable steps in that direction were taken by Burt and Schappert (2002), who used a split-sample experiment to validate pretest debriefing results, and by Forsyth et al. (2002), who compared cognitive appraisal and cognitive interviewing results. Questions remain about the relative effectiveness of various methods and the extent to which they improve data quality. Further studies are needed to address these questions so that questionnaire development and testing resources may be applied more effectively.

How can statistical organizations implement ambitious research studies to reduce measurement errors associated with questionnaires, particularly when many testing activities require additional contacts with business or institutional respondents? Admittedly, concern for respondent burden constrains questionnaire improvement research. Are we left in a quandary where research activities, which have the goal of reducing respondent burden through questionnaire improvements, themselves require short-term increases in respondent burden? Must emphasis in testing always be placed with large businesses that are already heavily burdened? We grant the critical importance of large businesses to aggregate statistics, but also recognize the preponderance of smaller businesses in the survey population. Routine cognitive testing with smaller businesses may well increase respondent friendliness for them and improve instruments for large businesses. Research is needed to ensure that cognitive pretesting with small businesses yields questionnaire improvements for large businesses as well.

We urge taking greater advantage of the recurring nature of many establishment surveys as a solution to the burden problem associated with questionnaire testing. Embedding research within recurring surveys presents an ideal opportunity to test and evaluate alternative formulations of data collection instruments, without adding to respondent burden. The marginal cost should be low, compared with undertaking additional data collection activities to conduct questionnaire design research. Indeed, the advent of electronic instruments offers cost-effective opportunities to test alternative questionnaire content along with electronic features. When the marginal cost is weighed relative to the increases in respondent burden associated with separate questionnaire research and the potential payoff in burden reduction from improved questionnaires, embedding research within recurring establishment surveys seems an obvious solution.

Most of the research identified in our literature review was conducted by government organizations or their contractors, and not by academics. (A few notable exceptions are studies by Dillman and by Manfreda and her colleagues.) Yet the response process and methodological issues could be studied in depth outside the routine government survey process. To expand the participation of academic researchers in establishment survey questionnaire development and testing, we encourage their collaboration with government organizations.

In the meantime, adoption of questionnaire development, evaluation, and testing methods by establishment surveys would be enhanced by a single comprehensive publication providing detailed descriptions of how to conduct these activities in the establishment setting. We believe that documenting current practices and their foundation will stimulate research interests and lead to improvements and innovations in questionnaire development and testing methods for establishment surveys.

ACKNOWLEDGMENTS

This chapter reports the results of research and analysis undertaken by U.S. Bureau of the Census staff and their collaborators. It has undergone a Census Bureau review more limited in scope than that given to official Census Bureau publications, and is released to inform interested parties of ongoing research and to encourage discussion of work in progress. The authors wish to thank the following for their substantial research assistance: Amy E. Anderson, Rebecca L. Morrison, and Kristin Stettler of the U.S. Bureau of the Census; JPSM intern Lori Parcel; Frances van den Enden of Statistics Canada; and Jacqui Jones and Sarah Williams of the Office for National Statistics, the United Kingdom. We also gratefully acknowledge helpful review comments from Karen Goldenberg, Allen Gower, Tony Hak, Nash Monsour, Laurie Schwede, Kristin Stettler, and our chapter editors, particularly Betsy Martin.

CHAPTER 20

Pretesting Questionnaires for Children and Adolescents

Edith de Leeuw and Natacha Borgers
Utrecht University

Astrid Smits
Statistics Netherlands

20.1 INTRODUCTION

Society is becoming more concerned with children's issues and children's rights. In most of the Western world, it is now recognized that the voices of younger children and adolescents should be heard and there is a demand for research that focuses on children as actors in their own right. As a consequence, survey researchers are realizing that information on children's opinions, attitudes, and behavior should be collected directly from the children; proxy reporting is no longer considered good enough *if* children can be interviewed themselves. Survey methodologists now focus on methods for designing questionnaires especially for children and adolescents and on methods for interviewing them. Official government agencies acknowledge children and adolescents as respondents and have developed and implemented special surveys for them (Scott, 1997). Also, academic research institutes and health organizations realize the need for accurate data collected directly from children and adolescents on their perspectives, actions, and attitudes (Greig and Taylor, 1999). Market research firms now acknowledge children and adolescents as special respondents and have guidelines for interviewing them (e.g., Esomar, 1999). However, relatively little is known about children and adolescents as respondents, and pretesting for this age group is a neglected issue (Blair, 2000).

Methods for Testing and Evaluating Survey Questionnaires, Edited by Stanley Presser, Jennifer M. Rothgeb, Mick P. Couper, Judith T. Lessler, Elizabeth Martin, Jean Martin, and Eleanor Singer
ISBN 0-471-45841-4

Children are not miniature adults. Their cognitive, communicative, and social skills are still developing as they grow older, and this affects their ability to answer survey questions (Borgers et al., 2000; Cynamon and Kulka, 2001; Zill, 2001). For surveys of adults, procedures to enhance response quality and the improvement of data collection methods are well documented (Biemer et al., 1991; Groves, 1989; Lyberg et al., 1997; Sudman et al., 1996). Still, even surveying adults is far from simple, and methodological studies have shown that adults may experience problems with certain questions, and that question characteristics affect the data quality in surveys (Krosnick and Fabrigar, 1997). Especially when questions are very complex and/or when information has to be retrieved from memory, adults have difficulty (Eisenhower et al., 1991; Tanur, 1992). Interviewing children and adolescents is both similar to and different from interviewing adults. With children as respondents, the same problems may be magnified, as a slight error (e.g., ambiguity) in the questionnaire may be more difficult to overcome or have a larger impact. Also, children may experience additional problems when responding to a question, and the questionnaire should be adapted to suit the cognitive, linguistic, and social competence of each age group. The usefulness of an answer to a question will depend on the age of the child and his or her verbal abilities, so pretesting questions for their suitability for specific age groups is highly advisable.

The age of 7 is a major turning point in the development of children. At this age, their language expands (Nelson, 1976), reading skills are acquired, and they start to distinguish different points of view (Gelman and Baillargeon, 1983; Selman, 1980). These are important prerequisites for the understanding of questions. With special care, children can be interviewed with structured questionnaires or complete self-reports from the age of 7 onward. At the age of 18, adolescents are generally treated as adults in surveys, as is reflected in definitions of adult populations for many surveys (e.g., ISSP).

In this chapter we discuss methods for pretesting questionnaires for respondents between 7 and 18 years old. We start with an integrative summary of empirical knowledge on the young as respondents. This section is organized around the major phases in child development and will serve as a conceptual framework for testing questionnaires for children and adolescents. We present guidelines for optimizing questionnaire testing methods for different age groups.

20.2 DEVELOPMENTAL INFLUENCES ON SURVEYING CHILDREN: REVIEW OF EMPIRICAL KNOWLEDGE

As children grow from infancy to adulthood, their thinking becomes more logical and their reasoning skills develop, memory and language develop, and social skills are acquired. Although there is considerable variation among children, depending on heredity, learning, experiences, and socioeconomic factors, consecutive stages can be discerned. The pioneer in child development research was Jean Piaget. Piaget's theory of developmental stages has provided the impetus for much psychological research and gives a useful framework for practical applications (Flavell et al., 1993). In the following sections, we discuss developmental issues for surveying children based on Piaget's stages, but amended with modern insights derived from information processing and sociocultural perspectives (see also Gray, 2002). One should always keep in mind that the stages presented should not be seen as sharply distinct categories, but rather, as a moving scale: There are differences within age groups, and there may be overlap between groups.

20.2.1 Middle Childhood (7 to 12)

Piaget (1929) saw the age of 7 as a major cognitive turning point; around this age children make the important transition from *preoperational* to the more advanced *concrete operational* period. Starting at age 7, children are better at logical, systematic thought using multiple pieces of information. Language skills develop further and reading skills are acquired. Children begin to learn about classifications and temporal relations, but still have problems with logical forms, such as negations. They become much more capable of perceiving underlying reality, despite superficial appearance (Flavell, 1985), but still may be very literal in interpreting words.

Consistent with Piaget's early view that young children have problems with logical negations and abstract thought, Holaday and Turner-Henson (1989) found that children in middle childhood have difficulties with "vague" words because they tend to interpret words literally. For example, offering vague quantifiers in questions about the frequency of behavior produces difficulties for children because they need clear definitions, especially in early middle childhood (7 to 10). For this age group, simple yes/no questions about doing something are better understood. Negatively formulated questions make the intended meaning ambiguous for children (as they do for adults) and should always be avoided in children's questionnaires. Younger children in middle childhood have particular difficulty with negatively phrased questions, while older children and adults experience less difficulty (Benson and Hocevar, 1985; Borgers and Hox, 2001; De Leeuw and Otter, 1995). To understand what is required, a child should also grasp the intended meaning of a question. As a result of their literal interpretation, the distance between the intended meaning and the literal meaning of words can cause serious problems for children in middle childhood. This is even more pronounced when *depersonalized* or indirect questions are used (Scott, 1997). A clear illustration is the observation made by Scott et al. (1995) during pretesting that in reaction to questions using the term "people my age," some children tried to guess the age of the interviewer before answering!

Another important factor for participating in questionnaire research is memory. In middle childhood the variety and effectiveness of memory strategies increase. Many studies have shown clear increases with age of the amount of information that can be kept and manipulated in working (short-term) memory (Swanson, 1999). Around age 10 to 11, the memory capacity of children is at the same level as adults (Cole and Loftus, 1987) and the constructive processes used by children seem to function much like those of adults (Kail, 1990). When questions are clear and concrete about the here and now, even young children (7 to 10) are able to give informative responses (Amato and Ochiltree, 1987). The still developing memory capacity also has consequences for the number and order of response alternatives. A limited number of response categories gives better results (Hershey and Hill, 1976). Holaday and Turner-Henson (1989) advise not more than three before the age of 10, but even with older children more than five is not advisable (Borgers et al., 2004). Scott et al. (1995) found good results using graphical representations (e.g., smiley faces) as response categories.

Retrospective questions pose extra problems for young children because of their still developing memory capacities. If the question is immediately recognizable and concerns salient and meaningful experiences (e.g., class outing, visit to pediatrician), even children in early middle childhood (7 to 10) can answer correctly, as their memory for salient issues is remarkable (Brainerd and Ornstein, 1991). However, several studies have shown that unreliable responses appear if these children are not involved or interested in the subject (Holaday and Turner-Henson, 1989; Vaillancourt, 1973). Younger children in particular are prone to construct scripts or event representations of familiar routines if they do not clearly recollect atypical events (Brainerd and Ornstein, 1991; Ceci and Bruck, 1993) or when more complex questions are asked (De Leeuw and Otter, 1995). Furthermore, there are developmental differences in reality monitoring, and under certain circumstances young children (early middle childhood) have more difficulty distinguishing between imagined events and those actually experienced (Ceci and Bruck, 1993; Johnson and Foley, 1984). This is corroborated by Saywitz (1987), who found that 8- and 9-year-olds tended to have less complete recall and more embellishments than did 11- to 12-year-olds.

Provided that extreme care is taken, diary-type methods can be used. The diary method is minimally demanding in terms of cognitive processes and memory, and uses the "here and now" type of question, which is especially appropriate for children (Amato and Ochiltree, 1987). Otter (1993) showed that using the diary method to measure 9-year-old children's leisure-time reading yielded good response quality, produced reliable and valid data responses, and was superior to self-administered paper-and-pencil questionnaires. Structured diaries were also used successfully to collect information about peer interactions of children in their final year at primary school (Ralph et al., 1997).

In addition to an age-related increase in working capacity, an age-related increase of processing speed has also been established. Kail (1993) found a steady decrease of reaction time and increase in processing speed with age; on six different tests, children of 8 and 9 took twice as long as adults, and children of 10 and 11 took 1.5 times as long. By age 17, they performed almost as fast as adults (Gray, 2002). Holaday and Turner-Henson (1989) therefore advise giving children more time to answer survey questions. One way to do this is to use longer introductions to a question. That has a positive effect on response quality, as shown by Borgers and Hox's (2002) finding that the number of words in introductions to questions was positively related to the reliability of children's responses.

Suggestibility has been a topic of much debate in the field of children's testimony in the past 20 years. For an overview, see Ceci and Bruck (1993) and the special issue of *Law and Contemporary Problems* (2002). There appear to be two aspects of suggestibility relevant for survey research (Bob Belli, personal communication): one is suggestibility resulting from cognitive factors, including potential alteration of memories for a past event, as discussed above. The other aspect is suggestibility that results from social and motivational factors, such as seeking to please the interviewer. According to Maccoby and Maccoby (1954), children as old as 8 years will assume that an adult knows everything already. In addition, they are afraid to say something wrong or foolish, especially in a situation that resembles school (Delfos, 2000). As a consequence, young children may react to the demand characteristics of the interview situation by responding in socially desirable ways (La Greca, 1990), or fall back on other response strategies, such as yea-saying (Maccoby and Maccoby, 1954). During middle childhood the structure of self-concept changes, and in late middle childhood (10 to 12) children start comparing themselves to others, and from approximately 10 years on, the effect of peers will be more present (Kohlberg and Puka, 1994). Furthermore, they become aware of the possibilities of putting on a facade and deceiving others intentionally (Selman, 1980). This is clearly illustrated by several methodological studies on children as respondents. Borgers and Hox (2001) reanalyzed questionnaire data from five studies and found that on sensitive questions the younger children had less item nonresponse than older children, while on nonsensitive issues this was reversed, indicating that older children prefer avoiding a socially undesirable answer. Van Hattum and De Leeuw (1999) found that a more private setting (CASI) resulted in fewer social desirable answers for children in late middle childhood.

20.2.2 Adolescence (12 to 18)

After the age of 11, children enter the stage of formal operations (Piaget, 1929). In this stage in early adolescence (roughly 12 to 16 years of age), cognitive functioning is well developed, including formal thinking, negations, and logic. There is a shift in emphasis from the real to the possible, from what is to what might be, and the young adolescent can manipulate ideas about hypothetical situations (Conger and Galambos, 1996).

At the beginning of adolescence memory capacity is fully developed and the constructive processes function much like those of adults. During adolescence memory processing increases rapidly, and by the age of 16 it approaches adult speed (Kail, 1993). Also, social skills are further developed. Selman (1980) calls this the stage of social and conventional system role taking (roughly 12 to 16), in which the young adolescent attempts to understand another person's perspective by comparing it to that of the social system in which he or she operates. Young adolescents in this age group are context sensitive and may have their own norms. After the age of 12, peers become more and more important, and numerous studies have shown that conformity to peers and peer pressure increases dramatically in early adolescence (Gray, 2002).

From 16 years onward, adolescents can be regarded as adults with respect to cognitive development and information processing. But resistance to peer pressure is very low and older adolescents have their own group norms and social norms. The social context of the survey (e.g., classroom, presence of siblings or friends, type and age of interviewer) can be important, especially in interaction with special topics (e.g., health, social networks). For example, in a drug survey among U.S. high school students, the more private data collection method worked best, resulting in more openness and increased reporting. Even the physical proximity (measured physical distance) of other students influenced the openness of answers (Beebe et al., 1998).

20.2.3 Summary

In surveying children, language ability is an important issue for the comprehension of questions. *Comprehension* is the first step in the question–answer process that has to be checked in pretesting questionnaires (cf. Tourangeau and Rasisnski, 1988). As reading and language skills are still developing in middle childhood (7 to 12), the understanding of words has to be checked very carefully for this group. Extra attention should be paid to complexity of wording, negations, and logical operators. As children can be very literal, depersonalized or indirect questions should be checked very carefully.

Memory and processing time is a second important issue. In middle childhood (7 to 12) both memory capacity and speed are still developing. Therefore, complexity of the question and number of response categories should be examined carefully in pretests. Retrospective questions may pose extra problems, and young children are prone to construct scripts of familiar routines if they do not clearly recollect events. In early adolescence (12 to 16) memory capacity is full grown, but memory speed is not. Even in this older age group, ample time for answering questions should be allowed.

In younger children, *suggestibility* is an important item. In early middle childhood (7 to 10), children have a tendency to please and are afraid of doing something wrong. This may result in more satisficing strategies and an inclination toward social desirability. In late middle childhood (10 to 12) children become less suggestible, but start to compare themselves with others. From the age of 12, peers become increasingly important, making adolescents increasingly sensitive to peer pressure and group norms. Sensitivity of topic and privacy of interview situation become important.

20.3 PRETESTING METHODS FOR QUESTIONNAIRES TO BE ADMINISTERED TO CHILDREN AND ADOLESCENTS

20.3.1 Setting the Stage: Survey Design Decisions

Designing and conducting quality surveys requires a careful decision process (e.g., Czaja and Blair, 1996; Lyberg et al., 1997). Designing surveys for children and adolescents is no exception; however, with young respondents some design issues are of extreme importance and warrant extra attention. These include the question of proxy reporting, mode issues, question wording, and consent.

Self-Report versus Proxy Before the age of 7, direct questionnaire research of children is not feasible, and one has to fall back on proxy interviews or on other forms of indirect data collection. Children younger than 7 do not appear to have sufficient meta memory skills to be questioned effectively and systematically (Memon et al., 1996) and experience severe problems in understanding more than very simple concepts (Riley et al., 2001). From the age of 7 on, children can be surveyed with structured questionnaires, and the older the child, the more reliable the answers (Borgers, 2003; Zill, 2001). Zill (2001) advises using an informed parent as informant for health-related issues until adolescence, and collecting information directly from a child on topics for which the child is the best informant. These include subjective phenomena, such as feelings, pain, but also questions on peer influence and peer behavior, and general questions in areas outside the scope of parents' knowledge. The latter is well illustrated by Blair (2000), who compared different protocols on children's food intake and checked these with validating information obtained through observation. Children aged 6 to 11 provided better information than their parents did. The main reason for the discrepancy was the faulty assumption by parents that the children had eaten the food taken with them to school (Blair, 2000). In general, the decision to rely on self or proxy reporting is made a priori on theoretical or practical grounds and differs from country to country and from topic to topic.[1] However, a pretest could provide useful data to guide this decision process. The study reported by Blair (2000) is a good example of this procedure.

Survey Mode Data on very young children are usually collected through observational and assessment studies performed by specially trained interviewers and through interviews with caretakers (Borgers et al., 1999; Zill, 2001). From 4 to 6 years of age, children can be interviewed, but not easily. The interview resembles a qualitative open interview with a topic list, the form is play and talk, and much attention should be given to nonverbal communication and communication of the rules and expectations (Delfos, 2000). Interviewing such young children is a special skill, which is outside the general frame of survey research. However, special handbooks on this topic have been published for counselors, social workers, and law officers. Although these books often focus on very sensitive topics such as sexual abuse, they give guidelines that are extremely useful for

[1]Population definitions for general surveys differ and start at either 18 or 15 years: Nordic countries such as Finland and Sweden have 18 as the lower age limit in official statistics, while the United Kingdom and the Netherlands use 15. In labor force surveys in Europe and the United States, persons 15 years and over are eligible as respondents. The International Social Survey (ISSP) uses 18 years as the lower limit; the European Social Survey (ESS) uses 15. For special surveys children as young as 10 (e.g., Level of Living in Sweden) or 12 (e.g., Crime Victimization Survey and Survey of Program Dynamics, United States) are eligible in official statistics, but permission is needed.

any interview with very young children (see, e.g., Aldridge and Wood, 1998, and Wilson and Powell, 2001). Other forms of special data collection techniques with young children are playing assessment, drawings, story completion, and puzzle tasks (for a description, see Greig and Taylor, 1999). Because at such a young age the child usually is not able to give detailed information on general background characteristics and facts about family, health, and schooling, caretakers are interviewed as proxies.

Starting at the age of 7, structured questionnaires can be used either during a survey interview or through self-completion. Which particular mode is chosen depends on design constraints, such as research topic and budget, and on the literacy of the intended population. From the age of 7 to 8 years old, educational researchers start to use simple self-administered questionnaires in the classroom. When literacy is a problem, a combination of methods is often used, with an instructor reading the questions aloud and the pupils recording their responses on a self-administered form (Borgers et al., 2000). Also, in individual or household surveys, a combination of methods can be used when asking sensitive questions of young respondents. For instance, Scott (1997) used a combination of a Walkman with prerecorded questions on tape and an anonymous self-administered questionnaire. If the budget allows, computer-assisted self-administrative methods, such as CASI or Audio-CASI, have advantages both in school surveys and in household surveys of young respondents (Hallfors et al., 2000). Children and adolescents are good respondents in computer-assisted surveys, and even ordinary schoolchildren as young as 8 years can successfully complete electronic questionnaires and enjoy the process (Van Hattum and De Leeuw, 1999).

A pretest can provide useful information to guide the mode decision. A good example is the study of Helweg-Larsen and Larsen (2001, 2002), who observed both standard mainstream and special education students 15 to 16 years old while they completed a pilot version of a Danish health survey. They found that the special education students took longer and read at such a slow rate that they lost grasp of what had just been asked in the text. Based on these observations and subsequent focus groups, the researchers decided to use Audio-CASI technology for their main study.

Consent One of the strictest codes for doing research with human subjects is the *Declaration of Helsinki* of the World Medical Association, set out to provide moral, ethical, and legal principles to biomedical researchers. Recent amendments of this declaration now include the issue of children and informed consent (Greig and Taylor, 1999; University of Essex, 2002). It states that "... when a subject is a minor, permission from the responsible relative replaces that of the subject in accordance with national legislation. Whenever the minor child is in fact able to give a consent, the minor's consent must be obtained in addition to the consent of the minor's legal guardian" (World Medical Association Declaration of Helsinki, paras. I.9, 11). Esomar, the world association for research professionals in opinion polling and market research, states in its guidelines that first of all, a researcher should conform to any relevant definitions in any national code of conduct and/or in national legislation, and second, that in the case of children under 14, explicit permission should be asked of a parent, guardian, or other person to whom the parent has conferred responsibility (Esomar, 1999). Legislation may vary from country to country regarding the age at which children can legally give their consent. For example, in the United Kingdom it is 16 (University of Essex, 2002), and as a consequence the British Market Research Society prescribes that consent of a parent or responsible adult must be obtained with children under 16 (Market Research Society, 2003). In Sweden, permission from parents is required until 18, even for social surveys such as the ESS (S. Svallfors, personal communication).

Still, permission of a parent or guardian is not enough. The declaration of Helsinki prescribes that the minor's consent must be given, too, if the minor is able to do this. Professional research organizations such as the Society for Research in Child Development also require that researchers inform the child about the study and obtain permission of the child in addition to the consent of the legal guardian (Goodwin, 2002). This implies that the information presented to the child should be given in clear language and at a level the child can understand. To verify this, a cognitive pretest of the wording and phrasing of the consent statement should take place for the relevant age groups. Research in this area is scarce, but an exception is the work of Abramowitz et al. (1995), who investigated the capacity of children in middle childhood (7 to 12) to give informed consent to participation in psychological research, using vignette descriptions followed by open interviews. Their main finding was that children could describe the purpose of the studies fairly accurately, and that the child's consent was not influenced by knowing whether parents had given their consent. However, many children had difficulties recounting the potential risks and benefits of the studies.

Wording of Questions and Response Choices When developing and evaluating questionnaires for children, a researcher should start by following the basic rules for general questionnaire construction and evaluation as outlined in handbooks such as Converse and Presser (1986), Dillman (1978, 2000a), Foddy (1996b), and Fowler (1995). These include good advice to use simple words, avoid ambiguity, ask one question at a time, and so on. But one has to do more. Methodological studies on adult populations have shown that adults sometimes experience problems with certain questions, and that question characteristics affect the data quality in surveys (cf. Krosnick and Fabrigar, 1997). Evidence for interaction effects between respondent characteristics and question characteristics has been found by Borgers and Hox (2001), De Leeuw and Otter (1995), Knäuper et al. (1997), and Schwarz et al. (1999). These studies show that the less cognitively sophisticated respondents are more sensitive to more difficult or cognitive demanding questions than the more cognitively sophisticated respondents, resulting in more item nonresponse and less reliable answers for respondents lower in cognitive ability.

With children as respondents, these problems are magnified. In addition, children experience specific problems when responding. Not only their cognitive, but also their communicative and social skills are still developing, and this affects different stages of the question–answer process. Therefore, special care should be given to the construction of questionnaires for children and adolescents. Pretesting of the questionnaire is certainly necessary to examine the adequacy of question wording and response options for different age groups. This is still a new field, and few publications about procedures and results are available. Levine and Huberman (2002) describe how they effectively used cognitive interviewing (think-aloud with probing) with children aged 9 to 14 to test questions from the U.S. National Assessment of Educational Progress. Hess et al. (1998a) describe similar positive experiences with adolescents aged 12 to 17 when pretesting the youth part of the U.S. Survey of Program Dynamics.

Design Decisions and Pretesting When surveying children, many design decisions may be guided or informed by using pretesting methods. A first step is consulting with experts in the field. The next step is evaluating the procedures and questionnaire using cognitive testing methods, using the intended respondents as informants. Cognitive pretests will enable the researcher to discover which wordings or questions are problematic for young respondents and why, thereby suggesting improvements in questionnaires for children. In the next section, well-known cognitive methods for pretesting with adults (e.g., Esposito and Rothgeb, 1997) are reviewed for usability with children. In addition, we discuss how these methods can be optimized for children.

20.3.2 Focus Groups

Different pretest methods have different strengths (Presser and Blair, 1994). The strength of focus groups is the interaction within the group; the participants stimulate each other to discuss topics and explain ideas (Morgan, 1997). As a consequence, a wide range of information can be gathered in a short time; however, this information is not always very detailed. Focus groups are useful for generating ideas and topics for questions, evaluating the data collection procedures planned, and evaluating the acceptability or sensitivity of certain topics, but for a detailed evaluation of the questions, in-depth interviews are more useful (Campanelli, 1997; Snijkers, 2002).

The usefulness of focus groups in the design phase of a survey is well illustrated by Scott et al. (1995), who conducted a series of six focus groups with children aged 11 to 15 in the United Kingdom. The decision to add a Young Person's Survey to the British Household Panel challenged the researchers to develop a way of interviewing children in their homes in privacy. Because of potential literacy problems, the researchers opted for prerecorded Walkman interviews with a paper self-completion response booklet. The goal of the focus groups was to help develop structured questions and to fine-tune the Walkman method. The focus groups took place in a neutral setting, the interviewer's home. Groups were separated by gender and by age groups (11 to 13 and 13 to 15) and lasted about two hours with a snack break at half time. Each focus group started with a general open discussion on health and health-related issues. This served as a warm-up but also provided information on the typical language use and on sensitivity of topics. This was followed by trying out formats for semistructured questions thought suitable for these age groups (e.g., response card with a range of smiley faces). Question formats were presented and discussed in the group. In the last phase of the focus group, Walkmans were handed out together with a short self-completion booklet. According to the researchers, the focus group discussions were very productive for identifying appropriate wordings, question formats, and response options for the development of the Young Person's Questionnaire. The Walkman test showed that children did not experience any technical problems when using a Walkman and provided useful feedback on voice type (Scott et al., 1995; see also Scott, 1997). A subsequent test of the redesigned procedure during the pilot phase of the Young Person's Survey was reported to be very successful (Scott, 1997).

Using focus groups of young persons in the design phase of special surveys may provide useful information, and although it is still in the pioneering phase, its use is growing. Different approaches may be used for different purposes. For instance, Spruyt-Metz (1999) used focus groups to pretest a self-administered questionnaire on health and risk behavior among Dutch high school students aged 12 to 17. She was interested primarily in question interpretation and the meaning of important concepts and used open-interview questioning. Cannell et al. (1992) used focus groups of U.S. adolescents to test the acceptability of health-related sensitive topics (e.g., cigarette smoking). They presented subjects with potential questions, and stimulated group discussion by giving specific probes on the understanding of the question, how one would react, and on whether or not one would answer it, or answer it truthfully.

A rather unorthodox but fruitful application of focus group techniques was employed by Watson et al. (2001) in New Zealand, who used *post pilot focus groups* to evaluate the usability of Multimedia-CASI techniques. Following completion of a questionnaire, students aged 12 to 18 participated in structured focus groups. Each group consisted of six to 10 students of the same gender and took about 40 minutes. Open-ended questions were used to stimulate discussions about available time, use of headphones and computers, but also question the difficulty and emotional burden of the questions. The focus groups revealed two important themes. First, the students were very positive about the multimedia computer interface, especially the audio component. In the eyes of the respondents, the computer made everything easier. The second perceived advantage was privacy. Students appreciated the computer but also emphasized how important it was that nobody else could read the screen.

Focus Groups with Children and Adolescents Compared to general adult focus groups, focus groups for children and adolescents appear to be more structured and more centered around specific tasks. Whether this is inherent for groups with children and young adolescents, or whether this is the result of the specific topics in the studies cited above, is unclear. The researchers do not describe in detail if and how the focus groups were adapted to the younger respondents. However, general publications about interviewing children (e.g., Delfos, 2000; Wilson and Powell, 2001) emphasize the importance of a well-designed protocol for open interview situations and the extreme importance of explaining clearly what is expected of the child. This is also stressed by Morgan et al. (2002), who wrote one of the first methodological articles about focus groups with children.

Although in most countries children and adolescents are acquainted with group discussions from classroom settings, they will not know what a *focus* group is and what its rules are. Therefore, it should be made very clear to them what is expected and also that a focus group is not school or a test situation. Also, during the focus group itself, the participants sometimes need to be reminded of the rules. For instance, Morgan et al. (2002) wrote simple rules on a flip chart in the beginning and left them on display during the entire session. Examples of these rules were: Everyone gets a chance to speak, speak one at a time; you do not have to put up your hand to talk (this is not school). Of course, explaining the rules is important when conducting focus groups with adults, too. But young respondents are still developing the cognitive and social skills for meta-communication (see also Section 20.2) and compared to focus groups for adults, the moderator has to pay more attention to meta-communication.

In general, many issues and good practices for focus groups with adults are common to conducting focus groups with children and with adults; it is a question of translating these good practices to the needs of younger age groups (Morgan et al., 2002). Through the setting and the explicit verbal and nonverbal behavior of the moderator, the researcher has to create a different interaction-stimulating environment for each age group. In the following paragraphs we discuss optimal focus group settings for different age categories, emphasizing the special needs of each group. We will not discuss the general rules for conducting good (adult) focus groups; for a thorough introduction we refer to Morgan (1997) and Stewart and Shamdasani (1990); for a quick overview, see Cheng et al. (1997) and American Statistical Association (1997). However, as certain topics, such as group size and homogeneity, are recurrent methodological issues in focus group setups for developing questionnaires (Bishoping and Dykema, 1999), we will comment explicitly on these topics.

Group Size Young children need more attention than older children, and as a general rule, the younger the participants, the smaller the group should be. For children in early middle childhood (ages 7 to 10) a group size of about five is optimal. To increase motivation and keep the attention of these young children, one moderator should constantly attend to motivating the children and keeping the conversation going. A second moderator will be necessary for general practical assistance in running a group of young children (see also Greig and Taylor, 1999; Morgan et al., 2002). More grown-ups in the room will disrupt the balance of power in the group, and it is advisable to have note takers in a separate room and to videotape the entire session for nonverbal cues and interactions (Annon, 1994).

In late middle childhood and early adolescence (ages 10 to 16) group sizes may range from 5 to 8 (Scott et al., 1995). A second moderator will no longer be needed for practical child-care issues and may be replaced by a note taker or observer. In late adolescence (16 to 18), group size may increase to 8 to 10 participants, only slightly less than in adult groups (cf. Bishoping and Dykema, 1999).

Group Homogeneity Group composition is an important consideration in focus groups. Homogeneity in age with small age bands (e.g., ages 7 and 8, 9 and 10) is recommended (Morgan et al., 2002). In early adolescence this is crucial, as the eldest will in general look down on the youngest, who has just left primary school. Therefore, age homogeneity should be strictly enforced, with the 12- and 13-year-olds separated from the older children (cf. Scott et al., 1995).

Whether or not groups should be homogeneous with respect to gender is age-dependent. Before the age of 10, gender homogeneity is not necessary, but in late middle childhood and early adolescence it is advisable (Greig and Taylor, 1999; Scott et al., 1995). In late adolescence much depends on the topic of the study and on culture. For instance, Spruyt-Metz (1999) varied the composition of focus groups of Dutch adolescents. She used both all-girl and all-boy groups, but also added mixed-gender groups to stimulate discussion. According to Spruyt-Metz (1999), having opposite sex members in the group may reduce acting-out behavior and make the group more task-oriented. Only for the adolescents of Turkish and Moroccan origin were the groups gender homogeneous, because of cultural taboos on discussing many of the topics in the protocol with members of the opposite sex. The findings of Bishoping and Dykema (1999) are helpful in deciding on gender homogeneity for focus groups with late adolescents and young adults (16+). They review extensively the importance of sociopsychological factors in focus groups for adults and conclude that sex segregation has negative effects, especially on disclosure of emotions and personal information, for men, while for women all-female groups enhance their input.

Scott et al. (1995) note that their focus groups were homogeneous in terms of socioeconomic status. But this could be country specific and dependent on the school system and whether or not there are large status differences between schools as there are in the United Kingdom.

For all age groups it is advised to avoid having close friends, or even classmates in one group, as this may have affect group dynamics. It may stimulate concentration lapses in younger children (Morgan et al., 2002) and inhibit open interactions (Scott et al., 1995). Especially in adolescence, when peer pressure is heavy (Gray, 2002), one should avoid selecting children from the same peer groups or school classes and preferably mix children from two or more schools.

Session Duration The younger the child, the shorter the attention span. In early middle childhood (7 to 10) the attention span is still limited, and this has consequences for the scheduling of a session. One should have short periods of discussion (around 20 minutes) alternated with play activities (Delfos,

2000). Morgan et al. (2002) used two 20-minute sessions separated by a short refreshment break; they also advise keeping the (tape) recorder running during the breaks to catch relevant remarks.

According to Delfos (2000), children 10 to 12 can have longer periods of discussion (30 to 45 minutes), alternated with refreshment breaks. Scott et al. (1995), who studied children aged 11 to 16, used focus groups that lasted approximately two hours. Although the attention span of these older children is longer, the moderator should carefully monitor the process and stimulate participation. Group discussion can be alternated with other activities, such as making lists of important points (Morgan et al., 2002), showing pictures, or having children handle survey material (Scott et al., 1995). Adolescents can handle discussion periods of one hour, after which a refreshment break is definitely needed. This is as long as most adult focus groups. Still, one has to remember that young adolescents are not adults. They need more time to think, as their mental processing speed is still lower (cf. Kail, 1993).

General Setting Notably with the younger children (7 to 10), the setting should be chosen with careful consideration of the demand characteristics of the room. The moderators should always be on the same eye level as the children (Annon, 1994; Delfos, 2000). Annon (1994) also notes that when a one-way mirror is used, it should not be on the same level as the children, as it may distract them. In setting the scene, it is also important to pay attention to the power balance. Morgan et al. (2002) explicitly chose an informal arrangement, in which all participants sat on soft mats on the floor in the middle of a pleasant light room in a community center. Furthermore, to reduce the hierarchical adult–child relationship, all used first names and all had colorful buttons with their names.

To promote group cohesion with these young children and to clearly communicate that interaction and participation are the goal of the session, group games are advised as warming up. Morgan et al. (2002) used a ball game to introduce the group members to each other; a ball was thrown to a group member, who had to state his or her name, favorite food, animal, and so on, and then throw the ball to another participant. This is also very useful to assess the cognitive and verbal development of the children and to tune into the child's language (Cares, 1999).

Similarly with children in late middle childhood (10 to 12), the setting should be chosen with consideration of the demand characteristics of the room, and the moderators should be on the same level as the children. However, one should avoid treating children this age as little ones, as they feel quite superior to the younger children in primary school. Warm-ups and informal introductions remain extremely important and age-related games play an important role in this. When moderators and children draw special name labels together, this helps to get acquainted and to reduce the authority imbalance (Hill et al., 1996). Nonverbal communication is an important part of controlling the group process and at regular times and after each subtopic, the moderator has to structure the session by summarizing and asking for additions from the children (Delfos, 2000).

For adolescents it is extremely important that the setting itself has no relationship at all with school or youth centers. It should be new and neutral territory for all, so that none of the adolescents is in a power advantage. Especially for the younger adolescents (12 to 16), careful monitoring of the group process is recommended, and shy adolescents should be encouraged. One way to do this is alternating the verbal discussions with other tasks. For instance, let each one individually write down what he or she thinks is important. The moderator can ask the quieter group members what they have written and so reduce dominations of the group by the more boisterous ones. Compared to adult focus groups, more time should be dedicated to warming up and acquainting the members with the rules and goals of a focus group. All focus groups are vulnerable to group pressure and conformity effects, but adolescents are more sensitive to peer pressure than younger children and adults. With adolescents, moderators have to be even more attentive to group processes, and give feedback when necessary. Finally, the moderators should realize that they themselves are *not* young (even if they are 22) and that fashions, music, and fads change very quickly (personal communication from M. Isacson) Moderators should never try to be one of the group, as in participant observation, and should never transcend their older adult identity (cf. Morgan et al., 2002).

20.3.3 In-Depth or Cognitive Interviews for Testing Questionnaires

Cognitive interviewing in the context of pretesting questionnaires is a form of in-depth interviewing used to find out what goes on in the head of a respondent when answering questions. The cognitive interview in questionnaire testing should not be confused with the cognitive interview in the context of law and child-witness literature. The cognitive interview of a child witness is a special structured interview taking the respondent step by step back to the event, and is explicitly designed to get more reliable reports on past events (e.g., Memon and Koehnken, 1992; Memon et al., 1996). To pretest questionnaires thoroughly, cognitive interviews are used to investigate the total question–answer process and discover sources of confusion and misunderstanding. This method is widely used as a pretest method to investigate the understanding of questions by *adults* and has proven to be successful in identifying potential problems in questions and in suggesting solutions for these problems (Campanelli, 1997; Presser and Blair, 1994; Willis et al., 1999b).

Potentially, cognitive pretesting of questions could also be a successful method with children and adolescents. It relies heavily on think-aloud procedures, which come very naturally to children. Young children often talk aloud in a noncommunicative manner during play or when performing tasks. According to the Russian developmental psychologists Vygotsky (Gray, 2002), this is a natural and necessary phase in the acquisition and internalization of language and verbal thought. Furthermore, think-aloud procedures are often used as an educational tool in primary and secondary schools, especially in teaching mathematics (Kraemer, 2002; P. Lynn, personal communication). Strangely enough, one of the first studies using cognitive testing procedures with young respondents (age 10 to 21) reported that think-aloud procedures were problematic and that most teenage respondents lacked the ability or the motivation to articulate their thought processes spontaneously (Stussman et al., 1993). Blair (2000) also reports problems using think-aloud protocols with young children (6 to 11). However, both studies gave standard think-aloud instructions for adults, and the procedures were not adapted for younger respondents. Stussman et al. (1993) suggest that traditional cognitive interviewing techniques need to be modified for the young, with more attention to nonverbal communications and more probes. In addition, Blair (2000) comments that more introduction and explanation are likely to be necessary for children to be good respondents.

Think-aloud procedures with young respondents can work well, as Hess et al. (1998b; see also Zukerberg and Hess, 1996) showed. They conducted cognitive interviews with adolescents aged 12 to 17 to evaluate question understanding, task difficulty, and question sensitivity for the youth questionnaire in the U.S. Survey of Program Dynamics. The researchers developed a detailed protocol beforehand that included probing questions. They report that during the interviews they found a greater need to probe than they typically do during cognitive interviews with adult respondents. This corroborates the conjecture of Stussman et al. (1993) that the young need more extensive probing.

Levine and Huberman (2002) also used think-aloud techniques successfully to test questions on background information from the U.S. National Assessment of Educational Progress questionnaires with children aged 9 and 13 to 14. Levine and Huberman (2002) developed a detailed protocol with special probes for the cognitive interviews, and interviewers were trained to use them. The young respondents were given a special instruction and explanation of the procedure. Each think-aloud was preceded by having the respondent read the specific question aloud. This facilitated the detection of language and comprehension problems and served as a warm-up for the think-aloud. During the think-aloud the young respondents were continuously encouraged in a neutral manner, and probes were used frequently.

Unique in the Levine and Huberman study is that validating information was available based on responses by parents and teachers, which enabled comparison of revised questions with original questions. It is encouraging that Levine and Huberman (2002) showed that revised questions had a lower error rate.

Cognitive Test Interviews with Children and Adolescents Using cognitive interviews for pretesting of children's questionnaires is possible and can result in worthwhile information, provided that the procedures are adapted to the special needs of children and adolescents. In the following paragraphs we discuss necessary adaptations to the general setup and protocol for in-depth interviews with adults. To accommodate different age groups, adaptations have to be made to all phases: arrival, introduction, start of the interview, interview proper, and ending (cf. Snijkers, 2002).

Arrival In early and late middle childhood (7 to 12) special attention has to be paid to this stage. The child will be accompanied be a parent, caretaker, or teacher, and both child and caretaker have to be welcomed and introduced to the interviewer, and time has to be taken to make the young child feel at ease. With children, the arrival stage includes many aspects of the introductory stage, too. Confidentiality and background information (why is the study done, etc.) have to be explained briefly to both parent and child. Therefore, some of the general procedures that with adult respondents are discussed in the introduction of the interview, are now introduced at the arrival stage when the parent or caretaker is still there [e.g., explaining videotaping, obtaining permission to record the session (both parent and child should give permission)]. In early adolescence, more often than not a caretaker will still accompany a child, and as a consequence, the arrival will take more time. With older adolescents, the situation more resembles the usual situation with adults. The arrival takes less time, with confidentiality and consent discussed during the introduction. However, in many countries, consent of a parent or caretaker is needed even for older adolescents (16 to 18) and should be obtained before the session.

Introduction For a successful cognitive laboratory interview the introduction is crucial. In general, one has to take more time to explain what the rules are and what is expected than with adults. The importance of this is illustrated by Presser et al. (1993), who asked youngsters preinterview questions on what a survey was. They found that neither younger (6 to 8), nor older children (9 to 11) had a clear idea what a survey was and what the goals and rules of a survey were. More explanation of question asking and answering is needed with children than with adults.

Starting the Interview Because the situation is completely new, the interviewer has to explain the procedures carefully, give clear examples, and practice the required tasks before the interview starts. For instance, one can rehearse thinking aloud using simple age-appropriate examples (e.g., a simple arithmetic task, a simple puzzle, sorting objects, etc). Extra time should be reserved for explanation and practice exercises, as part of a short training-phase before the real interview starts.

The Interview Itself In general, the same rules of thumb for duration are valid as for focus groups. However, the estimates given for focus groups are the maximum possible. Because of the lively nature and potential for interaction, focus groups are in general more relaxed and demand less concentration than does an individual in-depth interview. Especially with the youngest age group, one has to watch the child carefully and react to drops in attention.

Different interviewing techniques for different age groups are advised. Think-aloud is very natural for young children (7 to 10), who often still read aloud. Levine and Huberman (2002) explicitly asked 9- and 13-year-olds to start by reading the question aloud. Not only did this stimulate them to think aloud, it also provided clues for further probing. For example, when a child could not read or pronounce a word correctly, this could indicate a comprehension problem.

During the think-aloud the interviewer has to be continuously alert, reinforce the child, and start up the process if the child stops for a moment (ask "Why do you stop"; if tired/not concentrating, suggest a short break). Both Hess et al. (1998a) and Stussman et al. (1993) recommend that the interviewer probe more frequently than with adults, and it is advisable to prepare a probing protocol and train interviewers to use frequent probes (Levine and Huberman, 2002).

It is important to make sure that the child feels completely at ease. Although thinking aloud is quite natural for young children, they will *not* perform well when they feel uncomfortable or watched. Young children can be very open in a situation they trust, but become completely shy and introverted when they find themselves in an unknown situation (Scott, 1997). In some cases it is therefore better to have a parent or caretaker present at the interview. Only when a young child feels comfortable will he or she perform well.

Paraphrasing is a technique that should not be used with younger respondents. Especially in young middle childhood (7 to 10), paraphrasing a question will not work, since children this age tend to repeat a question literally.

Late adolescents (16 to 18) may feel very embarrassed when asked to do a think-aloud. But paraphrasing combined with direct probes (e.g., "What does this word mean?" "What do you think it means?") may give good results in this age group. For adolescents, it is important for the interviewer to reinforce them and reassure them that this is not a school test and that not the adolescent but the questionnaire is being evaluated! Adolescents often lack confidence and may be unsure about themselves and their performance. Reassurance and frequent reinforcement is far more important for this group than for adults (cf. Hess et al., 1998b).

20.3.4 Auxiliary Methods

Observation Monitoring of standardized interviews and self-administered questionnaire sessions is a relatively quick method that can provide useful additional information during field tests and pilot studies. Coding schedules developed for interviewing adults (e.g., Fowler and Cannell, 1996; Lessler and Forsyth, 1996; Oksenberg et al., 1991) are mainly for *verbal* behavior: for example, "interviewer reads verbatim," "interviewer deviates slightly," "respondent interrupts," "respondent asks clarification." Coding schedules for children should have more emphasis on *nonverbal* behavior, since children, especially younger children in middle childhood, will have more motor (movement) behavior. An example is provided by Presser et al. (1993), who developed and tested three interview protocols to measure daily food intake for children aged 6 to 11. They videotaped all test sessions and applied an extensive coding scheme with specific nonverbal codes for the child (e.g., head shaking, nodding, smiling) added to the standard verbal coding scheme of Oksenberg et al. (1991) for interviewer behavior. Presser et al. (1993) found that in the younger group, the interviewer deviated twice as much from verbatim reading of the questions as in the older group, and used more probes, indicating more problems in the question–answer process. They also found that younger children smiled about three times as much as older children. This could indicate that young children will smile or laugh to hide that they do not understand a question. However, the fact that it is possible to code overt children's nonverbal behavior reliably does not mean that the interpretation is necessarily clear. In the field of child interviews, there is little work on the interpretation of coded behaviors, and more research and development is necessary. The newly emerging field of usability testing with children (Hanna et al., 1997) is facing similar problems, forcing researchers to acquire more methodological knowledge about children as subjects (Markopoulos and Bekker, 2002).

There are few examples of systematic observation of children during pilot testing of self-administered questionnaires. Researchers generally only note down the time it takes to fill in a test or questionnaire, to acquire data to improve planning the major fieldwork. An exception is the work of Helweg-Larsen and Larsen (2001, 2002), who observed both standard mainstream and special education students, aged 15 to 16, while they completed a pilot version of a Danish health survey. The special education students who had learning problems took longer and read at such a slow rate that they lost grasp of what had just been asked in the text. It became apparent that students in special education, but also a number of mainstream students, experienced literacy problems.

Debriefing Interviewer and respondent debriefing studies have proved to be useful for studying response errors in survey data (e.g., Campanelli et al., 1991), and the observations of trained interviewers may provide worthwhile information on difficulties encountered in interviews with children. Until now this promising area has not been explored.

In a comparison of computer-assisted self-administered questionnaires with paper-and-pencil questionnaires in Dutch primary schools, Van Hattum and De Leeuw (1999) used a form of teacher debriefing in which teachers were asked about their experiences, the experiences of their pupils, and problems encountered during data collection. According to the teachers, asking sensitive questions (e.g., about bullying) by computer was less stressful than paper questionnaires for their young pupils (aged 9 to 12). Teachers also reported the problems their pupils had understanding several questions (e.g., meaning of certain words) but did not report any problems with the computer itself.

There are several examples of the use of respondent debriefing in surveys of children. Helweg-Larsen and Larsen (2001, 2002) in Denmark, and Watson et al. (2001) in New Zealand, added special debriefing questions at the end of computer-assisted questionnaires for adolescents. Topics included the computer interface, as well as privacy issues. Hess et al. (1998a) included debriefing questions in a field test of the youth questionnaire of the U.S. Census Survey of Program Dynamics. Like Scott (1997), they used a combination of Walkman and self-administered questionnaire, and at their debriefing focused on reactions to the audiocassette and privacy issues. Based on the debriefing results, the procedures were slightly modified to reduce repetition of the answer categories on the taped interview.

20.4 CONCLUSIONS

We discussed above various methods for pretesting questionnaires for children and adolescents. For the clarity of this chapter, we discussed each method separately, but this does not mean that in survey practice only one method should be used. In our opinion it is not either–or; the methods discussed in this chapter complement and reinforce each other and should be used in combination. This is clearly illustrated in the study of Presser et al. (1993; see also Blair, 2000), who used a variety of methods when developing interview protocols for food intake aimed at children aged 6 to 11. Besides think-aloud pretests, they compared different interview protocols and videotaped these for behavior coding. The same videotapes were also used as starting points in debriefing interviews. Data from all sources were combined to devise a new interview protocol for food intake. Another good example is the study by Reynes (2002; see also Reynes and Lorant, 2001), who used a combination of pretest methods when adapting the Buss and Perry Aggression Questionnaire to young French children aged 8 to 10. Experts were used to check the simplified vocabulary and sentence structure; the questionnaire was then pretested on 8-year-olds to make sure that all questions were understood; and in the final phase a pilot study was done on a large sample of 8- to 10-year-olds ($n = 500$) to check psychometric properties such as the reliability of the aggression scale. Hess et al. (1998a) used a similar procedure and combined the results of cognitive think-aloud interviews with those of a full field pretest to investigate potential problems in a self-administered questionnaire of adolescents (12 to 17) as part of the U.S. Survey of Program Dynamics.

Watson et al. (2001) and Helweg-Larsen and Larsen (2001, 2002) followed a slightly different procedure when pretesting health surveys for adolescents in New Zealand (12 to 18) and Denmark (15 to 16): After having completed the questionnaire in a pilot study, the respondent immediately took part in postpilot focus groups to investigate their experiences of the survey. Helweg-Larsen and Larsen (2001, 2002) also used systematic observation during the pilot.

Usually cognitive laboratory methods are used in a *pretest*, which is followed by a pilot or field test and the final study, but cognitive laboratory methods can also be useful as a *posttest* to gain insight into problems encountered during data collection or data analysis. Questionnaire test methods can be extremely useful after a survey is completed and when unexpected results are found, or in ongoing or longitudinal surveys. The goal of the questionnaire posttests is to identify sources of measurement errors encountered in the data. A prime example is the study of Jakwerth et al. (1999), who used standardized in-depth interviews to investigate reasons for the high item nonresponse rates reported over the years for the U.S. National Assessment of Educational Progress in achievement tests, for eighth graders (approximately 13 to 14 years).

Although in most research disciplines the instrumentation is checked, the methods vary. For example, in test development for educational research, an instrumentation phase is always included in which psychometric reliability and validity of the test are estimated on a large sample, while a cognitive pretest of the questionnaire is rarely employed. In survey research, cognitive pretests are being used increasingly and pave the way for the costly pilot phase. In our opinion a cognitive pretest should always be part of the test design stage. It is very cost-efficient and gives a thorough insight into what may be wrong with questions and test items and suggests ways to improve them.

ACKNOWLEDGMENTS

The views expressed in this chapter are those of the authors and do not necessarily reflect the policies of Statistics Netherlands. We thank all those dedicated researchers worldwide who shared their experiences with us and sent us examples of their questionnaires and procedures. We also thank Bob Belli, Sandra Berry, Janet Harkness, Kathy Heckscher, Joop Hox, Bärbel Knäuper, Betsy Martin, Jean Martin, and Stanley Presser for their suggestions and comments.

CHAPTER 21

Developing and Evaluating Cross-National Survey Instruments

Tom W. Smith
NORC

21.1 INTRODUCTION

As challenging as it is to develop questions, scales, and entire questionnaires within a monocultural context, the task becomes considerably more difficult in multicultural settings. Overlayering the standard need to create reliable and valid measures are the complications inherent in cross-cultural and cross-national differences in language, customs, and structure. Only by dealing with these challenges on top of the usual instrument-design issues can scientifically credible cross-national survey instruments emerge.

The basic goal of cross-national survey research is to construct questionnaires that are functionally equivalent across populations.[1] Questions need not only be valid, but must have comparable validity across nations. But the very differences in language, culture, and structure that make cross-national research so analytically valuable, seriously hinder achieving measurement equivalency.

Although the difficulty of establishing comparability is widely acknowledged, the challenge is more often ignored than met. A review of cross-national research (Bollen et al., 1993) found that "[m]ajor measurement problems are expected in macrocomparative research, but if one were to judge by current practices, one might be led to a different conclusion. Issues surrounding measurement are usually overlooked. ... Roughly three quarters of the books and articles do not

[1]On different types of equivalence, see Johnson (1998) and Knoop (1979).

Methods for Testing and Evaluating Survey Questionnaires, Edited by Stanley Presser, Jennifer M. Rothgeb, Mick P. Couper, Judith T. Lessler, Elizabeth Martin, Jean Martin, and Eleanor Singer
ISBN 0-471-45841-4

consider alternative measures or multiple indicators of constructs, whether measures are equally valid in the different countries, or the reliability of measures."

Considering the value of cross-national research, the importance of obtaining comparable measurements, and the frequent failure to take measurement seriously, there is an obvious need for improvement. This chapter contributes toward that goal by discussing (1) the development of equivalent questions in surveys, focusing on the question-asking and answer-recording parts; (2) response effects that contribute to measurement error in general and variable error structures across nations (e.g., social desirability, acquiescence bias, extreme response styles, don't knows and nonattitudes, neutral/middle options, question order, and mode); and (3) steps to enhance validity and comparability in cross-national surveys, including the form of source questions, translation procedures, and item development and pretesting.

21.2 QUESTION WORDINGS

Question wordings and their translation are "the weakest links" in achieving cross-national equivalence (Kumata and Schramm, 1956, p. 229). Questions have two parts: the body of the item presenting the substance and stimulus and the response scale recording the answers.

21.2.1 Question-Asking Part

First, there is the substantive meaning and conceptual focus of the question. The challenge is to achieve functional equivalence across versions of the questionnaire. One needs an optimal translation, but as important and difficult as this is, it is only part of the challenge. An optimal translation may not produce equivalency.

Even cognates between fairly closely related languages can differ substantially. For example, the concept *equality/egalité* is understood differently in the United States, English-speaking Canada, and French-speaking Canada (Cloutier, 1976). Similarly, for Spanish-speaking immigrants in the United States, *educación* includes social skills of proper behavior that are essentially missing from the more academic meaning of *education* in English (Greenfield, 1997).

A related problem occurs when a concept is easily represented by a word in one language and no word corresponds in another language. For example, a study of Turkish peasants (Frey, 1963, p. 348) concluded that "there was no nationally understood word, familiar to all peasants, for such concepts as 'problem,' 'prestige,' and 'loyalty.' . . ." Similarly, the Japanese concept of "giri" [having to do with duty, honor, and social obligation] has no "linguistic, operational, or conceptual corollary in Western cultures" (Sasaki, 1995, p. 102).

Besides language incompatibility, differences in conditions and structures hinder the achievement of functional equivalence. First, situational differences can interact with words that may have equivalent literal meanings but different social implications. As Bollen et al. (1993, p. 345) note: "Consider the young woman who has reached her family size goal. In the United States, if you ask such a woman whether it would be a problem if she were to get pregnant, she is likely to say yes. In Costa Rica, she may say no. This is because in Costa Rica, such a question may be perceived as a veiled inquiry about the likely use of abortion rather than a measure of commitment to a family size goal."

Also, structural differences mean that equivalent objects may not exist or that terms used to describe one object in one country describe something else in another country. For example, the American food-stamp program, which gives qualifying people script to purchase certain food, has no close equivalent in most other countries. In other cases, questions must ask not about the literal translation but about the functionally equivalent object. For example, most questions asking about the U.S. president would inquire about the German chancellor and the Israeli prime minister and not the German or Israeli president.

Variations in conditions and structures mean that the objects one asks about and how one asks about them differs across societies. This applies to behaviors and demographics as well as to attitudinal and psychological measures. For example, a study in Mali added to the standard U.S. occupational classifications of how jobs relate to data, people, and things a fourth dimension on relating to animals (Schooler et al., 1998). Similarly, items about spouses have to allow for multiple mates in Islamic and most African societies.

Basic demographics can be among the least compatible of variables. Some demographics must use country-specific terms for both questions and answers, as in the obvious example of region of residence, which uses country-specific units (e.g. "states" in the United States, "provinces" in Canada, "laender" in Germany), and of course the answers are unique geographic localities. Similarly, voting and party preference questions must refer to country-specific candidates and political parties.

Some demographic questions might be asked in either country-specific or generic, cross-country terms. For example, a generic education question might ask, "How many years of schooling have you completed?" A country-specific approach might ask about the highest degree obtained, the type of school attended, and/or the examination passed. The International Social Survey Program (ISSP) follows the latter course, judging that getting precise country-specific information on education is important. The former produces a simple, superficially equivalent measure but lumps together people educated in completely different educational tracks within a country. But the latter makes it necessary to compare unique country-specific educational categories across nations.

The problems of linguistic and structural equivalence increase the need for multiple indicators. Even with the most careful of translations, it is difficult to compare the distributions of two questions that employ abstract concepts and subjective response categories (Grunert and Muller, 1996; Smith, 1988b). It is doubtful that responses to the query "Are you very happy, pretty happy, or not too happy?" are precisely comparable across languages. Most likely the closest linguistic equivalent to "happy" will differ from the English concept in various ways, perhaps conveying different connotations and tapping related dimensions (e.g., satisfaction), but at a minimum probably expressing a different level of intensity. Similarly, the adjectives "very," "pretty," and "not too" are unlikely to have precise equivalents. Even when the English adjective "very" is consistently (and correctly) translated into the French "tres," it is not known if they cut the underlying happiness continuum at the same point.

To illustrate the extra need for multiple indicators in cross-national research, consider a scheme to compare French and Americans on psychological well-being:

A. A measure of general happiness
B. A measure of overall satisfaction
C. A scale of domain-specific satisfaction items

Franco-American comparisons on any one question or scale are suspect because of possible language ambiguities. Even the multiitem measure of domain-specific satisfaction would be insufficient since all items use the term *satisfaction*, with any nonequivalence compounded across items. Nor would the combination of the domain-specific and overall satisfaction help solve the problem since any disparity in the meaning of "satisfaction" across languages would merely be perpetuated. However, asking about how *happy/heureux* one is adds a question that is distinct from the satisfaction items and avoids correlated linguistic error from repeated terms. Similarly, the addition of the Bradburn affect-balance scale would have this same advantage since it asks about positive and negative emotions using largely different terminology.[2]

If linguistically distinct measures (i.e., items that use different terms, such as "satisfaction" and "happiness") are used, it is possible to reach unambiguous conclusions if the results across items are consistent (e.g., the French leading/trailing the Americans on all measures). With one measure it is impossible to know if observed differences (or nondifferences) are societal or merely linguistic. With two measures a consistent pattern on both items establishes a clear finding, but if the measures disagree, one may be societal and the other linguistic, and there is no basis to identify which is which. Three linguistically distinct measures of the same construct are desirable.[3] If all three agree, one has a clear, robust finding. If two agree and the third shows a different pattern, one has to be more cautious, but there is at least a "preponderance of evidence" toward one substantive interpretation of the cross-national differences. If all three results disagree (positive, negative, and no difference), no firm evidence about cross-national differences exists and further developmental work is needed. A similar approach is called *triangulation* (van de Vijver and Leung, 1997, p. 55; see also Przeworski and Teune, 1966, and Scheuch, 1989). As Jacobson et al. (1960, p. 218) noted: "However

[2]MacIntosh (1998a) argues that the affect-balance scale as used in the World Values Study was not comparable across nations. However, the point here is that the scale would not replicate measurement error associated with the terminology of the other scales.

[3]This does not refer to three single-item measures but to three linguistically distinct items or scales. In this example, the Bradburn scale has 10 items, and domain-specific satisfaction measures usually cover many different areas.

difficult it may be to deal with theoretical issues concerning psychological processes which intervene between observable stimuli and responses in intracultural studies, the cross-cultural research situation magnifies these problems and adds new ones."

21.2.2 Answer-Recording Part

Achieving equivalent response categories is as important as establishing the equivalency of the concepts and substance of questions is. Several solutions have been offered to increase the equivalency between responses to questions in cross-national research.

Nonverbal Scales

Numerical or other nonverbal scales are advocated by some (Fowler, 1993) on the belief that using fewer words increases equivalency. These scales include numerical instruments such as ratio-level magnitude-measurement scales, scalometers, feeling thermometers, and frequency counts. Nonnumerical nonverbal instruments include ladders, stepped mountains, and figures or symbols often used in psychological tests. Numerical scales are assumed to reduce problems by providing a universally understood set of categories with precise and similar meanings (e.g., 1, 2, 3 or 2:1) that do not require language labels for each response category. Similarly, visual questions and response scales using images are seen as reducing verbal complexity.

However, nonverbal approaches have their own problems. First, many numerical scales are more complex and difficult than verbal items. For example, the magnitude-measurement method assigns a base value to a reference object and other objects are rated by assigning values to them that reflect their ratio to the reference item (Hougland et al., 1992; Lodge, 1981; Lodge and Tursky, 1979, 1981, 1982; Lodge et al., 1975, 1976a,b). Robbing a store of $1000 might be selected as the reference crime and assigned a seriousness score of 100. People rate the seriousness of other crimes (e.g., jaywalking, homicide) as a ratio to the base of 100 preassigned to robbery. A serious problem with this technique is that many people (typically, 10 to 15% in the United States) are confused by this complex task and are unable to supply meaningful responses. The level of confusion probably would vary across countries, perhaps covarying with levels of numeracy.

Second, numerical scales are not as invariant in meaning and error free as their mathematical nature presupposes. Schwarz and Hippler (1995) found that people rate objects quite differently on a 10-point scale going from 1 to 10 than on one from −5 to −1 and +1 to +5. Also, on the 10-point scale, people routinely misunderstand what the midpoint of the scale is (Smith, 1994). In another example, the 101-point feeling thermometer is not actually used as such a refined measurement tool (Tourangeau et al., 2000; Wilcox et al., 1989). It is rare for more than about 10 to 20 values to be chosen by any respondents (mostly 10s and 25s/75s) and some people seem influenced by the temperature analogy and avoid very high ratings as "too hot."[4] No research establishes whether numerical scales are used consistently across nations.

Third, most societies have lucky and unlucky numbers which may influence numerical responses. Since the lucky and unlucky numbers vary across societies, the effects also differ.

Fourth, numerical scales only reduce the use of words in response scales and do not eliminate them. For example, the 10-point scalometer has to describe the dimension on which the objects are being rated (usually liking/disliking) and the scale's operation.

Finally, alternative numbering or grouping schemes influence the reporting of frequencies. Respondents are often unable or unwilling to provide an exact count and round off or estimate in various ways (Tourangeau et al., 2000), and there is no assurance that these are the same across societies.

Related problems occur with nonverbal nonnumerical questions and scales. Visual stimuli are not necessarily equivalent across cultures (Tanzer et al., 1995). For example, the color called "orange" in English is not clearly coded or distinctly labeled in Navajo. Because of this the Navajo do more poorly in matching objects that are "orange" (Jacobson et al., 1960). A second example comes from matrix items used in psychological testing, in which the missing element is placed in the bottom right corner (Tanzer, in press). This works for people using languages running from left to right and top to bottom. However, the matrix is wrongly oriented for Arab respondents, who read right to left and top to bottom. For them the missing element needs to be in the lower left corner.

Finally, visual stimuli must be replicated accurately across countries. The 1987 ISSP study on social inequality included a measure of subjective social stratification: "In our society there are groups which tend to be towards the top and groups which tend to be towards the bottom. Below is a scale that runs from top to bottom. Where would you place yourself on this scale?" There were 10 response categories, with 1 = top and 10 = bottom. This item was asked in nine countries. All countries showed a majority placing themselves toward the middle (4 to 7), but the Netherlands clearly was an outlier, with by far the fewest in the middle. Although most other differences appeared to reflect actual differences in social structure, the Netherlands' distinctive distribution did not match other measures of the inequality of Dutch society (e.g., income distributions, subjective class identifications) (Smith, 1988b).

Translation error was a suspected source of the deviation, but a check of the Dutch wording indicated that it was equivalent to the English and appropriate and clear in Dutch. It was then discovered that the scale displayed visually differed in the Netherlands from that employed elsewhere. The scale was to have 10 vertically stacked squares. The Dutch scale had 10 stacked boxes, but they formed a truncated pyramid, with the bottom boxes wider than those in the middle and top. Dutch respondents were apparently attracted to the lower boxes because they

[4]On climate as a factor in cross-national research (and possibly interacting with feeling thermometer), see Doob (1968).

were wider and were probably seen as indicating where more people were. This impact of differently arranged boxes was later verified experimentally (Schwarz et al., 1998).

21.2.3 Simple Response Scales

A second suggested solution, in a sense the opposite of the numerical approach, is the "keep-it-simple-stupid" approach. For surveys, this means using dichotomies. It is argued that yes/no, agree/disagree, and other antonyms have similar meanings and cutting points across languages. The argument is that it may be difficult to determine, because of language differences, just where someone is along a continuum, but relatively easy to measure on which side of a midpoint someone is.

But the assumption that dichotomies are simple and equivalent across societies is questionable. For example, "agree/disagree" in English can be translated into German in various ways with different measurement consequences (Mohler et al., 1998). Also, languages may disagree on the appropriateness of intermediate categories between dichotomies. For example, a "maybe" response may be encouraged or discouraged by a language.

Another drawback is loss of precision. Dichotomies only measure direction and not extremity and are likely to create skewed distributions. Moreover, it takes several questions using dichotomies to differentiate respondents as well as one item using five or seven scale points. (Of course, single items with multiple responses would usually not produce as valid or reliable a measure as multiple-item scales.)

21.2.4 Calibrating Response Scales

A third proposed solution calibrates response scales by measuring and standardizing the strength of the labels used. One procedure has respondents rate the strength of terms by defining each as a point on a continuum. There are three variants of this approach: First, one can rank the terms from weaker to stronger (or along any similar continuum) (Spector, 1976). This only indicates their relative position and not absolute strength or the distance between terms. Second, one can rate each term on a numerical scale (usually with 10 to 21 points) (Smith, 1997). This measures absolute strength and the distance between terms, and thus facilitates the creation of equal-interval scales. Finally, magnitude-measurement techniques locate each term on a ratio scale (Hougland et al., 1992; Lodge, 1981; Lodge and Tursky, 1979, 1981, 1982; Lodge et al., 1975, 1976a,b). This allows more precision than the numerical-scale approach.

Among the three variants, the middle appears most useful. The ranking method fails to provide the numerical precision needed to calibrate terms across languages. The magnitude-measurement technique does this but is much more difficult to administer and much harder for respondents to do. In addition, the extra precision that the magnitude-measurement procedure can provide over that achievable using a 21-point scale is probably unnecessary.

Studies show that (1) respondents can perform the required numerical-scaling tasks; (2) ratings and rankings are highly similar across different studies and populations; (3) test–retest reliability is high; and (4) variations in rating procedures yield comparable results. Thus, the general technique seems robust and reliable.[5]

Another approach for assessing the intensity of scale terms and response categories measures the distributions generated by different response scales (Hougland et al., 1992; Laumann et al., 1994; MacKuen and Turner, 1984; Orren, 1978; Sigelman, 1990; Smith, 1979). In an experimental, across-subjects design, one random group is asked to evaluate an object (e.g., one's happiness) with one set of response categories, and a second random group evaluates the same object with another set of response categories. Since the stimulus is constant and assignment is randomized, the number of people attracted to each category depends on the absolute location of each response category on the underlying continuum and the relative position of each of the scale points adopted. With some modeling over the true distribution, one calculates at what point each term cuts the underlying continuum (Clogg, 1982, 1984).

In a within-subjects design, people answer the same question two (or more) times, with different response categories being used (Orren, 1978). This allows the direct comparison of responses, but the initial evaluations may artificially influence later responses.

The advantage of the distributional approaches is that respondents do only what they are normally asked to do—answer substantive questions with a simple set of response categories. A disadvantage is that it is harder to evaluate many response terms, and thus these approaches are better suited for assessing an existing response scale than for evaluating a large number of terms that might be used to construct an optimal response scale. Also, the results depend on the precise underlying distribution and modeling adopted, and calculating the strength of terms is more complex than computing from direct respondent ratings.

The direct-rating approach yielded very promising results in a study of terms used in response scales in Germany and the United States (Mohler et al., 1998; Smith, 1997). Many response terms were highly equivalent in both countries, but some notable systematic differences also appeared. Bullinger (1995) obtained similar findings in his German/American study.

Besides the technical challenges that the approach demands, the major drawback is that separate methodological studies are needed in each language to establish the calibration. This is not something that every cross-national study can undertake. However, in theory once calibrations are determined, they could be used by other studies without extra data collection needed. Since the same response scales are used in many different substantive questions, a small number of carefully calibrated response scales could be used in many questions.

[5]However, vague frequency terms correspond to different absolute values, depending on the commonness or rarity of the event or behavior specified (Bradburn and Sudman, 1979; Schaeffer, 1991b).

21.3 RESPONSE EFFECTS

Cross-national comparability is also difficult to achieve because of differences in response effects (Hui and Triandis, 1985; Usunier, 1999). The special danger in cross-national surveys is that various error components may be correlated with nation, so that observed differences represent differences in response effects rather than in substance. Work by Saris (1998) across 13 cultural groups/nations indicates that measurement error is not constant. As he notes (p. 83): "Even if the same method is used, one can get different results due to differences in the error structure in different countries." Important cross-national sources of measurement variation include effects related to social desirability, acquiescence, extremity, no opinion, middle options, context/order, and mode.

21.3.1 Social Desirability

Social desirability effects distort responses (DeMaio, 1984; Johnson et al., 2000; Tourangeau et al., 2000). Image management and self-presentation bias lead respondents to portray themselves positively by overreporting popular opinions and actions and underreporting unpopular or deviant attitudes and behaviors.

Social desirability effects appear common across social groups but often differ in both intensity and particulars. First, the pressure to conform varies. Conformist societies would presumably have larger social desirability effects than individualist societies. Social desirability effects may interact with characteristics of respondents and interviewers, such as race/ethnicity, gender, social class, and age. In the United States, the most consistently documented interviewer effect is that people express more intergroup tolerance when being interviewed by someone of another race/ethnicity (Schuman et al., 1997). A similar effect was found in Kazakhstan involving Russian and Kazakh interviewers (Javeline, 1999).

Social desirability effects are likely to be greater when larger status/power differentials exist between interviewers and respondents, and these are likely to vary across nations. In developing countries, for example, interviewers tend to be members of educated elites, while in developed countries interviewers are typically of average status.

Moreover, the topics regarded as sensitive and the behaviors considered to be undesirable vary among cultures (Newby, 1998). For example, items about alcohol use are much more sensitive in Islamic countries than in Judeo-Christian societies. Similarly, cohabitation was commonplace in Sweden long before it became socially acceptable in the United States. Also, what it is legally permissible to ask varies. China now permits much survey research, but many political questions such as those about the Communist party are forbidden. To deal with social desirability effects, one can frame questions in less threatening ways, train interviewers to be nonjudgmental in asking items and responding to answers, and use modes that reduce self-presentation bias.

21.3.2 Acquiescence Bias

Acquiescence or yea-saying bias is the tendency for respondents to select affirming responses (Landsberger and Saavedra, 1967; Tourangeau et al., 2000). It is particularly likely to occur on agree/disagree and other items with clear affirming and rejecting responses. This bias can vary across cultures. Church (1987) found yea-saying to be particularly strong in the Philippines. Javeline (1999), using experiments with reverse-coded items in Kazakhstan, found high acquiescence overall, with an even higher level among the Kazakhs than among the Russians. Van Herk (2000) showed that the Greeks gave more positive responses than those in other European countries.

Acquiescence bias can be reduced by balancing scales so that the affirming response half the time is in the direction of the construct and half the time in the opposite direction (e.g., six agree/disagree items on national pride, with the patriotic response matching three agree and three disagree responses). In addition, formal reversals can be built into an instrument to catch yea-sayers (Bradburn, 1983; Javeline, 1999). Finally, alternative formats less susceptible to acquiescence bias, such as forced-choice items, can be employed (Converse and Presser, 1986; Krosnick, 1999). As Javeline (1999, p. 25) observed: "[M]embers of certain ethnic groups—in the name of deference, hospitality, or some other cultural norm [agree falsely] ... more frequently. ... [T]he fact that they do must be taken into account in designing questionnaires. We cannot change the respondents, so we must change our methods."

21.3.3 Extreme Response Styles

Some people are attracted to extreme categories (e.g., strongly agree, most important), whereas others avoid them and favor less extreme responses (e.g., agree, somewhat important). People tend to follow the extreme/nonextreme patterns regardless of the true strength of attitudes toward particular items, so the choice of categories may represent a response set rather than true opinions.

This tendency varies across racial and ethnic groups. Black students in the United States are more prone than white students to select extreme responses (Bachman and O'Malley, 1984), and Hispanics in the U.S. Navy use extreme categories more than do non-Hispanics (Hui and Triandis, 1989; see also Pasick et al., 1996).

Differences in the propensity to select extreme categories also appear in cross-national studies. Asians in general, and the Japanese in particular, avoid extreme responses (Chen et al., 1995; Chun et al., 1974; Hayashi, 1992; Lee and Green, 1991; N. Onodera, personal communication). Whether these differences are tied to cultural differences or explained by differences in other factors related to extremity preference, such as education, age, and income, is unknown (Greenfield, 1997; Greenleaf, 1992).

Various approaches to deal with extreme response styles exist. First, a multitrait multimeasurement design can determine if effects occur across instruments (Van Herk, 2000). Second, rankings rather than ratings have been proposed. This eliminates the possibility of an extreme response effect but forces respondents to complete a more difficult task, loses measurement differentiation, and assumes no ties across objects (Van Herk, 2000). A third approach argues that linguistic equivalency may have to be sacrificed to achieve functional equivalency on scales. Since the Japanese appear to be predisposed to avoid response categories with strong labels, some advocate that labels for Japanese categories be softened, so that "strongly agree" and "agree" would not be literally translated into Japanese, but rendered as equivalent to "agree" and "tend to agree" instead. Others suggest that the problem is a disconnect between translation equivalence and response-scale equivalence. For example, in a study of English, Chinese, and Japanese students, Voss et al. (1996) found that a number of terms used in typical survey responses and translated as equivalent were not rated as similar in intensity in quantitative comparisons. Thus, part of the difference may be due to mistranslation.

The issue is whether noncomparability in one aspect is needed to establish comparability on a more important basis. The general rule is to do things exactly the "same" across surveys. The challenge is identifying cases in which things that are the "same" really are not and an adjustment is needed to achieve equivalency.

Finally, steps can be taken during analysis. One can conduct analyses with items collapsed into dichotomies to see if this appreciably changes conclusions. For example, in an analysis of items on scientific knowledge on the ISSP, one summary scale using the 12 items merely counted the number of correct responses, while another used the five response categories (definitely true, probably true, can't choose, probably false, definitely false). The two scales produced similar results, suggesting robust findings (Smith, 1996).

21.3.4 "Don't Know" Responses and Nonattitudes

People have different propensities to offer opinions (Smith, 1984). "Don't knows" (DKs) are higher among the less educated across countries (Young, 1999), but even with education controlled, DK levels vary by country. Some of this cross-national variation is undoubtedly real, reflecting true differences in the level of opinionation, but some appears to come from different response styles. As Delbanco et al. wrote (1997, p. 71): "Attitudes about responding to surveys (e.g. a tendency for individuals to say they do not know an answer or to refuse to answer) may differ across countries." For example, Americans are more likely to supply personal income information than people in many European countries (Smith, 1991a). Also, there is apparently a greater willingness of people in some countries to guess about questions on scientific knowledge rather than admit ignorance by giving DK responses (Smith, 1996).

21.3.5 Neutral/Middle Options

Related to the issue of "no opinion" responses is whether neutral or middle options should be offered and what the impact of their inclusion or exclusion is. A no-middle-option question might ask one to "strongly agree, agree, disagree, strongly disagree," while a middle-option version might add "neither agree nor disagree." Research from several countries finds that providing ambivalent respondents with a clear response option produces more reliable results (O'Muircheartaigh et al., 1998; Smith, 1997).

21.3.6 Question Order

Context effects occur when questions asked previously influence responses to later questions (Smith, 1991b; Tourangeau, 1999; Tourangeau and Rasinski, 1988). The questions fail to remain independent of each other, and the prior stimuli or one's response joins with the stimuli from subsequent questions to affect the responses to subsequent items. Context effects undoubtedly operate across surveys everywhere, but experiments and detailed studies have apparently been done in only a few countries, and no coordinated cross-national experimental studies exist.

Even when the effects are consistent in their structure and nature, their impact may vary across countries. For example, Schwarz (1999a) has shown in Germany that the favorability ratings of politicians are influenced by the ratings of political figures preceding them. Asking first about a discredited and widely disliked politician leads to higher ratings of leaders in good standing who are asked about in subsequent questions. This effect would probably appear in other countries. If so, then in a cross-national survey of political leadership that asked about party leaders in some predetermined order (say, head of government, leader of largest opposition party, leader of next largest party, etc.), the ratings of some subsequent figures (e.g., opposition leaders) could easily vary artificially because of true differences in the status of leaders mentioned previously (e.g., the head of government). This, in turn, could lead to misinterpretations about the absolute popularity of the opposing leadership across nations.

Particularly likely to vary across cultures are context effects that are conditional on people's attitudes toward a preceding context-triggering question (Smith, 1991b). In the United States, asking first about government spending programs influences the percent saying that income taxes are too high. However, the effect depends on the popularity of the spending programs. Among those in the most pro-spending group, asking the spending items first *lowers* the percent saying that their income taxes are too high by 25 percentage points. But for those most against government spending, asking the spending items first *increases* the percent saying that their taxes are too high by 7 percentage points. Thus, even if the conditional context effects were similar across countries, the same net effect occurs only if the popularity of government programs was also comparable across countries.

21.3.7 Mode of Administration

Survey responses often differ by mode of administration (e.g., self-completion, in-person, telephone). Among the most consistent effects of mode is that more socially undesirable or sensitive behaviors (e.g., high alcohol consumption, criminal activity) are reported in self-completion modes than in interviewer-assisted modes (Hudler and Richter, 2001; Tourangeau et al., 2000). Keeping the mode constant will not solve the problem automatically, since mode may not have a constant impact across countries. For example, not only would showcards with words be of little use in low-literacy societies, but their inappropriate use would create greater differences between the literate and illiterate than a mode that did not interact with education and literacy so strongly.

In brief, various measurement effects influence survey responses. Sometimes these effects vary across countries. Of course, this does not mean that response effects always or even typically differ across groups and societies. A number of consistent results have been documented. For example, some social-desirability effects have been shown to be similar in Canada, the Netherlands, and the United States, telephone surveys produce lower-quality data in the same countries, and forbid/allow question variations have similar effects in Germany and the United States (Hippler and Schwarz, 1986; Scherpenzeel and Saris, 1997). But variable measurement effects remain a serious concern that researchers must look out for continually.

21.4 ENHANCING QUESTION COMPARABILITY

Various steps can be taken to enhance equivalence and achieve valid cross-national research. These include (1) cross-national cooperation over study and instrument design, (2) adopting a master questionnaire using question forms more conducive to reliable measurement and suitable for translation, (3) considering both emic and etic items (see Section 21.4.3), (4) following optimal translation procedures, (5) careful item development and pretesting, and (6) thorough documentation of surveys.

21.4.1 Cross-National Research Collaboration

Research imperialism or safari research in which a research team from one culture develops a project and instrument and imposes it rigidly on other societies should be avoided. As van de Vijver and Leung (1997, p. 12) have observed: “Many studies have been exported from the West to non-Western countries and some of the issues examined in these studies are of little relevance to non-Western cultures.” Instead, a collaborative multinational approach should be followed (Jowell, 1998; Schooler et al., 1998; Szalai, 1993; van de Vijver and Leung, 1997). For example, Sanders (1994, p. 518) noted: “One of its [the ISSP’s] greatest strengths is that a country can only be incorporated in the survey if a team of researchers from that country are available . . . to ensure that the translation of the core questions can be achieved without significantly altering their meaning. The potential problem of cross-national variation in meaning is accordingly minimized.” Another example of joint development involves a study of AIDS/HIV in three preliterate tribes in Mali. The research team consisted of American health and African specialists, Mali health researchers, and local tribal informants. They worked together to design and conduct a survey that performed well for the target populations (Schooler et al., 1998).

21.4.2 Question Form and Content

The first step in developing a questionnaire is to formulate items that avoid problematic constructions and make translations easier. Brislin (1986) offers 12 guidelines for making items more translatable:

1. Use short, simple sentences of less than 16 words. (But items can be of more than one sentence.)[6]
2. Employ active rather than passive voice.
3. Repeat nouns instead of using pronouns.
4. Avoid metaphors and colloquialisms.[7]
5. Avoid the subjunctive.
6. Add sentences to provide context to key items. Reword key phrases to provide redundancy.
7. Avoid adverbs and prepositions telling "where" or "when."
8. Avoid possessive forms where possible.
9. Use specific rather than general terms.
10. Avoid words indicating vagueness (e.g., "probably," "maybe," "perhaps").
11. Use wording familiar to the translators.
12. Avoid sentences with two different verbs if the verbs suggest different actions.

Other general rules about how to formulate questions have been developed for single-nation surveys, but are generally applicable across countries (e.g., Converse and Presser, 1986; Fowler, 1995; Sudman and Bradburn, 1982; van der Zouwen, 2000).

First, vague quantifiers (e.g., "frequently," "usually") should not be used. They have highly variable understandings both across respondents and question contexts (Miller et al., 1981).

Second, items with ambiguous or dual meanings should be avoided. Tanzer (forthcoming) noted that an internalized-anger item in the State-Trait Anger Expression Inventory ("I am secretly quite critical of others") could be understood as indicating internalized anger ("I keep my criticism of others to myself") or the indirect expression of anger ("I talk about other people behind their backs").

[6]Scherpenzeel and Saris (1997) find long questions superior but do not address sentence length. Pasick et al. (1996) oppose using compound sentences.

[7]Although metaphors often work well within a language, they are rarely translatable. For example, a General Social Survey item asks people to agree or disagree that "Right and wrong are not usually a simple matter of black and white; there are many shades of gray." This item was proposed for the ISSP but had to be dropped because other languages did not use the color metaphor. Similarly, it proved difficult to develop scales of psychological well-being in Hmong because items in the source language (English) used such terms as "downhearted" and "lighthearted," while the target language (Hmong) used different physiological metaphors, such as *nyuai siab* ("difficult liver"), for being depressed (Dunnigan et al., 1993).

In South Tyrol, German speakers tended to understand the item in the first sense, whereas Italian speakers favored the second.

Third, ambiguity also emanates from complex questions with more than one key element. Double-barreled questions are particularly problematic (Fowler, 1995; van der Zouwen, 2000).

Fourth, hypothetical and counterfactual questions should be avoided. People often lack coherent thoughts about imagined situations and may not even grasp the circumstances described (Fowler, 1995; van der Zouwen, 2000).

Fifth, terms should be simple and widely and similarly understood across all segments. One needs to avoid technical and jargon terms and aim for wordings suitable for general audiences. One wants to minimize interrespondent variability in comprehension. Words must not only be understood but comprehended similarly (Smith, 1989). When needed, definitions should be provided to clarify meanings (Converse and Presser, 1986; Pasick et al., 1996; Tourangeau et al., 2000).

Sixth, time references should be clear and precise (Fowler, 1995). For example, "Do you fish?" might be understood to mean "Have you ever gone fishing?" or "Do you currently go fishing?" It would be better to ask something like, "Have you gone fishing during the last 12 months?"[8]

Finally, some recommend avoiding the particularistic and using questions with a higher level of abstraction (Van Deth, 1999). Inglehart and Carballo (1997, p. 40) argue: "If we had asked questions about nation-specific issues, the cross cultural comparability almost certainly would have broken down. In France, for example, a hot recent issue revolved around whether girls should be allowed to wear scarves over their heads in schools (a reaction against Islamic fundamentalism). This question would have had totally different meanings (or would have seemed meaningless) in many other societies. On the other hand, a question about whether religion is important in one's life is meaningful in virtually every society on earth, including those in which most people say it is not." But other research indicates that people in general, and the less educated in particular, have more difficulty with abstract items than with concrete questions (Converse and Presser, 1986). In addition to these particular rules on constructing items, one should follow the rule that "more is better." Multiple indicators enhance scale reliability and reduce linguistic artifacts.

21.4.3 Emic and Etic Questions

Etic questions are items with a shared meaning and equivalence across cultures, and *emic questions* are items of relevance to some subset of the cultures under study. Suppose that one wanted cross-national data on political participation in general and contacting government officials in particular. In the United States, items on displaying bumper stickers, visiting candidate Web sites, and e-mailing

[8]On the difficulties of asking questions with time references (e.g., telescoping, forgetting curves) and means to deal with same (e.g., bounding and dating aids), see Tourangeau et al. (2000).

public officials would be relevant. In most developing countries, these would be rare to meaningless. Conversely, an item on asking a village elder to intervene with the government might be important in developing societies but have little relevance in developed nations. In such circumstances solutions include (1) using general questions that cover the country-specific activities within broader items, (2) asking people in each nation both the relevant and irrelevant participation items, or (3) asking a core set of common items (e.g., voting in local and national elections, talking to friends about politics), plus separate lists of country-specific political behaviors.[9]

Using general items is perhaps the least appropriate since the necessary loss of detail is usually a heavy price to pay, and general items may be too vague and sweeping. The relevant + irrelevant approach makes sense if the number of low-relevancy items is not too great and they are not so irrelevant that they are nonsensical or otherwise inappropriate. For example, the ISSP used this technique successfully in its study of environmental change, where items on personal car use were asked in all countries, even though ownership levels were quite low in some countries.

The emic/etic approach is useful if the common core is adequate for direct comparisons. For example, a study of obeisance to authority in the United States and Poland had five common items plus three country-specific items in Poland and four in the United States (Miller et al., 1981). This allowed direct cross-national comparisons as well as more valid measurement of the construct within countries (and presumably better measurement of how constructs worked in models).

Similarly, in developing the Chinese Personality Assessment Inventory, researchers found that important parts of Chinese personality did not match any dimension on standard Western scales (e.g., *ren quin*, or relationship orientation) and needed to be added (Cheung et al., 1996, p. 195). They noted: "[this] illustrates the importance of a combined emic–etic approach to personality assessment in non-Western cultures. ... The inclusion of relatively emic constructs are needed to provide a more comprehensive coverage of the personality dimension that are important to the local culture."

The emic–etic approach indicates that sometimes one needs to do things differently in order to do them equivalently (Przeworski and Teune, 1966).

21.4.4 Translation Procedures

Perhaps no aspect of cross-national survey research has been less subjected to systematic, empirical investigation than translation. Thoughtful pieces on how to do cross-national survey translations exist (Brislin, 1970, 1986; Harkness, 1999, 2001; Harkness and Schoua-Glusberg, 1998; Prieto, 1992; van de Vijver and Hambleton, 1996; Werner and Campbell, 1970). But rigorous experiments to test

[9]However, even identical actions, such as voting in the last national election, may not be equivalent. In some countries voting is legally mandatory, so it is not a meaningful measure of voluntary political activity. In other countries, elections are meaningless charades, so voting is not a meaningful measure of participating in a democracy.

the proposed approaches comparable to Schuman and Presser's work on question wording or Tourangeau and Schwarz's work on question order are lacking. Because of this the development of scientifically based translation has languished.

Translation is often wrongly seen as a mere technical step rather than as central to the scientific process of designing valid cross-national questions. Translation must be an integral part of the study design and not an isolated activity (Bullinger, 1995; Pasick et al. 1996). Pasick et al., (1996) describe translation as an integrated part of an eight-step process for designing a multilingual study. The steps are (1) conceptual development of topic, (2) inventory of existing items, (3) development of new questions, (4) question assessment through translation, (5) construction of a full draft questionnaire, (6) concurrent pretesting across all languages, (7) item revision, and (8) final pretesting and revisions. The key point is that changes in source and target language wordings occur at various points in the design process.

Achieving optimal translation begins at the design stage. Cross-national instruments should be designed by multinational teams of researchers who are sensitive to translation issues and take them into consideration during the design and development stages (Bullinger, 1995). They need to keep considering how each concept of interest can be measured in each language and society under study. Specifically, they should practice *decentering* (Harkness and Schoua-Glusberg, 1998; McGorry, 2000; Pasick et al., 1996; Potaka and Cochrane, 2002; Werner and Campbell, 1970), a process by which questions are formulated so that they are not anchored in one language but fit equally well in all applicable languages.[10] Of course, the problems of translation in general and decentering in particular multiply as the number of languages involved increases and as the linguistic and cultural differences between languages widen.

The various techniques for carrying out translations are discussed in greater detail in Chapter 22. They include the following approaches or some combination of them.

- Translation-on-the-fly, with multilingual interviewers doing their own translations when respondents do not understand the source language. This approach lacks standardization and quality control.
- The single-translator, single-translation approach, which is frequently used because it is quick, easy, and inexpensive, but is usually not recommended.
- Back-translation, involving (1) translation of questions in the source language into the target language by one translator, (2) retranslation back into the source language by a second translator, (3) comparison of the two source language questionnaires, and (4) when notable differences in the two appear, working with one or both translators to adjust the target language of the problematic questions. No assessment is made directly of the adequacy of the

[10]Decentering is not possible when a well-established scale developed in one language is being replicated across countries, but should be employed whenever new items and scales are being designed for a multilingual study.

target language questions (Blais and Gidengil, 1993), so a poorly worded item that back-translates successfully goes undetected.

- Parallel translation, involving (1) independent translations of questions in the source language into the target language by two translators, (2) comparison of the two translations, and (3) when found to differ appreciably, the two translators working with the developers of the source language questions to determine the reason for the variant translations (Bullinger, 1995). Compared to back translation, there is more emphasis on optimizing wording in the target language. It can also be done more quickly, since the two translations are done simultaneously rather than sequentially.
- Committee translation, with a team of translators and researchers discussing the meaning of items in the source language, possible translations in the target language, and the adequacy of the translations in the target language relating to such matters as level of complexity and naturalness as well as meaning (McGorry, 2000). This approach places the greatest emphasis on writing good questions, not just on translating words.

Although careful translation procedures are essential for developing equivalent items, they are insufficient alone. Quantitative methods should evaluate the results of the qualitative translation procedures. One approach is the direct evaluation of items. Bullinger (1995) describes a study in which two raters independently judged the difficulty of the wordings in the source language, then two other raters evaluated the quality of the translated items, and finally, two more raters assessed the back-translated items. This allowed both qualitative and quantitative evaluation of the translations, evaluations as to whether the items were of comparable understandability, and interrater reliability checks on the quantitative ratings.

Second, quantitative ratings of the terms used in response options can determine whether scale points are equivalent. Third, various statistical tests can assess the comparability of cross-national results (Ellis et al., 1989; MacIntosh, 1998b). Although usually applied at the analysis phase after data collection has been completed, they should be employed at the development stage. In particular, item-response theory (IRT) has been used to measure equivalency and even assess whether differences were due to translation errors or cultural differences. In a comparison of psychological scales using German and French data, 10% of the items tested as nonequivalent (Ellis et al. 1989). Excluding the nonequivalent items from the scales resulted in a major change in substantive interpretation. Using all items, the Germans rated lower than the French on self-actualization, but with only equivalent items used, no national differences appeared.[11] Others, however, find IRT tests too exacting and prefer other techniques, such as confirmatory or exploratory factor analysis (Ellis et al., 1989; MacIntosh, 1998b; Ryan et al., 1999).

[11]Nonequivalent items should not merely be discarded as flawed. As Ellis et al. (1989, p. 671) note: "[T]he non-equivalent items ... should be examined separately for potential clues of real cross-cultural differences."

Finally, these quantitative evaluation approaches can be combined. Items might be evaluated on various dimensions related both to language (e.g., clarity, difficulty) and substance (e.g., extremity, relevancy). These ratings could then be compared across languages (as in the rating of response options) and also correlated with results from pilot data using IRT or other techniques.

The various quantitative techniques should be used hand-in-hand with qualitative techniques. For example, in a German–American study of response options (Mohler et al., 1998), equivalent English and German terms for answer scales were developed by translators and then respondents rated the strength of the terms on the underlying dimensions (agreement/disagreement and importance). In almost all cases, the mean ratings of the German and English terms were the same, thereby validating translation equivalency (e.g., finding that "strongly agree" and its German translation were both rated similarly on a 21-point scale that ran from total and complete agreement to total and complete disagreement). In another German–American study (Ellis et al., 1989), an American verbal-reasoning question[12] was found by IRT testing not to be equivalent in Germany (i.e., it did not scale with other items in a comparable manner in both countries). Evaluation of the wording revealed that the difference occurred because in England and the United States, poodles are not considered as retrievers, but in Germany, poodles were originally bred as waterfowl retrievers and thus are part of the larger set. In both studies, qualitative assessment and quantitative measurement yielded consistent judgments about the equivalency (or nonequivalency) of response options or items.

It has been proposed that translation equivalence can be established by administering items in two languages to bilingual respondents. This approach is problematic because bilinguals understand and process language differently than monolinguals do (Blais and Gidengil, 1993; Ellis et al., 1989). Despite this serious impediment, useful evaluations can be gained by looking at how results compare across languages within societies. In a test of whether French Canadians were less supportive of democracy than English Canadians, Blais and Gidengil (1993) found that within and outside Quebec, both English and French Canadians interviewed in French were less supportive of elections than English and French

[12]The item was:

Instructions: Following are sets of facts and conclusions. In many cases, from reading the facts you can determine which conclusions are true and which are false. If the facts don't provide enough information to determine if a conclusion is true or false, answer "uncertain."

Facts:
All of the dogs in the park are retrievers.
Some of Ann's dogs are in the playground.
Cindy owns a poodle.
Most of Ann's dogs are poodles.

Conclusion:
All of Cindy's dogs are in the park.

TRUE FALSE UNCERTAIN

Canadians interviewed in English. Their statistical analysis showed that language, rather than culture, explained the differences in support for democracy.

Achieving item and scale equivalency is a challenging task, and optimal translations are essential for reaching this goal. Researchers should (1) make translations an integrated part of the development of studies; (2) utilize the best approaches, such as committee and combined translation, and (3) use quantitative methods to assess translations.

21.4.5 Pretesting and Related Questionnaire Development Work

Whereas pretesting is important in surveys conducted in a single country, its value greatly increases cross-nationally. Developmental work must establish that the items and scales meet acceptable standards of comprehension, reliability, and validity in each country and are comparable across countries (Krebs and Schuessler, 1986). As Hudler and Richter (2001, p. 19) have observed about cross-national research: "[I]t is very essential that the instrument is carefully designed and analyzed in a pretest." The pretesting should be "a team effort with multiple disciplines and preferably multiple cultures represented" (Pasick et al., 1996, p. 530).

Useful developmental and pretesting procedures include (1) cognitive interviews using think-aloud protocols in which respondents verbalize their mental processing of questions and computer-assisted concurrent evaluations (Bolton and Bronkhorst, 1996; Gerber and Wellens, 1997, 1998; Johnson et al., 1997; Krosnick, 1999; Prüfer and Rexroth, 1996; Tourangeau et al., 2000); (2) behavioral coding with the interviewer–respondent exchanges recorded, coded in detail, and then analyzed (Fowler and Cannell, 1996; Krosnick, 1999; Prüfer and Rexroth, 1996); and (3) conventional pretesting, including the use of probing (Converse and Presser, 1986; Fowler, 1995; Hudler and Richter, 2001).

Survey instruments may also be tested by (1) concurrent ethnographic analysis, in which the results from surveys and ethnographic studies are cross-validated (Gerber, 1999); (2) exemplar analysis, in which scales are assessed by asking people to describe what types of events would represent the response options (e.g., what would be an example of someone being completely satisfied with his or her job, somewhat dissatisfied, etc.) (Ostrom and Gannon, 1996); (3) conversational interviewing in which interviewers use alternative wordings to achieve better understanding (Conrad and Schober, 2000); (4) expert panels (Presser and Blair, 1994); and (5) the quantitative scaling of response options (Mohler et al., 1998; Smith, 1997).

Two major obstacles to effective developmental work in cross-national surveys are the absence of methodological studies of the various pretesting approaches and a general underutilization of pretesting. Few studies have systematically studied and compared pretesting methods (Presser and Blair, 1994; Willis and Schechter, 1997). Presser and Blair's comparison of pretest methods (conventional, cognitive, and behavior coding plus using an expert panel) found considerable differences in the number and nature of problems revealed by the different approaches. Expert panels identified the most problems, and behavioral coding was the most reliable. Conventional pretesting found the fewest analysis-related problems but the most interviewer-centered difficulties. Cognitive pretesting did the best in uncovering problems of understanding, but revealed almost none of the problems that interviewers had. Thus, the main lesson of their research is that multiple methods should be used.

There has been no similarly rigorous comparison of pretesting techniques cross-nationally. Studies on the use of cognitive pretesting in such varied countries as Australia, Belgium, Taiwan, and the United States (Foddy, 1998; Nuyts et al., 1997; Tien, 1999) all found this approach to be valuable. But the optimal combination of specific pretesting approaches in cross-national studies has not been established. Second, most cross-national studies fail to devote adequate time and resources to pretesting. A review of pretesting procedures used by the ISSP, surveys in bilingual countries (e.g., Canada, Belgium, and Switzerland), the World Fertility Survey (WFS), the European Social Survey, and other cross-national surveys found that resources allocated to pretests were usually severely limited and their usefulness notably curtailed. One common problem is that pretests are usually too small for anything but a qualitative assessment of whether items are working, and they are often limited to atypical, convenience samples, such as college students. A third problem is that most studies use only conventional pretesting, making only occasional use of cognitive pretesting, behavioral coding, think-alouds, and other advanced techniques. Most studies do not even document what their pretesting consisted of. For example, one major four-nation study reported only that "[e]xtensive pretesting was carried out to ensure the meaningfulness and cultural appropriateness of the questions" (Kohn et al., 1990, p. 968). Perhaps the most serious problem is that pretests are sometimes not allowed to play the important role in developing items for which they are designed. For example, while the WFS used larger pretests than usual (almost all with 100 or more cases), and even audiotaped many interviews (a good but rarely used procedure), the WFS's content was basically fixed before the pretests and revised little based on them (Cleland and Scott, 1987, pp. 32–33, 384). In other cases, the pretests are used to revise language in one country only and not to improve the instrument cross-nationally. Studies that allocate more time and effort to pretesting end up with better instruments (Bullinger, 1995; Perneger et al., 1999).

More methodological studies of pretesting are needed. Until such studies clarify the best practices, a few general guidelines based on what appear to be the best existing practices can serve as a guide: (1) the multipretesting procedures should be carried out across countries and languages with results evaluated by researchers expert in (a) the cultures and languages being investigated, (b) the substantive domains being studied, and (c) survey research methodology; (2) the pretesting and translating should be integrated and interactive processes; (3) the pretesting needs to be cross-national and not just within specific countries; and (4) the developmental process takes much more time and resources than for single-country monolingual studies and usually should involve multiple rounds of pretesting and larger samples.

21.4.6 Documentation

As Jowell (1998, p. 171) observed, good documentation and "detailed methodological reports about each participating nation's procedures, methods, and success rates ..." are essential. However, as Hermalin et al. (1985, p. 203) noted: "[M]aintenance and documentation are painstaking tasks for which little provision is made ..." Their work with the WFS found that some of the surveys no longer existed and that "the documentation for surviving surveys is often confused and incomplete." Although all phases of each survey from sampling to data processing need to be carefully recorded (Hudler and Richter, 2001), it is particularly important to include the original questionnaires used in each country so that they can be consulted to understand results across countries. The ISSP does this. Solid documentation is more than just a good practice that facilitates primary and secondary analysis. It enhances comparability from the start by forcing researchers to detail what procedures are being used in each country and how comparable they are.

21.5 CONCLUSIONS

The great challenge in cross-national survey research is that languages, social conventions, cognitive abilities, and response styles all vary across societies (Fiske et al., 1998). To obtain valid, equivalent measurement across cultures, measurement error from these sources must be minimized and equalized so that reliable and consistent substantive information emerges. Achieving this is difficult. The task of obtaining cross-national comparability is so complex and challenging that more effort is needed at all stages, from conceptualizing the research question to instrument development to survey analysis. But the benefits from cross-national research fully merit the extra efforts. As the Working Group on the Outlook for Comparative International Social Science Research has noted: "A range of research previously conceived of as 'domestic,' or as concerned with analytical propositions assumed invariant across national boundaries, clearly needs to be reconceptualized in the light of recent comparative/international findings" (Luce et al., 1989, p. 550). Unless a comparative perspective is adopted successfully, "models and theories will continue to be 'domestic' while the phenomena being explained are clearly not" (Luce et al., 1989, p. 564).

CHAPTER 22

Survey Questionnaire Translation and Assessment

Janet Harkness
Zentrum fur Umfragen Methoden und Analysen

Beth-Ellen Pennell
University of Michigan

Alisú Schoua-Glusberg
Research Support Services

22.1 INTRODUCTION

Survey translation is too often seen as a rather simple affair, not calling for great expenditure of time, expertise, or other resources. In this chapter we aim to demonstrate the relevance of survey translation quality to data quality. It illustrates that translating and assessing questionnaires is a complex undertaking that calls for proven procedures and protocols and cross-disciplinary, cross-cultural expertise.

Procedures for testing and assessment that are standard requirements for monolingual survey instruments are, surprisingly, not required for translated questionnaires. We address this oversight by first discussing examples of linguistic and cultural challenges faced when trying to produce translations that maintain equivalence of measurement across languages. The outline of practices and procedures that follows highlights other issues (e.g., whether to translate, consequences of "close" translation practices, and dealing with multiple languages). In the following sections we describe translation, assessment, pretesting, and documentation procedures and discuss language harmonization procedures.

Methods for Testing and Evaluating Survey Questionnaires, Edited by Stanley Presser, Jennifer M. Rothgeb, Mick P. Couper, Judith T. Lessler, Elizabeth Martin, Jean Martin, and Eleanor Singer
ISBN 0-471-45841-4

Questionnaires are usually translated to interview populations that cannot be interviewed in the language(s) already available. In translation jargon, we speak of translating out of a *source language* into a *target language*. Consequently, questionnaires that serve as the text for translation are called here *source questionnaires*. Questionnaires are translated in three main contexts: for cross-national survey projects, for within-country research in countries with several official languages, and for projects in which it is necessary to include populations that do not speak the majority language of a given country. The need for translations in all three contexts is growing. At the same time, no commonly accepted set of standards and procedures has been established in the survey research community either for translating questionnaires or for assessing the quality of translations produced. The view advanced here is that quality assurance for translated questionnaires calls for both statistical analysis of questionnaire performance and textual analysis of translation quality.

22.1.1 Effects of Translation

In the monocultural context, small changes in formulation or suboptimal design have been shown to affect respondents' understanding of the question asked or the accuracy of the measurement or count. Questionnaire designers go to considerable effort to try to ensure that the *intended* meaning of questions is also what respondents understand. In cross-cultural research, too, we can expect that small differences in formulation across languages can affect understanding and that inappropriate design or inappropriate translation can result in respondents not being asked what the researchers intended to ask.

Poor translation can rob researchers of the chance to ask the questions they want to ask across languages and cultures. However, we cannot always expect to notice from the data that translation problems have arisen. In addition, we cannot predict in advance what the effects of poor translation will be because this depends on what mistakes are made, the context in which they are made, and how much a given mistake or infelicity matters for a specific context. Examples discussed below illustrate different aspects of these points.

Mistranslations do, however, sometimes show up unmistakably in the data and may then be of such magnitude that one wonders how they happened. The following is an example from the first International Social Survey Programme (ISSP) survey in 1985. The British English source question formulation was as follows:

...Please show which you think should be allowed and which should not be allowed by ticking a box on each line.

...Organizing protest marches and demonstrations

(1. Definitely allowed/2. Probably allowed/3. Probably not allowed/4. Definitely not allowed)

Table 22.1 shows the distribution of answers among different countries for this question; the German and Austrian distributions for codes 1 and code 4 (bold in

Table 22.1 Marginals from the ISSP 1985 Codebook (Percent)

	Germany	Austria	Australia	Great Britain	United States	Italy
1. Definitely allowed	**10**	**11**	37	38	37	37
2. Probably allowed	21	22	32	32	29	31
3. Probably not allowed	39	40	17	12	19	19
4. Definitely not allowed	**31**	**28**	14	18	14	12

Source: From Mohler and Uher (2003).

table) differ substantially from those of other countries. The German and Austrian translations rendered the English "organizing protest marches and demonstrations" as "organizing protest marches *that obstruct/block traffic*" ("Protestmärsche organisieren *die den Verkehr behindern*"). The example also illustrates one of the dangers inherent in simply "borrowing" a translation, as the Austrian researchers did in this case.

In other cases, however, translation inadequacies may not be of prime importance for respondents' understanding of the question. The Italian translation of a set of English questions on family, women, and work (ISSP 1994) translated the first three items below with *madre* (mother) although "mother" occurs only in the first two questions of the set in English, while the third refers to "woman."

i. A working mother can establish just as warm and secure a relationship with her children as a mother who does not work.
ii. A pre-school child is likely to suffer if his or her mother works.
iii. All in all, family life suffers when the woman has a full-time job.

The translator may have copied *madre* automatically, not noticing the change to "woman." (Consistency and careful proofreading of translations are crucial.) Kussmaul (1995) discusses another possible cause: Translators sometimes read (and thus translate) what they expect from their own cultural perspective and not what is there in the source text. The translator could have automatically linked the mention of "woman" and "family" with the notion of (and word for) "mother." The slip might not greatly affect respondents' grasp of the issue if they shared the same perspective as the translator. At the same time, mentioning *madre* instead of *donna* (woman) makes it explicit that the family life referred to *is* one with children and, theoretically, could therefore affect responses.

22.2 QUESTIONNAIRE TRANSLATIONS: SURVEY NEEDS AND RESULTING CHALLENGES

With a few exceptions, both within-country research and cross-national research routinely begin translation work after the source questionnaire has been finalized in wording and design. In this chapter we therefore focus on challenges faced in translating existing questionnaires.

In deciding to translate a questionnaire rather than adapt questions, a major premise is that the source and translated questionnaires will ask the same questions (cf. "adapt" or "adopt and translate" in Harkness et al., 2003). Researchers often assume incorrectly that the less difference there is between source questions and translations in terms of language features, the more likely it is that the same questions are indeed asked. In discussing translation we must keep in mind that languages are essentially different in lexis (vocabulary), in structure, and in the functions language components have in one or the other language community. In targeting equivalence or comparability across translations, we need also to acknowledge the inevitability of difference across languages and cultures; translation necessarily changes texts.

Despite slight differences in their understanding of *translation* versus *adaptation*, Geisinger (1994), Hambleton and Patsula (1999), and Harkness et al. (2003) illustrate a growing recognition that keeping things the same is neither always possible nor always desirable. To date, however, a translated questionnaire is still often expected to be a rather close translation of the source, retaining the semantic and propositional content, the pragmatic meaning, as well as structural arrangements and the design and measurement properties of the source questionnaire.

Vocabulary (words), semantics, and pragmatic meaning of words and utterances do not match up neatly across languages. In the simplest case, one term in one language may cover less or more than what corresponding terms in another language cover. A health survey of Western design might thus pose questions distinguishing between physical and mental health but run into problems if translated closely for cultures, such as the Maori population in New Zealand, that do not distinguish between the two. In the following sections we present examples of classic challenges to matching meaning, structures, and measurement across questionnaire versions.

22.2.1 Matching Meaning

Misunderstanding within a language can be due to a variety of factors, not least of which is the intrinsically multiple nature of meaning. In the majority of instances, the context in which communication takes place, the shared common ground (Clark and Schober, 1992), and the surrounding discourse (co-text) will help clarify different types of ambiguity and avoid misunderstandings (cf. relevance theory; Sperber and Wilson, 1986). Even within a language, however, cultural diversity can increase the likelihood of misunderstanding. Ambiguity and resulting misunderstandings in source questionnaires create problems for translation.

Traditionally, linguistic ambiguity is discussed under the headings of *syntactical ambiguity* (e.g., "I really dislike answering machines" understood as "I dislike answering" or "I dislike the machines") and *lexical or semantic ambiguity*. Semantic ambiguity exists when one and the same word form can represent different lexemes (e.g., dictionary entries *ball*[1] and *ball*[2]). Both forms of ambiguity are found in questionnaires, partly because they provide fewer contextual cues to help disambiguate than many other text types. Respondents may not see any ambiguity and simply provide *their* interpretation of what is meant. Translators, in contrast, are trained to parse for meaning. When they spot ambiguities, they will want to know what to do. The following question (from a Likert-type battery) is from the 2002 European Social Survey (ESS): "Most asylum seekers are not in real fear of persecution at home." The source questionnaire carried a note (see Section 22.6) indicating that the question was intended to tap whether or not respondents felt asylum seekers were under the threat of persecution in their home countries. A Norwegian translation team produced two translations. One indicated that the asylum seekers were not very afraid, and the other that they were not really under the threat of persecution. Nuances of usage and idiom that make the second reading more salient to native British speakers left a trained group of nonnative speakers, despite the note, uncertain which was intended.

A more common problem for questionnaire translation is *semantic polysemy*. Polysemous words have a number of different but related meanings. Several of these may nominally fit one, and the same context and textual clues may not suffice to identify which lexical meaning is intended. An example from the "Frequently Asked Questions" for the American Time Use Survey reads: "People do a lot of important things for themselves, their families and their communities for which they don't get paid. This survey will help the government *value* those kinds of activities" (emphasis added).

The translation committee (see Section 22.4.1) producing the U.S. Spanish translation could not decide whether *value* meant "assign or determine the economic value" or "recognize the importance of" and consulted the survey designers, who identified "recognize the importance of" as the intended meaning. Where available at all, documentation on source questions is usually not at a level of detail sufficient to resolve translation issues. In contexts where questionnaire translators cannot ask for clarification, they necessarily decide themselves what is intended.

The suggestion is sometimes made that if an item is identified as ambiguous, translators should simply translate ambiguously. This implies that they should keep the same ambiguities and assumes that this can be done. Although ambiguity can sometimes be matched across languages, this is by no means always the case. Opinions also differ on whether or not maintaining ambiguity is advisable. Kussmaul (1995) suggests it is the translator's job to improve poor source texts in translation. Questionnaire designers, on the other hand, are likely to be keen that questions are translated, not "improved." At the very least, translators' documentation on the questionnaire and their translation output should include information on ambiguities noted and how these have been dealt with. By reporting back promptly, translators can help improve source questionnaires not yet finalized.

22.2.2 Matching Semantic Content and Measurement: The Example of Answer Scales

One of the easiest ways to demonstrate that languages are not isomorphic is to try to produce close translations of answer scales retaining semantic content and measurement properties across languages. Answer scales frequently combine negation and quantification, and these differ considerably across languages. A basic problem in trying to match up answer scales is, therefore, that the lexical and structural options available differ across languages. Very often, term-for-term equivalence is impossible. A well-constructed response scale in English might thus unavoidably end up in translation with labels that either change (or break) the scale design (cf. Harkness, 2003). Two brief examples illustrate the problems.

Coordinating elements in answer scales (italic below) bring together contrasting dimensions, as in *neither* good *nor* bad; *both* like *and* dislike). In translation these can lead to problems, as Henningsson et al. (1998, p. 3) document for isiZulu: "The neutral position is difficult to get across when the two opposed words have the same root in isiZulu. An adjective of this type [like 'satisfied'] is usually translated by a verb, in this case meaning 'to-be-satisfied,' and its negative by the same verb with a negative particle included, 'to-be-not-satisfied.' 'Neither... nor' is rendered by 'not... not' (*ngi... ngi*). Then, there is no difference between 'not to be satisfied' and 'to be not-satisfied.' If you are not satisfied, and not not-satisfied, there seems to be no position left."

We need not go far afield for our second example. In everyday contexts in English, it is not unusual to distinguish between "not agreeing" and "disagreeing." However, German has no expression that formally matches "disagree," only terms for "not agree" or "reject" are available. Thus English "agree–disagree" scales can only be translated as "agree–not agree" or "agree–reject." Research on "allow," "not allow," and "forbid" in English (Hippler and Schwarz, 1986) and on the German counterparts, *erlauben, nicht erlauben*, and *verbieten* (Hippler and Schwarz, 1986) indicates that respondents are more likely to endorse lack of affirmation ("not allow") than active negation ("forbid"). Similar effects might well hold for other sets of terms. Thus, methods research is needed to determine which term is best in German for "disagree." The "not agree" term (*nicht zustimmen*) and the "reject" term (*ablehnen*) are both used in German "homegrown" scales. (For research on ratings across the two languages, see Mohler et al., 1998.)

22.2.3 Matching Structures and Measurement: The Example of Gender

Human beings have sexes (genders); languages may or may not have *grammatical gender*. Grammatical gender assigns gender to nouns irrespective of whether they refer to animate or inanimate objects. Spanish and French have fairly elaborate systems of grammatical gender, Finnish has none, and English only vestigial remnants. The number of genders that a language has may also differ; German has three, French two.

While languages differ in whether they have grammatical gender, cultures differ in how important gender reference is. Where inclusive reference is a culturally salient issue (usually perceived as inclusive of women), solutions also differ. For example, English usage avoids gendered nouns ("chair" and "chairperson" are preferred to "chairman" and "chairwoman") and avoids using masculine forms when referring to men and women together. Politically correct Danish reference, in contrast, has the masculine forms for both sexes, avoiding the feminine forms (cf., "actress") formerly also used. Politically correct German, on the other hand, provides male and female forms side by side (cf., "policeman/policewoman").

Grammatical gender is of import for questionnaire translation because it may automatically assign a gender to someone talked about or talking, and this may change the stimulus respondents receive (cf., Harkness, 2003). In the CDC's Behavioral Risk Factor Surveillance System (BRFSS), as in many other health surveys in the United States, respondents are asked the following question: "Have you ever been told by a doctor, nurse or other health professional that you had asthma?" (*http://www.cdc.gov/brfss/pdf-es/2002brfss.pdf*)

A Spanish translation was required for U.S. fielding. In many forms of Spanish, it is normal to use the masculine form of words to refer to both men and women. Theoretically, we could expect in Spanish the masculine form for "doctor" and the masculine form for "nurse." However, possibly since very few nurses are male, the masculine form *enfermero* is still understood as "nurse who is a man." The U.S. Spanish translation, therefore, has the term for a male doctor and the term for a female nurse: "Alguna vez un doctor, enfermera o professional de la salud le ha dicho que tiene asma?" *http://www.cdc.gov/brfss/brfsques-qustionnaires-esp.htm*) In other languages, the issues change. In German, the term usually associated with "nurse" (*Krankenschwester*) is used only for female nurses, but a gender-neutral term, rather like "nursing staff member," could be used. German has no gender-neutral term for "doctor," however, and the masculine and feminine forms would both be used (*Arzt/Ärztin*) to achieve inclusive reference

Social realities may mean that the effects are negligible and, at least pragmatically, that item bias in translation can be discounted. At the same time, it is possible that some respondents could understand the Spanish version to mean that in answering, they should discount female doctors and male nurses. To date we have no tested evidence of the effects of such changes in reference. Research in other disciplines (on bilinguals) indicates that the grammatical gender of the forms used influences perceptions (e.g., Eberhard et al., 2002).

22.2.4 Matching Reference versus Meeting Social Norms: The Example of the "Tu/Vous" Distinction

Many languages have highly complex pronominal systems that take note of number, gender, age, kinship or in-group/out-group relationships, and status and prestige. In contrast, the personal pronoun system of present-day English contains few distinctions beyond singular/plural; first, second, or third person; subject/object use; and genitive forms.

Older forms of English had what is often called a *tu/vous distinction*, distinguishing between "you familiar" (thou, thee, thine) and "you nonfamiliar." Present-day English is, in fact, unusual among Indo-European languages in not having this distinction. Although *tu/vous* distinctions differ in structure and functions across languages, the *tu* form is typically used to indicate relational proximity and the *vous* form to indicate the reverse. The lack of this distinction in English source questions poses a special but persistent problem for questionnaire translation. Language-specific differences apart, adult users of languages with a *tu/vous* distinction address young children with the familiar *tu* form, irrespective of whether or not they know the children. They would tend to address young people they did not know (midteens) with the distal *vous* form, in acknowledgment of their nonchild status. When one and the same questionnaire is to be used for different age groups, this can become a translation or version problem.

Implementing *tu/vous* tailoring in translations is often more a problem of administrative logistics and costs than of how to translate. Appropriate individual versions can be produced on paper, tape, or computer. If planned for and funded, face-to-face interviewers could switch scripts on the spot; telephone interviewers could be cued by information on age and a respondent's voice and linguistic performance. At the same time, each added version increases costs. The U.S. National Household Survey on Drug Abuse, for example, included a Spanish version for youth-only modules in which the familiar pronoun *tú* and the corresponding verb form were used. However, providing age-tailored *tú/usted* audio-CASI tapes was not cost-effective given the relatively small number of Spanish monolingual youth interviewed.

22.3 CURRENT PRACTICE

The number of translations required to be able to field in one country can vary considerably, and this may affect the type of translation procedure able to be adopted. Some countries have several official languages and require written translations for these (e.g., Switzerland). Other countries may have large or key minority populations who do not speak the majority (or the official) language. The Philippine member of the ISSP, for example, regularly conducts ISSP surveys with questionnaires in five languages. (Ethnologue.com lists 169 living languages for the Philippine islands; some have only a handful of speakers: *http://www.ethnologue.com/show_country.asp?name=Philippines.*)

Unless coverage and written translations are obligatory, cost–benefit considerations will decide whether a population is included in a survey, and if it is, what form translation takes. In a review of barriers to the inclusion of language minorities in U.S. surveys, cost was seen as the most significant barrier (Li et al., 2001). In within-country research in particular, the financially cheapest solution is to exclude sample units who cannot use the language(s) already available. Not speaking the majority language goes hand in hand with social exclusion. Depending on the survey topic, it may therefore be appropriate to exclude barely integrated populations (e.g., recent immigrants in a study on national election campaign issues).

If minority-language speakers do need to be covered, funds may not permit a written translated questionnaire for every language. Various strategies then come into play. The Project on Human Development in Chicago Neighborhoods included a longitudinal survey of children and their primary caregivers. The first wave had written questionnaires in English, Spanish, and Polish. Households in which the child or caregiver could be interviewed in one of these languages were included in the study. Because of the longitudinal nature of the project, item nonresponse (all of the items for a particular household member) was accepted in the expectation that by the following wave, the person would speak enough English to participate. It became apparent in the first wave that the small number of monolingual Polish speakers did not justify the cost of a full Polish translation. As a result, a small number of key items were translated that Polish bilingual interviewers read out, recording the responses on an English questionnaire.

In multiple language contexts, interviewers might be equipped with five or six different language versions of questionnaires, but hardly with 15 or more. Faced with dealing with multiple languages and dialects in Indonesia, the WHO World Mental Health Initiative provided interviewers with a full translation in the national language, Bahasa Indonesia, and with translations of the key terms and items for multiple other languages, to be used alongside the majority language questionnaire. In the United States, the Census 2000 forms were principally self-completed. Although questionnaires were available in five languages other than English (Spanish, Chinese, Korean, Vietnamese, and Tagalog), language assistance guides were available in 49 other languages (de la Puente, 2002).

Sometimes bilingual interviewers are asked to provide ad hoc translations of the written questionnaire. Interviewer costs may well increase in this case, but the effort, time, material, and staff costs needed to produce written translations are saved. If recruiting bilingual interviewers is impractical (e.g., when it is difficult to predict what languages will be needed and how many respondents are involved), bilinguals are sometimes recruited ad hoc to act as *interpreters* on site (e.g., neighbors, workmates, children). When several language versions are being fielded, ad hoc translators (the interviewers) might translate from a translation that suited them better than the source questionnaire, thereby possibly increasing difference across interviews.

Such ad hoc procedures are usually defended on the grounds that the interviews are "better than no interview." Given the inevitable variance across performances and the lack of training and monitoring of those involved and given the care taken in monocultural contexts to ensure that interviewers *stick to their scripts*, the discrepancy in quality requirements for foreign language implementations is puzzling. In employing ad hoc procedures, a project basically abandons control over what is said and what is replied. Variation across interviews is unavoidable because on-site translators and interpreters, no matter how good, perform live.

22.3.1 Asking-the-Same-Question Approaches and Close Translation

Questionnaires should sound or read like questionnaires, not like translated questionnaires (cf. *covert translation*; House, 1977). If questions are translated too closely, because clients expect word faithfulness or because translators misjudge what is needed, respondents may be misled, asked a different question, or, at best, puzzled by stilted text. A stilted text is itself a message and a change of stimulus, in much the same way that a foreign accent is. Without being shown what "stay close to the original" should be, translators could focus on words rather than on the intended meaning of questions. Research on students of translation shows that inexperienced translators work more on the level of words than on the level of unit meaning, thereby increasing the likelihood of too close translation. Survey research, we note, often employs inexperienced people who translate infrequently and/or have little training.

Effects related to overly close translation fall into four broad categories. Several of these may apply for one and the same translation:

- Translators focus on meaning of words rather than meaning of questions.
- Respondents perceive a different question from that intended.
- Processing burden is increased for respondents.
- Translated questions do not sound natural.

Focus on Meaning of Words Rather Than Questions A Belgian translator was preparing to translate the question: "Do you think women should work full time, part time, or not at all if..." (followed by a set of family scenarios in a Likert-type question format (ISSP 2000). The question was already available in Dutch (Flemish), which is very similar to Belgian Flemish, and the Belgian translator had the Dutch translation. "Should" had been translated in the Dutch questionnaire in one of two ways, and the Belgian translator asked the project's translation advisor which of the two was correct for this question. One translation of "should" would indicate women have some kind of *duty or responsibility* to work, and the other would indicate that women should be *allowed to* work under the conditions specified. However, the English question aims to have respondents indicate whether they are in favor or not of women working under the conditions described. Thus neither of the translations proposed for "should" would capture the intended *sense of the question.*

Respondents Are Asked a Different Question A battery in the ESS 2002 requires respondents to indicate whether they consider themselves to be like a person described. One of the items (for male respondents) begins: "He looks for adventures...." A salient reading for the French close translation (*Il cherche les aventures*) is that the person is interested in finding and engaging in sexual adventures. Translated closely into German [*Er sucht (die) Abenteure*], one reading is that the person has lost certain adventures and is trying to find them again. A grammatically correct inclusion of a definite article in German (*die*) could also prompt listeners to ask "What/Which adventures are you talking about?"

Processing Is More Complex Phrases such as "if any" and "if at all" are useful phrases for survey questions in English, since they cover the eventuality that the question might not actually apply for some respondents: "How many times, *if at all*, have you done *X*?" (emphasis added). Translating this into Spanish produces a longer and more awkward text. Instead of a short and useful phrase, the translation has a full clause, is cumbersome, and potentially confusing: "*Si alguna vez hizo X, ¿cuántas veces lo ha hecho?*" ("If you have ever done *X*, how many times have you done it?")

Stilted Text for Respondents An ESS 2002 question measuring time spent on watching TV begins: "In an average week, how many hours do you spend watching. . . ." A close French translation "dans une semaine *moyenne*" was proposed. The idiomatic translation would be "dans une semaine *normale*," which is semantically closer to English "in a normal week." Close translation here would result in unidiomatic French. Obviously, idiomatic and proper use in one language cannot be expected to match up with idiomatic and proper use in another.

22.4 QUALITY PROCEDURES AND PROTOCOLS FOR TRANSLATION AND ASSESSMENT

In the next sections we outline briefly our recommendations on key procedures for translation and assessment.

Who should translate? Survey literature advocates that translators should be bilinguals, professional translators, people who understand empirical social science research, or some combination of these. None of these descriptions adequately assists the researcher in selecting a translator. Translating skills are more important than survey translation experience, and people will often need to learn on the job. Team approaches offer useful opportunities to train translators. Learning by simply doing (with no mentor) is unlikely to be successful, and translators who have had experience in translating questionnaires but were never actually trained may also prove difficult to (re-)train.

Performance and experience as well as official credentials are the best indicators for selecting a translator. The people most likely to be good questionnaire translators are people who are already good translators and who learn or are trained to become questionnaire translators. Informing them about the measurement components of questions and design requirements puts translators in an optimal position to produce good translations. They are also more likely to recognize problems and to be able to point out when a requirement cannot be met. Equipping them properly for the task with appropriate briefing and information on the source questionnaire (see Section 22.6) enables them to perform better (Gile, 1995; Gutknecht and Rölle, 1996; Hambleton, 1993; Holz-Mänttäri, 1984; Kussmaul, 1995; Wilss, 1996).

What tools do translators need? What translators produce as a translation depends on their ability and training, the quality of the source material, and the task specifications they receive. Translators need to be given support materials, example texts, and other information relevant for their part in producing instruments, including information on the target audience and administrative mode.

How many translators? In some traditional disciplines, such as literary translation, it is normal to use one translator. In others, such as software localizing, sharing work among translators is common. In survey work, the practice varies. However, survey translations should always be assessed. As soon as review or assessment is included, several people are involved. In approaches that specify "a translator," therefore, what is usually meant is that one person produces the first translation.

However, relying on one person to provide a questionnaire translation completely on his or her own can be problematic for a number of reasons. Working with several, on the other hand, may initially be complicated. An exchange of versions and views as part of the review process makes it easier to deal with regional variance, idiosyncratic interpretations, and inevitable oversights. Group discussion, including input from survey fielding staff, helps to identify comprehension problems for low-literacy populations and ambiguities more easily missed by someone working on his or her own (cf. Hambleton and Patsula, 1999). As described below, using more than one translator also allows for economical *translation splitting*.

22.4.1 Team Approaches

Different forms of team approaches have a long tradition in the translation sciences. Nida (an expert on Scripture translation) describes the general approach as "providing a greater balance of decision" (1964). Team approaches to survey translation and assessment have been found to provide a richer pool of options to choose from for translating items and a balanced critique of versions (Acquadro et al., 1996; Guillemin et al., 1993; Harkness and Schoua-Glusberg, 1998; McKay et al.,1996). The team can be thought of as a group with different talents and functions, bringing together the mix of skills and disciplinary expertise needed to produce an optimal version. Key members of the team need to have the cultural and linguistic knowledge required to translate appropriately in the required varieties of the target language. Collectively, members of the team also supply knowledge of the study, of questionnaire design, and of fielding processes (cf. Acquadro et al., 1996; McKay et al., 1996; van de Vijver and Hambleton, 1996).

The general framework for a team approach incorporates procedures of *translation, review*, and translation *approval*. Depending on the approach (see below), different groups may undertake each of these or individuals may fulfill several roles. There is broad general agreement on the skills and functions required for each role. The *translators* must be skilled practitioners who have received training on translating questionnaires. Translators generally translate out of the source language into their strongest language. (In most cases this is a person's "first" language.) *Reviewers* have at least as good translation skills as the translators but are familiar with questionnaire design principles as well as the study design and topic. One reviewing person with linguistic expertise, experience in translating, and survey knowledge is generally sufficient. If one person with these three areas of expertise is not available, two can cover the various perspectives. *Adjudicators* make the final decisions about which translation options to adopt. They understand the research subject, know about the survey design, and if not proficient in the languages involved, must be aided by a consultant who is.

Nida (1964) identifies three teams or committees, based on their primary function: consultative, review, or editorial. In his scheme, *consultative committees* consist of persons who *review and approve* the work but do not contribute directly to the translation. *Review committees* in Nida's scheme are comprised of people with substantive knowledge or other expertise. They may suggest changes in a translation based on their knowledge of the topic or the source questionnaire background and intent. Nida's *editorial committees* are those charged with creating the first draft of the translation. WHO's World Mental Health Initiative used an approach similar to Nida's: Two independent translations were submitted to an expert panel comprised of the original translators, substantive experts in the field (in this case, psychiatrists), and people familiar with previous versions of the instrument (the CIDI2.1). The panel reviewed problematic items and made recommendations for changes. Item alternatives were tested using in-depth interviews, focus groups, and formal pretests to resolve decisions when views differed.

In the Schoua-Glusberg committee model (Schoua-Glusberg, 1992), translation, review, and approval (adjudication) are interlinked processes. The adjudicator approving the translation is simultaneously the senior reviewer, who works with translators to arrive at a final version. The TRAPD team approach used in the European Social Survey sets out to accommodate multilingual contexts, multiple translations within one country, and language sharing across countries. TRAPD is an acronym for *translation, review, adjudication, pretesting, and documentation*, the five interrelated procedures that form the basis of the approach (Harkness, 2003). In this approach, adjudicators carry the final responsibility for translation decisions but need not have the translation skills required of reviewer and translators, nor possibly even expertise in all the required languages.

It is always necessary to "sign off" a final version. However, a different form of "sanctioning" a translation may also be necessary to satisfy constituency groups or political concerns. Such endorsements may be essential to receive government or community approval for a project. Examples can be found in surveys of indigenous minority populations, such as the Maori in New Zealand, or of immigrant populations, such as Latinos in the United States. The "committee" in this instance may consist of members of the constituent groups, government officials, substantive experts, and other stakeholders. These individuals may also provide cultural and ethnic insight and facilitate testing and administration by securing access to respondent pools. In a recent WHO study, some participating countries requested that the leading national substantive experts review and endorse the translated instrument before government approval would be given for the study.

22.4.2 Organizing Translations

Whether one or more translators are used, it is important to involve the translator(s) in the review process. Toward this, translators should be asked to keep notes of their translation queries, compromises, and any remaining problems. Templates can be used for this. When more than one translator is used, the actual work of translation can be organized in either parallel translations or a split translation.

In *parallel translations*, several translators make independent parallel translations of the same questionnaire (Acquadro et al., 1996; Brislin, 1980; Guillemin et al., 1993; Schoua-Glusberg, 1992). At a reconciliation meeting, translators and a translation reviewer go through the questionnaire discussing alternatives and agreeing on a version. The adjudicator may participate in the review process; otherwise, the version produced through discussion moves on to adjudication.

Task splitting can save time and effort (cf. Schoua-Glusberg, 1992). For *split translations,* the translation is divided up between translators in the alternating fashion used to deal cards in card games. This ensures that translators get an even spread of material from the questionnaire. Each translator translates his or her own section. Translators and the translation reviewer meet to discuss the translations and agree on a version. The adjudicator may be involved in the review; alternatively, the reviewed version moves on to adjudication.

Parallel translations are the more expensive option. However, the decision to have one translation, two parallel translations, or a split translation typically depends on more factors than cost alone. Split translations may be the obvious best choice for an experienced team used to working with one another and versed in note taking and consistency checks. For a group new to the procedures, or when translators are being trained, parallel translations may be better. An important key to successful team approaches using one translator seems to be unusual expertise of the translator and the reviewer/adjudicator (Montalbán, 2002).

22.4.3 Harmonization

Harmonization is the process of modifying translations across countries sharing a language in order to make one version suitable for use everywhere; for example, using the same Spanish instrument in Spain, Colombia, Mexico, and Puerto Rico or using the same French instrument in France and in the French-speaking parts of Switzerland, Canada, and Belgium. Translations completed in one country context are sometimes used in another without regard for cultural or country differences. Harmonization acknowledges these differences and attempts to find terms and phrases that are commonly understood across countries. When these cannot be found, documentation of the country differences becomes an important part of the process.

For the WHO World Mental Health Initiative, harmonization was undertaken for the Spanish translation for Spain and all Spanish-speaking Latin American participants. The process involved multiple meetings of all the participating country project directors (bilingual substantive experts) and the various translators. Phrases and items thought to be problematic in the English source document were first identified and alternative translations suggested. If consensus was reached, one translation was adopted. Where different versions were retained, the country-specific items were documented in a template. Countries sharing languages in the ESS 2002 produced a national version, then met to discuss harmonization. The same template format was used to document translations and harmonization decisions.

22.5 PRETESTING AND TRANSLATION ASSESSMENT PROCEDURES

Although pretests for translated questionnaires are not yet established practice everywhere (de la Puente, 2002), there is a growing general recognition of the need to pretest. In practical terms, introducing pretesting makes budget reallocations necessary, and reluctance to pretest can still be expected in contexts where multiple translated versions are needed.

It is useful to distinguish between procedures that assess the quality of translations as translations and those that assess how translated questions perform as instruments. Survey instrument evaluation must address both translation and performance quality.

Evaluations of the *translations* focus on issues such as whether the substantial content of a source question is captured in the translation, whether there are changes in pragmatic meaning (what respondents perceive as the meaning), and whether technical aspects are translated appropriately (e.g., linguistic *and* survey appropriateness of answer scales). Approaches combining translation, review, and adjudication are seen to be the most useful ways to evaluate and improve translation quality (see Section 22.4.1).

Assessments of *performance* can focus on how well translated questions work for the target population, how they perform in comparison to the source questionnaire, and on how data collected with a translated instrument compares with data collected with the source questionnaire. In the first case, assessment may indicate whether the level of diction is appropriate for the sample population, in the second, whether design issues favor one population over another, and in the third, whether response patterns for what is nominally "the same question" differ (or do not differ) in unexpected ways across instruments and populations.

Translation quality and performance quality are obviously linked, but good translation does not suffice to ensure that questions will function as desired in translation. Thus, well-translated questions may work better for an educated population than for a less well-educated population of the same linguistic group, either because the vocabulary is too difficult for the less-well educated or because the questions are less salient or meaningful for this group. Problems of question design, such as asking questions not salient to a target (second) population, should be addressed at the design level; they are difficult to resolve in terms of a translation. As testing literature points out, question formats also affect responses if the chosen format is culturally biased and more readily processed by respondents in one culture than in another (cf. Geisinger, 1994; Solano-Flores and Nelson-Barber, 2001; Tanzer, 2004).

22.5.1 Qualitative Techniques and Procedures

The *review and adjudication* procedures described earlier are key qualitative procedures for evaluating and refining translated questionnaires. Other qualitative procedures include *focus groups* and *cognitive interviews* with the target population, *field staff appraisals* of questionnaires, and *interviewer and respondent debriefings* after pilots and pretests. Less commonly used procedures are *think-alouds* and cognitive interviews with translators (Harkness, 1996). *Back translation* combines mechanistic steps with qualitative procedures.

Focus groups can be used to gain target population feedback on item formulation and how questions are perceived (Schoua-Glusberg, 1988). They are generally not suitable for assessment of entire (lengthy) questionnaires. To optimize their efficiency, materials pertinent for many items can be prepared (fill-in-the-blanks and multiple-choice tasks) and participants asked to explain terms and rate questions on clarity. At the same time, oral and aural tasks are more suitable than written when target population literacy levels are low or when oral/aural mode effects are of interest. Focus groups conducted to validate the Spanish translation of the U.S. National Health and Sexual Behavior Study (NHSB) revealed that participants did not know terms related to sexual organs and sexual behaviors considered unproblematic up to that point (Schoua-Glusberg 1989).

Cognitive interviews are an alternative means of collecting target population feedback, allowing problematic issues to be probed in depth. In the U.S. National Survey of Family Growth (Cycle VI), cognitive interviews conducted to validate the Spanish translation helped identify terms not well understood across participants and terms requiring alternative treatment for different groups of Spanish-speakers.

Interviewer and respondent debriefing sessions are valuable opportunities to collect interviewer and respondent feedback and probe comprehension of items or formulations. Debriefing sessions for the 1995 ISSP National Identity module in Germany revealed comprehension problems with terms covering ethnicity and confirmed cultural perception problems with questions about "taking pride" in being German.

Back translation is a procedure proposed originally to help researchers assess questionnaires in languages they could not read. The translated questionnaire is translated back into the source questionnaire language, and the two versions in the *source* language are then compared. On the basis of differences or similarities between the source questionnaire and the back-translated version, conclusions are drawn about the quality of the translated version, which itself is not evaluated (cf. Brislin, 1970, 1976, 1980; Sinaiko and Brislin, 1973; Werner and Campbell, 1970).

Weaknesses of back translation as an assessment tool have long been acknowledged, but there is growing recognition of its inadequacy as a systematic technique, in particular in comparison to textual appraisals of the translated target text (de la Puente, 2002; Geisinger, 1994; Hambleton and Patsula, 1999; Harkness, 2003; Schoua-Glusberg, 1997). We should also keep in mind that language consultants will be needed to make any translation emendations after assessment. Their presence obviates the need to resort to the rough assessment that basic back translation can provide. Even if reliable language consultants are unavailable, think-aloud protocols and probe interviews with translators have been shown to provide more information of value for improving translations (cf. Harkness, 1996). In general, a review-adjudication appraisal *of the translation* is to be preferred (e.g., Geisinger, 1994; Hambleton and Patsula, 1999).

22.5.2 Quantitative Techniques

Textual assessment of translation quality does not suffice to indicate whether questions will actually function as required across cultures; statistical analyses are required to investigate the measurement characteristics of items. Field pretesting options include split-ballot administrations to bilinguals, double administrations to bilinguals, pilot runs with the target population, and full field pretests (e.g., Hambleton, 1993; McKay et al., 1996; Sinaiko and Brislin, 1973). The central aim in testing items is to detect bias of different types that distort measurement systematically.

The emphasis placed on assessing translated instruments and the strategies employed differ across disciplines. In part, this is because instruments that are copyrighted and distributed commercially (as in health, psychology, and education) are also often evaluated extensively in pretests and after fielding. In addition, the type of instrument involved and the number of items covering a construct of interest may determine which assessment strategies are suitable. Some strategies call for a large number of items (e.g., item response theory); these are thus unsuitable for social science studies that tap a given construct or dimension with only one or two questions. Small sample sizes may rule out other strategies (such as multidimensional scaling and factor analysis), while others again are relatively unfamiliar in the social sciences (e.g., multitrait multimethod).

Statistical tests thus take various forms, depending on the characteristics of an instrument, the sample sizes available, and the focus of assessment. There is a large literature on the application of these to translated instruments, in particular in educational and psychological research. For general discussions, see Geisinger, 1994; Hambleton, 1993; Hambleton and Patsula, 1998; Hambleton et al., 2004; van de Vijver, 2003; van de Vijver and Hambleton, 1996; van de Vijver and Leung, 1997). Popular techniques used to explore measurement invariance and identify differential item functioning include variance analysis and item response theory (see, e.g., Allalouf et al., 1999; Budgell et al. 1995; Hulin, 1987; Hulin et al., 1982). Factor analysis (target factor rotation), confirmatory factor analysis, and multidimensional scaling can also be used to undertake dimensionality analyses (e.g., Fontaine, 2003; Reise et al., 1993; van de Vijver and Leung, 1997). Where scores are relevant (e.g., in credentialing tests) a design is needed to link scores on the source and target versions (cf. Geisinger, 1994).

One drawback for attitudinal and behavioral research is that the procedures referred to here call for an abundance of items or at least a multiple-item battery or scale. For the one-item case, item bias can be estimated using multitrait, multimethod procedures, as described in Scherpenzeel and Saris (1997) and Saris (2003). Post hoc analyses that examine translations on the basis of unexpected response distributions across languages are usually intended to help guide interpretation of results, not translation refinement (e.g., Braun and Scott, 1998). Caution is required in using such procedures for assessment because bias may also be present when differences in univariate statistics are not.

22.6 SURVEY DOCUMENTATION AS RELATED TO TRANSLATION

In a multiple language context, documentation is as critical to the translation process as it is to the final product. Several different stages and types of documentation are required, including documents for translator instructions, documents to ensure consistency across translations/translators, and documents that record the translation decision-making process. If the source questionnaire is still under development while translations are proceeding, documents are needed to track changes. Further documents are needed if the survey is computerized in order to integrate translations into the computer code. The final translation also needs to be documented for end users. We discuss each of these forms of documentation below.

22.6.1 Background Documentation for Translators

Questionnaires look simple but are carefully crafted tools of measurement. This is one reason why fairly short questions, such as "How many people, including children, live in this household?" are often accompanied by definitions of what counts as a household and what counts as "live in" or being a member of the household. In similar fashion, translators making a French translation would benefit from knowing whether the household definition centers on shared cooking and/or financial arrangements or has more to do with shared rooms and accommodation. The choice between different French terms (such as *ménage* and *foyer*) could lead to a different household composition count.

Annotations for translators in source questionnaires serve purposes similar to those of notes for interviewers. They are not intended to explain what words or phrases mean in ordinary terms but to provide information on what is meant in survey measurement terms. The ESS 2002 questionnaires carry modest annotations of this kind. Much of this information for translators could be produced as part of the design procedure and added to source questionnaires the way notes are provided for interviewers in a monolingual questionnaire in question-by-question or Q×Q specifications.

22.6.2 Translation Process Documents

In social science survey projects, documentation of translation decisions and difficulties is, to date, rare. One exception is the study documentation for the International Social Survey Programme (ISSP), which includes a short description of translation procedures and problems (e.g., Harkness et al., 2001). ISSP documentation is produced after the study has been completed. Experience with ISSP translation documentation indicates that records should be kept while the work is being done so that report writers do not need to rely on recollected accounts. With this in mind, translation protocol templates were developed for the ESS to facilitate *concurrent* translation documentation. Apart from the need to document for end users, reviewers and adjudicators may need to consult such notes in order to decide on the "final" or "best" choices. This documentation process can initially be time consuming but will make the review and adjudication process more efficient.

Monolingual and multilingual diagnostic instruments of long standing are frequently documented. For example, WHO's World Mental Health Initiative uses an expanded version of WHO's Composite International Diagnostic Interview (CIDI 2.1). The translated versions of the questionnaire are archived along with documentation of all the country-specific changes that occurred in the source document as a result of the translation process or other feedback. Harmonization decisions are also recorded across questionnaires for different versions of the same language.

22.6.3 Tracking Changes

It is important for consistency to track reoccurring sections across an instrument. Tracking documents generally start by listing all reoccurring phrases and words and their location in the questionnaire. As the translation proceeds, translation alternatives and then final translations are recorded. Once the final version is approved, it is a simple task to check that all instances of an item or section have been translated (or updated) consistently. Translator software tools can help automate this process.

In survey translation, the received wisdom has been to finish the source questionnaire before undertaking any translation activities in order to minimize version control and, presumably, costs (but see "advance translation"; Harkness, 2003). However, following this advice limits how the translation process can inform or improve the source document. Increasingly, major cross-national surveys are building in formal translation-testing mechanisms. To do so, a mechanism is needed to track and flag changes to the source questionnaire. Widely available document-sharing software can be used for this. In the World Mental Health Initiative, instrument development occurred over a four-year period. Each of the more than 27 participating countries fielded at different points in time over this period. Using a commercial Web-based product called *e-room*, documents were created that tracked every change to the source questionnaire by date, time, and the person making the change.

22.6.4 Documentation for Computerized Instruments

Translation documentation takes on another level of complexity if the instrument is computerized. Various approaches can be used to incorporate the translation, but ultimately it needs to be integrated with the computer code. When the integration step is undertaken, "hidden" phrases may need to be updated and translated. These hidden items include *fills* such as "he/she/they," interviewer instructions, error messages, and the like. These are generally translated after code integration and pose a new set of issues with regard to consistency across the questionnaire. Anticipating this material while constructing the source instrument can facilitate the process considerably.

22.6.5 Documenting Final Versions

Documentation of problems helps inform later versions of a study. In addition, if changes are made over time, records need to be available of the series of changes (or their absence) across translations. Secondary analysts can benefit from records of unavoidable differences or adaptations and from notes on mistakes found after a questionnaire was fielded. Countries sharing languages and harmonizing need to indicate, for example, where they do differ in formulation. Moreover, countries joining a project at a later date that share languages with countries already involved can also benefit from this documentation.

22.7 CONCLUSIONS

Developments are under way to establish recommended guidelines and procedures for survey translation for national and cross-national research (e.g., de la Puente, 2002; Hambleton, 1994; Hambleton, 2004; van de Vijver and Hambleton, 1996; and the International Testing Commission guidelines at *http://www.intestcom.org/itc_projects.htm*). We expect increasing cross-disciplinary exchange on issues that are central to translation to result in further improvements in tools and protocols for social science survey translations and assessment.

Many different fields have strengths from which cross-cultural survey procedures stand to benefit. Monocultural survey research has established reliable standards for testing monolingual instruments that can be adapted for multilingual work. The International Testing Commission has researched and published extensively on how to test measures and their translations for bias (*http://www.intestcom.org*). Cognitive survey research has explored the relevance of communication norms and pragmatically grounded theories of meaning (e.g., Sperber and Wilson, 1986) for understanding how respondents perceive and process survey information. Recent cognitive research illustrates how cultural framing shapes respondents' readings of questions (Braun, 2003) and how cross-cultural differences in perception and sensibility affect how respondents respond (Schwarz, 2003). These and further insights can gradually be applied to the translation of questionnaires in much the way that Katan (1999) suggests that we should deliberately capture culture in translation.

Linguistics can be expected to offer an ongoing refinement of our understanding of different types of meaning. Although only beginning to deal with questionnaire translation issues, translatology offers proven standards of recommended practice, refined conceptions of the goals of different types of translation, and details of procedures to be used to realize each. The field has sophisticated techniques for training translators, including training on electronic workbenches essential for some fields. Last, but not least, cross-cultural communication research has uncovered mechanisms that result in misunderstandings across cultures (e.g., Guirdham, 1999; Scollon and Scollon, 1995). Given more research, we can look forward to a time when these insights can be applied to improve procedures for producing questionnaires for multilingual fielding.

At the same time, it seems likely that increased cross-disciplinary exchange will also lead to changes in the nature of what counts as "asking the same questions" in translation. As recent discussions of respondent tailoring and mixed-mode implementations in monocultural research reflect (e.g., Dillman, 2000a), keeping things structurally and substantively the same is no longer automatically seen as the optimal route to take. In the long term we can expect that these developments will also affect cross-cultural designs and translations.

CHAPTER 23

A Multiple-Method Approach to Improving the Clarity of Closely Related Concepts: Distinguishing Legal and Physical Custody of Children

Nora Cate Schaeffer and Jennifer Dykema
University of Wisconsin–Madison

23.1 INTRODUCTION

The influence of question order on respondents' answers, particularly on answers to subjective questions, has been studied for decades. Nevertheless, there are relatively few studies that attempt to take advantage of what is known about these effects to manipulate question order in order to clarify concepts for respondents. As an example, when respondents are asked how many hours they *actually* worked in the preceding week, some report the number of hours they *usually* work instead. But if an instrument asks first about the number of hours the respondent usually works and then about the number of hours actually worked, the sequence of questions creates a contrast that appears to help clarify that the target concept for the second question is the actual number of hours worked (Rothgeb et al., 1991). Similarly, when respondents are asked questions about possible causes of AIDS, asking about transmitting AIDS by transfusion before asking about transmission by donation reduces the proportion incorrectly reporting that AIDS can be contracted by donating blood (Colasanto et al., 1992).

Methods for Testing and Evaluating Survey Questionnaires, Edited by Stanley Presser, Jennifer M. Rothgeb, Mick P. Couper, Judith T. Lessler, Elizabeth Martin, Jean Martin, and Eleanor Singer
ISBN 0-471-45841-4

Legal custody and *physical custody* are two closely related concepts that have been problematic in studies of separated families in the United States. When parents live apart from each other, either because of divorce or because they never married, it is common for courts to rule about two important aspects of the children's lives: (1) about legal custody or who has the right to make important decisions for the children; and (2) about physical custody or where the children should live.

Although these two concepts are important in the study of separated families in the United States, they are difficult to measure for several reasons. First, some of the terms used to describe custody arrangements in legal documents also refer to lay concepts, but the lay concepts may be somewhat different from the legal concepts. Second, the common language terms may be used routinely in ways that are more flexible or ambiguous than their legal homonyms, leaving question writers no choice but to use terminology that is inherently ambiguous and has no clear legal referent. For example, in everyday talk a *custodial parent* is the parent with whom the children live most of the time and the term *custody* may be understood to mean physical custody. However, the children's actual living arrangement may or may not have been ratified by the court and may or may not correspond to the court's order about placement (physical custody), and the noncustodial parent may or may not have the legal right to make decisions about the children (legal custody). Third, while most respondents are familiar with the concept of physical custody (the child must, after all, live somewhere), legal custody may not be salient except in some circumstances (e.g., when it is shared with a noncustodial parent). Fourth, the terminology used to describe these concepts has changed over time. What was once called *sole physical custody* is now more likely to be called *primary placement*, and the language with which respondents are most familiar may depend on when they had contact with the legal system.

In the portion of our project reported on here, our target concept is legal custody. This concept has been asked about in the Child Support Supplement (CSS) to the U.S. Current Population Survey (CPS); the CSS is administered approximately every other year and is one of the principal data sources policymakers use to monitor changes in characteristics of separated families. Helping respondents to report accurately about legal custody requires helping them to distinguish between legal and physical custody, for the reasons described earlier. The CSS was revised in the early 1990s in an attempt to improve the measurement of many survey concepts, including legal custody, in order to better monitor the impact of legal reforms. The development effort we report about takes the CSS as its starting point.

Our chapter presents a case study in which we report about efforts to improve the measurement of legal custody using the Parent Survey 3 (PS3), a telephone survey of mothers and fathers in divorce and paternity cases in Wisconsin. Our project used an array of techniques: (1) focus groups were used to explore the common terminology used in this domain; (2) cognitive interviews were used to evaluate questions about legal custody modeled on the CSS and then our revisions of those items; (3) a split-ballot experiment varied the position of the legal custody question in the survey to improve clarity; (4) respondents rated how sure they were about their legal custody; and (5) the interaction between the interviewer and respondent was coded using a detailed system of interaction coding.

Using the Court Record Database (CRD), which contains information about the PS3 respondents from public court records, we evaluate the quality of responses in the PS3. The overall study design offers several ways to evaluate our efforts: the accuracy of the response (by comparing responses in the PS3 and information from the court in the CRD), how sure the respondent was about her or his answer (from the PS3), the relationship between how sure the respondent was and her or his accuracy (using the PS3 and CRD), whether any differences between the two forms of the instrument are reflected in the interaction codes (from the interaction coding of PS3 responses), and whether the interaction between the interviewer and respondent can inform us about the accuracy of the respondent's answer (using PS3 interaction coding and CRD). Because the PS3 is based on the 1994 CSS, the PS3 should already show gains in accuracy over approaches used in surveys done a decade earlier. An earlier survey, the Parent Survey 2 (PS2), also sampled court cases from the CRD, and answers in the PS2 can be compared with the CRD to further assess the results of our development efforts.

In describing the stages of our testing program, we discuss key decisions made and explain why we chose specific versions of questions. The design does not allow us to test the specific contribution of focus groups, cognitive interviews, or interaction coding to the quality of the ultimate survey questions. Instead, we focus on the strength of a case study: the description of issues that arose in applying these methods and how they contributed to the decisions we needed to make in designing the questions. Unlike many case studies, however, this one has the advantage of having a criterion available for assessing its success.

23.2 TECHNIQUES USED IN THE DEVELOPMENT OF THE PARENT SURVEY 3

PS3 applied ideas and principles drawn from research in survey methodology to attempt to improve the accuracy with which respondents reported about a wide range of concepts. The 1994 CSS, revised from earlier years, provided the starting point for the instrument. We revised the CSS for the PS3 following a series of steps that was intended both to identify problems in the conceptualization and operationalization of concepts and to provide resources for solving those problems.

23.2.1 Focus Groups

The development of the PS3 began with focus groups using four key sample groups—divorced mothers, divorced fathers, mothers in paternity cases, and fathers in paternity cases—held between October and December 1994. Two focus groups were conducted with members of each of the four sample groups, but because of the number of topics that needed to be covered, custody was discussed in only six of the groups. Efforts were made to obtain economic and racial diversity within groups that ranged from 4 to 10 participants. Attempts were made to recruit from the CRD, and additional sources included newspaper advertisements, flyers distributed at public buildings and social service agencies, and a mailing to volunteers in a community service organization. Each participant was paid an incentive of $30. Each session had a male and female moderator, sessions were tape recorded, and rough transcripts were made.

23.2.2 Intensive Cognitive Interviews

Following the focus groups, we conducted three rounds of cognitive interviewing between May and December 1995 using similar methods as the focus groups to recruit participants. In each round, we attempted to interview one person from each sample group (e.g., a divorced mother, etc.), although our pool of recruits did not always permit that quota to be achieved. Interviews were tape recorded and item-by-item transcripts were produced.

Six interviewers (including the two authors) were trained in a full-day session that included completing an interview with another interviewer.[1] Then each interviewer completed a practice interview with a live respondent, and interviewers were given feedback based on review of the recorded interviews. Debriefings after each round of interviewing included an item-by-item review of problems, and comments were incorporated into the next round of revisions when appropriate. The debriefings themselves were not formally documented and their contribution to the final instrument cannot be described separately. Interviewers and respondents were matched by gender because of the sensitivity of the topic.

We decided to use more versions of the instrument and fewer respondents in each round to provide an opportunity for multiple revisions. Round 1 consisted of the six practice interviews and eight additional interviews and used questions modeled on the 1994 CSS. Rounds 2 and 3 used different versions of the instrument that were informed by the results of the preceding round. For round 2, two versions of the instrument, round 2X and round 2Y, were fielded simultaneously; nine respondents were interviewed with version X and seven with version Y. Eight participants were interviewed in round 3. The number of responses available for inspection varies because of skip patterns or problems with tape recording.

23.2.3 Parent Survey 3

Although we adopted the stance of the CSS and focused on the measurement of legal custody, our developmental interviewing suggested that *custody* was

[1]During the course of interviewing, the performance of one interviewer was found to be below standard and another's work was also deficient; the final round of interviewing used only the three best interviewers.

ambiguous, even when modified by *physical* or *legal.* It appeared that some parents reported about the more familiar concept—physical custody—even when they were asked about legal custody. The final survey items attempted to improve the clarity of the definition of legal custody by using the order of questions to create a contrast. By placing the question about physical custody before that about legal custody, we offered respondents the opportunity to report about the more familiar concept first; we expected that the distinction between physical and legal custody would be clearer when respondents had already answered about physical custody. The PS3 included a split-ballot experiment to evaluate the effect of this approach. After answering questions about legal and physical custody, respondents rated how sure they were about their reports.

To obtain a criterion to use in assessing the accuracy of responses, the PS3 used a reverse record-check study design and sampled court records from the 1989 to 1992 cohorts of the CRD within two counties in Wisconsin. The CRD includes abstracts of initial court actions and subsequent updates to create ongoing histories for court cases with minor children that petitioned for divorce or paternity; this allows us to determine whether the court ever awarded joint legal custody up to the time of the interview. Two counties were included to increase the diversity of the sample. Computer-assisted telephone interviews were conducted from May to December 1997 (using 1996 as reference period, when appropriate). Parents from 797 court cases were sampled and interviews were conducted with 344 mothers and 222 fathers by the Letters and Sciences Survey Center at the University of Wisconsin. The response rate was 47% for mothers and 31% for fathers. Because it had been some time since many of the cases had been to court, the predominant reason for nonresponse was that parents could not be located or contacted (approximately 80% of the nonresponders); approximately 20% of the nonresponders refused. Analyses of the CRD omit 14 respondents from the PS3 who were not matched in the CRD.

23.2.4 Interviewer–Respondent Interaction Coding Data

Of the 566 PS3 interviews, 529 were recorded on audiotape. Twenty interviews were not taped because the respondent denied permission to be recorded; 17 were lost due to recording errors. We coded behaviors that interviewers and respondents exhibit during the survey interview and that have been studied over a number of investigations (Cannell and Robison, 1971; Cannell et al., 1968; Mathiowetz and Cannell, 1980; Morton-Williams, 1979) along with some new behaviors (e.g., pauses, elaborations). Coders were instructed to code up to three exchange levels and were given detailed instructions for applying the codes. A question–answer sequence began with the interviewer's reading of the survey question and ended when the interviewer started the next question. Within a sequence, a new exchange level began each time the interviewer spoke.

Ten persons, most of whom had interviewing experience, were trained to do the coding. To assess intercoder reliability, a sample of 57 cases, selected from a randomly generated list, was coded independently by two coders and measures of interrater agreement were produced. These cases included a total of 6791 administrations of the questions, which serve as the unit of analysis for the reliability analysis. Interrater agreement is assessed with kappa statistics which provide the ratio of the difference between the observed and expected levels of agreement to the proportion of agreement that is unexplained (see Fleiss, 1981):

$$\kappa = (P_{\text{obs}} - P_{\text{exp}})/(1 - P_{\text{exp}})$$

Values of kappa greater than 0.75 indicate excellent agreement beyond chance; values between 0.40 and 0.75 indicate fair to good agreement; and values below 0.40 indicate poor agreement (Fleiss, 1981, p. 218, citing Landis and Koch, 1977). All of the behaviors fall within the range fair to excellent.

23.3 FOCUS GROUP DISCUSSIONS OF LEGAL AND PHYSICAL CUSTODY

In several of the focus groups, the first question about custody was "What is custody?" without specifying "legal" or "physical" (see Figure 23.1). Both interpretations appeared in the groups: For example, in one group with divorced mothers, one response was that *custody* referred to physical custody, but other participants interpreted it differently: "In Wisconsin, it means legal primarily" and "custody is different from placement." In the groups with fathers, most initial

---

Now I'm going to change the subject. The next topic I'd like to talk with you about has to do with something related to child support, that is, where children live.

- First, let's talk about something very general. Can you tell me, from your point of view, what does it mean for a child to "live" with a parent?
- And let's start by defining some terms. What is custody?
  - What is physical custody? What does *physical custody* mean?
  - What is legal custody? What does *legal custody* mean?
  - Are they the same or different?
  - What is joint physical custody?
  - What is joint legal custody?
- How does a parent get physical custody?
  - Does the parent have to go to court?
- How does a parent get legal custody?
  - Does the parent have to go to court?
- *Concepts*: noncustodial parent
nonresident parent

Figure 23.1 Section about custody from guide used for Focus Group 4.

interpretations focused on physical placement. The range of responses can be illustrated with these comments (FG7-DivFa):[2]

> "Who has legal responsibility over the child."
>
> "The legal definition would probably be primary placement."
>
> "Who has physical custody. We have joint custody over the kids but she has physical custody."

Several themes emerged. One is that parents may use similar terminology to describe both court-ordered arrangements and their actual practice without signaling that the referent had changed:

> "I have total custody of my children, but if I'm not around for a couple of days and they get sick or something, he better take care of them immediately. When I don't have them and he's got them, he has custody of them. When they're with me, I have custody of them, even though, in the eyes of the court, I have sole custody of the children. When he has them, he has sole custody of them, I don't want to hear from him." (FG1-DivMo)

Note that "total custody" at first might appear to indicate that the court has ordered the mother both legal custody and placement; subsequently, "custody" appears to refer to where the children are at a given moment and to the decision-making authority that goes with having the children in one's charge.

The fact that the parent with the children must make decisions about them, which is a potential source of ambiguity for legal and physical custody, is a theme in several comments:

> "I think that in theory there's probably a big difference between physical and legal custody, but in practice there's not much difference." (FG2-PatMo)
>
> "They say there's joint custody but then they say there has to be a primary custodian. ... I think the system is dealing with a double definition of custody that they haven't resolved." [What does custody mean?] ... I guess the primary physical custodian aspects of it." (FG3-DivFa)

Another interpretation was that "legal" custody referred to having the court ratify an agreement about physical custody or placement:

> "Physical custody is where those kids are. Legal custody the court said whoever supposed to have the kid." (FG5-DivMo)

[2]In the transcripts from the focus groups and cognitive interviews, there is an identification number for the respondent or group and an abbreviation that indicates the sample status of the respondent or focus group. The key for the sample status is: Div = divorce case, Pa = paternity case, Mo = mother, Fa = father. Interviewers' probes and comments are shown in square brackets in an abbreviated form. In some places the transcripts have been edited for the purpose of confidentiality and readability.

Nevertheless, other comments distinguished between physical and legal custody in the way required by the survey concepts, for example:

> "I have placement which is physical custody. But we have joint custody on the decisions . . . [Is that legal custody?] That's joint legal custody. [Would you call it that?] Would I call it legal custody? Yeah, I would call it joint legal custody." (FG5-DivMo)

Substantively, the focus groups suggested that caution was needed in using the following terminology:

> "Custody" by itself had many possible interpretations: who had the right to make decisions about the children, where the children actually were at a given time, where the children actually were most of the time, and where the court said the children should live.
>
> "Legal custody" without "joint" was particularly vulnerable to being interpreted as referring to an arrangement about physical custody that had been approved by the court. "Joint legal custody" appeared to be relatively (although not perfectly) clear as a way to refer to shared decision-making authority.
>
> "Physical custody" was frequently understood to refer to where the court said the children should live, but it could also refer to where the children actually live, or where they are at a given moment.
>
> Some participants thought that "placement" referred to a situation in which the court placed the children with someone other than their parents, but many participants used "physical custody" and "placement" as synonyms.
>
> "Joint physical custody" was commonly understood as an arrangement in which the child lived with both parents, but participants varied in what amount of time with each parent was implied by this arrangement.

Methodologically, several problems arise in using material from and drawing lessons from the focus groups:

1. When participants offered a variety of interpretations (as they often did), they sometimes appeared to be following conversational conventions that required "new" contributions rather than "redundant" contributions (Grice, 1975).
2. The group discussion made it difficult to control the context in which a term was presented to participants. For example, in reading the transcripts of the focus groups we have the impression that "joint physical custody" and "joint legal custody" were more likely to be conflated when they were presented as a pair, but it is difficult to mass evidence in support of that impression.
3. By themselves, the focus groups offer little guidance about how to word questions that might communicate survey concepts clearly to most respondents.

23.4 INTENSIVE COGNITIVE INTERVIEWS ABOUT LEGAL AND PHYSICAL CUSTODY

We had two main goals for the portion of the cognitive interviews concerned with legal custody: (1) to determine whether respondents' interpretations of key concepts conformed with analytic goals (e.g., whether respondents' definitions of legal custody matched the analytic definition) and (2) to examine respondents' justifications for their answers so that we could determine whether they were accurately classifying themselves according to analytic concepts. To achieve these goals we used an initial request that the respondent think-aloud followed by structured (scripted) probes that were designed to assess the respondent's understanding of the key concepts (see panel A of Figure 23.2).

We adopted these two approaches because we quickly discovered that some questions could be explored fruitfully using the think-aloud technique and others could not. For example, respondents were usually able to think out loud when calculating the amount of child support they were paid in the preceding year. But when we asked respondents to classify themselves (e.g., with respect to legal custody), many respondents simply "knew" which category they belonged in, and provided brief initial answers that did not contain enough detail to assess whether or not they were interpreting the survey question as we intended. Consequently, we relied heavily on structured probes as a strategy to meet our goals, and our interviews might be best described as intensive cognitive interviews (Esposito and Rothgeb, 1997; Willis et al., 1991).

For the questions about custody, we found that respondents were most likely to think out loud spontaneously in answering the survey question when they had some uncertainty about the question, as exemplified by the following example about joint physical custody:

> R303 (DivFa): Um, well yes they did, they called it joint custody. But uh, but um, um, let's see ... The question is joint custody, I mean joint physical custody in my head, joint custody to me was that if you had, if there were legal questions or medical questions that, that both parents had to agree on before decisions were made. Um, as far as like joint physical custody, er, I don't even know if that means anything unless you're with [inaudible] unless you're with your kids half the time, if you live in the same time, which they don't, so, so now that I forgot the question ... legal custody, joint legal custody and joint physical custody. I guess joint legal custody means you have a say in the decisions, you know, one may go to school, and you know, are they going to get, uh, hepatitis B vaccination And then uh, and then joint physical custody means that if you wanted to you could, I mean, in my case you could press it to where you can have them 50% of the time and hopefully you wouldn't screw them up psychologically down the road. That's another story.

Structured probes had several advantages: (1) topics are covered in the same way in all interviews (e.g., every respondent provided definitions of key concepts

Panel A: Question about Legal Custody from Round 3 of Cognitive Interviewing

504. *Joint legal custody of a child means that both parents have the legal right to help make decisions about the child, for example about medical care or education.*

Did a court, judge, or divorce decree ever give you and (CHILD)'s (father/mother) joint legal custody?

Probes: *YOW.*

What did "joint legal custody" mean to you?

What term would you use to describe the arrangement you have about decisions for your (child/children)- would you call it joint legal custody, joint custody or would you use some other term? **IF OTHER TERM:** What term would you use?

IF YES: Did you ever have a different legal arrangement about legal custody?

Confidence probes:

How certain are you that you that a court, judge, or divorce decree (did/did not) give you and (CHILD)'s (father/mother) joint legal custody: not at all certain, slightly certain, pretty certain, very certain, or extremely certain?

What did you mean when you said you were [ANSWER]?

Panel B: Illustrative Answers to Question about Legal Custody in Round 3 of Cognitive Interviewing

R403 (PaMo): No. [YOW.] It means is, it would mean does he have any right to say anything about S. [What: What did 'joint legal custody' mean to you?] That, um, he's, we would both be, um, the people that make decisions about S. [What term would you use for that arrangement you have about decisions, um, and not you in particular but just in general, would you call this arrangement that we've described joint legal custody, joint custody or some other term?] Joint legal custody seems like a good one. [How certain are you that a court, judge, or divorce decree (did/did not) give you and (CHILD/the children)'s (father/mother) joint legal custody?]. Extremely certain. [And what did you mean when you said that you were extremely certain?] That I'm positive. [Okay.]

R409 (PaMo): No. [YOW.] Meaning does he have legal custody to, uh, um, make decisions and see her. [Um-hum.] No, he does not. [What did 'joint legal custody' mean to you?] Legal custody meaning, um, can that person have the over 90% visitation. Or, you know, um, actual living arrangement with that child. [What term would you use to describe the arrangement you have about decisions for your child. Would you call it joint legal custody, joint custody or would you use some other term?] Um, I guess, cooperative living, something like that, something a little, instead of, I don't know. [Okay.] [How certain are you that a court, judge, or divorce decree (did/did not) give you and (CHILD/the children)'s (father/mother) joint legal custody?] Slightly. Slightly. [What did you mean when you said you were slightly certain?] Uh, I just meant that I'm not really sure if I have exact, you know, if I have the legal custody. [Um-hum.] I'm really not sure. [Um-hum.] I just know that I am the custodial parent but I'm not really sure if they've, you know, um, ordained me to be, you know, this legal parent.

Figure 23.2 Illustrative question, structured probes, and answers from round 3 of cognitive interviewing.

such as legal custody in their own words); (2) analysis was facilitated because the probes and the answers to them could be located easily in the transcripts; and (3) the standard probes helped to guide and control the behavior of interviewers who had little experience conducting cognitive interviews.

When many probes were used in the practice interviews, enough time and discussion had elapsed that the respondent's original answer to the survey question began to seem remote. Based on that experience and advice from other researchers, we used a maximum of four probes. The majority of our probes were of three types: those asking respondents to repeat the question in "your own words" (YOW), to define what a concept "means to you" (MTY), and to say why they chose the answer they did (ANSWER).

Probes that ask the respondent to "replay" all or part of the survey question can be thought of as on a continuum. On one end are probes that ask the respondent to repeat the question, and these may be most useful when one is interested in seeing whether the respondent can retain the elements of the question in short-term memory. On the other end of the continuum are questions that ask respondents for the "gist" of the question. Such probes attempt to discover the "pragmatic" question that the respondent has constructed from the target survey question. Because we were most concerned with the interpretation of the survey question, we used probes (e.g., "Can you use your own words to tell me what that question is asking?") that focused on the respondent's reconstruction of the question.

We also relied heavily on concept probes (e.g., "What does 'joint physical custody' mean to you in this question?") to assess whether respondents shared our definitions of key concepts. Although some respondents simply pointed to a previous answer when asked to state the question in their own words, most provided definitions. These definitions are particularly helpful in examining whether the conceptual distinctions that respondents make mirror analytic distinctions. Similarly, although some respondents replied to the "Why did you choose [ANSWER]?" probe by saying something like "because he does," most provided a justification such as "because that was the order" or "I have been allowed and have seen my son" that lets us assess whether the respondent classified himself or herself accurately given the goals of the question.

There is very little guidance in the literature about how to analyze the data from cognitive interviews. We adopted the following approach. Transcriptions were organized by survey question and, within question, by type of respondent. For each respondent we documented the respondent's initial answer to the survey question and then the respondent's answer to each structured probe separately. Panel B in Figure 23.2 shows examples from round 3. From the transcriptions, reports were written for most items that included the following information: a background section that outlined the population for whom the question was relevant (e.g., all mothers and fathers) and the purpose of the question; a history section that contained the wording of the question from previous rounds of testing and summarized relevant issues (if applicable); a section that highlighted possible revisions of the question to test in the next round of cognitive interviewing (or the PS3 survey for the final round of interviewing); and finally, a decision about the wording for the question.

23.4.1 Round 1: Testing Questions Modeled on the Revised Child Support Supplement

The question used in the April 1990 CSS included the ambiguous phrase "joint custody": "Does the child(ren)'s father have visitation privileges, joint custody, or neither?" The version used in 1992 was more precise and had instructions that said to "mark all that apply," but could still be heard as asking about "legally awarded joint physical custody": "Does (CHILD)'s (father/mother) have joint legal custody, visitation privileges, neither?" The version used in April 1994 was the starting point for our testing: "Joint legal custody of a child means that both parents have the right to help make decisions about the child. Did a court or judge EVER give you and (CHILD)'s (father/mother) joint LEGAL custody?"

One of the authors was involved with development of the 1994 CSS, and the focus groups and interviews conducted as part of that earlier process indicated the potential for confusion between physical and legal custody. The 1994 CSS provided the model for the question that was used in round 1 of cognitive interviewing. In round 1, the question about joint legal custody appeared last in a series of four questions: two questions about visitation privileges, a question about joint physical custody, and then the question about joint legal custody. We followed the CSS in focusing analytic interest on joint legal custody; a question about physical custody was used only to highlight the contrast between physical and legal custody. Because misunderstandings could appear at either question, we discuss both below.

All of the 13 respondents (one interview was not tape recorded) asked whether they have joint physical custody answer "no," which is not surprising because this was relatively uncommon in Wisconsin at the time these parents' cases went to court. Ten of the respondents provide an adequate definition of joint physical custody (that it involves the child spending time with each parent).[3] The answers of the remaining three respondents suggest potential problems. R101 clearly describes joint legal custody instead of joint physical custody:

> R101 (DivMo): No. [And what does joint physical custody mean to you?] Means we both have say so. [And that means . . . what does that mean?] That party cannot make all the decisions, that it's both parties together. And in our case, the judge says, it cannot be done, because we can't talk to each other.

Instead of answering the survey question, R201 provides a report of potentially relevant information (see Schaeffer and Maynard, 2002), which could lead to her being miscoded as "yes" if the interviewer did not understand the survey concept; but this interviewer repeats the survey question and receives a negative answer that appears to be accurate. The respondent's definition of joint physical custody includes joint legal custody as a component.

> R201 (DivMo): Um we had joint custody with me having placement but there were times when CHILD went to stay with his dad. Maybe not for school year purposes but stay with his dad. [Interviewer repeated question] No. [YOW] Whether or not we both had physical custody of him at the same time. [Joint physical custody mean?]

[3]Only three of these 10 respondents mention the role of the court in their definition, but because we were asking about physical custody only in the service of improving our measurement of legal custody, we were less concerned with this component of the definition.

> Meaning that you've got joint custody but the physical placement is with one parent or the other or both parents and you would rotate that from either every few months or every few weeks whatever you decided on. As it turned out ours was, um I had physical placement.

Like some other respondents, R201 uses "joint custody" instead of "joint legal custody" for that concept. The phrase "joint custody," although not common, was used often enough that we were concerned that it might lead to confusion on the part of interviewers or respondents.

When we asked about joint legal custody, most of the initial answers to the question are immediate, but at least two respondents give signs of uncertainty (R205: "to the best of my recollection, no"; R221: "I don't think so. I have to go back and look at the court order"). An additional five respondents provide answers that are incorrect in whole or part: R102 "didn't realize" that joint legal custody has to do with decisions and then says that "custody ... means living with," but that "it sounds like it also means like giving medical consent and things like that." The definitions of joint legal custody offered by R105 ("we both have custody of her ... neither one of us would have to make child support payments") and R106 ("if he's got the right to have her, too"), both sound as though they are referring to physical rather than legal custody. R205 notes that physical and legal custody sound "the same to me."

The first round of interviewing led us to several conclusions: For some respondents, the concepts of joint physical and joint legal custody are so closely connected that they might have difficulty adopting the distinction; the currency of the phrase "joint custody" for "joint legal custody" could lead to misunderstandings by respondents without any evidence of the misunderstanding being available to the interviewer, and interviewers might misinterpret the phrase "joint custody" when respondents used it; "legal custody" is likely to be less familiar to respondents than physical custody; some respondents would (reasonably) answer "no" to the question about legal custody if their arrangement about (physical) custody had not been made through the court. Thus, although "physical" and "legal" custody might be a contrasting pair, the contrast between "joint physical" and "joint legal" custody did not seem to be sharp enough.

23.4.2 Round 2: Testing Questions Revised after Round 1

Informed by the results of round 1, we fielded two versions of the series of questions in round 2. In round 2X we used the same order as round 1, preceding the question about joint legal custody with one about joint physical custody. Round 2Y (which was very similar to the final version in Figure 23.3) began with the more familiar concept of "primary" physical custody or primary placement. To both versions we added an introduction that alerted respondents that we were about to ask about two related concepts and that "it is easy to confuse them." In addition, we added terminology about "placement" to the questions about physical custody.

	Order of Questions		
	Final	Initial	Wording
Introduction (preceded whichever question appeared first in the series)			*The next questions are about the relationship between (CHILD) and (his/her) (father/mother).*
Noncustodial Parent Has Visitation Privileges	1	6	*A parent's right to see a child is sometimes called visitation privileges. Does (CHILD)'s (father/ mother) have the right to see (CHILD), whether or not (he/she) actually does?*
Visitation Privileges Made Legal	2	7	*Were the visitation privileges or right to see the child ever made legal by a court, judge, or in a divorce decree?*
Primary Physical Custody	3	3	Introduction read in the Final Form only: *I'm going to ask about physical custody and legal custody. It's easy to confuse them, so I'm going to read some definitions.* *Sometimes children live with one parent for most of the time. This is called primary physical custody or primary placement.* *Did a court, judge, or divorce decree ever give you or (CHILD)'s (father/mother) primary physical custody or primary placement of (CHILD)?*
Sure of Answer About Primary Physical Custody	4	4	*How sure are you that a court, judge, or divorce decree (did/did not) give you or (CHILD)'s (father/ mother) primary physical custody or primary placement of (CHILD): not at all sure, slightly sure, pretty sure,very sure, or extremely sure?*
Joint Physical Custody	5	5	*IF NO PRIMARY PLACEMENT: Sometimes children live with each parent for part of the time. This is called joint physical custody or shared placement.* *Did a court, judge, or divorce decree ever give you and (CHILD)'s (father/mother) joint physical custody, shared physical custody, or shared placement of (CHILD)?*
Joint Legal Custody	6	1	Introduction read in the Initial Form only: *I'm going to ask about legal custody and physical custody. It's easy to confuse them, so I'm going to read some definitions.* *Joint legal custody of a child means that both parents have the legal right to help make decisions about the child, for example about medical care or education.* *Did a court, judge, or divorce decree ever give you and (CHILD)'s (father/mother) joint legal custody?*
Sure of Answer About Joint Legal Custody	7	2	*How sure are you that a court, judge, or divorce decree (did/did not) give you and (CHILD)'s (father/mother) joint legal custody: not at all sure, slightly sure, pretty sure, very sure, or extremely sure?*

Figure 23.3 Question wording and ordering of questions about legal and physical custody in Parent Survey 3.

The answers to the question about joint physical custody in round 2X continue to show problems. Two of the nine respondents answer "yes," but one is probably incorrect (R308: "I don't think they used the word joint physical custody, but they did say joint custody ... as long as I give her 24-hour notice I can come and get the child"); the other begins similarly, but in the course of thinking out loud realizes his mistake (R303: "yes they did, they called it joint custody ... but uh, but um, um, let's see the question is joint custody, I mean joint physical custody ..."). R305 provides a report that could indicate some confusion about the distinction between physical and legal custody: "A judge gave us joint custody, but I have custodial custody." But when asked to define joint physical custody, at least seven respondents provide a definition that is basically correct and the other two are roughly correct.

In round 2Y, six of seven parents say "yes" to the question about primary placement without any signs of hesitation. The remaining respondent answers that "we have joint custody with physical placement with the mother" but is not properly probed to select "yes" or "no." One other respondent mentions "joint custody" but also contrasts it with placement. All the respondents provide an adequate definition of primary physical custody.

In reading the reports, it seems to us that the difference between round 2X and round 2Y is visible in answers to the subsequent question about legal custody, our focal question. Four of the seven respondents in round 2X express uncertainty or change their answers, as illustrated by this response:

> R306 (DivFa-X): Um, I'm not sure. I'm not sure if there was joint legal custody um, I would have to get out you know the papers and see if that's something that's specified. I mean, I would assume that we have some sort of, you know, joint decision making as far as [Can't hear R] things go, but I can't—you said medical, you know, what if they were sick, you know, what would be decided upon, surgery and things and I'm sure that's, you know, we would do things like that. Whether it's a legal thing or not, I'm not positive.

In contrast, all the respondents in round 2Y answer "yes" or "no" initially, and only one elaborates on his "yes."

In summary, most respondents in both versions (13 altogether) give adequate definitions of joint legal custody; the probe is skipped incorrectly for two respondents, and one respondent says that he would not have known what joint legal custody was without the definition in the question.

23.4.3 Round 3: Testing Questions Revised after Round 2X/Y

For the final round of testing, we adopted the approach taken in round 2Y (with some minor modifications); primary physical custody or placement appeared to be clearer to respondents than joint physical custody and so seemed likely to provide a clearer contrast with joint legal custody. Of the seven respondents answering about joint legal custody in round 3, the three mothers say "yes" and the four fathers "no," and all initial responses appear unproblematic. Five of the respondents give adequate definitions of joint legal custody when asked to do so. The remaining two respondents provide adequate paraphrases of the question about legal custody, but when probed for a definition of the term, give problematic or ambiguous definitions.

Overall, the responses contain considerably fewer indications of problems than those in earlier rounds of cognitive interviewing. It appears that using an introduction to alert respondents that it is easy to confuse legal and physical custody, and asking about the familiar concept of primary physical custody first clarifies the meaning of joint legal custody. Nevertheless, we wanted to test these impressions more formally, so we included a split-ballot experiment in the PS3.

23.5 PARENT SURVEY 3: SPLIT-BALLOT EXPERIMENT, RECORD CHECK, AND INTERACTION CODING

The experiment in the PS3 varied the position of the target question about joint legal custody (see Figure 23.3 for wording of questions). The form we refer to as "final" followed the order that we had used in round 3 of cognitive testing: Questions about visitation privileges were followed by questions about primary physical custody and, in the final position, the question about joint legal custody (and the subsequent rating of sureness). The form we refer to as "initial" put the question about in the joint legal custody first; only the introduction served to alert respondents that legal custody might be confused with another concept. We anticipated that those who received the final form would report about joint legal custody more accurately than those who received the initial form because answering about physical custody first would make it clearer that legal custody was a different concept.[4] However, because of the introduction—the effect of which we did not test—we did not expect question order to have a large effect.

Our study offers several methods for evaluating the manipulation of question order, in addition to a comparison of the two forms in the split-ballot experiment. To assess whether our revisions affected the accuracy of answers, we examine both the agreement between the survey response and the CRD and respondents' assessments of their accuracy. We use interaction coding to examine whether the two forms differ in how well interviewers were able to maintain standardization or with respect to behaviors by respondents that might indicate that they had difficulty understanding the questions or answering accurately; this also allows us to address debates about the effects of standardization on accuracy (see Dykema et al., 1997). In addition to these assessments, the CRD allows us to examine

[4]The questions we used in the PS3 are longer than questions that are typically used in many surveys. Previous research has suggested that longer questions may provide higher-quality data than shorter questions, perhaps because they better clarify survey concepts or because they give respondents more time to think (see, e.g., Blair et al., 1977, and Marquis et al., 1972). Although anecdotal evidence suggests that interviewers often complain about long questions, our training included a discussion of the rationale for longer questions, and the interviewers rarely complained about this at debriefings during question development and pretesting. But see the analysis of interaction codes later.

the accuracy of respondents' beliefs about their own accuracy and whether the interaction coding is a guide to the accuracy of respondents' answers. Such an assessment is important because ratings of sureness and interaction coding can be used even when record checks are not possible.

23.5.1 Split-Ballot Experiment

The only item for which there was a significant ($p < 0.05$) difference between the initial and final forms in the PS3 is the question about how sure respondents are that their answer about joint legal custody was accurate.[5] More respondents are extremely sure of their answers when the question about joint legal custody is final (78.2%) than when it appears first (69.8%) (other items not shown). We next examine whether or not the placement of the question about legal custody affects how accurate respondents actually are.

23.5.2 Record Check: Comparison of Survey Responses and the CRD

The data in Table 23.1 show that the correspondence between the PS3 and the CRD is the same for both forms in the split ballot experiment ($p > 0.05$ for G^2 test for three-way interaction). Approximately 49.8% of the PS3 sample have joint legal custody according to the CRD (calculated from Table 23.1), and the overall proportion of cases in which reports of having or not having joint legal custody match the CRD is very similar for the two forms, 83.4 to 85.6%. To evaluate our approach more systematically, we examine the proportion of those with joint legal custody in the CRD who report joint legal custody in the PS3 (the sensitivity of the questions) and the proportion of those without joint legal custody in the CRD who do not report it in the PS3 (the specificity of the questions) (Fleiss, 1981). For the two forms, the proportions of those with joint legal custody in the CRD who report

Table 23.1 PS3 and PS2 Survey Reports of Joint Legal Custody by CRD Reports of Joint Legal Custody (and Form), All Respondents[a]

	PS3				PS2	
	Final Form: CRD Joint Legal Custody		Initial Form: CRD Joint Legal Custody		CRD Joint Legal Custody	
Survey Report of Joint Legal Custody	No	Yes	No	Yes	No	Yes
No	72.3	7.0	75.5	2.9	93.2	29.8
Yes	27.7	93.0	24.5	97.1	6.8	70.2
n	112	129	159	140	1280	517

[a]Table omits respondents who said "don't know" or "refused" for the PS2. In PS3, the interaction with form is not significant ($p > 0.05$).

[5]Because the distribution of the responses about sureness is highly skewed, so that the categories at lower levels of sureness are quite sparse, we collapsed the item to have two categories. Results were very similar when we examined the unrecorded item and a version with three categories.

accurately in the PS3 are 93 and 97%, and the proportions of those in the CRD without joint legal custody who report accurately are 72 and 76%. Our approach appears to have fairly high sensitivity, but its specificity is less satisfactory. Thus, the questions appear to be quite good at identifying those with joint legal custody, but they do less well in helping those who do *not* have joint legal custody to recognize how to classify themselves with respect to the survey concept.

The strategy used in the PS3 to measure legal custody was quite different from that used in previous surveys. But because the results from the two forms used in the PS3 are relatively similar, the PS3 alone offers little guidance in assessing the success of the strategy used in the PS3 as compared with other common approaches. An earlier survey, the 1989 PS2, offers a comparison that is helpful in roughly evaluating how well the revisions in the PS3 succeeded. The PS2 also interviewed a sample of divorce and paternity cases (from earlier cohorts in the CRD) by telephone. There are many differences between the PS2 and the PS3, including differences in the cohorts and counties sampled from the sample frame of Wisconsin court cases, the terminology used by the court to describe custody and placement, the relative proportions of divorce and paternity cases in the population, and the response rate (see Dykema and Schaeffer, 2000, for details about the PS2). Nevertheless, we provide the comparison as a baseline for evaluating our efforts in the PS3.

Like the question used in the 1990 CSS, the PS2 questions were intended to be about legal custody but do not actually make the survey concept clear—either to the respondent or the interviewer—and the concepts may have been less distinct at that time. The PS2 question (shown below) includes an instruction to the interviewer in capital letters saying the question is not about where the children actually live; this instruction is not clarifying, however, because where the children actually live is distinct from both legal custody and the placement ordered by the court:

> (L. 68) "My next questions are about the *agreement* you and the child(ren)'s (father/mother) had about custody in 1988. When I ask about legal court-ordered agreement, I mean any agreement that went through a court."
>
> "First, did you have a *legal*, court-ordered agreement about custody of (CHILD/the children) in 1988?"
>
> We want to know about the *LEGAL CUSTODY ARRANGEMENT*, even if it is different from where the children actually lived. This question is *NOT* about where the children actually lived.
>
> (L.69) Write names of all noncircled court-order children on blanks.
>
> "According to your court-ordered agreement, who had legal custody of (the/each) child in 1988? Was it: you, the other parent, someone else, or did you have joint custody?"

Calculations based on Table 23.1 indicate that joint legal custody was much less common among respondents in the PS2 (approximately 29%) than in the PS3 sample (approximately 50%). Overall, the percentage of cases in the PS2 in

Table 23.2 Match between CRD and PS3 Reports of Joint Legal Custody by Sureness by Form[a]

	Final Form		Initial Form	
How Sure	Not Matched	Matched	Not Matched	Matched
Very or less	21.6	21.6	56.1	26.6
Extremely or exact	78.4	78.4	43.9	73.4
n	37	190	41	248

[a]Table omits 7 respondents who broke off the interview before answering the question about joint legal custody, 5 respondents with missing information from the CRD, 18 respondents who said "don't know" or refused the stem question about joint legal custody, and 6 respondents who said "don't know" or refused to answer the question about their confidence about having/not having joint legal custody. Three-way interaction is significant ($p < 0.01$).

which reports of having or not having joint legal custody match the CRD (87%) is very similar to the PS3 (85%). However, the similarity in the overall proportion of incorrect responses in the two surveys masks a substantial difference in the types of errors made in the two surveys. Approximately 7% of those without joint legal custody in the CRD and 30% of those with joint legal custody in the CRD report incorrectly in the PS2—the reverse of the situation for the PS3. Thus, compared to the PS2, the PS3 has higher sensitivity, but lower specificity.

Table 23.2 shows the level of sureness for those whose survey reports about joint legal custody did and did not match the CRD, separately for each form. When the question about joint legal custody is in the final position, there is no relationship between accuracy and how sure the respondent is: Approximately 78% of respondents are extremely sure, whether or not their PS3 report actually matches the CRD. When the question about legal custody is in the initial position, however, respondents for whom the two sources match are much more likely than those for whom the two data sources do not match to be extremely sure that their PS3 answer is accurate (G^2 for test of the interaction = 5.38, $p < 0.02$). It is possible that in the initial form, respondents must do more cognitive work in order to answer the question, and respondents may base their assessment of sureness on their level of effort. In the final form, the sequence of questions may itself create a sense of sureness that is based on having heard a question on the topic of physical custody before being asked about legal custody rather than on a level of effort.

23.5.3 Interviewer–Respondent Interaction Coding Data

To date, studies have examined only a few aspects of the potential contribution of interaction coding to the testing and evaluation of survey instruments. There is evidence that interaction coding is reliable in detecting respondent and interviewer behaviors that may indicate problems in a survey question (Presser and Blair, 1994, p. 87). In their analysis of questions about health care utilization, Dykema et al. (1997) found that when respondents qualified an answer or frequently exhibited any of the behaviors that were coded, their answers were consistently (and sometimes significantly) less accurate. In a single case, interrupting the initial reading of an item was associated with more accurate answers. For seven of 10 items, a substantive change in the question during the initial reading was associated with more accurate answers, although the effect was significant for only one item. However, the same unexpected relationship was found for several items using measures that summarized the question-asking behavior of the interviewer. In addition, some behaviors of respondents such as giving adequate or qualified answers are associated with reliability in their answers (Hess et al., 1999; see also Mathiowetz, 1998). Our case study allows us to ask two questions about the validity of interaction coding as a methodology for evaluating survey questions: Using the split-ballot experiment, we can ask whether the interaction reflects or provides insight into the similarities and differences already observed in the two versions of these items. Using the record-check component of our study we ask which of the behaviors, if any, predict accuracy. We expect the interaction codes (see Table 23.3) for the two forms to be highly similar given the similarities between the two forms, but possibly to show some trace of the interaction among form, accuracy, and the sureness rating in Table 23.2.

Table 23.3 shows the proportion of times a code was assigned for primary physical custody, joint legal custody, and the sureness rating for joint legal custody. Major changes occur less frequently for the two questions about visitation privileges (the proportion of readings with major changes ranges from 0.05 to 0.16; results not shown) than for the three questions about legal or physical custody status (range is 0.13 to 0.36). The two questions that begin the two forms—visitation privileges and joint legal custody, respectively—have a major change more often when they initiate the series than when they appear in the penultimate position (results shown only for joint legal custody). The difference in the proportion of cases with a major change in joint legal custody when that question appears in the initial position with the introduction (0.37) versus the final position without the introduction (0.14) suggests that the majority of the changes might have occurred in the introduction.

For the questions about primary physical and joint legal custody, the codes for the interviewer's reading of the questions present mirror images across instrument forms: For example, whichever one of the two is read before the other is subject to a major change about one-third of the time probably because of the sentence introducing the contrast. Similarly, whichever of the two precedes the other is more likely to evoke an expression of feeling from the interviewer. When the question about physical custody immediately follows that about visitation (final form), the question about physical custody is more likely to be interrupted (0.11) and to evoke an expression of feeling from the respondent (0.05) than when it precedes the question about visitation (0.06 and 0.01, respectively), but there is no accompanying significant increase in follow-up behavior by the interviewer.

Summary measures (not shown) that provide the mean of the proportion of items in the section for which the code was assigned indicate no significant differences ($p < 0.05$) in the incidence of the codes between the final and initial

Table 23.3 Proportion of Select Interviewer–Respondent Interaction Codes Evaluated at the Question for PS3 Questions About Primary Physical and Joint Custody, by Form (Order)[a]

		Primary Physical Custody		Joint Legal Custody		Sure of Answer About Joint Legal Custody	
		Final (3)	Initial (3)	Final (6)	Initial (1)	Final (7)	Initial (2)
Interviewer question asking							
Exact	Reads question exactly.	**0.26**	**0.55**	**0.56**	**0.26**	**0.43**	**0.56**
Exact with repairs	Reads question with correction.	0.05	0.08	**0.15**	**0.07**	*0.04*	*0.09*
Slight change	Adds/deletes 1 to 3 words that do *not* alter the question's meaning.	**0.34**	**0.23**	**0.14**	**0.31**	0.08	0.08
Major change	Adds/deletes 4 or more words *or* add/deletes 1 or more words that alter the question's meaning.	**0.34**	**0.13**	**0.14**	**0.37**	**0.43**	**0.27**
Not asked	Skips applicable question.	0.01	0.00	**0.02**	**0.00**	0.02	0.01
Interviewer follow-up behaviors							
No follow-up behaviors	No follow-up behaviors.	0.45	0.46	0.62	0.59	**0.48**	**0.57**
Adequate follow-up only	Follow-up behavior conforms to principles of standardization (e.g., neutral probe).	0.40	0.40	*0.29*	*0.36*	0.27	0.27

(continued)

Table 23.3 *(continued)*

		Primary Physical Custody		Joint Legal Custody		Sure of Answer About Joint Legal Custody	
		Final (3)	Initial (3)	Final (6)	Initial (1)	Final (7)	Initial (2)
Any inadequate follow-up	Follow-up behavior does not conform to principles of standardization (e.g., leading probe).	0.15	0.15	0.08	0.05	**0.25**	**0.16**
Other interviewer behaviors							
Elaboration	Includes a clarifying phrase.	<0.01	0.01	<0.01	<0.01	0.02	0.01
Verification	Includes a verification of previously offered information.	0.01	0.01	0.01	<0.01	0.02	<0.01
Stress problem	Fails to stress capitalized word or phrase.	0.37	0.43	**0.61**	**0.51**	**0.38**	**0.50**
Laughter	Laughs.	0.03	0.02	*0.02*	*0.05*	**0.08**	**0.04**
Responds with feeling	Conveys feelings about the respondent or survey.	**0.13**	**0.02**	**0.01**	**0.10**	0.05	0.05
Respondent question answering[b]							
Codable answer	Response answers question/verification/probe and can be coded into response categories/format.	0.59	0.60	0.84	0.80	**0.65**	**0.77**
Implicitly codable answer	Response answers question/verification/probe but does not match response	0.16	0.20	0.07	0.05	**0.29**	**0.12**

Uncodable answer	Response either does not answer question/ verification/probe *or* it cannot be coded into response categories/ format.	0.18	0.13	0.06	0.09	0.04	0.07
Other respondent behaviors							
Qualification	Includes a qualifier such as "probably" or "about."	0.07	0.04	0.04	0.06	<0.01	0.01
Seeks clarification	Requests repeat of the question or clarification of a term.	0.05	0.06	0.02	0.03	0.02	0.03
Elaboration	Provides information in addition to a codable answer.	0.14	0.11	0.05	0.07	0.04	0.04
Interruption	Interrupts initial reading of the question.	**0.11**	**0.06**	0.08	0.10	**0.31**	**0.20**
Pause	Pauses/hesitates 2 seconds or more.	0.10	0.09	0.06	0.09	0.02	0.04
Don't know	Offers a don't know response.	0.03	0.04	0.03	0.02	**0.00**	**0.02**
Refusal	Refuses to answer.	0.00	<0.01	0.00	0.00	0.00	0.00
Laughter	Laughs.	*0.05*	*0.02*	0.03	0.04	*0.11*	*0.06*
Responds with feeling	Conveys feelings about the survey.	**0.05**	**0.01**	0.02	0.02	0.02	0.02

(*continued*)

Table 23.3 ***(continued)***

		Primary Physical Custody		Joint Legal Custody		Sure of Answer About Joint Legal Custody	
		Final (3)	Initial (3)	Final (6)	Initial (1)	Final (7)	Initial (2)
Exchange levels							
One exchange level	Interaction unfolded over one level.	0.45	0.46	0.62	0.59	**0.47**	**0.57**
Two exchange levels	Interaction unfolded over two levels.	0.28	0.32	0.27	0.27	**0.44**	**0.30**
Three exchange levels	Interaction unfolded over three levels.	0.13	0.13	0.07	0.07	0.05	0.06
Four or more exchange levels	Interaction unfolded over four levels.	*0.13*	*0.09*	0.05	0.08	0.04	0.06
n		238	282	237	286	224	278

[a] Significance tests are from two-tailed t-tests. Differences of $p < 0.05$ are shown in boldface type and differences of $p < 0.10$ are shown in italic.
[b] Proportions for codable, implicitly codable, and uncodable answers do not add to 1.00; differences reflect the fact that some respondents provided "other" responses.

forms taken as a whole. However, for three of the behaviors, the incidence in the section is greater in the final than the initial form (although these differences are of borderline significance, $p < 0.10$): The interviewer makes a major departure in reading the question by making a major change or not asking the question, the interviewer expresses a feeling, and the respondent interrupts.

There are some other form differences for the two questions about how sure respondents are about their answers (the results for the sureness rating for joint physical custody are not shown in Table 23.3). For the rating of sureness about joint legal custody, both forms have similar levels of uncodable answers, but the composition of codable answers is different for the two forms: Codable answers are less frequent when the question is last in the series (0.65) than when it is second (0.77), but more answers are implicitly codable in the former case (0.29) than in the latter (0.12). The slightly higher incidence of "don't know" answers in the initial form is consistent with the overall lower level of sureness expressed by respondents who provided a rating (discussed earlier). That, and the lower frequency of interruptions, inadequate follow-up, interviewer laughter, implicitly codable answers, and two exchange levels when the joint legal custody questions precede the physical custody questions (initial form) may be indirect evidence of the process that generates the statistical interaction among the variables observed in Table 23.2.

The results in Table 23.4 address whether or not the interaction codes are associated with the accuracy of responses.[6] The odds ratios express the percentage increase (when greater than 1) or decrease in the log odds of agreement between the CRD and PS3 with respect to joint legal custody. In contrast to the findings of Dykema et al. (1997), we find that interviewers who make major departures from the wording of the question—either by changing the wording of the question or by simply not asking the question—have significantly lower odds of obtaining an accurate response than do interviewers who read the question exactly or with only slight changes. Producing a codable answer in the first exchange shows a large positive association with accuracy. Accuracy is lower in those situations where follow-up is required than where it is not, but the reduction appears more substantial when the follow-up does not follow the rules of standardization: There is a 50% reduction in the odds of reporting accurately when all the follow-up is adequate by the rules of standardization, and a 75% reduction when any of the follow-up is inadequate. Many other features of the interaction at the question about joint legal custody are associated significantly with reduced accuracy: a qualified answer, a pause before answering, laughter by the respondent, and an increasing number of exchange levels. These behaviors are recognizable as those that occur (sometimes together) when the respondent is

[6]For each behavior, we first examined whether the behavior interacted with form in its effects on accuracy. These tests were hampered by the low incidence of many of the behaviors. Of the models for which we could obtain stable estimates, two interaction codes (interviewer question asking and the respondent giving a codable answer) interacted significantly with form; however, the effect of the behavior in those models differed from those shown in Table 23.5 only in the size of the effects, not their direction or significance, so the models in Table 23.5 pool both the final and initial forms.

Table 23.4 Joint Legal Custody: Regression Equations of Accuracy in Survey Reports on Interviewer and Respondent Interaction Codes, Evaluated at the Question[a]

	Odds Ratios[b]	Standard Error
Interviewer question asking		
Major change/not asked [vs. exact/exact with repair/slight change][c]	0.59*	0.15
Interviewer follow-up behaviors		
[No follow-up behaviors]	—	
Adequate follow-up only	0.50**	0.12
Any inadequate follow-up	0.20**	0.08
Other interviewer behaviors		
Elaboration [vs. none]	0.23	0.32
Verification [vs. none]	0.91	1.03
Stress Problem [vs. none]	0.83	0.19
Laughter [vs. none]	0.73	0.43
Feeling [vs. none]	0.64	0.27
Respondent question answering		
[Uncodable/other answer]	—	
Codable answer	4.80**	1.39
Implicitly codable answer	1.44	0.67
Other respondent behaviors		
Qualification [vs. none]	0.22**	0.09
Seeks clarification [vs. none]	0.35+	0.21
Elaboration [vs. none]	0.59	0.24
Interruption [vs. none]	0.91	0.36
Pause [vs. none]	0.21**	0.07
Don't know [vs. none]	0.04**	0.03
Refusal [vs. none]	*d*	
Laughter [vs. none]	0.37*	0.18
Feeling [vs. none]	0.60	0.41
Exchange levels		
One to four exchange levels	0.61**	0.07

[a] Results are from logistic regression equations on 505 cases. Cell entries show the odds ratios ($\exp^b$) and standard errors.

[b] $^*p < 0.05$; $^{**}p < 0.01$; $^+p < 0.10$.

[c] Omitted category is shown in brackets.

[d] Variable is dropped from the model because of too little variation and/or too few cases.

having difficulty providing a codable answer (see, e.g., Schaeffer and Maynard, 2002). We cannot determine whether the inaccuracy associated with these behaviors can be reduced by improved instrument design or interviewer training or whether the concept of joint legal custody is impervious to such techniques, but our analysis confirms earlier research in finding that some features of the interaction can be used as indicators of measurement error (see also Hess et al., 1999; Mathiowetz, 1998).

23.6 CONCLUSIONS

What was the overall success of our efforts to improve the measurement of joint legal custody, and how did each of the developmental methods we used appear to contribute to the success (or failure) of those efforts? Although the PS2 is different in many ways from the PS3, the PS2 results suggest that the proportion of positives in the CRD that were correctly identified was relatively low for a survey question that simply referred to "custody," as did the PS2 and the 1990 CSS. It appears that the PS3 approach is more sensitive than that of the PS2, but it unexpectedly has reduced specificity; that is, it is less successful than the PS2 at eliciting correct answers from parents who do not have joint legal custody.

Our focus group discussions indicated that an unmodified reference to "custody," like that in the PS2, is ambiguous; it is plausible that some false negatives in the PS2 were from respondents who answered about joint physical—not joint legal—custody. Our effort to identify those with joint legal custody more accurately, which is relevant for the increasing proportion of the population who actually have joint legal custody, was successful. The material from the focus groups provides grounds for speculating about how false positives could arise (e.g., a parent might report that they have joint legal custody because the noncustodial parent must make decisions about the child when the child is in his or her care). But a review of the answers to the cognitive interviews does not provide enough clear examples of possible false positives to provide a basis for speculating about how the development process might have been improved to provide a more even-handed outcome. It is possible that we were simply more focused on reducing false negatives and neglected to guard against false positives, for example by using follow-up questions to determine more precisely what respondents meant when they reported that they had joint legal custody. Because both versions of the question sequence that we tested in our split-ballot experiment show the same pattern—high sensitivity and lower specificity—it is tempting to speculate that the shared introduction to the series may have contributed to both the success and the failure of our revision.

Because it is a case study, the present analysis does not allow us to give a precise assessment of the contribution of each of the developmental methods we applied to the overall results. The focus groups suggested that respondents were likely to use language in ways that were flexible and, from our point of view, imprecise and highly ambiguous. The groups identified a broad range of possible sources of error but gave little guidance in choosing among them or in assessing how much the dynamics of the focus groups themselves might have contributed to what we observed (see Bischoping and Dykema, 1999). The cognitive interviews, despite their small number, appeared to provide useful information about the respondents' interpretations of questions. To be most useful, cognitive interviewers must be certain to probe respondents to select a category and to pursue the possibility of both false positives and false negatives. In our case, we should probably have probed respondents more extensively about the legal basis for their answers to explore the possibility of false positives more fully. Our finding that the relationship between respondents' sureness about their survey answers and their actual accuracy may depend on the sequence of survey questions suggests a reason to be cautious about seeking reassurance about validity from the self-assessments of respondents.

Of the methods used here, interaction coding is the only one to examine the behavior of the interviewer, so it provides information very different from that provided by the focus groups or cognitive interviews. Situations that deviate from the paradigmatic exchange in which the interviewer reads a question and the respondent answers (Schaeffer and Maynard, 1996) are associated with reduced accuracy, as indicated by the effects of making major changes in administering the question and multiple exchange levels. In addition, when the interviewer needs to follow up the respondent's answer, the outcome is likely to be reduced accuracy, more so when the follow-up behavior is inadequate. Finally, the results of our split-ballot experiment and the comparison with court records highlight the importance of complementing enhanced development efforts with evaluation. This evaluation highlights that development efforts should explicitly consider the contribution of both false positives and false negatives to error, and develop methods to reduce both types of error.

ACKNOWLEDGMENTS

We wish to thank James Esposito and Rachel Caspar for their comments and Sheri Meland for her research assistance. This research was supported by National Institutes of Health grant HD31042 to Nora Cate Schaeffer and Judith A. Seltzer. Computing was provided by the Center for Demography and Ecology, which receives core support from the Center for Population Research of the National Institute for Child Health and Human Development (HD-05876). Additional support was provided by grants to Nora Cate Schaeffer from the Graduate School of the University of Wisconsin–Madison. Some of the data analyzed here were collected under contracts between the Institute for Research on Poverty and the State of Wisconsin, Department of Health and Social Services. We are grateful for the assistance of staff at the Institute for Research on Poverty, which receives support from the Office of the Assistant Secretary for Planning and Evaluation, U.S. Department of Health and Human Services. Opinions and conclusions are those of the authors.

CHAPTER 24

Multiple Methods for Developing and Evaluating a Stated-Choice Questionnaire to Value Wetlands

Michael D. Kaplowitz, Frank Lupi, and John P. Hoehn
Michigan State University

24.1 INTRODUCTION

In this chapter we present a case study of an iterative multiple-method approach to survey design and evaluation. Three types of focus groups, as well as cognitive interviews, were used to develop a survey instrument. The case study focuses on two aspects of the experience. First, the case illustrates how the information from the group discussions and cognitive interviews can be used iteratively to design an environmental valuation questionnaire. Second, and more generally, the case presents insights into the relative strengths and weaknesses of group discussions and cognitive interviews for evaluating a questionnaire. The questionnaire in this case study was designed to elicit values associated with wetlands.

Environmental valuation, sometimes referred to as *nonmarket valuation*, is a field of economics aimed at estimating economic values for changes in environmental and ecosystem services (Freeman, 1993). Information on the economic benefits of environmental quality is used in cost–benefit analyses of environmental policies (Arrow et al., 1996) and legal cases involving natural resource damages (Ward and Duffield, 1992). Because environmental and natural resources are not typically traded on markets, economists have developed survey methods for environmental valuation based on individuals' preferences. *Stated-choice surveys*, sometimes referred to as *choice experiments* (Opaluch et al., 1993) or

Methods for Testing and Evaluating Survey Questionnaires, Edited by Stanley Presser, Jennifer M. Rothgeb, Mick P. Couper, Judith T. Lessler, Elizabeth Martin, Jean Martin, and Eleanor Singer
ISBN 0-471-45841-4

conjoint analysis (Gan and Luzar, 1993), have been widely applied in market research (Louviere, 1991), transportation economics (Louviere et al., 2001), development economics (Rubey and Lupi, 1997), and environmental valuation (Adamowicz et al., 1993; Boxall et al., 1996; Mackenzie, 1993; Opaluch et al., 1993; Swallow et al., 1998). This case study focuses on the development of a stated-choice questionnaire for valuing wetland protection (Louviere et al., 2001).

Stated-choice questionnaires typically present respondents with information about the attributes (e.g., size, type, quality, cost) of particular environmental goods or services. The questionnaires also inform respondents about the choice context and implications of possible trade-offs. They then ask respondents to choose between alternative bundles of goods and services (see Figure 24.1). In using such questionnaires, all respondents receive identical information about the

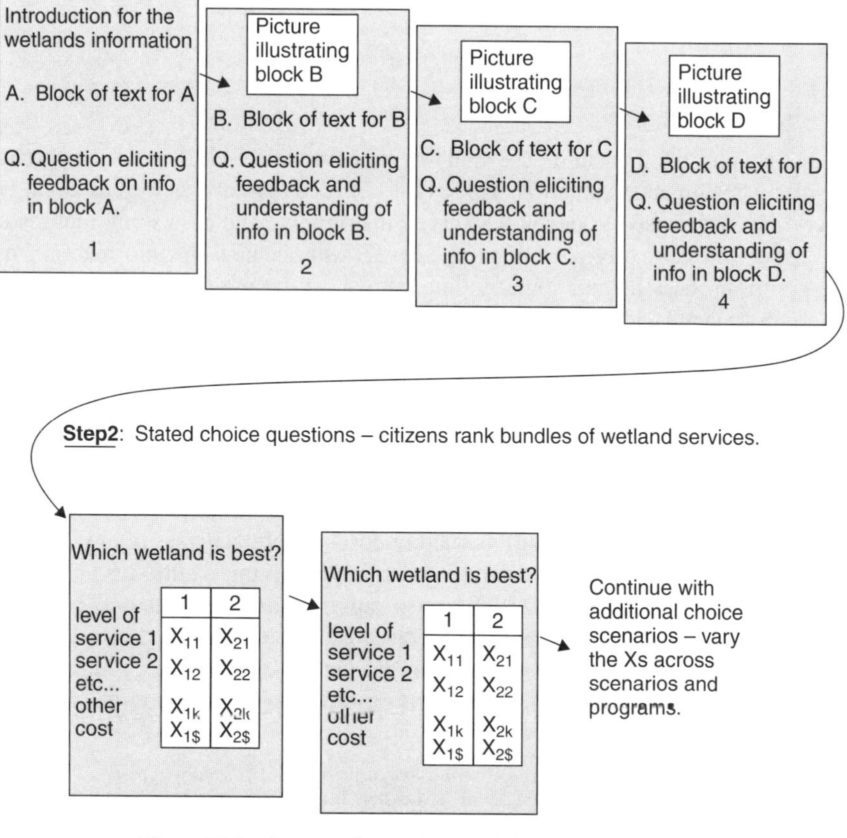

Figure 24.1 Common format for stated-choice questionnaires.

attributes, policies, and choice context. For example, pages 1–4 in Figure 24.1 illustrate how information on wetland science and policy might be presented to respondents in an interactive and uniform manner. However, the levels of attributes of the goods and services being studied in a stated choice questionnaire are varied across respondents based on an experimental design. That is, respondents receive the same information treatments but make choices among alternative scenarios with differing attribute levels. In Figure 24.1 this would be accomplished by varying the "Xs" on the figure's pages 5–6 based on an experimental design. Respondents are asked to select their preferred "outcome" for several pairs of attribute arrays. That is, in Figure 24.1, respondents would be asked to select their preferred "wetland" (1 or 2) among the choices on page 4, their preferred "wetland" (1 or 2) on page 5, and so on. Statistical analysis of the attribute trade-offs implicit in respondents' choices reveals underlying economic values associated with goods and services. The analysis of individuals' informed trade-offs reveals the public's value for the environmental and natural resource services in question (see, e.g., Lupi et al., 2002).

24.1.1 The Design Challenge

Unlike market goods, environmental and natural resources (e.g., ecosystems) are often complex and not widely understood by the public (Barbier, 1994; Costanza et al., 1989; Schwarz, 1997). The researchers' task is to design a stated choice survey that will inform the general public about ecosystem functions and policies in an unbiased, realistic, and easily understood manner. Furthermore, the survey instrument needs to place respondents in a believable context for making realistic and informed choices among alternative scenarios (see Arrow et al., 1993; Mitchell and Carson, 1989). The challenge is to build all these elements into a friendly, inviting survey that "create[s] respondent trust and perceptions of increased rewards and reduced costs for being a respondent" (Dillman, 2000a, p. 27).

Despite increasing use of natural resource valuation surveys (Carson et al., 1994a), the literature on their design is comparatively thin. The difficulty of designing environmental and natural resource valuation questionnaires has been recognized for some time (Carson and Mitchell, 1993; Carson et al., 1998; Mitchell and Carson, 1989). "Producing a good [stated choice] survey instrument requires substantial development work" (Carson, 2000, p. 1415). However, the resource valuation literature is relatively silent regarding the design, evaluation, and testing of survey instruments. While a few prominent environmental valuation studies (e.g., Exxon Valdez oil spill) have been recognized for their questionnaire development (Carson et al., 1994b, 1998), these studies tend to be associated with high-profile, well-financed research efforts in support of significant natural resource damage litigation.

Environmental and natural resource economists have reported some use of qualitative methods during the design, evaluation, and testing of questionnaires for valuing environmental resources (e.g., Boyle et al., 1994; Carson et al., 1994b;

Chilton and Hutchinson, 1999; Mitchell and Carson, 1989). This follows increased use of cognitive methods as tools for developing questionnaires in other fields; foremost among these is cognitive interviewing (Tourangeau et al., 2000). Carson (2000) recommends using focus groups and in-depth interviews to determine the plausibility and understandability of environmental goods and services and the scenarios presented to respondents. However, qualitative interviews and focus groups may provide different but complementary information for environmental valuation (Kaplowitz and Hoehn, 2001). Cognitive interviews may place respondents in a setting that facilitates their sharing of (sensitive) resource information they otherwise might not share in a group (Kaplowitz, 2000).

24.1.2 Wetland Ecosystems and Policy

The questionnaire developed in this case focused on trade-offs that the public is willing to make when it comes to protection of inland freshwater wetland ecosystems. Wetlands are transitional ecosystems that occupy a spectrum between land and water ecosystems [National Research Council (U.S.), Committee on Characterization of Wetlands, 1995]. Types of wetlands include bottomland swamps, tidal marshes, cattail marshes, vernal ponds, fens, and bogs. Wetlands provide a range of ecological and biogeochemical functions, such as water storage, maintenance of surface and groundwater flows, biochemical cycling, and maintenance of characteristic habitats. In the United States, a wetland protection policy of "no net loss" seeks to stem the loss of wetlands. To operationalize the "no net loss" policy, state and federal governments require mitigation (i.e., replacement) of destroyed wetlands through the creation, restoration, or protection of equivalent wetlands in the area [National Research Council (U.S.), Committee on Mitigating Wetland Loss, 2001]. However, even in instances where wetland acreage is maintained, the quality of wetlands and their ability to provide services is often diminished (Dahl, 2000). Stated choice techniques provide a means for quantifying the public's preferences for possible trade-offs between wetland quantity and quality that typically arise with mitigation decisions.

24.1.3 Methods Used in Case Study

This study used both focus groups and cognitive interviews to design and evaluate a wetlands stated choice questionnaire. Figure 24.2 illustrates the iterative multiple-method approach employed. Focus groups were used in several distinct ways. First, a series of focus groups with participants sharing characteristics of likely respondents was used to help researchers conceptualize, contextualize, and frame questions as well as identify appropriate language and cognitive obstacles for survey design. Second, a focus group was conducted with a panel of wetland scientists and regulatory officials to ensure the appropriateness of the thrust as well as the wetland science and policy of the questionnaire. Third, a series of focus groups with typical Michigan residents were conducted to evaluate and identify the relative strengths and weaknesses of two alternative draft questionnaires. The challenge of using focus groups for instrument evaluation led the

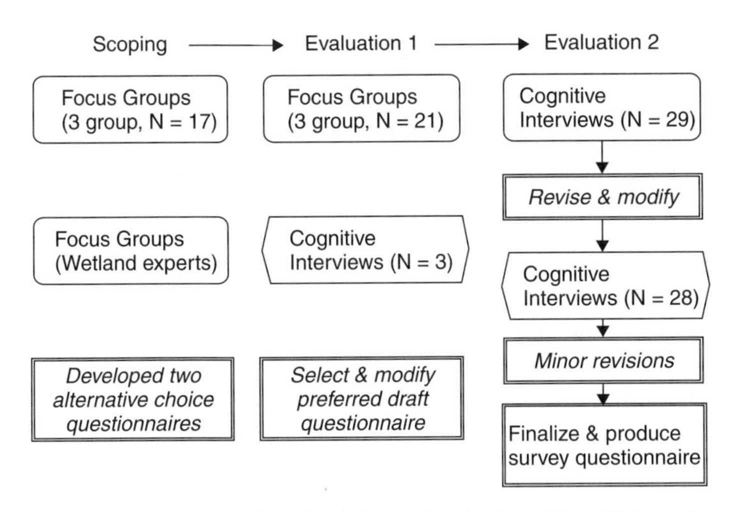

Figure 24.2 Iterative questionnaire design and evaluation with multiple methods.

researchers to conduct several cognitive interviews to evaluate the alternative questionnaires. The success of these cognitive interviews for evaluating the draft instruments resulted in the exclusive use of additional cognitive interviews for the next planned wave of questionnaire design and evaluation. It should be noted that the evaluation focus groups and cognitive interviews took place directly following participants' self-administration of the same version of the draft survey instruments. These mixed methods are presented and discussed in chronological order. The chapter concludes with a discussion of the relative merits (and limitations) of the methods as means of developing and evaluating a questionnaire.

24.2 SCOPING AND QUESTIONNAIRE DEVELOPMENT

24.2.1 Focus Groups with Likely Respondents

Initially, the investigators used a series of focus groups with randomly recruited participants from mid-Michigan to learn from members of the general public about their uses, perceptions, and understanding of wetlands, wetland types and services, and wetland policy. These focus groups helped scope out the territory and concepts for the questionnaire, so we call them "scoping" focus groups. The focus group size was conditioned by the desire for them to be "small enough for everyone to have opportunity to share insights and yet large enough to provide diversity of perceptions" (Krueger, 1994, p. 17). Following the maxim that one needs as many sessions as long as nothing new is being learned in the last session (Maxwell, 1996; Morgan, 1997), the researchers planned to conduct enough of these initial focus groups until no major new information was revealed. In all, three were held, with five, six, and eight participants respectively. The scoping focus group sessions followed a detailed discussion guide to lead respondents through several topics, including natural resources of importance to people, prior knowledge of wetlands, cognizance of wetland policies, and reaction to wetland replacement scenarios.

Participants for the focus groups (and later, cognitive interviews) were recruited from the general population of adults in the vicinity of Lansing, Michigan, using random telephone recruitment. Potential participants were contacted initially using randomly selected telephone numbers from local telephone directories. Trained recruiters followed carefully crafted recruitment scripts to invite potential participants to the sessions. Potential participants were asked to participate in a group discussion of "natural resource issues in Michigan." Potential participants were also asked a series of brief questions to screen out those persons with advanced knowledge of wetlands or potential conflicts that might unduly influence the sessions. Potential participants were not told that they would be discussing wetlands. In keeping with generally accepted focus group procedures, participants received a small ($40) honorarium for their participation.

These focus group sessions were organized for evenings on the campus of Michigan State University (MSU). Each focus group took roughly two hours. The sessions were held in a special focus group facility at MSU. The same trained focus group moderator used the same specially prepared discussion guide for all three focus groups. The moderator followed the guide and used nondirective prompts to encourage participants to participate and elaborate their responses. The detailed discussion guide was used to lead respondents through several topics and to (1) learn what natural resources were important to people, (2) explore people's level of prior knowledge concerning wetlands, (3) gather information concerning people's knowledge of wetland types, (4) explore people's knowledge of public policies relating to wetlands, (5) learn people's opinions of the importance of certain wetland functions, (6) evaluate how people process given wetland definitions/pictures, and (7) examine people's reaction to a particular wetland replacement scenario. These seven components are conceptually in line with the elements required for designing an effective wetland stated choice questionnaire.

The focus groups were both video and audio recorded. In addition, the investigators and the moderator kept written notes. Additionally, worksheets used by respondents to identify their top natural resource issues and rank the wetland functions were also collected and analyzed. The data collected were subsequently analyzed iteratively making use of a grounded-theory approach (Strauss and Corbin, 1990). The goal of such analyses is not to produce simple counts of things, but to "fracture" the data and rearrange it into categories that facilitate understanding and comparisons of the data (Maxwell, 1996; Strauss and Corbin, 1990). This approach helped the researchers extract and derive the major ideas and themes from the focus groups (Krueger, 1994).

The scoping focus groups helped the researchers learn how the general public thought about wetlands, identify information gaps, and frame the task of designing information treatments to convey wetland ecosystem science and policy information. The scoping focus groups revealed that many Michigan residents have some experience and familiarity with wetlands and some knowledge of wetland services. Respondents generally understood that wetlands support a variety of functions, especially habitat for plants and animals. As one respondent pointed out: "Fish, turtles, muskrats, and you know deer, deer live in the swamp, they go in the swamp to drink the water, so you get all sorts of animals in the swamp, like bears ... it's their kind of a refuge, for all the animals in the world."

These scoping focus groups also revealed the surprising but widely held misperception that "trees don't grow in wetlands" and that "wetlands kill trees." These "dead tree" comments occurred in all of the scoping focus group discussions. In all instances when a participant raised this misperception, the other participants did not refute the statements. What makes this misperception especially interesting is that in Michigan, lowland hardwood wetlands and lowland conifer wetlands make up more than two-thirds of Michigan's wetlands (Comer, 1996). This misconception was addressed by explicitly including wooded wetlands information as part of the subsequent draft questionnaires. See Hoehn et al. (2001) for more details about the scoping focus groups and their findings.

24.2.2 Focus Group with Subject-Matter Experts

After learning from Michigan residents about their use, (mis)understanding, and perceived value for wetland services, a focus group was convened with a panel of wetland scientists and state regulatory officials. This two-hour semistructured group discussion focused on project-related ecological and economic concepts, the scoping focus group results, and future project plans. Unlike the scoping focus groups with the general public, this session was generally informal, with one of the researchers taking the lead in guiding the discussion. The session began with a brief presentation about the research project, potential research questions, and the findings from the scoping focus groups. The researchers took written notes of the session and later used these notes to, among other things, follow up on specific points with individual experts.

The session with the subject-matter experts helped the researchers to (1) clarify pertinent wetland ecosystem information, (2) understand statewide wetland protection and mitigation policies, and (3) identify valuation needs that complement current regulatory approaches. Vetting possible information treatments, choice contexts, and survey design elements with these experts helped ensure accuracy and appropriateness of the wetland ecosystem information for the survey. Topics discussed included wetland functions and their measurement, wetland mitigation, mitigation ratios, mitigation banks, public perceptions, wetland pictures and diagrams, possible choice scenarios, and recommendations on additional scientific resources. This expert group discussion helped create a science-based wetland questionnaire and identify key gaps in scientific techniques for assessing wetland mitigation. For example, while wetland ecosystems provide many ecological functions and societal services, the experts believed it essential to understand the importance survey respondents place on habitat quality (e.g., Brinson, 1993; Brinson et al., 1995). Since this meshed with the scoping focus group findings on

Wetlands Scorecard #2:
How do the Drained and Restored Wetlands Compare?

Wetland Features	Drained Wetland	Restored Wetland
Is it marsh, wooded, or a *mix* of marsh and woods?	Marsh	Wooded
How large is it?	12 acres	18 acres
Is it open to public?	Yes	Yes
Are there trails and nature signs?	No	No
How good is the habitat for different species?		
Amphibians and reptiles like frogs and turtles	Good	Excellent
Small animals like raccoon, opossum, and fox	Good	Good
Songbirds like warblers, waxwing, and vireo	Excellent	Good
Wading birds like sandpiper, heron, or crane	Good	--
Wild flowers	--	Good

What do the habitat ratings mean?

Excellent: The wetland habitat supports these species in better than average numbers and variety; a casual observer is very *likely to see a variety* of these species.

Good: The wetland habitat supports these species in average numbers and variety; a casual observer is *likely to see a few* of these species.

--: The wetland habitat supports these species in very small numbers or not at all; a trained observer is *unlikely to find any* of these species.

Wetland Case #2

The scorecard on the left page compares the natural features of the drained and restored wetlands. The rows in the table describe different features and habitats of the two wetlands. The box at the bottom of the scorecard explains the habitat ratings.

In your opinion, is the restored wetland good enough to offset the loss of the drained wetland in Case #1? (Circle the letter next to your decision)

a. Yes, the restored wetland offsets the loss of the drained wetland

b. No, the restored wetland does not offset the loss of the drained wetland

c. Too close to call

d. Not sure

The Fine Print:
The drained and filled wetlands...
...are common wetland types.
...do not contain any rare species or rare habitat.
...are the same in terms of features not mentioned in the scorecard.

Figure 24.3 Sample of the choice question pages showing the wetland alternatives.

the importance of habitat to participants, subsequent research efforts focused on a survey addressing changes in wetland habitat.

24.2.3 Draft Instruments

Based on information from the scoping focus groups, the subject-matter expert focus group, and the literature, two alternative draft instruments were prepared. Both of these draft instruments presented information about wetland ecosystems, wetland habitat, and wetland restoration in easy-to-understand information treatments. These stated-choice instruments followed standard design by presenting pairwise alternatives to respondents in a tabular form (Louviere et al., 2001). As Figure 24.3 illustrates, respondents were informed about the features and habitat attributes of a "drained" and "restored" wetland and then asked whether the "restored wetland" was adequate compensation for the "drained wetland." Two draft questionnaires using this basic approach were developed by the researchers for evaluation. These two questionnaires differed primarily in the context of the choices that respondents were asked to make. One instrument asked respondents to compare a "drained wetland" with its attributes to a "restored wetland" with a different set of attributes. The other instrument asked respondents to compare two restored wetlands, one that provided the legally "required" amount of restoration and another that provided "extended restoration" at a higher cost.

24.3 QUESTIONNAIRE EVALUATION PHASE 1

24.3.1 Evaluation Focus Groups

To evaluate the feasibility and utility of either or both of the draft instruments and to gauge the level of understanding of potential survey respondents, the researchers organized a second series of focus groups. Participants for the evaluation focus groups were recruited following the procedures outlined for the scoping focus groups. The three evaluation focus groups had eight, six, and seven participants, respectively. The group discussions began with the participants each taking the same draft, self-administered questionnaire. The groups were moderated by a professional focus group facilitator. The moderator followed a structured discussion guide that was designed to elicit participants' general impressions and difficulties with the questionnaires and then focus participants' discussion on specific aspects of the questionnaire (e.g., "Wetland Scorecard," wetland policy information). When all the participants had completed the questionnaire, the moderator began the sessions by asking participants, "What did you think about the booklet?" After the general impressions were elicited, the group discussion followed the discussion guide that built on much of Fowler's (1995) presurvey evaluation question-and-answer process. Respondents were asked to describe "out loud" how they made their choice between alternative wetlands (a retrospective think-aloud approach). Respondents were probed to explain what they thought about while making their choices and invited to "think aloud" to describe their thought processes. The moderator solicited comments on the choice table's design, ease of use, and information adequacy. Participants were also asked to define terms contained in the questionnaire and describe the wetland information they read in the booklet and they relied on to make their choice. Similarly, respondents were asked to explain the wetland habitat rating scales used in the instrument as well as to share any uncertainties or confusions. The focus groups were audio and video recorded; the researchers observed from behind a one-way mirror, and the moderator took notes. Transcripts from the sessions as well as the researchers' notes and observations were used in subsequent analyses.

Challenge of Staying on the Task of Evaluation It appears that the same group dynamics that make focus groups so useful for revealing certain types of information make it challenging, to say the least, for groups to stay focused on the task of evaluating a specific survey instrument and its components. For example, in our evaluation focus groups, there was a tendency for the group and its individuals toward digressions. Conversational digressions sometimes followed very specific and targeted probes from the moderator. In this section we provide some examples of the types of digressions that occurred in the evaluation focus groups.

Session Start Digression Away from Evaluation It was clear from the outset that it would be difficult to help the group stay on the task of instrument evaluation. Instead of answering the moderator's initial probe about how they found the survey instrument, participants often used their first opportunity to speak to voice their political views and personal experiences. In one instance, participants started their session by discussing issues relating to "government control." This type of session start digression was evident in all three group sessions. Another type of session start digression observed by the researchers was discussion of respondents' personal experiences with wetlands instead of their experiences with the instrument itself. It appeared that when one group member pointed out his or her recollection of a local issue related to the subject matter (e.g., a local legal battle concerning wetland protection and development) in the initial response to the moderator's first probe about the instrument, other group members readily and immediately contributed their similar recollections, which then led to group conversation off the task of instrument evaluation.

EXAMPLE (EVALUATION FOCUS GROUP 1)

Probe: "What did you think of the booklet?"

R1: "I was just thinking about all of the development in the area I live."

R4: "I remember when the mall was first proposed to be built ..." (R4 describes the wetlands that were there)

R6: "or the pressure on Lake Lansing ..." (R6 speaks about how a local lake used to be, including its surrounding wetlands).

Tendency toward the "Big Picture" Another phenomenon that the focus group sessions seemed to reveal was a tendency for participants to avoid detailed analysis of the survey instrument by, instead, offering general observations about larger, society-wide issues. These big picture digressions away from the task of instrument evaluation took the form of persistent talk in the sessions on topics such as property rights; conversations about how, in the past, the loss of public wetlands was not adequately replaced; and discussions about public willingness to support activities such as education. The tendency of group participants to think about and discuss big picture items made it difficult to direct focus group members' attention to the task of evaluating the particular survey instrument. These big picture comments were volunteered in response to specific probes by the moderator about the instrument as well as a respondent's reaction to hearing another participants' observations about the instrument.

Empathetic, Not Focused Feedback Participants in the survey evaluation sessions also tended to respond to other participants' comments in empathetic ways intended, we believe, to be helpful and supportive of others. These sorts of feedback may be one reason focus groups are so useful for other research tasks. However, in our case, such empathetic feedback did not seem useful and on the task of survey instrument evaluation. Every group session had participants providing empathetic "answers" to respondents' questions and moderator's probes. While one participant may have been reflecting and forming a response to a probe, fellow group members seemed anxious to fill the apparent void with what they perhaps believed to be helpful suggestions. These dynamics made it difficult for us to learn what many participants retained and understood from their experience with the survey booklet.

EXAMPLE (EFG 1)

R16: "Why do we have wetlands? Why are we restoring them?"

Probe: (Moderator looking directly at R16) "What did it say in the booklet?" (Four separate respondents ignore the moderator's specific probe of R16 and provide their ideas and information to R16.)

Probe: (Moderator again looking directly at R16): "So that wasn't clear, R16, why we worry about wetlands after reading the booklet?" (Again, this probe was answered by two respondents other than R16.)

Balancing Points of View Similar to our observation of empathetic responses offered by participants to the probes of others, we often found focus group participants responding in ways that seemed to defend their particular point of view. That is, it seemed that many participants took it as their role to provide an alternative perspective or to keep their point of view on the table.

EXAMPLE (EFG 2)

R9: "It doesn't state whether its public or private land that these wetlands are on. ... It affects the property owner. It has to do with property owners' rights." (R9 gives an extended description of a Florida farmer whose land was declared wetland.)

Later in the session, after R9 has discussed private property issues twice:

R9: "[in making a wetland choice] I'm assuming the government is doing this with their own land." (R9 describes his choice thought process.)

R14: (After R9 finishes talking) "I want to weigh in on the private ownership issue. I don't think private ownership trumps the common good of the world. And so I think that when an owner owns a piece of land, whatever it is, that owner is responsible for proper stewardship of whatever that land is. That was there before anyone owned it. So I don't see the government as a kind of overburdening burden on a person for compliance, but rather as a guideline that helps people in this case be good stewards of the land."

R9: ". . . private ownership, private property is why this nation is great." (There are three more such exchanges; each time after R9 comments on property rights).

Although it might be possible for such digressions to be helpful for evaluating some types of instruments, this point/counterpoint dynamic too often detracted from our task of instrument evaluation.

Challenge of Learning Group Members' Individual Experiences As foreshadowed by the discussion of difficulties staying on task, the focus groups format made it difficult to learn about particular individuals' experiences with the self-administered survey instrument. This challenge of getting feedback on the individual experience, the problems as well as the strengths of the survey's component parts, manifests itself in a number of ways. For example, there was a tendency for group members to answer probes intended for others; participants frequently offered up "yeah, me too's"; and the retrospective think-alouds fell short on revealing individuals' experiences and difficulties with the instrument. Unlike the benefit of group dynamics we experienced in our focus groups for exploratory research, the group dynamics in the evaluation focus groups detracted from our ability to probe for individuals' experiences and input regarding the survey instrument.

Validating Comments The evaluation focus groups were filled with instances where one respondent offered his or her observation about the instrument or their experience going through the instrument which was then greeted with a chorus of "yeah, me too." The difficulty of such "yeahs" is that we are unsure whether these collective agreements are an indication of genuinely shared experiences with the instrument or rather are "good neighbor"-type conversational responses, the type of responses that may conform with what cognitive psychologists call the "logic of conversation" (e.g., Sudman et al., 1996). These responses do not necessarily reveal individuals' actual experience with the survey instrument.

EXAMPLE (EFG 1)

R8: "I thought the questions were too easy . . . it was like if you know nothing, here are the facts."

R4: "Pretty objective questions; I didn't feel very emotional."

Probe: "R8 mentions that the questions were easy. I'm wondering how other people found the survey?"

R6: "Yes, easy."

R3: "Oh yes, easy."

R1: "I didn't think it was difficult."

This tendency toward a "me too" effect also happened with comments about possible problems with the instrument. For example, at one point a respondent complained that the survey was unclear on the location of the restored wetland. This resulted in the group discussing (and then complaining about) the same thing, even though they had not previously mentioned it. This "me too" effect makes it hard to tell if there is a real problem with the instrument or if participants could see a potential problem once it had been raised by someone.

Think-Aloud Shortcomings As part of the group instrument evaluation, we asked respondents to take the moderator through what they were thinking about as they went through the choice elements of the questionnaire. However, the results raised several concerns about the efficacy of using think-alouds in groups. As the other observations and concerns about group dynamics illustrate, it was generally difficult to learn how individuals thought about and processed the information in the questionnaire. Furthermore, the group setting seemed to inhibit participants from sharing or admitting to difficulties with the instrument.

Despite difficulties, the think-alouds in the group setting did result in some participant comments on elements of the choice as well as the reason for some participants' selection of an alternative.

EXAMPLE (EFG 3)

R19: "I kept flipping back and forth in the booklet to make sure I understood what it was that was being lost."

R20: "I did what you said [to R19]. I compared the left to the right. And its going from a good habitat for birds to a poor. Why pay to make it worse?"

Nevertheless, the think-alouds suffered from problematic group dynamics. In response to a specific probe about "trails," participants volunteered observations suggesting trails are a good thing. These comments were "balanced" by other participants who pointed out that more trails meant more people to litter and destroy the resource.

EXAMPLE (EFG 3)

Probe: How did you think about that problem [wetland choice 1]?

R18: "I voted yes. It's open to the public, has a few trails."

R19: "No, I voted no. I don't want it open to the public. There's lots of wetlands open to the public."

(A lengthy discussion ensues about pros and cons of public access. Five separate participants weigh in, despite the moderator's probes to determine how this affected choices.) Thus, it seems that an "observation" volunteered by a person during the think-aloud portion of the group session resulted in digressions instead of revealing the participants' own decision-making process.

24.3.2 Opportunistic Cognitive Interviews

Although the researchers had planned to rely solely on focus groups in the first phase of questionnaire evaluation, their experience with the focus groups (described above), coupled with the opportunity to use "extra" focus group participants, prompted the researchers to conduct several cognitive interviews to evaluate the draft questionnaires. These cognitive interviews allowed the researchers to have respondents evaluate the "preferred" draft questionnaires (see below) in a setting more closely approximating an individual, self-administered questionnaire. These cognitive interviews were conducted by individual researchers using the same basic format as the focus groups—respondents first took the self-administered questionnaire and the interview followed the focus group discussion guide. They were conducted in parallel with the evaluation focus groups, and the participants were drawn from the same pool of recruits. Recall that the target focus group size for the evaluation sessions was seven or eight participants. These interviews took advantage of the additional participants that arrived for the group session, which amounted to one respondent in the first group and two from the last group. While the script used was the same as that for the focus groups, the cognitive interviews tended to be more open and somewhat less formal. The intent was for respondents talk about how they answered questions and made the wetland choices, as well as to probe them for understanding.

Despite their small number, the cognitive interviews were quite successful, and the major concerns that were generated about the draft instruments echoed those raised in the evaluation focus groups. Furthermore, the cognitive interviewers found few difficulties keeping the interviewees on the task of instrument evaluation. Moreover, respondents seemed to share their thought processes freely during the think-aloud portion of the cognitive interviews. Consequently, the researchers decided that the subsequent phase of questionnaire evaluation should abandon the use of focus groups and should concentrate instead on cognitive interviews for questionnaire evaluation.

24.3.3 Implications for Survey Development

The evaluation focus groups and several cognitive interviews helped in identifying a preferred questionnaire version for further development and evaluation as well as provided specific insights into questionnaire formatting, information treatments, and other questionnaire design elements. Respondents' explanations of their stated choices (e.g., indications that when habitat is worse, the number of acres needs to increase) revealed that respondents generally understood the tasks they were being asked to perform and generally accepted the reasonableness of the choice context. The evaluation focus groups and the several cognitive interviews revealed respondents to be more comfortable with the wetland comparisons set in a context of comparing the attributes of a destroyed wetland to those of a single restored wetland (rather than the version comparing alternative restored wetlands). That is, the first round of questionnaire evaluation revealed that respondents had such difficulty with one of the questionnaire versions that it was dropped from further consideration.

Two aspects of the difficulty with the "expanded restoration" instrument became clear. First, in order to state a choice about a wetland restoration proposal, respondents expressed their desire to know about the attributes of the wetland that was lost. That is, respondents had difficulty assessing whether the "expanded restoration" scenario was worth it, or even a reasonable trade, without knowing the features of the drained wetland. While discussed at length in evaluation focus groups, some quotes illustrate this point:

R19: "Doesn't say what you give up."

R17: "We don't know what was destroyed."

Second, the "expanded restoration" choice context was unclear to some respondents in light of existing wetland mitigation policy. The "expanded restoration" version raised many questions about paying for such a policy within the existing legal framework. The following excerpts illustrate some of these difficulties with the "expanded restoration" choice context:

R21: "What are we talking about here? There is a law in place that takes care of that end of it [impairments to existing wetlands]."

R18: "I was wondering who was responsible for all this repairing of the wetland. Is it the taxpayer? It was saying something about taxpayers."

The confusion seemed to stem from providing information in the expanded restoration version of the survey about the legal requirements for wetland mitigation along with information about potential additional taxes to support wetland restoration beyond that required by law. The version comparing drained wetlands to restored wetlands did not cause these confusions and was selected for further development and testing.

Both the evaluation focus groups and the opportunistic cognitive interviews yielded some consistent feedback on potential changes to the questionnaire. For example, respondents voiced reservations about answering the wetland restoration choice question because of fears that rare species or rare wetland types might be involved. To address these concerns, a "fine print section" was added to the questionnaire, near the choice questions, to reiterate key aspects of the choice context explained earlier in the survey booklet (e.g., wetlands with rare species are not subject to mitigation). It was also observed that respondents did not seem to have any difficulty completing three wetland choice scenarios, so the survey instrument was revised to increase the number of choice scenarios to five per survey. Being able to present each respondent with additional choice scenarios increased the amount of data that could be collected with a given budget.

24.4 QUESTIONNAIRE EVALUATION PHASE 2

24.4.1 Planned Evaluation Cognitive Interviews

After the researchers revised and modified the preferred draft questionnaire, it was decided to conduct two more rounds of cognitive interviews to evaluate the questionnaire and determine whether respondents understood the wetland law and policy information in the instrument, accepted as plausible policy context presented, and were able to make informed trade-offs between alternatives. Two rounds of cognitive interviews were conducted. These cognitive interviews were performed by graduate students specifically recruited, trained, and screened by the researchers. These interviewers received written material on conducting cognitive interviews; they received instruction from the researchers; they each conducted "practice" interviews using the draft questionnaire and discussion guide with several respondents randomly intercepted and recruited at the university's union building and international center; and they each conducted at least one cognitive interview with a randomly selected respondent while being observed by the researchers. While all potential cognitive interviewers were paid for their time to be trained, only those deemed acceptable by the researchers were hired to conduct the questionnaire evaluation cognitive interviews reported here and relied upon by the researchers. In the first round of evaluation cognitive interviews, 29 interviews were conducted in two days. Feedback from the first round of interviews was used to modify the survey questionnaire prior to the second round. The second round of evaluation cognitive interviews took place two weeks later, with the interviewers conducting 28 cognitive interviews in two days.

For both rounds of these evaluation cognitive interviews, respondents were recruited from the general population of mid-Michigan using the same methods outlined above for the focus groups. These respondents shared the same demographic characteristics as the general population of mid-Michigan. Respondents came to a central meeting facility on the Michigan State University Campus. All participants first completed the same version of the draft questionnaire. Upon completion of the questionnaire, participants were escorted across the hall to a semiprivate office where they were "interviewed" by one of the specially trained cognitive interviewers or one of the researchers. The cognitive interviews followed a detailed interview script that focused the interview on the key aspects of the questionnaire. The interviewers took detailed notes during the session and completed a postinterview evaluation.

Staying on Task While the group evaluation sessions presented the challenge of keeping participants on the task of evaluating the survey instrument, this challenge did not dominate the cognitive interview evaluations. In general, participants in the cognitive interviews were focused, cooperative, and energetic in providing interviewers with feedback about the survey instrument and their thought processes and strategies for answering.

Session Starts The vast majority of the cognitive interviews began with respondents staying on the task set forth by the interviewer. After a general introduction, the cognitive interviews began with the interviewer stating, "Let's talk about the booklet. So what do you think? Do you have any overall impressions or comments on the booklet?" Overwhelmingly, most of the interviewees' initial comments and subsequent discussions were about the instrument, not general concerns. The data show that 84% of participants started their sessions on task.

EXAMPLES

On-task

II 7: "It was well done and easy to follow. The questions' difficulty increased with the cases. I wanted more specific location information to help make decisions."

II 52: "It was clear and concise. Not hard to follow. No questions harder than any others but I had some difficulty deciding how I felt about replacing mixed wetlands with a single type of wetland, and vice versa."

EXAMPLE

Off-task

II 56: "It was interesting. I have mixed emotions. My house is between wetlands. My neighbor drained his wetland into my woods ... now I can't fill my wetlands in the front yard to sell ...."

Digressions During the cognitive interviews, it was reported to be fairly easy to keep respondents on task and to direct them back from their digressions off task to the task of instrument evaluation. For example, well after the interview was under way, interviewers gave the participants the following prompt: "Let's talk about case 1. Tell me a little about how you made your decision." In response to this probe, only four of the 57 respondents (7%) took the conversation off task. That is, 93% of cognitive interviewees responded by sharing their decision-making process for case 1 with the interviewer.

EXAMPLES

On-task

II 59: "I compared the condition before with the restored condition to see if restoration supplied everything that was there before. Acres were important, but if it was in place, it was important to me to note the habitat quality."

II 35: "The table made it clear, as it presented the restored wetland in the same manner as the drained wetland. I saw they were the same type, obviously in different locations. The replaced had more acres, more was better for amphibians."

Off-task

II 2: "I would call a troubleshooter to help me decide."

Learning Individuals' Experiences The cognitive interviews provided a useful context to learn about how the individual respondents understood, interpreted, and used the information contained in the survey booklet.

Saliency of Ratings The draft instrument made use of a habitat rating scale (see Figure 24.3) to describe the characteristics of various wetlands. From the focus groups it remained unclear whether the habitat ratings were understood, accepted, and useful to all respondents. This was addressed toward the end of the cognitive interviews, where respondents were asked: "Tell me in your own words about what the habitat ratings mean? What does ... [poor] mean?" In response, 52 respondents (91%) replied with feedback that reflected either their recall of the exact language of the instruments' rating scale or a scale in their own words that reasonably reflected the instruments' choice categories. Specifically, 36 respondents (63%) recalled and used language and terms that were provided in the instrument's descriptions, despite not having the survey at hand when this question was asked. An additional 16 respondents (28%) shared their own scale or heuristic for the habitat ratings that evidenced the ability to appreciate differing levels of species present in the alternative wetland scenarios.

Think-Alouds The cognitive interviews provided respondents with a setting in which they could share their individual thought processes, decision-making strategies, and difficulties. Of particular interest was learning from respondents about how they thought about and processed the information contained in the wetland choice questions. Most respondents revealed that they noticed and considered the different levels of attributes presented to them.

EXAMPLE

Prompt: "How did you use habitat ratings?"

II 52: "I was looking for equal or clear replacement of the loss. Ideally, impact could be made better, but not necessary. All categories of equal value? I looked for maintenance or improvement in significance of the five categories. Not always true in all of the cases or categories."

The think-aloud portion of the cognitive interview evaluations also resulted in individuals admitting to ignoring and misunderstanding information. The one-on-one setting seemed to make some people more comfortable sharing their feedback. So we were able to learn from the same prompt both how respondents understood and negotiated the survey instrument successfully and unsuccessfully.

EXAMPLE

Prompt: "How did you use habitat ratings?"

II 19: "I didn't use them."

II 35: "I didn't bother to look at that part of the page ... I just used common sense."

24.4.2 Implications for Survey Development

The cognitive interviews revealed that for the most part, respondents were comfortable with the questionnaire. The cognitive interviews also revealed that some people were not perceiving the full context and consequences of their wetland choice accurately. That is, they told the interviewers that they felt that by choosing not to accept a restoration scenario it would mean that the original wetland would get destroyed. To address this misperception, a new stand-alone page was added to the survey to introduce the choice setting briefly and to articulate more clearly the consequences of the wetland choice response options.

After two waves of cognitive interviews, the interviewers, their notes, and the debriefing forms seemed to support the notion that the revised survey questionnaire passed Fowler's test criteria (1995, p. 152). That is, the questions were consistently understood by respondents, and the questions posed tasks that respondents could perform. These criteria are also consistent with the design requirements established in Arrow et al. (1993). Moreover, no systematic or major misperceptions were revealed. As a result, no further individual cognitive interviews were conducted.

24.5 DISCUSSION

As a result of our use of multiple methods, including our use and analysis of videotapes from the focus groups and detailed written records of each cognitive interview, we were able to observe and analyze objectively the utility of focus groups as a design tool and an evaluation method. Furthermore, we documented and analyzed the results from the 57 cognitive interviews. As the preceding sections demonstrate, we learned a great deal about the appropriateness of our instrument design with the iterative use of multiple methods. Furthermore, and perhaps more important for the reader, we also discovered that the various methods appear to have some strong relative strengths and weaknesses in our case. In this section we use our experience, data, and analysis to discuss some observed strengths and weakness of the methods as instrument evaluation tools.

We must start with the observation that our evaluation focus group moderators were not very successful in keeping the participants focused on the task of survey instrument evaluation. We readily concede that others may be better skilled than our moderators. However, we did use professional focus group moderators with significant experience as well as a detailed focus group guide that targeted the instrument and its various components. Furthermore, after we observed the difficulties of keeping the first evaluation focus group on the survey evaluation task, we switched moderators for the next evaluative focus group. Although the experience of the first evaluation group helped us stress the importance of keeping the group on the task of instrument evaluation for the subsequent moderator, switching moderators did not cure the problematic group dynamics and digressions in the subsequent evaluative focus groups.

The general environment and setting of the evaluative focus groups and the cognitive interviews also seemed to have had an impact on participants' ability to stay on task. In the cognitive interviews, each respondent took the self-administered questionnaire by themselves and later joined an interviewer in a private space (e.g., empty office) for their "discussion about the booklet." The cognitive interview environment appeared to support participants staying on task, recalling their experience going through the instrument as well as sharing their think-aloud thought processes. In contrast, the participants in the evaluative focus groups seemed to easily mistake the goal of the session. Probes to individuals in the group evaluations designed to bring them back to the task of evaluating the survey instrument did not overcome conversational norms and a tendency to interact with and on behalf of others.

In terms of task comprehension, both types of evaluation methods provided evidence of participants' understanding of the wetland scenarios in the survey. For example, both methods revealed people who when making their choice stated that they considered the similarities and differences of the alternative scenarios. However, the fact that a few focus group respondents noticed differences in the attribute levels in the choice questions did not tell us what other members of the group understood. In contrast, it was readily apparent in the cognitive interviews that all respondents noticed the differing levels of attributes across the five pairwise wetland mitigation scenarios. Furthermore, the cognitive interviews provided richer evidence for a larger share of participants (everybody) on their comprehension of the wetland trade-off task. It should also be noted that completed survey data were looked at and subsequently analyzed econometrically to see whether participants' answers were logically sensible as indicators of respondents' having understood the task, and the results were highly supportive.

The focus group evaluations did provide an opportunity for some respondents to identify important issues and difficulties with the survey instrument, as well as reasonable approaches to solving them. For example, the groups were helpful to the researchers for identifying the more promising version of the survey instrument. However, the researchers found themselves trying to guard against unduly underweighting or overweighting feedback and information revealed during the evaluative focus groups. For example, after one focus group it appeared that property rights was a major concern that needed to be addressed. However, subsequent analysis of data from that session as well as the other focus groups and cognitive interviews revealed the property rights issue to have been a big concern to only one respondent. Nonetheless, we did make some beneficial changes to the draft questionnaire as a result of the focus groups. For example, we added "fine print" to address concerns about the mitigation of wetlands that were rare or contained rare species. Having an iterative process allowed us the opportunity to make changes knowing that we would be able to test and evaluate the efficacy and appropriateness of these changes.

Our experience showed that respondents' discussion of difficulties with the survey instrument seemed easier and more readily accessible in the cognitive interviews. In fact, the focus group participants did not raise many personal difficulties with the instrument during the evaluation focus groups. This result does seem to parallel some previous research comparing focus groups and cognitive interviewers (Kaplowitz, 2000). Cognitive interviews appear to be the preferable evaluation methods for getting at embarrassing and potentially sensitive information.

Our experience also revealed some important differences and similarities regarding the resource demands for the two evaluation methods. The focus groups and the cognitive interviews both require researchers to develop detailed interview guides for survey instrument evaluation. Doing so ensures that the interviewer/moderator will at least try to focus the attention of subjects on the evaluation issues of importance to the researchers. Similarly, both of these two methods can require similar amounts of recruiting effort to arrange for participants to attend the sessions. Although the cognitive interviews may require more interviewer time and preparation for the researcher than focus groups, the cognitive interviews do require a smaller time commitment from potential participants than do focus groups. In some cases, this may make it easier to recruit for cognitive interviews, especially if the cognitive interviewer is flexible about the times and locations of the interviews. The two methods also differ in the time requirements for data analysis as well as the richness of the data they generate. That is, focus groups may reveal more rich data on a broader range of subjects than the data yielded by a cognitive interview with a single respondent focused on the evaluation of a particular instrument. The analysis of a two-hour videotape and transcript from a focus group session of seven participants requires different data analysis skills and resources than does the analysis of seven cognitive interview forms. Also, the two methods do differ in the amount of time and other resources needed to train and pay interviewers/moderators. A typical focus group needs one moderator conducting a session with participants. To get input from the same number of individuals, cognitive interviewing would require one moderator for separate sessions or several moderators simultaneously. However, our experience suggests that to get comparable information, the number of cognitive interviews need not be as large as the number of focus group participants.

24.6 CONCLUSIONS

The purpose of this chapter was to present a case study on an iterative, multiple-method approach to survey design and evaluation in the context of environmental valuation. The survey instrument that was developed used a stated choice approach to examine the effect that different levels of wetland attributes such as acres and habitat quality would have on people's choices regarding wetland restoration alternatives. The iterative approach allowed us to make ongoing adjustments to the survey questionnaire based on the interview feedback. Overall, we found that respondents were quite capable of the survey task of making wetland choices, despite the potential complexity of wetland ecosystems and wetland mitigation policy.

The case highlights our use of mixed methods to develop and evaluate the questionnaire. A key aspect is the comparison of group and cognitive interviews for evaluation of a questionnaire. Although the case study reported is by no means a definitive study on the relative merits of focus groups and cognitive interviews as survey instrument evaluation methods, some clear lessons do emerge. It is also recognized that many projects do not have the resources to conduct the large number of cognitive interviews that we were able to do. However, a smaller number of cognitive interviews conducted in an iterative manner would probably be similarly beneficial. Of course, when time and other resources are limited, researchers must make trade-offs among various design, evaluation, and implementation considerations. It is our recommendation that questionnaire evaluation should probably include more than one iteration of cognitive interviews.

Our experience confirms that focus groups are an excellent tool for scoping and designing survey questionnaires. We learned about several wetland issues and information gaps that needed to be addressed in the design of the survey. In using focus groups to evaluate draft survey instruments, we found group sessions capable of identifying major concerns and difficulties with the instrument. However, for our evaluation needs, focus groups were not ideal tools for evaluating draft survey instruments. We found that for the purposes of questionnaire evaluation, focus group dynamics too often tended to yield rich conversations not germane to the task of instrument evaluation. Conversely, the cognitive interviews yielded a detailed set of information for virtually all respondents. Additionally, interviewers found it easy to help respondents stay on the task of survey instrument evaluation. With the cognitive interviews, we were more confident that we learned of each person's opinions and experiences with the instrument without the complications of the group process. Our experience using cognitive interviews to evaluate survey instruments suggests their value as a questionnaire evaluation tool.

CHAPTER 25

Does Pretesting Make a Difference? An Experimental Test

Barbara Forsyth
Westat

Jennifer M. Rothgeb
U.S. Bureau of the Census

Gordon B. Willis
National Cancer Institute

25.1 INTRODUCTION

In this chapter we present results from research designed to determine (1) whether questionnaire pretesting results predict actual problems encountered in survey data collection, and (2) whether survey administration is facilitated or survey outcomes are improved using revisions based on pretesting results. The research reported here was conducted in two phases. In phase 1 we used several pretesting techniques to test a set of survey items and to develop revised questions. In phase 2 we conducted a telephone survey using a split-sample experiment to administer both the original and the revised questions. We explore whether results from the phase 1 pretesting research predict problems observed when the original questions are administered in the phase 2 telephone survey. We also examine whether question revisions developed based on pretest results produce improved survey outcomes.

25.1.1 Background

Questionnaire pretesting is standard practice for several U.S. government statistical agencies and other organizations involved in designing or conducting national

Methods for Testing and Evaluating Survey Questionnaires, Edited by Stanley Presser, Jennifer M. Rothgeb, Mick P. Couper, Judith T. Lessler, Elizabeth Martin, Jean Martin, and Eleanor Singer
ISBN 0-471-45841-4

surveys. Pretesting methods that are commonly used include expert review, cognitive interviewing, behavior coding, and respondent debriefing. The ability to make good, informed decisions about pretest standards and pretest practices is enhanced by data that address methodological questions such as the following:

- Which pretesting methods are most effective for identifying questionnaire problems?
- Which pretesting methods are most useful for providing information to fix questionnaire problems?
- How does the set of effective methods differ depending on survey characteristics or pretest purposes?
- What is the most effective way to combine sets of pretesting methods to address particular pretest goals?

Researchers have taken different approaches to answering questions such as these. Willis et al. (1999a) noted that one way to distinguish these research approaches is according to the criteria for methods evaluation. Following Willis et al. we identify three general approaches to methods evaluation.

- *Exploratory research* compares pretest methods in terms of their effectiveness for detecting unexpected questionnaire problems.
- *Confirmatory research* compares pretesting methods in terms of their effectiveness for confirming or disconfirming questionnaire problems that are suspected based on other results.
- *Reparatory research* compares pretesting methods in terms of their effectiveness for suggesting revisions that improve survey outcomes.

Exploratory and confirmatory research focus on how well pretest techniques *detect* questionnaire problems. Reparatory research focuses on how effectively pretest techniques identify ways to *improve*• questionnaire items once problems have been identified.

Exploratory research is relatively common. The designs typically make direct comparisons between different pretesting methods in terms of the numbers and types of problems identified when the methods are applied to a constant set of survey materials (e.g., Campanelli, 1997; DeMaio et al., 1993; Oksenberg et al., 1991; Presser and Blair, 1994). A small number of exploratory studies have focused on comparing variants of a particular method—for example, alternative approaches for adding probe questions to cognitive interview protocols or alternative behavior coding schemes (e.g., Conrad and Blair, 2001; Davis and DeMaio, 1993; Edwards et al., 2002; Foddy, 1996a).

Confirmatory research is less common than exploratory research. The most prevalent confirmatory design involves assessing relations between pretesting results and survey results, especially survey measures assumed to be related to data quality (e.g., Davis and DeMaio, 1993; Willis and Schechter, 1997). Generally, these studies focus on predictions from pretests using individual pretesting methods. We have identified no confirmatory studies examining combinations of pretesting methods that are typical of actual pretesting practice. One major purpose of the research reported here is to extend the typical confirmatory design to explore predictions of survey outcomes based on results obtained by applying a sequence of pretesting methods similar to those used in actual practice.

Reparatory research is rare. We've identified two studies where researchers used split-sample field test designs to compare survey results from questions revised based in part on pretesting research activities with survey results from unrevised questions (Lessler et al., 1989; Turner et al., 1992a). In both studies, the reparatory results are difficult to interpret because questionnaire revisions were based on additional information beyond the pretest results. A second major purpose of the research reported here is to conduct reparatory research to determine whether questionnaire revisions made solely in response to pretesting results improve survey administration and/or survey outcomes.

25.1.2 Objectives

The research reported here was conducted in two phases. In phase 1 we used expert review, questionnaire appraisal, and cognitive interview methods to pretest three sets of survey items. One goal of the phase 1 research was to compare the three pretesting methods in terms of the numbers and types of potential problems identified. In phase 1, three organizations used all three of the methods. As a consequence, we were also able to examine agreement across organizations. Those findings are reported in Rothgeb et al. (2001). A second goal of the phase 1 pretesting research was to develop indicators of questionnaire problems to use for predicting phase 2 survey outcomes. We developed a qualitative problem classification scheme for this purpose. The classification scheme is described below (Section 25.2.2). A third goal of the phase 1 pretesting research was to develop recommendations for questionnaire revisions expected to improve survey outcomes in phase 2.

In phase 2 we conducted a split-sample field experiment using a random-digit-dial (RDD) telephone survey. Household cases were assigned randomly to either a control questionnaire or an experimental questionnaire. The control questionnaire included the items pretested in phase 1. The experimental questionnaire included comparable items that were revised based on the pretest evaluation.

We designed the telephone survey field experiment to answer two research questions.

- *Research question 1*: Do pretesting results from phase 1 predict problems in the control condition of the phase 2 field experiment?
- *Research question 2*: Do questionnaire revisions based on phase 1 pretest findings improve survey outcomes in the experimental condition of the phase 2 field experiment?

Research question 1 is confirmatory. Results from the phase 2 field experiment are used to confirm or disconfirm suspected questionnaire problems identified in phase 1 pretesting. Research question 2 is reparatory. Results from the phase 2 field experiment are used to determine whether questionnaire revisions based on phase 1 pretesting results produced more effective items.

25.2 DESIGN

25.2.1 Phase 1: Questionnaire Pretesting Research Design

In selecting a pretest design, we were interested in including both the experimental factors directly of interest and additional design factors that would enhance the generalizability of our results. The design factors we incorporated were:

- *Pretesting methods.* We chose to focus on three pretesting methods: informal expert review, questionnaire forms appraisal, and cognitive interviewing.
- *Survey organization.* Researchers from each of three survey research organizations conducted pretest activities. In addition to enhancing generalizability, we included research organization as a design factor so we could explore organizational differences. (Those results are reported in Rothgeb et al., 2001.)
- *Pretest experience.* A senior methodologist with considerable pretest research experience led the pretesting team at each organization. Each team consisted of two additional pretest researchers. The organizations aimed to select one pretest researcher with moderate pretesting experience and one with relatively little pretesting experience. Our aim was to include a mix of experience levels within each organization to enhance generalizability.
- *Questionnaire content.* We pretested a total of 83 questionnaire items selected from three survey questionnaires. We selected survey topics that the pretest researchers had relatively little experience with, including questions about (1) household telephone expenses and vehicles owned, from the U.S. Bureau of the Census's 1998 Consumer Expenditure Survey; (2) use of alternative transportation modes, from the U.S. Department of Transportation's 1995 National Public Transportation Survey; and (3) attitudes toward environmental issues from the U.S. Environmental Protection Agency's 1999 Urban Environmental Issues Survey.
- *Pretest method sequence.* Each pretest researcher used all three pretest methods. We selected a single pretest method order that seemed to reflect common practice and to minimize undesirable carryover effects. Each researcher completed an informal expert review first, followed by the questionnaire forms appraisal and then cognitive interviews.

We selected a Latin square design for conducting phase 1 pretest research activities. Under this design, each researcher conducted an informal expert review with one set of pretest items, a questionnaire forms appraisal with a second set of pretest items, and three cognitive interviews with the third set of pretest items. The

Latin square design ensured that each item was tested under all three pretesting techniques and by each organization. However, each individual staff member reviewed an item under only one of the three pretesting methods.

We analyzed the phase 1 results by comparing the number of problems detected by each pretesting method and by each organization. The pretest "problem scores" ranged from 0 to 9 (based on evaluation by three organizations × three pretesting techniques). An item's problem score was 0 when no organization identified a problem with the item based on any pretest method. An item's problem score was 9 when all three organizations identified problems with the item under all three pretest methods. Problem scores between these two extremes reflected disagreements across organizations, across pretest methods, or both.

Results from phase 1 indicated that there was little variation in the numbers or types of problems identified across participating research organizations; the organizations seemed to use similar criteria to identify and label questionnaire problems encountered in pretesting. Further, although the pretesting techniques varied in terms of the numbers of problems they identified, from a qualitative perspective all three were found to focus mainly on problems related to question comprehension and communication.

For purposes of the research reported here, we used the phase 1 problem scores to select the most problem-prone items. From the 83 items tested, we selected 12 items with problem scores of 8 or above to include in the phase 2 field experiment. The 12 items selected were very problematic, as is clear from the following examples:

Example item 1: Is local bus service available in your town or city? (Include only services that are available for use by the general public for local or commuter travel, including dial-a-bus and senior citizen bus service. Do not include long-distance buses or those chartered for specific trips.)

Example item 2: First, I'm going to read you a list of different issues that may or may not occur in your community. . . . I am going to read the list of issues and I want you to tell me how high or low a priority each is in the community. Use a scale of 1 to 10, with 1 meaning "very low priority" and 10 meaning "very high priority."

a. Depletion of the water table

25.2.2 Phase 2: Questionnaire Design

The control questionnaire included original versions of the 12 items selected, and the experimental questionnaire included revised versions of the 12 items selected. Our research purposes required that we revise questions based solely on pretest results. To meet this goal, we reviewed all notes gathered through the phase 1 pretest and analysis, using a problem classification coding scheme (CCS) to document question problems identified during phase 1 pretesting.

Table 25.1 shows the CCS problem categories. The CCS consists of a hierarchy of 28 codes. At the highest level of the hierarchy, the codes are grouped under

Table 25.1 Problem Classification Coding Scheme (CCS)

1. Comprehension and communication
    - Interviewer difficulties
        1. Inaccurate instructions
        2. Complicated instruction
        3. Difficult to administer
    - Question content
        4. Vague topic/term
        5. Complex topic
        6. Topic carried over from earlier question
        7. Undefined term(s)
    - Question structure
        8. Transition needed
        9. Unclear respondent instruction
        10. Question too long
        11. Complex, awkward syntax
        12. Erroneous assumption
        13. Several questions
    - Reference period
        14. Carried over from earlier question
        15. Undefined
        16. Unanchored or rolling
2. Memory retrieval
    17. Shortage of cues
    18. High detail required or information unavailable
    19. Long recall period
3. Judgment and evaluation
    20. Complex estimation
    21. Potentially sensitive or desirability bias
4. Response selection
    - Response terminology
        22. Undefined term(s)
        23. Vague term(s)
    - Response units
        24. Responses use wrong units
        25. Unclear what response options are
    - Response structure
        26. Overlapping categories
        27. Missing categories
5. Other
    28. Something else

the familiar headings of the traditional four-stage cognitive response model: problems in comprehension and communication, retrieval from memory, judgment and evaluation, and response evaluation (e.g., Tourangeau, 1984). Within each of the four stages there are midlevel categories of problems, and the lowest-level codes provide the most detailed descriptions of question problems identified during phase 1 pretesting.

The three authors jointly assigned CCS codes to pretest results for the 12 items selected. We applied the CCS to each item a total of nine times: once for each combination of pretest method and research organization. We selected CCS codes collaboratively and assigned as many codes as we agreed applied to the documented problems. A total of 257 problems were identified across the nine separate evaluations of the 12 most problem-prone items. These problems involved 28 unique problem codes. [Details of this analysis are provided in Rothgeb et al., (2001).]

We used the CCS codes and testers' notes to revise items to address the specific problems identified by our pretest research activities. For example, if the CCS codes indicated that pretest respondents had problems understanding a question because it used "undefined terminology," we used testers' notes to identify terms that caused problems for pretest respondents and developed a revised question that addressed only the documented terminology problem(s). Pretesting typically identified multiple problems with each problem-prone question. Consequently, revisions generally addressed multiple design problems.

We faced four general challenges as we developed question revisions.

- *Identifying question objectives.* We didn't have specific objectives for most of the items tested. As a result, we frequently had to agree on preliminary assumptions about question objectives before we could develop item revisions. This is probably not typical of most questionnaire revision where revision includes discussions between substantive and methodological experts to clarify and refine question objectives.
- *Revising items with multiple problems.* The CCS results identified multiple problems for all 12 items pretested. We chose to develop revisions addressing all problems identified. This feature of our design influences how we interpret analytic results. Differences in outcome measures between the control and experimental question versions cannot be linked to any one specific change. Rather, differences must be attributed to the combination of revisions selected.
- *Simplifying complex items.* Several of the original problem-prone items were identified as too complex. Effective revision depended on decomposing these items into two or more simpler items. As a consequence, we had to select analytic strategies that assess experimental effects when there is a many-to-one correspondence between the experimental and control question versions.
- *Developing items for CATI interviews.* Some of the original questions came from paper-and-pencil questionnaires. We had to design them for

administration in a CATI instrument without introducing extraneous changes that would interfere with our pretest predictions and analytic conclusions.

Of course, we also had to contend with more common design issues related to time constraints and allocated space within the questionnaire. Our interview content represented just one of four experiments included in the survey design. Our portion of the control questionnaire contained a total of 26 items. These 26 items included the selected 12 items that are the focus of our methodological experiment, and 14 additional items and transitional instructions included to establish and maintain interview flow. Our portion of the experimental questionnaire consisted of 44 items. These 44 items included 26 that were revised versions of 12 pretested items, and 18 additional items and transitional passages included for interview flow.

We included the original and the revised versions of the 12 items selected as part of two versions of an omnibus survey questionnaire. The omnibus survey was conducted by the Census Bureau in August and September 2000. Details of the split-sample field test design and methodology are provided in Section 25.3.

25.3 METHODOLOGY

25.3.1 Phase 2 Data Collection

The Census Bureau's omnibus Questionnaire Design Experimental Research Survey (QDERS) was conducted in August and September 2000. QDERS interviews were conducted by telephone from one of the Bureau's telephone interviewing facilities. The survey used RDD sampling procedures and computer-assisted telephone interview (CATI) survey instruments. The RDD sample represented households in the continental United States, and the sample consisted of 10,000 telephone numbers randomly assigned to one of the two questionnaire versions. For each eligible sample household, interviewers identified one adult household member to serve as the household respondent based on eligibility and willingness to participate. With the respondent's permission, the interviews were audiotaped. The interviews lasted approximately 15 minutes. Both versions of the QDERS questionnaire included seven sets of questions. The three sets of questions of interest here are on the topics of telephone expenditures, transportation, and attitudes about the environment. The other topics covered in the questionnaires were health insurance, home mortgages, income, and basic household demographics.

Interviewers completed interviews over a four-week period. The interview staff consisted of 24 experienced telephone interviewers, split randomly into two groups. During the first two-week period, one group of interviewers trained on and administered one version of the questionnaire. The other group of interviewers trained on and administered the other questionnaire version. At the end of the first two-week data collection period, interviewers were retrained on the alternative version and conducted their remaining interviews with a second half-sample. We selected this approach for staffing because we wanted all interviewers to administer both questionnaires, and we wanted to minimize interference. This approach produced one "pure" debriefing session for each questionnaire version.

Interviewers completed interviews in 1862 households. Using accepted response rate calculation guidelines (American Association for Public Opinion Research, 2000), the set of 1862 interviews represents a response rate between 42 and 55%. There were no differences between the two questionnaire conditions in terms of household response, household nonresponse, or interview refusal rates.

25.3.2 Dependent Measures of Survey Outcomes

We selected three sets of measures as indicators of phase 2 survey administration and as potential measures of data quality: item nonresponse rates under the two questionnaire versions, behavior coding results for the two questionnaire versions, and interviewer ratings collected as part of the study's interviewer debriefing activities. Behavior coding and interviewer ratings are not direct measures of data quality. Instead, they are measures of questionnaire flow and interviewer opinions that predict quality measures (e.g., Hess and Singer, 1995). Measures of item nonresponse are traditionally accepted as indicators of survey data quality (e.g., Groves, 1989; Hox et al., 1991; Turner et al., 1992a). All three outcome measures are useful because methodologists would generally agree that decreased item nonresponse, fewer problematic behavior codes, and improved interviewer ratings are signs of a successfully revised questionnaire.

Item Nonresponse Rates We computed "don't know" and refusal frequencies separately for each item in each questionnaire version. Item refusal frequencies were uniformly low, so we combined "don't know" and refusal into a single nonresponse frequency for each item. We computed item nonresponse rates by dividing each item nonresponse frequency by the total number of respondents expected to answer that item. In both questionnaire versions, a few items could be skipped based on earlier responses. Respondents who skipped an item were eliminated from the nonresponse computations for that item.

We also computed an index of nonresponse for each respondent for the purpose of comparing the control and experimental questionnaire versions. This index was computed by dividing the respondent's total number of nonresponses by the number of items administered to the respondent. We computed the mean subject nonresponse index by averaging the individual indices across respondents.

Behavior Coding All of the telephone interviews were recorded on audiotape. We selected a random sample of 98 cases from each questionnaire version for behavior coding. The staff who conducted the behavior coding was not involved in any other study activities. The behavior codes documented five interviewer behaviors:

- Question read exactly or with a slight change in wording
- Major change in question wording

- Question worded as verification
- Question mistakenly omitted
- Other follow-up behavior (e.g., question repeated, probed for clarification, provided clarification).

The behavior codes documented five respondent behaviors:

- Adequate response
- Qualified response
- Inadequate response
- Interruption
- Request for clarification or repetition

Behavior coders assigned codes for up to two interviewer–respondent interactions. Analyses presented here focus on the first behaviors coded during the first interaction for each item in each coded interview.

We used the codes assigned to interviewer and respondent behaviors to develop two behavior coding problem indicators for each item in each interview. The interviewer problem indicator was 1 if an assigned code indicated that the interviewer had difficulty administering the item; otherwise, the interviewer problem indicator was 0. The respondent problem indicator was 1 if an assigned code indicated that the respondent had difficulty understanding or answering the item, or 0 if no code was assigned indicating a respondent problem. We computed frequency distributions for the interviewer and respondent problem indicators by questionnaire version.

Interviewer Ratings Interviewers participated in debriefing sessions at the end of both two-week data collection periods, just after completing interviews with a single questionnaire version. At both time points, interviewers administering the control and experimental questionnaire versions were debriefed separately. At each debriefing session, interviewers completed an item rating task for the questionnaire they had just finished administering. They reviewed the questionnaire items independently and rated each in terms of how often it caused problems for them as interviewers and also in terms of how often it caused problems for respondents. Interviewers used a three-point scale to indicate that an item (1) caused no problems, (2) caused some problems, or (3) caused a lot of problems. Also, interviewers wrote comments about the types of interviewer and respondent problems they experienced with each item, when applicable.

We computed frequency distributions for the three rating categories by item and questionnaire version, separately for the interviewer and respondent problem ratings. We combined frequencies for ratings of "some problems" and "a lot of problems" to make presentations parallel across the three sets of survey outcome variables.

25.4 RESULTS

25.4.1 Research Question 1: Do Pretesting Results from Phase 1 Predict Survey Outcomes in the Control Condition of the Phase 2 Field Experiment?

We begin by describing the summary measures we developed to reflect the phase 1 pretest results and to predict phase 2 survey outcomes. All analyses in this section examine data collected using the control questionnaire version.

The CCS codes in Table 25.1 describe two general sets of pretesting problems: interviewer problems (CCS codes 1 through 3) and respondent problems (CCS codes 4 through 27). We developed two summary measures of pretest problem severity for each item by counting the number of CCS codes assigned to the item during the course of pretesting, separately for the codes describing interviewer problems and for the codes describing respondent problems. We used these problem severity measures to classify each item in two ways: first, as low, moderate, or high in interviewer problems, and second, as low, moderate, or high in respondent problems. These classifications were the basis for predicting interviewer and respondent problems observed during the phase 2 survey data collection.

Recall that the 12 items of interest here were selected because they were the most problem-prone. The least problematic item in this set had a total of 13 CCS codes assigned to it (combined across interviewer and respondent problem codes). The most problem-prone item had a total of 36 CCS codes assigned to it. Therefore, the low, medium, and high classifications of interviewer and respondent problem severity actually represent a relatively narrow range of pretest problem severity. Our results may not apply to less problem-prone items.

Interviewer Problems We gathered two outcome measures of interviewer problems with the control questionnaire version during the phase 2 survey data collection: behavior-coded interviewer problems observed for each item and interviewer ratings of problems they experienced with each item. Table 25.2 shows the behavior coding results, and Table 25.3 shows the interviewer rating results. The top panels in Tables 25.2 and 25.3 show results for items expected to have low numbers of interviewer problems based on the phase 1 pretesting results. The middle panels in both tables show results for items expected to have moderate numbers of interviewer problems based on phase 1 results, and the bottom panels show results for items expected to have high numbers of interviewer problems.

As expected based on the phase 1 pretest results, the proportion of interviewer problems documented by behavior coders is small in the top panel of Table 25.2 and larger in the bottom panel of Table 25.2. Chi-square tests with directional post hoc tests indicated no significant differences in interviewer problems between items identified as low and items identified as moderate in interviewer problems based on phase 1 pretest results. When these two sets of items are combined, the behavior coding results show significantly more behavior-coded interviewer problems for items with high pretest interviewer problems than for items with low or moderate pretest interviewer problems.

Table 25.2 Interviewer Problems Identified in Phase 2 Behavior Coding by Pretest Interviewer Problem Severity[a]

	Percent of Behavior-Coded Interviews		
Pretest Interviewer Problem Severity	No Interviewer Problems	One or More Interviewer Problems	Total n
Low (0 codes assigned)	96.1	3.9	180
Moderate (1–2 codes assigned)	97.2	2.8	668
High (4–6 codes assigned)	89.5	10.5[b]	277

[a] Chi square = 24.94, $p < 0.01$.
[b] One-tailed $p < 0.001$.

Table 25.3 Interviewer Problems Identified by Phase 2 Interviewer Ratings by Pretest Interviewer Problem Severity[a]

	Percent of Rated Items		
Pretest Interviewer Problem Severity	No Interviewer Problems	Some or a Lot of Interviewer Problems	Total n
Low (0 codes assigned)	93.2	6.8	44
Moderate (1–2 codes assigned)	87.6	12.3	154
High (4–6 codes assigned)	78.8	21.2[b]	66

[a] Chi square = 5.174, $p < 0.10$.
[b] One-tailed $p < 0.05$.

Interviewer ratings in Table 25.3 show a similar pattern. Chi-square analyses with directional post hoc tests indicated no significant differences in interviewer ratings of interviewer problems between items identified as low and items identified as moderate in interviewer problems based on phase 1 pretest results. When these two sets of items are combined, the interviewer rating results show significantly more rated interviewer problems for items with high pretest interviewer problems than for items with low or moderate pretest interviewer problems.

Respondent Problems We gathered two outcome measures of respondent problems with the control questionnaire version during the phase 2 survey data collection: behavior-coded respondent problems observed for each item and interviewer ratings of apparent respondent problems for each item. Table 25.4 contains the behavior coding results and Table 25.5 contains the interviewer rating results.

Table 25.4 Respondent Problems Identified in Phase 2 Behavior Coding by Pretest Respondent Problem Severity[a]

	Percent of Behavior-Coded Interviews		
Pretest Respondent Problem Severity	No Respondent Problems	One or More Respondent Problems	Total n
Low (12–16 codes assigned)	89.1	10.9	341
Moderate (20–21 codes assigned)	76.8	23.2[b]	379
High (29–30 codes assigned)	57.1	42.9[b]	170

[a] Analysis is based on 10 items for which explicit responses were required. Two items were excluded from this analysis because they were instructions that required no explicit response. Chi square = 67.9, $p < 0.001$.
[b] One-tailed $p < 0.001$.

Table 25.5 Respondent Problems Identified by Phase 2 Interviewer Ratings by Pretest Respondent Problem Severity[a]

	Percent of Rated Items		
Pretest Respondent Problem Severity	No Respondent Problems	Some or a Lot of Respondent Problems	Total n
Low (12–16 codes assigned)	95.4	4.6	88
Moderate (20–21 codes assigned)	90.8	9.2	109
High (29–30 codes assigned)	69.2	30.8[b]	65

[a] Analysis is based on 12 items, including two sets of instructions that required no explicit response because interviewers were able to rate these items. Chi square = 25.156, $p < 0.001$.
[b] One-tailed $p < 0.001$.

The top panels in Tables 25.4 and 25.5 show results for items expected to have low numbers of respondent problems based on the phase 1 pretesting results. The middle and lower panels in both tables show results for items expected to have moderate and high numbers of respondent problems based on phase 1 results, respectively.

The general patterns of results in Tables 25.4 and 25.5 are as expected based on pretest results. In Table 25.4, the proportion of respondent problems documented by behavior coders increases consistently with increased pretest problem severity. Chi-square tests with directional post hoc tests indicated significantly more behavior-coded respondent problems for items with high pretest respondent problems than for items with moderate respondent problems. Also, behavior-coded respondent problems were significantly higher for items with moderate pretest respondent problems than for items with low pretest respondent problems.

In Table 25.5, chi-square tests with directional post hoc tests indicated no significant differences in rated respondent problems between items with moderate and low pretest respondent problems. When these two sets of items are combined, the interviewer rating results show significantly more rated respondent problems for items with high pretest respondent problems than for items with moderate or low pretest respondent problems.

Item Nonresponse We used a subset of CCS codes for predicting patterns of nonresponse in the phase 2 survey. We hypothesized that item nonresponse would be related to pretest problems with memory recall (CCS codes 17 through 19 in Table 25.1) and item sensitivity (CCS code 21 in Table 25.1). We computed a measure of pretest recall and sensitivity problem severity for each item by counting the number of recall and/or sensitivity-oriented CCS codes assigned to the item. Based on this measure, we classified items as low or high in pretest recall and sensitivity problem severity. This classification was the basis for predicting item nonresponse rates. Because of the way the items clustered, we did not identify a category of items with "moderate" recall and/or sensitivity problem severity.

Table 25.6 contains overall rates of nonresponse by recall and sensitivity problem severity. In Table 25.6, overall item nonresponse is lower for items identified in phase 1 pretesting as having relatively few recall and sensitivity problems. Item nonresponse is significantly higher for items identified as having more problems related to recall and sensitivity.

Summary Results from behavior coding, interviewer ratings, and nonresponse rates consistently indicated that problems observed during pretesting do predict

Table 25.6 Phase 2 Item Nonresponse for Control Questionnaire Items by Pretest Recall and Sensitivity Problem Severity[a]

	Item Nonresponse	
Pretest Recall and Sensitivity Problem Severity	Percent of Items Administered	Total n
Low (0–3 codes assigned)	1.0	4313
High (5–11 codes assigned)	8.8[b]	4303

[a] Analysis is based on 10 items for which explicit responses were required. Two items were excluded from this analysis because they were instructions that required no explicit response.
[b] One-tailed $p < 0.001$.

problems observed when the same items are administered in the field.

- Items with relatively many interviewer problems during pretesting also have relatively many behavior-coded and rated interviewer problems in the field.
- Items with relatively many respondent problems during pretesting also have relatively many behavior-coded and rated respondent problems in the field.
- Items with relatively many recall and sensitivity problems during pretesting also have relatively high nonresponse rates in the field.

Thus, information from phase 1 pretesting was consistently useful for detecting problems that were confirmed by survey results. In the next section we present results for assessing the repairs made based on the phase 1 pretest results.

25.4.2 Research Question 2: Do Questionnaire Revisions Made Based on Phase 1 Pretest Findings Improve Survey Outcomes in the Experimental Condition of the Phase 2 Field Experiment?

We address research question 2 by examining differences between the control and the experimental questionnaire versions in the survey outcome measures selected. These analyses focus on 10 of the 12 items discussed in Section 25.4.1. The experimental versions for two of the control questionnaire items were incorrectly programmed in the experimental version of the CATI instrument. We excluded these items from our analyses because the data are not comparable.

Questionnaire revisions often involved decomposing one complicated item in the control questionnaire version into several simpler items in the experimental questionnaire version. These revisions influenced our analytic approach. Because of skip instructions, sample sizes could vary across the experimental questionnaire items that we compared to a single control questionnaire item. In most of these comparisons, our analyses compare results from a single control questionnaire item with results from multiple experimental questionnaire items, proportionally weighted according to their sample sizes. We used equal weights to compute the mean subject index of nonresponse.

Item Nonresponse Table 25.7 contains item nonresponse rates and mean subject indexes of nonresponse for the control and experimental questionnaire versions. For both measures, overall nonresponse was significantly lower for the experimental treatment questionnaire than for the control questionnaire, although the differences are small in magnitude.

Behavior Coding Table 25.8 contains results for behavior-coded interviewer problems and behavior-coded respondent problems for the control and experimental questionnaire versions. Directional and nondirectional tests of the proportions in Table 25.8 revealed no significant differences between the two questionnaires

Table 25.7 Phase 2 Measures of Nonresponse for Control and Experimental Questionnaire Versions

	Control Questionnaire		Experimental Questionnaire	
	Percent of Administered Items That Were Unanswered	Total Number of Items or Number of Respondents	Percent of Administered Items That Were Unanswered	Total Number of Items or Number of Respondents
Item nonresponse	5.9[a]	6898	4.7[a]	10,520
Mean subject index	6.4[a]	915	4.8[a]	930

[a] One-tailed $p < 0.001$.

Table 25.8 Interviewer and Respondent Problems Identified in Phase 2 Behavior Coding for Control and Experimental Questionnaire Versions[a]

	Control Questionnaire		Experimental Questionnaire	
	Percent of Behavior-Coded Interviews	Total n	Percent of Behavior-coded Interviews	Total n
One or more interviewer problems	4.4	939	6.9	1291
One or more respondent problems	23.6	705	25.9	1072

[a] One- and two-tailed tests indicate no significant differences between questionnaire versions.

in terms of either behavior-coded interviewer problems or behavior-coded respondent problems.[1]

Interviewer Ratings Table 25.9 contains results for rated interviewer problems and rated respondent problems for the control and experimental questionnaire versions. Interviewer ratings of interviewer problems yielded an unexpected result. Interviewer ratings identified *more* interviewer problems with the experimental questionnaire than with the control questionnaire. This difference was significant according to chi-square tests, but the direction was opposite that expected based on the pretest findings, so it was not significant under our directional post

[1] Item-level analyses indicated that two items were largely responsible for the absolute increase in behavior-coded interviewer problems with the experimental questionnaire version. The original versions of both items were easy for interviewers to read but difficult for respondents to understand based on pretest results. Apparently, revisions made to enhance communication also increased interviewer reading problems. Behavior-coded respondent problems increased with the experimental questionnaire version for seven items and decreased for two. Once averaged over items, the difference was not significant.

Table 25.9 Interviewer and Respondent Problems Identified by Phase 2 Interviewer Ratings for Control and Experimental Questionnaire Versions

	Control Questionnaire		Experimental Questionnaire	
	Percent of Rated Items	Total n	Percent of Rated Items	Total n
Some or a lot of interviewer problems	11.0	218	24.8[a]	314
Some or a lot of respondent problems	55.3	217	45.3[b]	300

[a]Chi square = 15.884, $p < 0.001$.
[b]One-tailed $p < 0.05$.

hoc tests. Interviewer ratings of respondent problems identified fewer respondent problems with the experimental questionnaire than with the control questionnaire. Chi-square analyses with directional post hoc tests indicated that this difference was significant.

Summary Results from analyses of item nonresponse, behavior coding, and interviewer ratings indicate that the question revisions we made had mixed effects on survey outcomes.

- Item revisions reduced item nonresponse, but the improvement was small.
- Item revisions had no effect on the types of interviewer and respondent problems identified by behavior coding.
- Item revisions repaired the types of respondent problems identified by interviewer ratings.
- Item revisions did not repair the types of interviewer problems identified by interviewer ratings and may have made these interviewer problems worse.

25.5 CONCLUSIONS

We close by summarizing our key findings, discussing their implications, and then suggesting some directions for future research.

25.5.1 Findings Related to Problem Detection

Our initial research question asked whether pretesting results based on expert review, questionnaire appraisal, and cognitive interviews predict actual problems in field survey outcomes. Our findings suggest that they did. Questions that pretesting identified as particularly problematic for interviewers elicited more inappropriate behavior-coded interviewer behaviors than less problematic items.

Interviewers also rated these items as causing more problems for them than did the less problematic items.

Questions that pretesting identified as posing particularly large problems for respondents also elicited more uncodable responses, more respondent requests for clarification, and/or more respondent interruptions compared with less problematic items, based on behavior coding analyses. Findings based on interviewer ratings of respondent problems were consistent with this trend. Finally, items that pretesting identified as posing large problems related to memory and estimation elicited more nonresponse than did less problematic items.

25.5.2 Findings Related to Problem Repair

The second major research question we posed asked whether questionnaire revisions based on pretest results yield improved survey outcomes. Based on our results, the answer is unclear. We obtained some positive evidence, but it certainly was not pervasive. The revised items in the experimental questionnaire produced a very small improvement in nonresponse and a larger improvement as assessed through interviewer ratings of respondent problems. Thus, there is some evidence that our pretest results served a reparative function, at least from the respondent's perspective. At the same time, there were no differences between the two questionnaire versions in terms of behavior-coded respondent problems. So the results are mixed.

We had little evidence that question modifications served to improve questions from the point of view of the interviewer. Our revisions had no effect on interviewer problems reflected in behavior coding, although behavior coding also suggested that these problems were generally not large to begin with. Based on interviewer ratings, our question revisions may have made interviewer problems worse. Initially, we chose to work with a very problematic set of items. Item-level analyses suggested that solving respondent problems required steps that made the revised items difficult for interviewers to administer. The increase in interviewer ratings of interviewer problems with the experimental questionnaire version makes it clear that there were limits to our ability to use the pretest findings to make revisions that consistently improved all of the survey outcome measures selected.

25.5.3 Study Strengths, Limitations, and Possible Explanations

Our study design had two important strengths. First, unlike previous confirmatory studies, our design used a split-sample experiment to examine the confirmatory power of a combination of pretesting methods. In this sense, our design mimicked an important feature of actual pretesting practice. Second, unlike previous reparatory studies, we made revisions to the questionnaire based only on pretest findings. As a result, any observed differences between items in the control and experimental questionnaires can be attributed to decisions made based on pretest results.

It is unfortunate that we did not strengthen the sensitivity of our basic design by purposefully including less problematic items in the control questionnaire version. Including less problem-prone items would allow us to assess the confirmatory and reparatory effects of pretesting for "pretty good" items in addition to the "pretty bad" and "very bad" items studied here. We believe that the contrast would be informative.

We were not particularly successful in improving questions that pretesting identified as problematic. We can think of a few possible reasons for our lack of success.

Possibility 1: The identified problems were insurmountable We attempted to eradicate flaws in the very worst questions that we found. Perhaps this goal was simply not possible. In many cases, we attempted to decompose one difficult question into several simpler ones. In these cases, concepts may have been too complex or too undeveloped to fix with revised wordings. It may be necessary to revisit measurement objectives or develop a new question-asking strategy to fix these items.

Possibility 2: The modifications were flawed A major function of pretesting is to point the way toward question improvement. However, this is not an automatic process. It requires a proficient questionnaire designer. It may be that the pretesting methods we employed were effective and uncovered real problems but that the redesign phase was deficient because we selected ineffective revisions.

For reasons that are probably obvious, this is not our preferred conclusion. The people involved in developing our revised items all have considerable experience pretesting questionnaires in a variety of contexts and roles. It seems implausible to us to maintain that we were very capable when using pretest results to find problems, but incapable of identifying obvious solutions that were indicated by the same pretest results.

Possibility 3: The survey outcome measures were flawed In the absence of objective measures of data quality, it can be difficult to assess question functioning, data quality, or the effects of pretesting (Willis et al., 1999a). Although we selected outcome measures that are generally thought to serve as indirect measures of data quality, there is no assurance that they act as reliable proxies of validation measures. Nonresponse is generally recognized as a gross measure of data quality. Interviewer ratings are subjective. In addition, interviewer ratings are probably biased toward identifying items with obvious administration difficulties and away from identifying items with more subtle problems that affect analysis and interpretation. Behavior coding is most effective for detecting overt problems that are easily observable.

We believe that the latter point is especially significant. Pretesting methods such as cognitive interviewing are designed to investigate covert problems. For example, intensive probing might be used to determine whether respondents interpret a term as intended. Pretesting methods such as behavior coding can serve as effective validation measures for cognitive interviewing only when we expect that covert problems in comprehension, recall, or response selection will produce overt, codable behavior. We can imagine question modifications that markedly improve question understanding without improving behavior coding results.[2]

It seems reasonable to hypothesize that in some cases, the pretest methods studied here successfully confirmed problems and led to effective repairs, but these successes were undetected in the survey outcome measures we studied. Better measures of response accuracy would shed light on this possible explanation for our lack of reparatory success. For example, in the current questionnaires we could measure accuracy of reported telephone bill amounts either by accessing telephone company billing records or by asking respondents to send us copies of their most recent telephone bills. Similarly, we could evaluate the accuracy of reported access to public transportation by contacting local bus companies.

Possibility 4: Pretesting alone is ineffective Of course, it may be that we were not able to repair question defects because the pretesting methods we used do not always suggest "fixes" that actually alleviate the inherent problems. Less radically, our findings may suggest that it is insufficient simply to pretest and make modifications before fielding a survey questionnaire. Additional pretesting with question revisions may be needed to ensure effective repairs. Pretesting revisions in field settings may be particularly important. The common practice of conducting cognitive interviewing as a series of iterative "rounds" clearly adheres to this recommendation. Our already complicated pretest design would not allow the added complexity of iterative pretesting.

A related thought is that our pretesting protocol may not have been entirely effective. The selected combination of methods we used (expert review, followed by appraisal, followed by cognitive interviewing) was effective for confirmatory purposes, and at least somewhat effective for reparatory purposes. However, we do not know how the selected pretest combination compares with others, nor can we say anything about the contributions the individual pretesting methods made to the overall effectiveness of the combination. Designs that would allow comparisons with other combinations or with a specific pretest method would require larger staffs and more resources than those available to us.

25.5.4 Recommendations for Further Research

First, we'd like to see reparatory studies that use more direct and more sensitive measures of data quality, especially studies that undertake some type of

[2]An example from our item-level analyses illustrates. Our behavior coding results indicated few respondent problems with a control questionnaire item on automobile use. ("Is it used for business?") We observed considerably more behavior coded respondent problems for the revised item in the experimental questionnaire version. ["In the past 12 months, was (FILL CAR) used *at all* as part of a job or business? Do not include commuting to work."] Pretesting indicated that the original item had comprehension problems. We believe that the revision represents a reasonable move toward improvement. The trends for these two items led us to wonder about the types of problems detected by our behavior coding results.

validation. For example, one could develop and pretest questions about specific health insurance coverages. A reparatory study of the pretest methods selected could include procedures for gathering general health plan information to validate eventual survey responses. In this way it should be possible to disentangle two sets of issues:

- *Administration quality*— the degree to which the questionnaire flows well, avoids negative interviewer reactions, and reduces the frequencies of problematic behavior codes.
- *Data quality*— the degree to which the questionnaire elicits reliable, accurate, and unbiased information from respondents.

We hope that pretesting positively influences both sets of issues.

Second, we think it would be useful to develop reparatory designs that allow more precise testing for effects of particular revisions on survey outcomes and data quality. The problem-prone items we selected to study gave us a variety of potential problems to repair. Early on, we chose to develop revisions that would address as many of the documented problems as possible. We believe that this approach reflects common pretesting practice. As a result, we think our findings represent the types of effects to expect in practice, where the objective is to fix as many problems as possible.

This approach to revision affords less experimental control. We cannot draw specific conclusions about the reasons *why* particular revisions we made were more or less effective. A more effective reparatory design might use a mix of revision strategies. For example, one might include some experimental questions revised to fix multiple problems and others that reflect one incomplete but unconfounded revision.

Third, we recommend additional research that includes a mix of more and less problem-prone items in control and experimental questionnaire versions. This type of mix is essential for determining that pretesting demonstrates both sensitivity and specificity. In other words, we want to know that our methods effectively identify items with true problems (sensitivity). We also want to know that our methods effectively identify items with no problems (specificity). This type of item mix would provide two additional advantages. A mix of problem severities would allow us to assess whether different levels of problem severity are particularly easy or difficult to repair. Also, a mix of problem severities would allow us to study trade-offs between problem detection and repair. For example, it may be that severe problems that are easy to detect are also especially difficult to repair. Conversely, it may be easiest to fix items that are not obviously problematic. Thus, a simple wording change may fix an important problem uncovered only by intensive pretesting methods.

Fourth, we recommend more research using iterative pretest designs where pretesting for problem diagnosis is followed by preliminary repairs, further pretesting, and additional repairs before fielding. Because of the subjective nature of questionnaire design, the "fixes" for a set of problems constitute a new stimulus.

This is particularly true for items identified as having severe problems. We believe that additional rounds of pretesting are wise before fielding major revisions on a large scale. Studies of the types of changes made across iterative pretests might shed light on the apparently subjective questionnaire revision process.

We'll close with one last observation. We believe that it's difficult to find problems in survey questions. It's more difficult to fix them. Even more difficult is demonstrating that a repair is in fact an improvement.

References

Aaker, D., Bagozzi, R., and Carman, J., 1980, On using response latency to measure preference, *Journal of Marketing Research* 17: 237–244.

Abramowitz, R., Freedman, J., Henry, K., and van Brunschot, M., 1995, Children's capacity to agree to psychological research: knowledge of risks and benefits and voluntariness, *Ethics and Behavior* 5: 25–48.

Acquadro, C., Jambon, B., Ellis, D., and Marquis, P., 1996, Language and translation issues, in B. Spilker (ed.), *Quality Life and Pharmacoeconomics in Clinical Trials*, 2nd ed. Philadelphia: Lippincott-Raven.

Adamowicz, W., Louviere, J., and Williams, M., 1993, Combining revealed and stated preference methods for valuing environmental amenities, *Journal of Environmental Economics and Management* 26: 271–292.

Aldridge, M., and Wood, J., 1998, *Interviewing Children: A Guide for Childcare and Forensic Practitioners*. New York: Wiley.

Allalouf, A., Hambleton, R., and Sireci, S., 1999, Identifying the sources of differential item functioning in translated verbal items, *Journal of Educational Measurement* 36: 185–198.

Alwin, D., and Krosnick, J., 1991, The reliability of survey attitude measurement: the influence of question and respondent attributes, *Sociological Methods and Research* 20: 139–181.

Amato, P., and Ochiltree, G., 1987, Interviewing children about their families: a note on data quality, *Journal of Marriage and Family* 49: 669–675.

American Association for Public Opinion Research, 2000, *Standard Definitions: Final Dispositions of Case Codes and Outcome Rates for Surveys*. Ann Arbor, MI: AAPOR.

American Statistical Association, 1993, *Proceedings of the International Conference on Establishment Surveys*. Alexandria, VA: ASA.

American Statistical Association, 1997, *What Are Focus Groups?* ASA series "What Is a Survey." Alexandria, VA: ASA, Section on Survey Research Methods.

Methods for Testing and Evaluating Survey Questionnaires, Edited by Stanley Presser, Jennifer M. Rothgeb, Mick P. Couper, Judith T. Lessler, Elizabeth Martin, Jean Martin, and Eleanor Singer
ISBN 0-471-45841-4

American Statistical Association, 2000, *Proceedings of the 2nd International Conference on Establishment Surveys*. Alexandria, VA: ASA.

Anderson, A., Nichols, E., and Pressley, K., 2001, Usability testing and cognitive interviewing to support economic forms development for the 2002 economic census, *Proceedings of the Statistics Canada Symposium 2001: Achieving Data Quality in a Statistical Agency: A Methodological Perspective*.

Andrews, F., 1984, Construct validity and error components of survey measures: a structural modeling approach, *Public Opinion Quarterly* 48: 409–442.

Andrusiak, G., 1993, Re-engineering of industry statistics: maintaining relevance in trying times, *Proceedings of the International Conference on Establishment Surveys*. Alexandria, VA: American Statistical Association, pp. 724–728.

Annon, J., 1994, Recommended guidelines for interviewing children in case of alleged sexual abuse, *IPT Journal* 6. Retrieved May 28, 2002 at *http://www.ipt-forensics.com/journal/volume6/j6_3_2.htm*.

Aronson, E., Wilson, T., and Brewer, M., 1998, Experimentation in social psychology, in D. Gilbert, S. Fiske, and G. Lindzey (eds.), *Handbook of Social Psychology*. Boston: McGraw-Hill.

Arrow, K., Solow, R., Leamer, E., Portney, P., Rader, R., and Schuman, H., 1993, Report of the NOAA Panel on Contingent Valuation, *Federal Register* 58: 4601–4614.

Arrow, K., Cropper, M., Eads, G., Hahn, R., Lave, L., Noll, R., Portney, P., Russell, M., Schmalensee, R., Smith, V., and Stavins, R., 1996, Is there a role for benefit–cost analysis in environmental, health, and safety regulation? *Science* 272: 221–222.

Austin, J., and Delaney, P., 1998, Protocol analysis as a tool for behavior analysis, *Analysis of Verbal Behavior* 15: 41–56.

Azocar, F., Arean, P., Miranda, J., and Munoz, R., 2001, Differential item functioning in a Spanish translation of the Beck Depression Inventory, *Journal of Clinical Psychology* 57: 355–365.

Babyak, C., Gower, A., Mulvihill, J., and Zaroski, R., 2000, Testing of the questionnaires for Statistics Canada's Unified Enterprise Survey, *Proceedings of the 2nd International Conference on Establishment Surveys*. Alexandria, VA: American Statistical Association, pp. 317–326.

Bachman, J., and O'Malley, P., 1984, Black–white differences in response styles, *Public Opinion Quarterly* 48: 491–509.

Baddeley, A., 1986, *Working Memory*. Oxford: Clarendon.

Bagozzi, R., and Yi, Y., 1991, Multitrait–multimethod matrices in consumer research, *Journal of Consumer Research* 17: 426–439.

Baker, R., and Wojcik, M., 1992, Interviewer and respondent acceptance of CAPI, in *Proceedings of the Bureau of the Census Annual Research Conference*. Washington, DC: U.S. Bureau of the Census.

Barbier, E., 1994, Valuing environmental functions: tropical wetlands, *Land Economics* 70: 155–173.

Barnum, C., 2002, *Usability Testing and Research*. New York: Allyn & Bacon.

Bassili, J., 1993, Response latency versus certainty as indexes of the strength of voting intentions in a CATI survey, *Public Opinion Quarterly* 57: 54–61.

Bassili, J., 1995, Response latency and the accessibility of voting intentions: what contributes to the accessibility and how it affects vote choice, *Personality and Social Psychology Bulletin* 21: 686–695.

Bassili, J., 1996, The how and why of response latency measurement in telephone surveys, in N. Schwarz and S. Sudman (eds.), *Answering Questions*. San Francisco: Jossey-Bass.

Bassili, J., and Fletcher, J., 1991, Response-time measurement in survey research: a method for CATI and a new look at nonattitudes, *Public Opinion Quarterly* 55: 331–346.

Bassili, J., and Krosnick, J., 2000, Do strength-related attitude properties determine susceptibility to response effects? New evidence from response latency, attitude extremity, and aggregate indices, *Political Psychology* 21: 107–132.

Bassili, J., and Scott, B., 1996, Response latency as a signal to question problems in survey research, *Public Opinion Quarterly* 60: 390–399.

Bates, N., 1992, The Simplified Questionnaire Test (SQT): results from the debriefing interviews, unpublished report, U.S. Bureau of the Census, August 18.

Bates, N., and Good, C., 1996, An evaluation of the 1995 test census integrated coverage measurement (ICM) interview: results from behavior coding, *Proceedings of the ASA Section on Survey Research Methods*. Alexandria, VA: American Statistical Association.

Bates, N., and Morgan, M., 2002, Non-interview rates for selected major demographic household surveys: 1990–2001, unpublished report, U.S. Bureau of the Census, Demographic Surveys Division, U.S. Bureau of the Census, August 2.

Beatty, P., 1997, A review of "Answering questions: methodology for determining cognitive and communicative processes in survey research," N. Schwarz and S. Sudman (eds.), *Journal of Official Statistics* 13: 435–438.

Beatty, P., undated, Classifying questionnaire problems: five recent taxonomies and one older one, unpublished manuscript, Office of Research and Methodology, National Center for Health Statistics.

Beatty, P., and Schechter, S., 1998, *Questionnaire Evaluation and Testing in Support of the Behavioral Risk Factor Surveillance System (BRFSS), 1992–1998*, Cognitive Methods Working Paper Series, No. 26. Hyattsville, MD: National Center for Health Statistics.

Beatty, P., Schechter, S., and Whitaker, K., 1996, Evaluating subjective health questions: cognitive and methodological investigations, *Proceedings of the ASA Section on Survey Research Methods*. Alexandria, VA: American Statistical Association, pp. 956–961.

Beaule, A., Blackburn, Z., Falconer, L., Foster, A., Hansen, S., and Yang, Q., 1997, *CAI Survey Instrument and System Testing*. Ann Arbor, MI: University of Michigan.

Beebe, T., Harrison, P., McRae, J., Jr., Anderson, R., and Fulkerson, J., 1998, An evaluation of computer-assisted self-interviews in a school setting, *Public Opinion Quarterly* 62: 623–632.

Beekman, A., Deeg, D., van Tilburg, T., Smit, J., Hooijer, C., and Van Tilburg, W., 1995, Major and minor depression in later life: a study of prevalence and risk factors, *Journal of Affective Disorders* 36: 65–75.

Beizer, B., 1990, *Software Testing Techniques*. New York: Van Nostrand Reinhold.

Belli, R., Schwarz, N., Singer, E., and Talarico, J., 2000, Decomposition can harm the accuracy of behavioral frequency reports, *Applied Cognitive Psychology* 14: 295–308.

Belson, W., 1981, *The Design and Understanding of Survey Questions*. Aldershot, U.K.: Gower.

Benson, J., 1998, Developing a strong program of construct validation: a test anxiety example, *Educational Measurement: Issues and Practice* 17: 10–22.

Benson, J., and Hocevar, D., 1985, The impact of item phrasing on the validity of attitude scales for elementary school children, *Journal of Educational Measurement* 22: 231–240.

Berl, J., Lewis, G., and Morrison, R., 1976, Applying models of choice to the problem of college selection, in J. Carroll and J. Payne (eds.), *Cognition and Social Behavior*. Hillsdale, NJ: Lawrence Erlebaum.

Berry, S., and O'Rourke, D., 1988, Administrative designs for centralized telephone survey centers: implications of the transition to CATI, in R. Groves, P. Biemer, L. Lyberg, J. Massey, W. Nicholls, and J. Waksberg (eds.), *Telephone Survey Methodology*. New York: Wiley.

Bethlehem, J., and Hundepool, A., 2002, On the documentation and analysis of electronic questionnaires, presented at the International Conference on Questionnaire Development, Evaluation, and Testing, Charleston, SC.

Bickart, B., and Felcher, E., 1996, Expanding and enhancing the use of verbal protocols in survey research, in N. Schwarz and S. Sudman (eds.), *Answering Questions*. San Francisco: Jossey-Bass.

Biderman, A., 1980a, *Report of the Workshop on Applying Cognitive Psychology to Recall Problems of the National Crime Survey*, Washington, DC, September 17–18.

Biderman, A., 1980b, Crime-circumscribed versus broader-net screening approaches. item 361 in Crime Survey Research Consortium Teleconference, August 7.

Biderman, A., 1981a, Cue specificity, time reference, mnemonics, and semantics, item 581 in Crime Survey Research Consortium Teleconference, June 4.

Biderman, A., 1981b, Further reflections on the recollection problem, item 633 in Crime Survey Research Consortium Teleconference, September 21.

Biderman, A., Cantor, D., Lynch, J., and Martin, E., 1986, *Final Report of the National Crime Survey Redesign Program*. Washington, DC: Bureau of Social Science Research.

Biemer, P., 1988, Measuring data quality, in R. Groves, P. Biemer, L. Lyberg, J. Massey, W. Nicholls, and J. Waksberg (eds.), *Telephone Survey Methodology*. New York: Wiley.

Biemer, P., 1997, *Dual Frame NHIS/RDD Methodology and Field Test: Analysis Report*, final analysis report for CDC.

Biemer, P., 2001, Nonresponse bias and measurement bias in a comparison of face to face and telephone interviewing, *Journal of Official Statistics* 17: 295–320.

Biemer, P., in press, Analysis of classification error for the revised Current Population Survey employment questions, to appear in *Survey Methodology*.

Biemer, P., and Bushery, J., 2001, Application of Markov latent class analysis to the CPS, *Survey Methodology* 26: 136–152.

Biemer, P., and Fecso, R., 1995, Evaluating and controlling measurement error in business surveys, in B. Cox, D. Binder, B. Chinnappa, A. Christianson, M. Colledge, and P. Kott (eds.), *Business Survey Methods*. New York: Wiley.

Biemer, P., and Forsman, G., 1992, On the quality of reinterview data with applications to the current population survey, *Journal of the American Statistical Association* 87: 915–923.

Biemer, P., and Stokes, S., 1991, Approaches to the modeling of measurement errors in surveys, in P. Biemer, R. Groves, L. Lyberg, N. Mathiowetz, and B. Sudman (eds.), *Measurement Errors in Surveys*. New York: Wiley.

Biemer, P., and Tucker, C., 2001, Estimation and correction for underreporting errors in expenditure data: a Markov latent class modeling approach, presented at the International Statistical Meetings, Seoul, Korea.

Biemer, P., and Wiesen, C., 2002, Latent class analysis of embedded repeated measurements: an application to the National Household Survey on Drug Abuse, *Journal of the Royal Statistical Society*, Series A, 165: 1.

Biemer, P., Groves, R., Lyberg, L., Mathiowetz, N., and Sudman, S., 1991, *Measurement Errors in Surveys*. New York: Wiley.

Biemer, P., Woltman, H., Raglin, D., and Hill, J., 2001, Enumeration accuracy in a population census: an evaluation using latent class analysis, *Journal of Official Statistics* 17: 129–149.

Birch, S., Schwede, L., and Gallagher, C., 1998, *Juvenile Residential Facility Census Questionnaire Redesign Project: Results from Phase 2 Cognitive Interviewing Testing*, prepared by the U.S. Bureau of the Census for the Department of Justice.

Birch, S., Schwede, L., and Gallagher, C., 1999, *Juvenile Residential Facility Census Questionnaire Redesign Project: Results from Phase 3 Mail-out Test Analysis*, prepared by the U.S. Bureau of the Census for the Department of Justice.

Birch, S., Schwede, L., and Gallagher, C., 2001, *Juvenile Probation Survey Development Project: Results of Phase 1 Exploratory Interviews*, prepared by the U.S. Bureau of the Census for the Department of Justice.

Bischoping, K., and Dykema, J., 1999, Toward a social psychological program for improving focus group methods of developing questionnaires, *Journal of Official Statistics* 15: 495–516.

Blair, E., and Burton, S., 1987, Cognitive processes used by survey respondents to answer behavioral frequency questions, *Journal of Consumer Research* 14: 280–288.

Blair, E., Sudman, S., Bradburn, N., and Stocking, C., 1977, How to ask questions about drinking and sex: response effects in measuring consumer behavior, *Journal of Marketing Research* 14: 316–321.

Blair, J., 2000, Assessing protocols for child interviews, in A. Stone, J. Turkkan, C. Bachrach, J. Jobe, H. Kurtzman, and V. Cain (eds.), *The Science of Self-Report: Implications for Research and Practice*. Mahwah, NJ: Lawrence Erlbaum.

Blair, J., and Presser, S., 1993, Survey procedures for conducting cognitive interviews to pretest questionnaires: a review of theory and practice, *Proceedings of the ASA Section on Survey Research Methods*. Alexandria, VA: American Statistical Association.

Blair, J., Menon, G., and Bickart, B., 1991, Measurement effects in self vs. proxy responses: an information-processing perspective, in P. Biemer, R. Groves, L. Lyberg, N. Mathiowetz, and S. Sudman (eds.), *Measurement Errors in Surveys*. New York: Wiley.

Blais, A., and Gidengil, E., 1993, Things are not always what they seem: French–English differences and the problem of measurement equivalence, *Canadian Journal of Political Science* 26: 541–555.

Bless, H., Bohner, G., Hild, T., and Schwarz, N., 1992, Asking difficult questions: task complexity increases the impact of response alternatives, *European Journal of Social Psychology* 22: 309–312.

Bollen, K., Entwisle, B., and Alderson, A., 1993, Macro-comparative research methods, *Annual Review of Sociology* 19: 321–351.

Bolton, R., 1993, Pretesting questionnaires: content analyses of respondents' concurrent verbal protocols, *Marketing Science* 12: 280–303.

Bolton, R., and Bronkhorst, T., 1996, Questionnaire pretesting: computer-assisted coding of concurrent protocols, in N. Schwarz and S. Sudman (eds.), *Answering Questions: Methodology for Determining Cognitive and Communicative Processes in Survey Research.* San Francisco: Jossey-Bass.

Bond, T., and Fox, C., 2001, *Applying the Rasch Model: Fundamental Measurement in the Human Sciences.* Mahwah, NJ: Lawrence Erlbaum.

Borgers, N., 2003, *Questioning Children's Responses.* Amsterdam: TT-Publikaties.

Borgers, N., and Hox, J., 2001, Item non-response in questionnaires with children, *Journal of Official Statistics* 17: 321–335.

Borgers, N., and Hox, J., 2002, Reliability of responses in questionnaire research with children, unpublished manuscript.

Borgers, N., de Leeuw, E., and Hox, J., 1999, Surveying children: cognitive development and response quality in questionnaire research, in A. Christianson (ed.), *Official Statistics in a Changing World.* Stockholm: SCB.

Borgers, N., de Leeuw, E., and Hox, J., 2000, Children as respondents in survey research: cognitive development and response quality, *Bulletin de Methodologie Sociologique* 66: 60–75.

Borgers, N., Hox, J., and Sikkel, D., 2004, Response effects in surveys on children and adolescents: the effect of number of response options, negative wording, and neutral midpoint, *Quality & Quantity* 38: 17–33.

Bosley, J., Conrad, F., and Uglow, D., 1998, Pen CASIC: design and usability, in M. Couper, R. Baker, J. Bethlehem, C. Clark, J. Martin, W. Nicholls II, and J. O'Reilly (eds.), *Computer-Assisted Survey Information Collection.* New York: Wiley.

Boxall, P., Adamowicz, W., Swait, J., Williams, M., and Louviere, J., 1996, A comparison of stated preference methods for environmental valuation, *Ecological Economics* 18: 243–253.

Boyle, K., Desvousges, W., Johnson, F., Dunford, R., and Hudson, S., 1994, An investigation of part–whole biases in contingent valuation studies, *Journal of Environmental Economics and Management* 27: 64–83.

Bradburn, N., 1983, Response effects, in P. Rossi, J. Wright, and A. Anderson (eds.), *Handbook of Survey Research.* New York: Academic Press.

Bradburn, N., and Sudman, S., 1979, *Improving Interview Methods and Questionnaire Design.* San Francisco: Jossey-Bass.

Brainerd, C., and Ornstein, P., 1991, Children's memory for witnessed events: the developmental backdrop, in J. Doris (ed.), *The Suggestibility of Children's Recollections.* Washington, DC: American Psychological Association.

Braun, M., 2003, Communication and social cognition, in J. Harkness, F. van de Vijver, and P. Mohler (eds.), *Cross-Cultural Survey Methods.* New York: Wiley.

Braun, M., and Scott, J., 1998, Multidimensional scaling and equivalence: is having a job the same as working?, in J. Harkness (ed.), *Cross-Cultural Survey Equivalence,* ZUMA-Nachrichten Spezial 3. Mannheim, Germany: ZUMA.

Brenner, M., 1982, Response effects of "role-restricted" characteristics of the interviewer, in W. Dijkstra and J. van der Zouwen (eds.), *Response Behaviour in the Survey-Interview*. London: Academic Press.

Briggs, C., 1986, *Learning How to Ask: A Sociolinguistic Appraisal of the Role of the Interview in Social Science Research*. Cambridge: Cambridge University Press.

Brinson, M., 1993, *A Hydrogeomorphic Classification of Wetlands*. Greenville, NC: U.S. Army Corps of Engineers.

Brinson, M., Hauer, F., Lee, L., Nutter, W., Rheinhardt, R., Smith, R., and Whigham, D., 1995, *Guidebook for Application of Hydrogeomorphic Assessment to Riverine Wetlands*. Vicksburg, MS: Waterways Experiment Station, U.S. Army Corps of Engineers.

Brislin, R., 1970, Back-translation for cross-cultural research, *Journal of Cross-Cultural Research* 1: 185–216.

Brislin, R., 1976, Introduction, in R. Brislin (ed.), *Translation: Applications and Research*. New York: Gardner.

Brislin, R., 1980, Translation and content analysis of oral and written material, in H. Triandis and J. Berry (eds.), *Handbook of Cross-Cultural Psychology*, Vol. 2. Boston: Allyn & Bacon.

Brislin, R., 1986, The wording and translation of research instruments, in W. Lonner and J. Berry (eds.), *Field Methods in Cross-Cultural Research*. Thousand Oaks, CA: Sage.

Brown, A., Hale, A., and Michaud, S., 1998, Use of computer assisted interviewing in longitudinal surveys, in M. Couper, R. Baker, J. Bethlehem, C. Clark, J. Martin, W. Nicholls, and J. O'Reilly (eds.), *Computer Assisted Survey Information Computer Assisted Survey Information Collection*. New York: Wiley.

Budgell, G., Raju, N., and Quartetti, D., 1995, Analysis of differential item functioning in translated assessment instruments, *Applied Psychological Measurement* 19: 309–321.

Bullinger, M., 1995, German translation and psychometric testing of the SF-36 health survey: preliminary results from the IQOLA project, *Social Science Medicine* 41: 359–366.

Bureau, M., 1991, Experience with the use of cognitive methods in designing business survey questionnaires, *Proceedings of the ASA Section on Survey Research Methods*. Alexandria, VA: American Statistical Association, pp. 713–717.

Burns, E., Carlson, L., French, D., Goldberg, M., Latta, R., and Leach, N., 1993, Surveying an uncharted field, *Proceedings of the International Conference on Establishment Surveys*. Alexandria, VA: American Statistical Association, pp. 37–44.

Burnside, R., 2000, Towards best practice for the design of electronic data capture instruments, presented at the Statistics Methodology Advisory Committee meeting, Australian Bureau of Statistics, Canberra, Australia, November.

Burnside, R., and Farrell, E., 2001, Recent developments at the ABS in electronic data reporting by businesses, *Proceedings of the Statistics Canada Symposium 2001: Achieving Data Quality in a Statistical Agency: A Methodological Perspective*.

Burr, M., Levin, K., and Becher, A., 2001, Examining Web vs. paper mode effects in a federal government customer satisfaction study, presented at the Annual Conference of the American Association for Public Opinion Research, Montreal, Quebec, Canada.

Burt, C., and Schappert, S., 2002, Evaluation of respondent and interviewer debriefing techniques on questionnaire development methods for health provider-based surveys, presented at the International Conference on Questionnaire Development, Evaluation, and Testing Methods, Charleston, SC.

Burton, S., and Blair, E., 1991, Task conditions, response formulation processes, and response accuracy for behavioral frequency questions in surveys, *Public Opinion Quarterly* 55: 50–79.

Cahalan, M., Mitchell, S., Gray, L., Chen, S., and Tsapogas, J., 1994, Recorded interview behavior coding study: national survey of recent college graduates, *Proceedings of the ASA Section on Survey Research Methods*. Alexandria, VA: American Statistical Association.

Campanelli, P., 1997, Testing survey questions: new directions in cognitive interviewing, *Bulletin de Methodologie Sociologique* 55: 5–17.

Campanelli, P., Martin, E., and Creighton, K., 1989a, Respondents' understanding of labor force concepts: insights from debriefing studies, *Proceedings of the 5th Annual Research Conference*. Washington, DC: U.S. Bureau of the Census.

Campanelli, P., Rothgeb, J., and Martin, E., 1989b, The role of respondent comprehension and interviewer knowledge in CPS labor force classification, *Proceedings of the ASA Section on Survey Research Methods*. Alexandria, VA: American Statistical Association.

Campanelli, P., Martin, E., and Rothgeb, J., 1991, The use of respondent and interviewer debriefing studies as a way to study response error in survey data, *The Statistician* 40: 253–264.

Campbell, D., and Fiske, D., 1959, Convergent and discriminant validation by the multitrait multimethod matrices, *Psychological Bulletin* 56: 81–105.

Campbell, D., and Stanley, J., 1963, *Experimental and Quasi-experimental Designs for Research*. Boston: Houghton Mifflin.

Cannell, C., and Kahn, R., 1953. The collection of data by interviewing, in L. Festinger and D. Katz (eds.), *Research Methods in the Behavioral Sciences*. New York: Dryden.

Cannell, C., and Kahn, R., 1968, Interviewing, in G. Lindzey and E. Aronson (eds.), *The Handbook of Social Psychology*. Reading, MA: Addison-Wesley.

Cannell, C., and Robison, S., 1971, Analysis of individual questions, in J. Lansing, S. Withey, and A. Wolfe (eds.), *Working Papers on Survey Research in Poverty Areas*. Ann Arbor, MI: Institute for Social Research.

Cannell, C., Fowler, F., and Marquis, K., 1968, The influence of interviewer and respondent psychological and behavioral variables on the reporting in household interviews, *Vital and Health Statistics, No. 26*, Data Evaluation and Methods Research. Washington, DC: U.S. Department of Health and Human Services.

Cannell, C., Marquis, K., and Laurent, A., 1977, A summary of studies of interviewing methodology, *Vital and Health Statistics*, No. 2, Data Evaluation and Methods Research. Rockville, MD: U.S. Department of Health, Education, and Welfare.

Cannell, C., Miller, P., and Oksenberg, L., 1981, Research on interviewing techniques, in S. Leinhardt (ed.), *Sociological Methodology*. San Francisco, CA: Jossey-Bass.

Cannell, C., Camburn, D., Dykema, J., and Seltzer, S., 1992, *Applied Research on the Design and Conduct of Surveys of Adolescent Health Behaviors and Characteristics*. Ann Arbor, MI: Institute for Social Research, University of Michigan.

Cantor, D., and Phipps, P., 1999, Adapting cognitive techniques to establishment surveys, *Proceedings of CASM II Seminar*. Hyattsville, MD: National Center for Health Statistics, pp. 74–78.

Cantril, H., 1944, *Gauging Public Opinion*. Princeton, NJ: Princeton University Press.

Cantril, H., and Fried, E., 1944, The meaning of questions, in H. Cantril (ed.), *Gauging Public Opinion*. Princeton, NJ: Princeton University Press.

CARES, 1999, Preschool children: notes from a clinical response to child abuses, presented at CARES Northwest Conference 1999. Retrieved May 28, 2002 at *http://www.geocities.com/wellesley/8819.html*.

Carlson, L., Preston, J., and French, D., 1993, Using focus groups to identify user needs and data availability, *Proceedings of the International Conference on Establishment Surveys*. Alexandria, VA: American Statistical Association, pp. 300–308.

Carroll, J., 1997, Human–computer interaction: psychology as a science of design, *Annual Review of Psychology* 48: 61–83.

Carson, R., 2000, Contingent valuation: a user's guide, *Environmental Science and Technology* 34: 1413–1418.

Carson, R., and Mitchell, R., 1993, The issue of scope in contingent valuation, *American Journal of Agricultural Economics* 75: 1263–1267.

Carson, R., Wright, J., Carson, N., Alberini, A., and Flores, N., 1994a, *A Bibliography of Contingent Valuation Studies and Papers*. La Jolla, CA: Natural Resource Damage Assessment, Inc.

Carson, R., Hanemann, W., Kopp, R., Krosnick, J., Mitchell, R., Presser, S., Ruud, P., and Smith, V., 1994b, *Prospective Interim Lost Use Value due to DDT and PCB Contamination in the Southern California Bight*. La Jolla, CA: Natural Resource Damage Assessment, Inc.

Carson, R., Hanemann, W., Kopp, R., Krosnick, J., Mitchell, R., Presser, S., Ruud, P., and Smith, V., 1998, Referendum design and contingent valuation: the NOAA panel's no-vote recommendation, *Review of Economics and Statistics* 80: 484–487.

Caspar, R., and Couper, M., 1997, Using keystroke files to assess respondent difficulties with an ACASI instrument, *Proceedings of the ASA Section on Survey Research Methods*. Alexandria, VA: American Statistical Association.

Ceci, S., and Bruck, M., 1993, The suggestibility of the child witness: a historical review and synthesis, *Psychological Bulletin* 113: 403–439.

Center for IT Accommodation, 1998, *http://www.section508.gov*. Washington, DC: U.S. General Services Administration.

Centers for Disease Control and Prevention, 1995, Health related quality of life measures, United States, *Morbidity and Mortality Weekly Report* 44.

Centers for Disease Control and Prevention, 1998, Self-reported frequent mental distress among adults, United States, 1993–1996, *Morbidity and Mortality Weekly Report* 47: 325–331.

Chen, C., Lee, S., and Stevenson, H., 1995, Response style and cross-cultural comparisons of rating scales among East Asian and North American students, *Psychological Science* 6: 170–175.

Chen, W., and Thissen, D. 1997, Local dependence indexes for item pairs using item response theory, *Journal of Educational and Behavioral Statistics* 22: 265–289.

Cheng, L., Choi, C., Easley, D., and Jackson, J., 1997, Focus groups, Indiana University continuous learning project. Retrieved May 28, 2002 at *http://universe.indiana.edu/clp/rf/focusg.htm*.

Cheung, F., Leung, K., Fan, R., Song, W., Zhang, J., and Zhang, J., 1996, Development of the Chinese Personality Assessment Inventory, *Journal of Cross-Cultural Psychology* 27: 181–199.

Chilton, S., and Hutchinson, W., 1999, Exploring divergence between respondent and researcher definitions of the goods in contingent valuation studies, *Journal of Agricultural Economics* 50: 1–16.

Christianson, A., and Tortora, R., 1995, Issues in surveying businesses: an international survey, in B. Cox, D. Binder, B. Chinnappa, A. Christianson, M. Colledge, and P. Kott (eds.), *Business Survey Methods*. New York: Wiley.

Chromy, J., Finkner, A., and Horvitz, D., 2004, Survey design issues in planning and early implementation of NAEP, in L. Jones and I. Olkin (eds.), *The Nation's Report Card: Evolution and Perspectives*. Bloomington, IN: Phi Kappa Delta International.

Chun, K., Campbell, J., and Yoo, J., 1974, Extreme response style in cross-cultural research: a reminder, *Journal of Cross-Cultural Psychology* 5: 465–480.

Church, A., 1987, Personality research in a non-Western setting: The Philippines, *Psychology Bulletin* 102: 272–292.

Citro, C., and Michael, R., 1995, *Measuring Poverty: A New Approach*. Washington, DC: National Academy Press.

Clark, H., and Schober, M., 1992, Asking questions and influencing answers, in J. Tanur (ed.), *Questions about Questions: Inquiries into the Cognitive Bases of Surveys*. New York: Russell Sage.

Clayton, R., and Werking, G., 1998, Business surveys of the future: the World Wide Web (WWW) as a data collection methodology, in M. Couper, R. Baker, J. Bethlehem, C. Clark, J. Martin, W. Nicholls, and J. O'Reilly (eds.), *Computer-Assisted Survey Information Collection*. New York: Wiley.

Clayton, R., Rosen, R., McCarthy, W., and Kennedy, J., 2000a, Progress and projections in computer assisted data collection at the Bureau of Labor Statistics, *Proceedings of the 2nd International Conference on Establishment Surveys*. Alexandria, VA: American Statistical Association, pp. 393–402.

Clayton, R., Searson, M., and Manning, C., 2000b, Electronic data collection in selected BLS establishment programs, *Proceedings of the 2nd International Conference on Establishment Surveys*. Alexandria, VA: American Statistical Association, pp. 493–448.

Cleland, J., and Scott, C., 1987, *The World Fertility Survey: An Assessment*. Oxford: Oxford University Press.

Clogg, C., 1977, *Unrestricted and Restricted Maximum Likelihood Latent Structure Analysis: A Manual for Users*, Working Paper 1977-09. University Park, PA: Populations Issues Research Office, Pennsylvania State University.

Clogg, C., 1982, Using association models in sociological research: some examples, *American Journal of Sociology* 88: 114–134.

Clogg, C., 1984, Some statistical models for analyzing why surveys disagree, in C. Turner and E. Martin (eds.), *Surveying Subjective Phenomena*. New York: Russell Sage.

Clogg, C., and Eliason, S., 1985, Some common problems in log-linear analysis, *Sociological Methods and Research* 16: 8–14.

Cloutier, E., 1976, Les Conception Américaine, Canadienne–Anglaise, et Canadienne–Française l'idée d'égalité, *Canadian Journal of Political Science* 9: 581–604.

Cohany, S., Polivka, A., and Rothgeb, J., 1994, Revisions in the Current Population Survey, *Employment and Earnings*, 13–37.

Cohen, B., Zukerberg, A., and Pugh, K., 1999, Improving respondent selection procedures in establishment surveys: implications from the Schools and Staffing Survey (SASS), presented at the Annual Conference of the American Association for Public Opinion Research, St. Petersburg Beach, FL.

Cohen, J., 1960, A coefficient of agreement for nominal scales, *Educational and Psychological Measurements* 20: 37–46.

Colasanto, D., Singer, E., and Rogers, T., 1992, Context effects on responses to questions about AIDS, *Public Opinion Quarterly* 56: 515–518.

Cole, C., and Loftus, E., 1987, The memory of children, in S. Ceci, M. Toglia, and D. Ross, (eds.), *Children's Eyewitness Memory*. New York: Springer.

Collins, M., Sykes, W., Wilson, P., and Blackshaw, N., 1988, Nonresponse: the UK experience, in R. Groves, P. Biemer, L. Lyberg, J. Massey, W. Nicholls, and J. Waksberg (eds.), *Telephone Survey Methodology*. New York: Wiley.

Colosi, R., 2001, *An Analysis of Time Stamp Data: Person-Based Interviewing vs. Topic-Based Interviewing*. Washington, DC: U.S. Bureau of the Census, December 10.

Comer, P., 1996, *Wetland Trends in Michigan since 1800: A Preliminary Assessment*. Lansing, MI: Michigan Natural Features Inventory.

Comijs, H., Dijkstra, W., Bouter, L., and Smit, J., 2000, The quality of data collection by an interview on the prevalence of elder mistreatment, *Journal of Elder Abuse and Neglect* 12: 57–72.

Conger, J., and Galambos, N., 1996, *Adolescence and Youth: Psychological Development in a Changing World*. New York: Longman.

Conrad, F., and Blair, J., 1996, From impressions to data: increasing the objectivity of cognitive interviews, *Proceedings of the ASA Section on Survey Research Methods*. Alexandria, VA: American Statistical Association.

Conrad, F., and Blair, J., 2001, Interpreting verbal reports in cognitive interviews: probes matter, *Proceeding of the ASA Section on Survey Research Methods*. Alexandria, VA: American Statistical Association.

Conrad, F., and Schober, M., 2000, Clarifying question meaning in a household telephone survey, *Public Opinion Quarterly* 64: 1–28.

Conrad, F., Brown, N., and Cashman, E., 1998, Strategies for estimating behavioural frequency in survey interviews, *Memory* 6: 339–366.

Conrad, F., Blair, J., and Tracy, E., 2000, Verbal reports are data! A theoretical approach to cognitive interviews, Office of Management and Budget, *Proceedings of the 1999 Federal Committee on Statistical Methodology Research Conference*, pp. 317–326.

Conrad, F., Couper, M., Tourangeau, R., and Baker, R., 2003, Use and non-use of clarification features in Web surveys, presented at the American Association for Public Opinion Research Annual Conference, Nashville, TN.

Converse, J., and Presser, S., 1986, *Survey Questions: Handcrafting the Standardized Questionnaire*. Thousand Oaks, CA: Sage.

Cook, T., and Campbell, D., 1979, *Quasi-experimentation: Design and Analysis Issues for Field Settings*. Chicago: Rand McNally.

Cooke, D., and Michie, C., 1997, An item response theory analysis of the Hare Psychopathy Checklist revised, *Psychological Assessment* 9: 3–14.

Corby, C., 1984, *Content Evaluation of the 1977 Economic Censuses (DE-2)*, Statistical Research Report 84/29. Washington DC: U.S. Bureau of the Census.

Corby, C., 1987, Content evaluation of the 1982 economic censuses: petroleum distributors, in *1982 Economic Censuses and Census of Governments: Evaluation Studies*. Washington DC: U.S. Bureau of the Census, pp. 27–50.

Cork, D. L., Cohen, M. L., Groves, R. M., and Kalsbeek, W. (eds.), 2003, *Survey Automation: Report and Workshop Proceedings*. Washington, DC: National Academies Press.

Corten, I., Saris, W., Coenders, G., Van der Veld, W., Aalberts, C., and Kornelis, C., 2002, Fit of different models for multitrait–multimethod experiments, *Structural Equation Models* 9: 213–233.

Cosenza, C., 2001, Standardized cognitive testing: will quantitative results provide qualitative answers? *Proceedings of the ASA Section on Survey Research Methods*. Alexandria, VA: American Statistical Association.

Cosenza, C., 2002, Not your grandparent's cognitive testing: exploring innovative methods in cognitive evaluations of questions, presented at the International Conference on Questionnaire Development, Evaluation, and Testing, Charleston, SC.

Costanza, R., Farber, S., and Maxwell, J., 1989, Valuation and management of wetland ecosystems, *Ecological Economics* 1: 335–361.

Couper, M., 1999, The application of cognitive science to computer assisted interviewing, in M. Sirken, D. Herrmann, S. Schechter, N. Schwarz, J. Tanur, and R. Tourangeau (eds.), *Cognition and Survey Research*. New York: Wiley.

Couper, M., 2000, Usability evaluation of computer-assisted survey instruments, *Social Science Computer Review* 18: 384–396.

Couper, M. P., and Burt, G., 1993, The impact of computer-assisted personal interviewing (CAPI) on interviewer performance: the CPS experience, *Proceedings of the Section on Survey Research Methods*. Alexandria, VA: American Statistical Association.

Couper, M., and Burt, G., 1994, Interviewer attitudes toward computer-assisted personal interviewing, *Social Science Computer Review* 12: 35–54.

Couper, M., and Groves, R., 1992, The role of the interviewer in survey participation, *Survey Methodology* 18: 263–278.

Couper, M., and Hansen, S., 2002, Computer-assisted interviewing, in J. Gubrium and J. Holstein (eds.), *Handbook of Interview Research*. Thousand Oaks, CA: Sage.

Couper, M., and Nicholls, W., 1998, The history and development of computer assisted survey information collection methods, in M. Couper, R. Baker, J. Bethlehem, C. Clark, J. Martin, W. Nicholls, and J. O'Reilly (eds.), *Computer-Assisted Survey Information Collection*. New York: Wiley.

Couper, M., and Schlegel, J., 1998, Evaluating the NHIS CAPI instrument using trace files, presented at the Annual Meeting of the American Association for Public Opinion Research.

Couper, M., Hansen S., and Sadosky S., 1997, Evaluating interviewer performance in a CAPI survey, in L. Lyberg, P. Biemer, M. Collins, E. de Leeuw, C. Dippo, N. Schwarz, and D. Trewin (eds.), *Survey Measurement and Process Quality*. New York: Wiley.

Couper, M., Baker, R., Bethlehem, J., Clark, C., Martin, J., Nicholls, W., II, and O'Reilly, J., 1998, *Computer Assisted Survey Information Collection*. New York: Wiley.

Couper, M., Beatty, P., Hansen, S., Lamias, M., and Marvin, T., 2000, CAPI design recommendations, report submitted to the Bureau of Labor Statistics, June.

Cox, B., and Chinnappa, B., 1995, Unique features of business surveys, in B. Cox, D. Binder, B. Chinnappa, A. Christianson, M. Colledge, and P. Kott (eds.), *Business Survey Methods*. New York: Wiley.

Cox, B., Elliehausen, G., and Wolken, J., 1989, Surveying small businesses about their finances, *Proceedings of the ASA Section on Survey Research Methods*. Alexandria, VA: American Statistical Association, pp. 553–557.

Cox, B., Binder, D., Chinnappa, B., Christianson, A., Colledge, M., and Kott, P., 1995, *Business Survey Methods*. New York: Wiley.

Cronbach, L., 1951, Coefficient alpha and the internal structure of tests, *Psychometrika* 6: 297–334.

Crutcher, R., 1994, Telling what we know: the use of verbal report methodologies in psychological research, *Psychological Science* 5: 241–244.

Crysdale, J., 2000, Harmonizing survey content: integrating Canada's Annual Survey of Manufactures into the Unified Enterprise Survey, presented at the 2nd International Conference on Establishment Surveys, Buffalo, NY.

Cynamon, M., and Kulka, R., 2001, Collecting data from children and adolescents, in M. Cynamon and R. Kulka (eds.), *Proceedings of the 7th Conference on Health Survey Research Methods*. Hyattsville, MD: U.S. Department of Health and Human Services.

Czaja, R., and Blair, J., 1996, *Designing Surveys: A Guide to Decisions and Procedures*. Thousand Oaks, CA: Pine Forge Press.

Czajka, J., 1983, Subannual income estimation, in M. David (ed.), *Technical, Conceptual, and Administrative Lessons of the Income Survey Development Program*. New York: Social Science Research Council.

Dahl, T., 2000, *Status and Trends of Wetlands in the Conterminous United States, 1986 to 1997*. Washington, DC: Fish and Wildlife Service, U.S. Department of the Interior.

Davidsson, G., 2002, Cognitive testing of mail surveys at Statistics Sweden, presented at the International Conference on Questionnaire Development, Evaluation, and Testing Methods, Charleston, SC.

Davis, W., and DeMaio, T., 1993, Comparing the think-aloud interviewing technique with standard interviewing in the redesign of a dietary recall questionnaire, *Proceedings of the ASA Section on Survey Research Methods*. Alexandria, VA: American Statistical Association.

Davis, W., DeMaio, T., and Zukerberg, A., 1995, *Can Cognitive Information Be Collected Through the Mail? Comparing Cognitive Data Collected in Written versus Verbal Format*, Working Paper in Survey Methodology SM95/02. Washington, DC: U.S. Bureau of the Census.

de la Puente, M., 2002, A blueprint for obtaining high quality data from non–English speaking households: translation guidelines, pretesting standards, and related research, presented to the U.S. Bureau of the Census Advisory Committee.

de Leeuw, E., and Otter, M., 1995, The reliability of children's responses to questionnaire items: question effects in children questionnaire data, in J. Hox, B. van der Meulen, J. Janssens, J. ter Laak, and L. Tavecchio (eds.), *Advances in Family Research*. Amsterdam: Thesis Publishers, pp. 251–258.

Delbanco, S., Lundy, J., Hoff, T., Parker, M., and Smith, M., 1997, Public knowledge and perceptions about unplanned pregnancies in three countries, *Family Planning Perspectives* 29: 70–75.

Delfos, M., 2000, *Luister je wel naar mij? Gespreksvoering met kinderen tussen vier en twaalf jaar* [Are You Really Listening to Me? Interviewing Children between 4 and 12]. Amsterdam: SWP, WESP-Publicatiereeks.

DeMaio, T., 1983a, *Approaches to Developing Questionnaires: Statistical Policy Working Paper 10*, Subcommittee on Questionnaire Design, Federal Committee on Statistical Methodology. Washington, DC: Office of Management and Budget.

DeMaio, T., 1983b, *Results of the 1980 Applied Behavior Analysis Survey or What People Do with Their Census Forms*, Preliminary Evaluation Results Memorandum 61. Washington, DC: U.S. Bureau of the Census, October 26.

DeMaio, T., 1984, Social desirability and survey measurement: a review, in C. Turner and E. Martin (eds.), *Surveying Subjective Phenomena*. New York: Russell Sage.

DeMaio, T., and Jenkins, C., 1991, Questionnaire research in the census of construction industries, *Proceedings of the ASA Section on Survey Research Methods*. Alexandria, VA: American Statistical Association, pp. 496–501.

DeMaio, T., and Rothgeb, J., 1996, Cognitive interviewing techniques: in the lab and in the field, in N. Schwarz and S. Sudman (eds.), *Answering Questions: Methodology for Determining Cognitive and Communicative Processes in Survey Research*. San Francisco: Jossey-Bass.

DeMaio, T., Mathiowetz, N., Rothgeb, J., Beach, M., and Durant, S., 1993, *Protocol for Pretesting Demographic Surveys at the Census Bureau*, Washington, DC: U.S. Bureau of the Census.

Dijkstra, W., 1999, A new method for studying verbal interactions in survey interviews, *Journal of Official Statistics* 15: 67–85.

Dijkstra, W., 2002, *Sequence Viewer 3.0*. Amsterdam: Department of Social Research Methods, Vrije Universiteit.

Dijkstra, W., and Van der Zouwen, J., 1982, *Response Behaviour in the Survey-Interview*. London: Academic Press.

Dillman, D., 1978, *Mail and Telephone Surveys: The Total Design Method*. New York: Wiley.

Dillman, D., 2000a, *Mail and Internet Surveys: The Tailored Design Method*. New York: Wiley.

Dillman, D., 2000b, Procedures for conducting government-sponsored establishment surveys: comparison of the total design method (TDM), a traditional cost-compensation model, and tailored design, *Proceedings of the 2nd International Conference on Establishment Surveys*. Alexandria, VA: American Statistical Association, pp. 469–471.

Dillman, D., 2000c, Progress in using the World Wide Web to conduct establishment surveys: comments on three important papers, *Proceedings of the 2nd International Conference on Establishment Surveys*. Alexandria, VA: American Statistical Association, pp. 469–471.

Dillman, D., Sinclair, M., and Clark, J., 1993, Effects of questionnaire length, respondent-friendly design, and a difficult question on response rates for occupant-addressed census mail surveys, *Public Opinion Quarterly* 57: 289–304.

Dillman, D., Singer, E., Clark, J., and Treat, J., 1996a, Effects of benefits appeals, mandatory appeals, and variations in statements of confidentiality on completion rates for census questionnaires, *Public Opinion Quarterly* 60: 3.

Dillman, D., Jenkins, C., Martin, B., and DeMaio, T., 1996b, *Cognitive and Motivational Properties of Three Proposed Decennial Census Forms*, Technical Report 96-29. Pullman, WA: Social and Economic Sciences Research Center.

Dillman, D., Carley-Baxter, L., and Jackson, A., 1999, *Skip Pattern Compliance in Three Test Forms: A Theoretical and Empirical Evaluation*, Technical Report 99-01. Pullman, WA: Social and Economic Sciences Research Center.

Dillman, D., Caldwell, S., and Gansemer, M., 2000, Visual design effects on item nonresponse to a question about work satisfaction that precedes the Q-12 agree–disagree items, unpublished manuscript.

Dillon, A., 1987, Knowledge acquisition and conceptual models: a cognitive analysis of the interface, in D. Diaper and R. Winder (eds.), *People and Computers III*. Cambridge: Cambridge University Press.

Dippo, C., 1997, Survey measurement and process improvement: concepts and integration, in L. Lyberg, P. Biemer, M. Collins, E. de Leeuw, C. Dippo, N. Schwarz, and D. Trewin (eds.), *Survey Measurement and Process Quality*. New York: Wiley.

Dippo, C., Chun, Y., and Sander, J., 1995, Designing the data collection process, in B. Cox, D. Binder, B. Chinnappa, A. Christianson, M. Colledge, and P. Kott (eds.), *Business Survey Methods*. New York: Wiley.

Dodds, J., 2001, Experiences with Internet reporting on e-business surveys, presented at the 34th International Field Directors and Technologies Conference, Montreal, Quebec, Canada.

Doob, L., 1968, Tropical weather and attitude surveys, *Public Opinion Quarterly* 32: 423–430.

Doyle, P., 2002, AAPOR roundtable: improving income measurement, *Proceedings of the ASA Section on Survey Research Methods*. Alexandria, VA: American Statistical Association.

Doyle, P., Martin, E., and Moore, J., 2000, Methods panel to improve income measurement in the survey of income and program participation, *Proceedings of the ASA Section on Survey Research Methods*. Alexandria, VA: American Statistical Association, pp. 953–958.

Draisma, S., 2000, RESPONSE: a simulation model for question answering in survey interviews, dissertation, Vrije Universiteit, Amsterdam.

Drasgow, F., and Parsons, C., 1983, Applications of unidimensional item response theory models to multidimensional data, *Applied Psychological Measurement* 7: 189–199.

Dumas, J., and Redish, J., 1994, *A Practical Guide to Usability Testing*. Norwood, NJ: Ablex.

Duncan, O., 1984, Rasch measurement in survey research: further examples and discussion, in C. Turner and E. Martin (eds.), *Surveying Subjective Phenomena*, Vol. 2. New York: Russell Sage.

Dunnigan, T., McNall, M., and Mortimer, J., 1993, The problem of metaphorical nonequivalence in cross-cultural survey research: comparing the mental health statuses among refugee and general population adolescents, *Journal of Cross-Cultural Psychology* 24: 344–365.

Dutka, S., and Frankel, L., 1991, Measurement errors in business surveys, in P. Biemer, R. Groves, L. Lyberg, N. Mathiowetz, and S. Sudman (eds.), *Measurement Errors in Surveys*. New York: Wiley.

Dykema, J., and Schaeffer, N. C., 2000, Events, instruments, and reporting errors, *American Sociological Review* 65: 619–629.

Dykema, J., Lepkowski, J. M., and Blixt, S., 1997, The effect of interviewer and respondent behavior on data quality: analysis of interaction coding in a validation study, in L. Lyberg, P. Biemer, M. Collins, E. de Leeuw, C. Dippo, N. Schwarz, and D. Trewin (eds.), *Survey Measurement and Process Quality*. New York: Wiley.

Eargle, J., 2000, Sample loss rates through wave 12 for the 1996 panel, unpublished memorandum, U.S. Bureau of the Census, May 23.

Eberhard, K., Scheutz, M., Targowski, K., and Spies, J., 2002, The effects of grammatical gender associations in one's native language on the processing of word representations in a second language: evidence for interactive processing of lexical representations in bilingual memory, poster presented at the 15th Annual CUNY Conference on Human Sentence Processing, New York.

Edwards, B., Bittner, D., Edwards W., and Sperry, S., 1993, CAPI effects on interviewers: a report from two major surveys, *Proceedings of the Bureau of the Census Annual Research Conference*. Washington, DC: U.S. Bureau of the Census.

Edwards, W., and Cantor, D., 1991, Towards a response model in establishment surveys, in P. Biemer, R. Groves, L. Lyberg, N. Mathiowetz, and S. Sudman (eds.), *Measurement Errors in Surveys*. New York: Wiley.

Edwards, W., Narayanan, V., Fry, S., Catania, J., and Pollack, L., 2002, A comparison of two behavior coding systems for pretesting questionnaires, presented at the 57th Annual Conference of the American Association for Public Opinion Research, St. Petersburg, FL.

Eisenhower, D., Mathiowetz, N., and Morganstein, D., 1991, Recall error: sources and bias reduction techniques, in P. Biemer, R. Groves, L. Lyberg, N. Mathiowetz, and S. Sudman (eds.), *Measurement Errors in Surveys*. New York: Wiley.

Eldridge, J., Martin, J., and White, A., 2000, The use of cognitive methods to improve establishment surveys in Britain, *Proceedings of the 2nd International Conference on Establishment Surveys*. Alexandria, VA: American Statistical Association, pp. 307–316.

Ellis, B., Minsel, B., and Becker, P., 1989, Evaluations of attitude survey translations: an investigation using item response theory, *International Journal of Psychology* 24: 665–684.

Embretson, S., and Reise, S., 2000, *Item Response Theory for Psychologists*. Mahwah, NJ: Lawrence Erlbaum.

Epstein, J., Barker, P., and Kroutil, L., 2001, Mode effects in self-reported mental health data, *Public Opinion Quarterly* 65: 529–549.

Ericsson, K., 2002, Towards a procedure for eliciting verbal expression of non-verbal experience without reactivity: interpreting the verbal overshadowing effect within the theoretical framework for protocol analysis, *Applied Cognitive Psychology* 16: 981–987.

Ericsson, K., and Simon, H., 1980, Verbal reports as data, *Psychological Review* 87: 215–251.

Ericsson, K., and Simon, H., 1984, *Protocol Analysis: Verbal Reports as Data*. Cambridge, MA: MIT Press.

Ericsson, K., and Simon, H., 1993, *Protocol Analysis: Verbal Reports as Data*, 2nd ed. Cambridge, MA: MIT Press.

Ericsson, K., and Simon, H., 1998, How to study thinking in everyday life: contrasting think-aloud protocols with descriptions and explanations of thinking, *Mind, Culture, and Activity* 5(3): 178–186.

Erikson, J., 2002, Coherence analysis as a tool for questionnaire evaluation in enterprise statistics, presented at the International Conference on Questionnaire Development, Evaluation, and Testing, Charleston, SC.

Esomar, 1999, Guideline on interviewing children and young people, retrieved November 8, 2003 at *http://esomar.nl/guidelines/interviewing_children_99.htm.*

Esposito, J., 2002, Iterative, multiple-method questionnaire evaluation research: a case study, presented at the International Conference on Questionnaire Development, Evaluation, and Testing, Charleston, SC, November 15–17.

Esposito, J., and Rothgeb, J., 1997, Evaluating survey data: making the transition from pretesting to quality assessment, in P. Lyberg, L. Biemer, M. Collins, E. de Leeuw, C. Dippo, N. Schwarz, and D. Trewin (eds.), *Survey Measurement and Process Quality*. New York: Wiley.

Esposito, J., Rothgeb, J., Polivka, A., Hess, J., and Campanelli, P., 1992, Methodologies for evaluating survey questions: some lessons from the redesign of the Current Population Survey, presented at the International Conference on Social Science Methodology, Trento, Italy.

European Social Survey, 2002, *European Social Survey Round 1: Report of the First Year*. London: NatCen.

Everitt, B., and Haye, D., 1992, *Talking about Statistics: A Psychologist's Guide to Data Analysis*. New York: Halsted Press.

Fazio, R., 1990, A practical guide to the use of response latency in social psychological research, in C. Hendrick and M. Clark (eds.), *Research Methods in Personality and Social Psychology*. Thousand Oaks, CA: Sage.

Fienberg, S., and Tanur, J., 1987, Experimental and sampling structures: parallels diverging and meeting, *International Statistical Review* 55: 75–96.

Fienberg, S., and Tanur, J., 1989, Combining cognitive and statistical approaches to survey design, *Science* 243: 1017–1022.

Finch, J., 1987, Research note: the vignette technique in survey research, *Sociology* 21: 105–114.

Fisher, R., and Geiselman, R., 1992, *Memory-Enhancing Techniques for Investigative Interviewing: The Cognitive Interview*. Springfield, IL: Thomas.

Fisher, S., and Adler, R., 1999, Using iterative cognitive testing to identify response problems on an establishment survey of energy consumption, presented at the Annual Conference of the American Association for Public Opinion Research, St. Petersburg Beach, FL.

Fisher, S., Goldenberg, K., O'Brien, E., Rosen, R., Stinson, L., and Willimack, D., 2000, Using cognitive methods to improve questionnaires for establishment surveys: a demonstration, presented to the Federal Economic Statistics Advisory Council, December 14. Washington, DC: U.S. Bureau of Labor Statistics.

Fisher, S., Goldenberg, K., O'Brien, E., Tucker, C., and Willimack, D., 2001a, Measuring employee hours in government surveys, presented to the Federal Economic Statistical Advisory Council, June 7. Washington, DC: U.S. Bureau of Labor Statistics.

Fisher, S., Frampton, K., and Tran, R., 2001b, Pretesting the Survey of Respirator Uses and Practices: cognitive and field testing of a new establishment survey, *Proceedings of the Annual Meeting of the American Statistical Association*. Alexandria, VA: ASA.

Fiske, A., Kitayama, S., Markus, H., and Nisbett, R., 1998, The cultural matrix of social psychology, in D. Gilbert, S. Fiske, and G. Lindzey (eds.), *The Handbook of Social Psychology*, Vol. 2. Boston: McGraw-Hill.

Fiske, S., and Ruscher, J., 1989, On-line processes in category-based and individuating impressions: some basic principles and methodological reflections, in J. Bassili (ed.), *On-Line Cognition in Person Perception*. Mahwah, NJ: Lawrence Erlbaum, pp. 141–174.

Flannery, W., Reise, S., and Widaman, K., 1995, An item response theory analysis of the general and academic scales of the self-description questionnaire II, *Journal of Research in Personality* 29: 168–188.

Flavell, J., 1985, *Cognitive Development*. Upper Saddle River, NJ: Prentice Hall.

Flavell, J., Miller, P., and Miller, S., 1993, *Cognitive Development*. Upper Saddle River, NJ: Prentice Hall.

Fleishman, J., Spector, W., and Altman, B., 2002, Impact of differential item functioning on age and gender differences in functional disability, *Journal of Gerontology: Social Sciences* 57B(5): S275–S284.

Fleiss, J., 1981, *Statistical Methods for Rates and Proportions*. New York: Wiley.

Fletcher, J., 2000, Two-timing: politics and response latencies in a bilingual survey, *Political Psychology* 21: 27–53.

Foddy, W., 1996a, The in-depth testing of survey questions: a critical appraisal of methods, *Quality and Quantity* 30: 361–370.

Foddy, W., 1996b, *Constructing Questions for Interviews and Questionnaires: Theory and Practice in Social Research*. Cambridge: Cambridge University Press.

Foddy, W., 1998, An empirical evaluation on in-depth probes used to pretest survey questions, *Sociological Methods and Research* 27: 103–133.

Folstein, M., Folstein, S., and McHugh, P., 1975, Mini mental state: a practical method for grading the cognitive state of patients for the clinician, *Journal of Psychiatric Research* 8: 102–109.

Fontaine, J, 2003, Multidimensional scaling, in J. Harkness, F. J. R. van de Vijver, and P. Ph. Mohler (eds.), *Cross-Cultural Survey Methods*. New York: Wiley, pages 235–246.

Forsman, G., and Schreiner, I., 1991, The design and analysis of reinterview: an overview, in P. Biemer, R. Groves, L. Lyberg, N. Mathiowetz, and S. Sudman (eds.), *Measurement Errors in Surveys*. New York: Wiley.

Forsyth, B., 1990, A summary of agency interviews, unpublished manuscript, Research Triangle Institute, Research Triangle Park, NC.

Forsyth, B., and Lessler, J., 1991, Cognitive laboratory methods: a taxonomy, in P. Biemer, R. Groves, L. Lyberg, N. Mathiowetz, and S. Sudman (eds.), *Measurement Errors in Surveys*. New York: Wiley.

Forsyth, B., Lessler, J., and Hubbard, M., 1992, Cognitive evaluation of the questionnaire, in C. Tanur, J. Lessler, and J. Gfroerer (eds.), *Survey Measurement of Drug Use: Methodological Studies*. Rockville, MD: National Institute on Drug Abuse, pp. 12–52.

Forsyth, B., Levin, K., and Fisher S., 1999, Test of an appraisal method for establishment survey questionnaires, *Proceedings of the ASA Section on Survey Research Methods*. Alexandria, VA: American Statistical Association, pp. 145–149.

Forsyth, B., Weiss, E., and Anderson, R., 2002, A comparison of appraisal and cognitive interview methods for testing organizational survey questionnaires, presented at the International Conference on Questionnaire Development, Evaluation, and Testing Methods, Charleston, SC, November 13–17.

Fowler, F., Jr., 1992, How unclear terms affect survey data, *Public Opinion Quarterly* 56: 218–231.

Fowler, F., Jr., 1993, *Survey Research Methods*. Thousand Oaks, CA: Sage.

Fowler, F., Jr., 1995, *Improving Survey Questions: Design and Evaluation*. Thousand Oaks, CA: Sage.

Fowler, F., Jr., and Cannell, C., 1996, Using behavioral coding to identify cognitive problems with survey questions, in N. Schwarz and S. Sudman (eds.), *Answering Questions: Methodology for Determining Cognitive and Communicative Processes in Survey Research*. San Francisco: Jossey-Bass.

Fowler, F., and Mangione, T., 1990, *Standardized Survey Interviewing*. Thousand Oaks, CA: Sage.

Fowler, F., and Roman, A., 1992, *A Study of Approaches to Survey Question Evaluation*, report produced for the U.S. Bureau of the Census, February 25.

Fox, J., 2001, Usability methods for designing a computer-assisted data collection instrument for the CPI, *Proceedings of the Federal Committee on Statistical Methodology Research Conference*, Arlington, VA, pp. 107–112.

Fracasso, M., 1989, Reliability and validity of response categories for open-ended questions in the Current Population Survey, *Proceedings of the ASA Section on Survey Research Methods*. Alexandria, VA: American Statistical Association.

Francoz, C., 2002, Review of the French industrial R&D survey, presented at the International Conference on Questionnaire Development, Evaluation, and Testing Methods, Charleston, SC, November 13–17.

Frazis, H., and Stewart, J., 1996, Keying errors caused by unusual response categories: evidence from the current population survey, unpublished manuscript, Office of Employment Research and Program Development, U.S. Bureau of Labor Statistics.

Freedman, H., and Mitchell, B., 1993, The development of surveys of waste management: the Canadian experience, *Proceedings of the International Conference on Establishment Surveys*. Alexandria, VA: American Statistical Association, pp. 52–61.

Freedman, S., and Rutchik, R., 2002, Information collection challenges in electric power and natural gas, *Proceedings of the Annual Meeting of the American Statistical Association*, Alexandria, VA.

Freeman, A., 1993, *The Measurement of Environmental and Resource Values*. Washington, DC: Resources for the Future.

Frey, F., 1963, Survey of peasant attitudes in Turkey, *Public Opinion Quarterly* 27: 335–355.

Friedenreich, C., Courneya, K., and Bryant, H., 1997, The lifetime total physical activity questionnaire: development and reliability, unpublished manuscript, Division of Epidemiology, Alberta Cancer Board, Canada.

Fu, H., Darroch, J., Henshaw, S., and Kolb, E., 1998, Measuring the extent of abortion underreporting in the 1995 National Survey of Family Growth, *Family Planning Perspectives* 30: 128–133, 138.

Gallagher, C., and Schwede, L., 1997, *Facility Questionnaire Redesign Project: Results from Phase 1 Unstructured Interviews and Recommendations for Facility-Level Questionnaire*, prepared by the U.S. Bureau of the Census for the Department of Justice.

Gallhofer, I., and Saris, W., 2000, Formulierung und Klassifizierung von Deutschen Fragen, *ZUMA Nachrichten* 46: 43–73.

Gan, C., and Luzar, E., 1993, A conjoint analysis of waterfowl hunting in Louisiana, *Journal of Agricultural and Applied Economics* 25: 36–45.

Gaul, B., 2001, Recent developments in electronic data collection at the U.S. Census Bureau, *Proceedings of the Statistics Canada Symposium 2001: Achieving Data Quality in a Statistical Agency: A Methodological Perspective*.

Geisinger, K. F., 1994. Cross-cultural normative assessment: translation and adaptation issues influencing the normative interpretation of assessment instruments, *Psychological Assessment* 16(4): 304–312.

Gelman, R., and Baillargeon, R., 1983, A review of some Piagetian concepts, in P. Mussen (ed.), Handbook of Child Psychology, Vol. 3, 4th ed. New York: Wiley.

Geoffrey, M., and Peel, D., 2000, *Finite Mixture Models*. New York: Wiley.

Gerber, E., 1990, Calculating residence: a cognitive approach to household membership judgements among low income Blacks, Report prepared for the U.S. Bureau of the Census, Washington, DC: U.S. Bureau of the Census.

Gerber, E., 1994, *The Language of Residence: Respondent Understandings and Census Rules*. Center for Survey Methods Research, U.S. Bureau of the Census.

Gerber, E., 1999, The view from anthropology: ethnography and the cognitive interview, in M. Sirken, D. Herrmann, S. Schechter, N. Schwarz, J. Tanur, and R. Tourangeau (eds.), *Cognition and Survey Research*. New York: Wiley.

Gerber, E., and DeMaio, T. J., 1999, Probing strategies for establishment surveys, presented at the Annual Conference of the American Association for Public Opinion Research, St. Petersburg Beach, FL.

Gerber, E., and Wellens, T., 1997, Perspectives on pretesting: "cognition" in the cognitive interview?, *Bulletin de Methodologie Sociologique* 55: 18–39.

Gerber, E., and Wellens, T., 1998, The conversational analogy, forms literacy, and pretesting in self-administered questionnaires, presented to the International Sociological Association, Montreal, Quebec, Canada.

Gerber, E., Keeley, C., and Wellens, T., 1997, Census rules and rostering decisions: a vignette study, *Proceedings of the ASA Section on Survey Research Methods*. Alexandria, VA: American Statistical Association.

Gergen, K., 1973, Social psychology as history, *Journal of Personality and Social Psychology* 26: 309–320.

Gile, D., 1995, *Basic Concepts and Models for Interpreter and Translator Training*. Amsterdam: John Benjamins.

Goddard, G., 1993, How not to collect fire statistics from fire brigades, *Proceedings of the International Conference on Establishment Surveys*. Alexandria, VA: American Statistical Association, pp. 107–110.

Goldenberg, K., 1994, Answering questions, questioning answers: evaluating data quality in an establishment survey, *Proceedings of the ASA Section on Survey Research Methods*. Alexandria, VA: American Statistical Association, pp. 1357–1362.

Goldenberg, K., 1996, Using cognitive testing in the design of a business survey questionnaire, *Proceedings of the ASA Section on Survey Research Methods*. Alexandria, VA: American Statistical Association, pp. 944–949.

Goldenberg, K., and Phillips, M., 2000, Now that the study is over, what did you really tell us? Identifying and correcting measurement error in the job openings and labor turnover survey pilot test, *Proceedings of the 2nd International Conference on Establishment Surveys*. Alexandria, VA: American Statistical Association.

Goldenberg, K., and Stewart, J., 1999, Earnings concepts and data availability for the current employment statistics survey: findings from cognitive interviews, *Proceedings of the ASA Section on Survey Research Methods*. Alexandria, VA: American Statistical Association, pp. 139–144.

Goldenberg, K., Butani, S., and Phipps, P. 1993, Response analysis surveys for assessing response errors in establishment surveys, *Proceedings of the International Conference on Establishment Surveys*. Alexandria, VA: American Statistical Association, pp. 290–299.

Goldenberg, K., Levin, K., Hagerty, T., Shen, T., and Cantor, D., 1997, Procedures for reducing measurement errors in establishment surveys, *Proceedings of the ASA Section on Survey Research Methods*. Alexandria, VA: American Statistical Association, pp. 994–999.

Goldenberg, K., Gomes, A., Manser, M., and Stewart, J., 2000, Collecting all-employee earnings data in the Current Employment Statistics survey, presented at the Joint Statistical Meetings of the American Statistical Association, Indianapolis, IN.

Goldenberg, K., Anderson, A., Willimack, D., Freedman, S., Rutchik, R., and Moy, L., 2002a, Experiences implementing establishment survey questionnaire development and testing at selected U.S. government agencies, presented at the International Conference on Questionnaire Development, Evaluation, and Testing Methods, Charleston, SC.

Goldenberg, K., Willimack, D., Fisher, S., and Anderson, A., 2002b, Measuring key economic indicators in U.S. government establishment surveys, presented at the International Conference on Improving Surveys, Copenhagen, Denmark.

Goodman, L., 1974, Exploratory latent structure analysis using both identifiable and unidentifiable models, *Biometrika* 61: 215–231.

Goodwin, C., 2002, *Research in Psychology; Methods and Design*. New York: Wiley.

Gower, A., 1994, Questionnaire design for business surveys, *Survey Methodology* 20: 125–136.

Gower, A., and Nargundkar, M., 1991, Cognitive aspects of questionnaire design: business surveys versus household surveys, *Proceedings of the U.S. Bureau of the Census Annual Research Conference*, pp. 299–312.

Graesser, A., Kennedy, T., Wiemer-Hastings, P., and Ottati, V., 1999, The use of computational cognitive models to improve questions on surveys and questionnaires, in M. Sirken, D. Herrmann, S. Schechter, N. Schwarz, J. Tanur, and R. Tourangeau (eds.), *Cognition and Survey Research*. New York: Wiley.

Graesser, A., Wiemer-Hastings, K., Kreuz, R., Wiemer-Hastings, P., and Marquis, K., 2000a, QUAID: a questionnaire evaluation aid for survey methodologists, *Behavior Research Methods, Instruments, and Computer* 32: 254–262.

Graesser, A., Wiemer-Hastings, K., Wiemer-Hastings, P., and Kreuz, R., 2000b, The gold standard of question quality on surveys: experts, computer tools, versus statistical indices, *Proceedings of the ASA Section on Survey Research Methods*. Alexandria, VA: American Statistical Association, pp. 459–464.

Granquist, L., 1995, Improving the traditional editing process, in B. Cox, D. Binder, B. Chinnappa, A. Christianson, M. Colledge, and P. Kott (eds.), *Business Survey Methods*. New York: Wiley.

Gray, P., 2002, *Psychology*. New York: Worth.

Gray, W., and Salzman, M., 1998, Damaged merchandise? A review of experiments that compare usability evaluation methods, *Human–Computer Interaction* 13: 203–361.

Greatbatch, D., Luff, P., Heath C., and Campion, P., 1993, Interpersonal communication and human–computer interaction: an examination of the use of computers in medical examinations, *Interacting with Computers* 5: 193–216.

Greenfield, P., 1997, You can't take it with you: why ability assessments don't cross cultures, *American Psychologist* 52: 1115–1124.

Greenleaf, E., 1992, Measuring extreme response style, *Public Opinion Quarterly* 56: 328–351.

Greig, A., and Taylor, J., 1999, *Doing Research with Children*. London: Sage.

Grice, H., 1975, Logic and conversation, in P. Cole and J. L. Morgan (eds.), *Syntax and Semantics*. San Diego, CA: Academic Press.

Griffiths, G., and Linacre, S., 1995, Quality assurance for business surveys, in B. Cox, D. Binder, B. Chinnappa, A. Christianson, M. Colledge, and P. Kott (eds.), *Business Survey Methods*. New York: Wiley.

Griffiths, J., 2001, Household income screening procedures in the Survey of Income and Program Participation (SIPP) Methods Panel, *Proceedings of the ASA Section on Government Statistics*. Alexandria, VA: American Statistical Association.

Groves, R., 1989, *Survey Errors and Survey Costs*. New York: Wiley.

Groves, R., 1996, How do we know what we think they think is really what they think? in N. Schwarz and S. Sudman (eds.), *Answering Questions*. San Francisco: Jossey-Bass.

Groves, R., 2002, Challenges in the future of scientific surveys, keynote address presented at the International Field Directors and Technologies Conference, Clearwater Beach, FL, May 19–22.

Groves, R., and Couper, M., 1998, *Nonresponse in Household Interview Surveys*. New York: Wiley.

Groves, R., Berry M., and Mathiowetz, N., 1980, Some impacts of computer assisted telephone interviewing on survey methods, *Proceedings of the ASA Section on Survey Research Methods*. Alexandria, VA: American Statistical Association.

Groves, R., Cantor, D., Couper M., Levin, K., McGonagle, K., and Singer, E., 1997, Research investigations in gaining participation from sample firms in the Current Employment Statistics program, *Proceedings of the ASA Section on Survey Research Methods*. Alexandria, VA: American Statistical Association, pp. 289–294.

Grunert, S., and Muller, T., 1996, Measuring values in international settings: are respondents thinking "Real" life or "ideal" life, *Journal of International Consumer Marketing* 8: 169–185.

Guenzel, P., Berckmans, T., and Cannell, C., 1983, *General Interviewing Techniques*. Ann Arbor, MI: Survey Research Center, University of Michigan.

Guggenmoos-Holzmann, I., and Vonk, R. 1998, Kappa-like indices of observer agreement viewed from a latent class perspective, *Statistics in Medicine* 17(8): 797–812

Guillemin, F., Bombardier, C., and Beaton, D., 1993, Cross-cultural adaptation of health-related quality of life measures: literature review and proposed guidelines, *Journal of Clinical Epidemiology* 46: 1417–1432.

Guirdham, M., 1999, *Communicating across Cultures*. London: Macmillan.

Gutknecht, C., and Rölle, L., 1996, *Translating by Factors*. Albany, NY: State University of New York Press.

Haberman, S., 1979, *Analysis of Qualitative Data: New Developments*. San Diego, CA: Academic Press.

Hagenaars, J., 1988, Latent structure models with direct effects between indicators: local dependence models, *Sociological Methods and Research* 16: 379–405.

Hak, T., and van Sebille, M., 2002, Het respons proces bij bedrijfsenquêtes [The response process in establishment surveys], internal report. Rotterdam/Voorburg, The Netherlands: Erasmus Research Institute of Management, Erasmus University/Centraal Bureau voor de Statistiek (Statistics Netherlands).

Hallfors, D., Khatapoush, S., Kadushin, C., Watson, K., and Saxe, L., 2000, A comparison of paper vs. computer-assisted self interview for school alcohol, tobacco, and other drug surveys, *Evaluation and Program Planning* 23: 149–155.

Hambleton, R., 1993, Translating achievement tests for use in cross-national studies, *European Journal of Psychology Assessment* (Bulletin of the International Test Commission) 9: 57–68.

Hambleton, R., 1994, Guidelines for adapting educational and psychological tests: a progress report, *European Journal of Psychology Assessment* (Bulletin of the International Test Commission) 10: 229–244.

Hambleton, R., 2004, Issues, designs, and technical guidelines for adapting tests in multiple languages and cultures, in R. Hambleton, P. Merenda, and C. Spielberger (eds.), *Adapting Educational and Psychological Tests in Cross-Cultural Assessment*. Mahwah, NJ: Lawrence Erlbaum.

Hambleton, R., and Patsula, L., 1998, Adapting tests for use in multiple languages and cultures, *International and Interdisciplinary Journal of Quality-of-Life Measurement* 45: 153–171.

Hambleton, R., and Patsula, L., 1999, Increasing the validity of adapted tests: myths to be avoided and guidelines for improving test adaptation practices, *Journal of Applied Testing Technology* 1: 1–30.

Hambleton, R., Robin, F., and Xing, D., 2000, Item response models for the analysis of educational and psychological test data, in H. Tinsley and S. Brown (eds.), *Handbook of Applied Multivariate Statistics and Mathematical Modeling*. San Diego, CA: Academic Press, pp. 553–585.

Hambleton, R., Merenda, P., Spielberger, C., 2004, *Adapting Educational and Psychological Tests in Cross-Cultural Assessment*. Mahwah, NJ: Lawrence Erlbaum.

Hanna, L., Risden, K., and Alexander, K., 1997, Guidelines for usability testing with children, *Interactions* 4: 9–14.

Hansen, M., Hurwitz, W., and Pritzker, L., 1964, The estimation and interpretation of gross differences and the simple response variance, in C. Rao (ed.), *Contributions to Statistics*. Calcutta: Pergamon Press, pp. 111–136.

Hansen, S., Couper, M., and Fuchs, M., 1998, Usability evaluation of the NHIS CAPI instrument, *Proceedings of the ASA Section on Survey Research Methods*. Alexandria, VA: American Statistical Association.

Hansen, S., Beatty P., Couper M., Lamias, M., and Marvin, T., 2000, The effect of CAI screen design on user performance: results of an experiment, report submitted to the U.S. Bureau of Labor Statistics, August.

Harkness, J., 1996, Thinking aloud about survey translation, presented at the International Sociological Association Conference on Social Science Methodology, Colchester, Essex, England.

Harkness, J., 1999, In pursuit of quality: issues for cross-national survey research, *International Journal of Social Research Methodology* 2: 125–140.

Harkness, J., 2001, Questionnaire development, adaption, and assessment for the ESS, presented to the International Conference on Quality in Official Statistics.

Harkness, J., 2003, Questionnaire translation, in J. Harkness, F. van de Vijver, and P. Mohler (eds.), *Cross-Cultural Survey Methods*. New York: Wiley.

Harkness, J., and Schoua-Glusberg, A., 1998, Questionnaires in translation, in J. Harkness (ed.), *Cross-Cultural Survey Equivalence*, ZUMA-Nachrichten Spezial 3. Mannheim, Germany: ZUMA.

Harkness, J., Langfeldt, B., Scholz, E., and Klein, S., 2001, *ISSP Study Monitoring, 1996–1998*, Reports to the ISSP General Assembly on Monitoring Work Undertaken for the ISSP by ZUMA, Germany. Mannheim, Germany: ZUMA.

Harkness, J., van de Vijver, F., and Johnson, T., 2003, Questionnaire design in comparative research, in J. Harkness, F. van de Vijver, and P. Mohler (eds.), *Cross-Cultural Survey Methods*. New York: Wiley.

Harley, M., Pressley, K., and Murphy, E., 2001, 2002 economic electronic style guide, *Proceedings of the Statistics Canada Symposium 2001: Achieving Data Quality in a Statistical Agency: A Methodological Perspective*.

Hart, J., 1965, Memory and the feeling-of-knowing experience, *Journal of Educational Psychology* 56: 208–216.

Hayashi, E., 1992, Belief systems, the way of thinking, and sentiments of five nations, *Behaviormetrics* 19: 127–170.

Hays, R., Morales, L., and Reise, S., 2000, Item response theory and health outcomes measurement in the 21st century, *Medical Care* 38(9 Supplement): II-28 to II-42.

Heeringa, S., Hill, D., and Howell, D., 1995, *Unfolding Brackets for Reducing Item Nonresponse in Economic Surveys*, Health and Retirement Study Working Paper 94-029. Ann Arbor, MI: Survey Research Center, Institute for Social Research, University of Michigan.

Heerwegh, D., and Loosveldt, G., 2002, Describing response behavior in Web surveys using client side paradata, presented at the International Workshop on Web Surveys, Mannheim, Germany.

Helweg-Larsen, K., and Larsen, H., 2001, Experiences from a pilot study: the potential for conducting a national questionnaire study on the well-being of adolescent school children, with a particular focus on the sexual encounters with adults, presented to the Danish Ministry of Social Affairs, prior to the national surveys on youth, well being with a focus on CSA.

Helweg-Larsen, K., and Larsen, H., 2002, A nation-wide survey conducted among 15–16 year old Danish school children that included questions on sensitive topics, presented at the International Conference on Improving Surveys (ICIS-2002), Copenhagen.

Henningsson, B., 2001, An enlightened client, *Proceedings of the QUEST 2001 Workshop*. Washington, DC: U.S. Bureau of the Census, pp. 73–77.

Henningsson, B., Mdluli, P., Näsholm, H., and Polfeldt, T., 1998, Some problems in questionnaire translation between English and isiZulu, presented at the 3rd Conference on Methodological Issues in Official Statistics, Stockholm, Sweden.

Heberlein, T., and Baumgartner, R., 1978, Factors affecting response rates to mailed questionnaires: a quantitative analysis of the published literature, *American Sociological Review* 43: 447–462.

Hermalin, A., Entwisle, B., and Myers, L., 1985, Some lessons from the attempt to retrieve early KAP and fertility surveys, *Population Index* 51: 194–208.

Hershey, M., and Hill, D., 1976, Positional response set in pre-adult socialization surveys, *Social Science Quarterly* 56: 707–714.

Hess, J., and Singer, E., 1995, The role of respondent debriefing questions in questionnaire development, *Proceedings of the ASA Section on Survey Research Methods*. Alexandria, VA: American Statistical Association, pp. 1075–1080.

Hess, J., Rothgeb, J., Zukerberg, A., Richter, K., LeNestrel, S., and Moore, K., 1998a, Teens talk: are adolescents willing and able to answer survey questions? *Proceedings of the ASA Section on Survey Research Methods*. Alexandria, VA: American Statistical Association.

Hess, J., Rothgeb, J., and Zukerberg, A., 1998b, Developing the Survey of Program Dynamics Survey Instruments, SM 98/07. Washington, DC: U.S. Bureau of the Census.

Hess, J., Singer, E., and Bushery, J., 1999, Predicting test–retest reliability from behavior coding, *International Journal of Public Opinion Research* 11(4): 346–360.

Hess, J., Moore, J., Pascale, J., Rothgeb, J., and Keeley, C., 2001, The effects of person-level versus household-level questionnaire design on survey estimates and data quality, *Public Opinion Quarterly* 65: 574–584.

Hill, D., 1987, Response errors around the seam: analysis of change in a panel with overlapping reference periods, *Proceedings of the ASA Section on Survey Research Methods*. Washington, DC: American Statistical Association, pp. 210–215.

Hill, M., Laybourn, A., and Borland, M., 1996, Engaging with children about their emotions and well-being: methodological considerations, *Children and Society* 10: 129–144.

Hippler, H., and Schwarz, N., 1986, Not forbidding isn't allowing: the cognitive basis of the forbid–allow asymmetry, *Public Opinion Quarterly* 50: 87–96.

Hippler, H., Schwarz, N., and Sudman, S., 1987, *Social Information Processing and Survey Methodology*. New York: Springer-Verlag.

Hochstim, J., 1967, A critical comparison of three strategies of collecting data from households, *Journal of the American Statistical Association* 62: 976–989.

Hoehn, J., Kaplowitz, M., and Lupi, F., 2001, *Wetland Uses and Functions as Perceived by Mid-Michigan Residents: Qualitative Research Results*. East Lansing, MI: Agricultural Economics Department, Michigan State University.

Holaday, B., and Turner-Henson, A., 1989, Response effects in surveys with school-age children, *Nursing Research (Methodology Corner)* 38: 248–250.

Holm, S., 1979, A simple sequentially rejective multiple test procedure, *Scandinavian Journal of Statistics* 6: 65–70.

Holstein, J., and Gubrium, J., 1995, *The Active Interview*. Thousand Oaks, CA: Sage.

Holz-Mänttäri, J., 1984, Sichtbarmachung und Beurteilung translatorischer Leistungen bei der Ausbildung von Berufstranslatoren [The elucidation and evaluation of translation performances in translator training], in W. Wilss and G. Thome (eds.), *Die Theorie des Übersetzens und ihr Aufschlußwert für die Übersetzungs- und Dolmetschdidaktik*. Tübingen, Germany: Narr.

Hougland, J., Johnson, T., and Wolf, J., 1992, A fairly common ambiguity: comparing rating and approval measures of public opinion, *Sociological Focus* 25: 257–271.

House, C., and Nicholls, W., II, 1988, Questionnaire design for CATI: design objectives and methods, in R. Groves, P. Biemer, L. Lyberg, J. Massey, W. Nicholls, and J. Waksberg (eds.), *Telephone Survey Methodology*. New York, Wiley.

House, J., 1977, *A Model for Translation Quality Assessment*. Tübingen, Germany: Narr.

Houtkoop-Steenstra, H., 2002, Questioning turn format and turn-taking problems in standardized interviews, in D. Maynard, H. Houtkoop-Steenstra, N. Schaeffer, and J. Zouwen (eds.), *Standardization and Tacit Knowledge: Interaction and Practice in the Survey Interview*. New York: Wiley.

Hox, J., deLeeuw, E., and Kreft, I., 1991, The effect of interviewer and respondent characteristics on the quality of survey data: a multilevel model, in P. Biemer, R. Groves, L. Lyberg, N. Mathiowetz, and S. Sudman (eds.), *Measurement Errors in Surveys*. New York: Wiley.

Hu, L., and Bentler, P., 1999, Cutoff criteria for fit indexes in covariance structure analysis: conventional criteria versus new alternatives, *Structural Equation Models* 6: 1–55.

Hudler, M., and Richter, R., 2001, *Theoretical and Methodological Concepts for Future Research and Documentation on Social Reporting in Cross-Sectional Surveys*, Reporting Working Paper 18. Vienna: Lazarsfeld-Gessellschaft fuer Sozialforschung.

Hui, C., and Triandis, H., 1985, The instability of response sets, *Public Opinion Quarterly* 49: 253–260.

Hui, C., and Triandis, H., 1989, Effects of culture and response format on extreme response style, *Journal of Cross-Cultural Psychology* 20: 296–309.

Hui, S., and Walter, S., 1980, Estimating the error rates of diagnostic tests, *Biometrics* 36: 167–171.

Hulin, C., 1987, A psychometric theory of evaluations of item and scale translation: fidelity across languages, *Journal of Cross-Cultural Psychology* 18: 115–142.

Hulin, C., Drasgow, F., and Komocar, J., 1982, Applications of item response theory to analysis of attitude scale translations, *Journal of Applied Psychology* 6: 818–825.

Hunt, S., Sparkman, R., and Wilcox, J., 1982, The pretest in survey research: issues and preliminary findings, *Journal of Marketing Research* 19: 269–273.

Inglehart, R., and Carballo, M., 1997, Does Latin America exist? And is there a Confucian culture? A global analysis of cross-cultural differences, *Political Science and Politics* 30: 34–46.

Jabine, T., Straf, M., Tanur, J., and Tourangeau, R., 1984, *Cognitive Aspects of Survey Methodology: Building a Bridge between Disciplines*. Washington, DC: National Academy Press.

Jacobson, E., Kumata, H., and Gullahorn, J., 1960, Cross-cultural contributions to attitude research, *Public Opinion Quarterly* 24: 205–223.

Jakwerth, P., Stancavage, F., and Reed, E., 1999, *An Investigation of Why Students Do Not Respond to Questions*, NAEP Validity Studies. Palo Alto, CA: American Institute for Research.

Javeline, D., 1999, Response effects in polite cultures: a test of acquiescence in Kazakhstan, *Public Opinion Quarterly* 63: 1–28.

Jay, G., Belli, R., and Lepkowski, J., 1994, Quality of last doctor visit reports: a comparison of medical records and survey data, *Proceedings of the ASA Section on Survey Research Methods*. Alexandria, VA: American Statistical Association, pp. 362–367.

Jeavons, A., 2002, Paradata: concepts and applications, presented at Net Effects4, Barcelona, Spain.

Jenkins, C., 1992, Questionnaire research in the Schools and Staffing Survey: a cognitive approach, presented at the Joint Statistical Meetings of the American Statistical Association, Boston.

Jenkins, C., and Dillman, D., 1997, Towards a theory of self-administered questionnaire design, in L. Lyberg, P. Biemer, M. Collins, L. Decker, E. de Leeuw, C. Dippo, N. Schwarz, and D. Trewin (eds.), *Survey Measurement and Process Quality*. New York: Wiley-Interscience.

Jobe, J., and Herrmann, D., 1996, Implications of models of survey cognition for memory theory, in D. Herrmann, C. McEvoy, C. Herzog, P. Hertel, and M. Johnson (eds.), *Basic and Applied Memory Research: Practical Applications*. Mahwah, NJ: Lawrence Erlbaum.

Jobe, J., and Mingay, D., 1989, Cognitive research improves questionnaires. *American Journal of Public Health*, 79: 1053–1055.

Jobe, J., and Mingay, D., 1991, Cognition and survey measurement: history and overview, *Applied Cognitive Psychology* 5: 175–192.

Johnson, M., and Foley, M., 1984, Differentiating fact from fantasy: the reliability of children's memory, *Journal of Social Issues* 40: 33–50.

Johnson, M., Shively, W., and Stein, R., 1999, Contextual data and the study of elections and voting behavior: connecting individuals to environments, presented at the Future of Election Studies Conference, University of Houston.

Johnson, T., 1997, Social cognition and responses to survey questions among culturally diverse populations, in L. Lyberg, P. Biemer, M. Collins, E. de Leeuw, C. Dippo, N. Schwarz, and D. Trewin (eds.), *Survey Measurement and Process Control*. New York: Wiley.

Johnson, T., 1998, Approaches to equivalence in cross-cultural and cross-national survey research, in J. A. Harkness (ed.), *Cross-Cultural Survey Equivalence*, ZUMA Nachrichten Spezial 3. Mannheim, Germany: ZUMA.

Johnson, T., van de Vijver, F., Harkness, J., and Mohler, P., 2000, The effects of cultural orientations on survey response: the case of individualism and collectivism, Paper presented to the International Conference on Logic and Methodology, Cologne, Italy.

Jorgensen, P., 1995, *Software Testing: A Craftsman's Approach*. London: CRC Press.

Jowell, R., 1998, How comparative is comparative research? *American Behavioral Scientist* 42: 168–177.

Juster, F., and Smith, J., 1997, Improving the quality of economic data: lessons from the HRS and AHEAD, *Journal of the American Statistical Association* 92: 1268–1278.

Juster, F., and Smith, J., 1998, Enhancing the quality of data on income and wealth: recent developments in survey methodology, prepared for the 25th General Conference of the International Association for Research in Income and Wealth, Cambridge, August 23–29.

Kahneman, D., 1973, *Attention and Effort*. Upper Saddle River, NJ: Prentice Hall.

Kail, R., 1990, *The Development of Memory in Children*. New York: W.H. Freeman.

Kail, R., 1993, The role of global mechanism in developmental change in speed of processing, in M. Howe and R. Pasnak (eds.), *Emerging Themes in Cognitive Development*. New York: Springer.

Kalton, G., and Miller, M., 1991, The seam effect with Social Security income in the survey of income and program participation, *Journal of Official Statistics* 7: 235–245.

Kalton, G., Lepkowski, J., Montanari, G., and Maligalig, D., 1990, Characteristics of second wave nonrespondents in a panel survey, *Proceedings of the ASA Section on Survey Research Methods*. Alexandria, VA: American Statistical Association, pp. 462–467.

Kaplowitz, M., 2000, Statistical analysis of sensitive topics in group and individual interviews, *Quality and Quantity: International Journal of Methodology* 34: 419–431.

Kaplowitz, M., and Hoehn, J., 2001, Do focus groups and personal interviews reveal the same information for natural resource valuation? *Ecological Economics* 36: 137–147.

Katan, D., 1999, *Translating Cultures*. Manchester, U.K.: St. Jerome Publishing.

Katz, D., 1940, Three criteria: knowledge, conviction, and significance, *Public Opinion Quarterly* 4: 277–284.

Kennickell, A., 1997, *Using Range Techniques with CAPI in the 1995 Survey of Consumer Finances*, Survey of Consumer Finances Working Paper. Washington, DC: Board of Governors of the Federal Reserve System.

Kenny, D., and Kashy, D., 1992, Analysis of the multitrait–multimethod matrix by confirmatory factor analysis, *Psychological Bulletin* 112: 165–172.

Kinder, J., and Baird, D., 2000, Data capture initiatives and the collection of business data in the Office for National Statistics, *Proceedings of the 2nd International Conference on Establishment Surveys*. Alexandria, VA: American Statistical Association, pp. 1256–1261.

Kingston, N., and Dorans, N., 1985, The analysis of item-ability regressions: an exploratory IRT model-fit tool, *Applied Psychological Measurement* 9: 281–288.

Kinsey, S., and Jewell, D., 1998, A systematic approach to instrument development in CAI, in M. Couper, R. Baker, J. Bethlehem, C. Clark, J. Martin, W. Nicholls, and J. O'Reilly (eds.), *Computer-Assisted Survey Information Collection*. New York: Wiley.

Kirk, J., and Miller, M., 1986, *Reliability and Validity in Qualitative Research*. Thousand Oaks, CA: Sage.

Kirk, R., 1968, *Experimental Design: Procedures for the Behavioral Sciences*. Pacific Grove, CA: Brooks/Cole.

Knäuper, B., Belli, R., Hill, D., and Herzog, A., 1997, Question difficulty and respondents' cognitive ability: the effect on data quality, *Journal of Official Statistics* 13: 181–199.

Knoop, J., 1979, Assessing equivalence of indicators cross-national survey research: some practical guidelines, *International Review of Sport Sociology* 14: 137–156.

Kohlberg, L., and Puka, B., 1994, *Kohlberg's Original Study of Moral Development: The Development of Modes of Moral Thinking and Choice in the Years 10 to 16*. New York: Garland.

Kohn, M. L., Atsushi, N., Schoenbach, C., Schooler, C., and Slomczynski, K., 1990, Position in class structure and psychological functioning in the United States, Japan, and Poland, *American Journal of Sociology* 95: 964–1008.

Költringer, R., 1995, Measurement quality in Austrian personal interview surveys, in W. Saris and A. Münnich (eds.), *The Multitrait–Multimethod Approach to Evaluate Measurement Instruments*. Budapest, Hungary: Eötvös University Press, pp. 207–224.

Koriat, A., Goldsmith, M., and Pansky, A., 2000, Toward a psychology of memory accuracy, *Annual Review of Psychology* 51: 481–537.

Kornhauser, A., 1951, Constructing questionnaires and interview schedules, in M. Jahoda, M. Deutsch, and S. Cook (eds.), *Research Methods in Social Relations: Part Two*. New York: Dryden.

Kraemer, J., 2002, *Évaluer pour mieux comprendre les enfants et améliorer sa pratique* [Better Understanding of Children and Improvement of Educational Daily Practice through Evaluation]. Arnhem, The Netherlands: National Center for Evaluation of Education, CITO.

Krebs, D., and Schuessler, K., 1986, *Zur Konstruktion von Einstellungsskalen im Internationalen Vergelich*, ZUMA-Arbeitsbericht 86/01. Mannheim, Germany: ZUMA.

Krosnick, J., 1991, Response strategies for coping with the cognitive demands of attitude measures in surveys, *Applied Cognitive Psychology* 5: 213–236.

Krosnick, J., 1999, Survey research, *Annual Review of Psychology* 50: 537–567.

Krosnick, J., and Fabrigar, L., 1997, Designing rating scales for effective measurement in surveys, in L. Lyberg, P. Biemer, M. Collins, E. de Leeuw, C. Dippo, N. Schwarz, and D. Trewin (eds.), *Survey Measurement and Process and Process Quality*. New York: Wiley.

Krosnick, J., and Fabrigar, L., Forthcoming, *Designing Questionnaires to Measure Attitudes*. New York: Oxford University Press.

Krueger, R., 1994, *Focus Groups: A Practical Guide for Applied Research*. Thousand Oaks, CA: Sage.

Kumata, H., and Schramm, W., 1956, A pilot study of cross-cultural meaning, *Public Opinion Quarterly* 20: 229–238.

Kussmaul, P., 1995, *Training the Translator*. Amsterdam: John Benjamins.

Kuusela, H., and Paul, P., 2000, A comparison of concurrent and retrospective verbal protocol analysis, *American Journal of Psychology* 113: 387–404.

Kydoniefs, L., 1993, The Occupational Safety and Health Survey, *Proceedings of the International Conference on Establishment Surveys*. Alexandria, VA: American Statistical Association, pp. 99–106.

Kydoniefs, L., and Stinson, L., 1999, Tapping data users to compare and review surveys, *Proceedings of the ASA Section on Survey Research Methods*. Alexandria, VA: American Statistical Association, pp. 968–972.

LaBarbera, P., and MacLachlan, J., 1979, Response latency in telephone interviews, *Journal of Advertising Research* 19: 49–55.

Labillois, T., and March, M., 2000, Cost-recovery business surveys: helping policy makers acquire information required to deal with newly arising issues, *Proceedings of the*

2nd International Conference on Establishment Surveys. Alexandria, VA: American Statistical Association, pp. 1250–1255.

Laffey, F., 2002, Business survey questionnaire review and testing at Statistics Canada, presented at the International Conference on Questionnaire Development, Evaluation, and Testing Methods, Charleston, SC.

La Greca, A., 1990, Issues and perspectives on the child assessment process, in A. La Greca (ed.), *Through the Eyes of the Child: Obtaining Self-Reports from Children and Adolescents.* Boston: Allyn & Bacon.

Landis, J., and Koch, G., 1977, The measurement of observer agreement for categorical data, *Biometrics* 33: 159–174.

Landsberger, H., and Saavedra, A., 1967, Response set in developing countries, *Public Opinion Quarterly* 31: 214–229.

Lansdale, M., and Ormerud, T., 1994, *Understanding Interfaces: A Handbook of Human–Computer Interaction.* London: Academic Press.

Lansing, J., Ginsberg, G., and Braaten, K., 1961, *An Investigation of Response Error.* Urbana, IL: Bureau of Economic and Business Research, University of Illinois.

Lashley, K., 1923, The behavioristic interpretation of consciousness II, *Psychological Bulletin* 30: 329–353.

Laumann, E., Gagnon, J., Michael, R., and Michaels, S., 1994, *The Social Organization of Sexuality: Sexual Practices in the United States.* Chicago: University of Chicago Press.

Law and Contemporary Problems, 2002, Special Issue on Children, 65: 1.

Lazarsfeld, P., 1950, The logical and mathematical foundation of latent structure analysis, in L. Stouffer, L. Guttman, E. Suchman, P. Lazarsfeld, S. Star, and J. Claussen (eds.), *Measurement and Prediction.* Princeton, NJ: Princeton University Press.

Lazarsfeld, P., and Henry, N., 1968, *Latent Structure Analysis.* Boston: Houghton Mifflin.

Leach, N., 1999, Using cognitive research to redesign federal questionnaires for manufacturing establishments, presented at the Joint Statistical Meetings of the American Statistical Association, Baltimore.

Lee, C., and Green, R., 1991, Cross-cultural examination of the Fishbein behavioral intentions model, *Journal of Business Studies* 2: 289–305.

Lepkowski, J., and Couper, M., 2002, Nonresponse in the second wave of longitudinal household surveys, in R. Groves, D. Dillman, J. Eltinge, and R. Little (eds.), *Survey Nonresponse.* New York: Wiley.

Leslie, T., 1996, *1996 National Content Survey Results*, Internal DSSD Memorandum 3. Washington, DC: U.S. Bureau of the Census.

Leslie, T., 1997, Comparing two approaches to questionnaire design: official government versus public information design, *Proceedings of the American Statistical Association.* Alexandria, VA: ASA, pp. 336–341.

Lessler, J., and Forsyth, B., 1996, A coding system for appraising questionnaires, in N. Schwarz and S. Sudman (eds.), *Answering Questions. Methodology for Determining Cognitive and Communicative Processes in Survey Research.* San Francisco: Jossey-Bass.

Lessler, J., Tourangeau, R., and Salter, W., 1989, Questionnaire design research in the cognitive research laboratory, *Vital and Health Statistics*, Series 6, No. 1, DHHS Publication PHS-89-1076. Washington, DC: U.S. Government Printing Office.

Lessler, J., Caspar, R., Penne, M., and Barker, P., 1993, Developing computer assisted interviewing (CAI) for the National Household Survey on Drug Abuse, *Journal of Drug Issues* 30: 9–34.

Lessler, J., Weeks, M., and O'Reilly, J., 1994, Results from the National Survey of Family Growth Cycle V Pretest, *Proceedings of the ASA Section on Survey Methods Research*. Alexandria, VA: American Statistical Association.

Levi, M., and Conrad, F., 1995, A heuristic evaluation of a World Wide Web prototype, *Interactions* 3: 4.

Levine, R., and Huberman, M., 2002, What types of survey items can elicit valid responses from fourth and eighth grade students, presented at the International Conference on Questionnaire Development, Evaluation, and Testing, Charleston, SC; see also *http://nces.ed.gov/pubsearch/pubsinfo.asp?pubid=200119* and *http://nces.ed.gov/pubsearch/pubsinfo.asp?pubid=200206*, both retrieved September 26, 2002.

Levinsohn, J., and Rodriguez, G., 2001, Automated testing of Blaise questionnaires, presented at the 7th International Blaise Users Conference, Washington, DC, September.

Li, R., McCardle, P., Clark, R., Kinsella, K., and Berch, D., 2001, *Diverse voices—Inclusion of Language-Minority Populations in National Studies: Challenges and Opportunities*. Bethesda, MD: National Institute on Aging and National Institute of Child Health and Human Development.

Lodge, M., 1981, *Magnitude Scaling: Quantitative Measurement of Opinions*. Thousand Oaks, CA: Sage.

Lodge, M., and Tursky, B., 1979, Comparisons between category and magnitude scaling of political opinion employing SRC/CPS items, *American Political Science Review* 73: 50–66.

Lodge, M., and Tursky, B., 1981, On the magnitude scaling of political opinion in survey research, *American Journal of Political Science* 25: 376–419.

Lodge, M., and Tursky, B., 1982, The social–psychological scaling of political opinion, in B. Wegener (ed.), *Social Attitudes and Psychophysical Measurement*. Mahwah, NJ: Lawrence Erlbaum.

Lodge, M., Cross, D., Tursky, B., and Tanenhaus, J., 1975, The psychological scaling and validation of a political support scale, *American Journal of Political Science* 19: 611–649.

Lodge, M., Cross, D., Tursky, B., Tanenhaus, J., and Reeder, R., 1976a, The psychophysical scaling of political support in the "real world," *Political Methodology* 3: 159–182.

Lodge, M., Tanenhaus, J., Cross, D., Tursky, B., Foley, M., and Foley, H., 1976b, The calibration and cross-modal validation of ratio scales of political opinion in survey research, *Social Science Research* 5: 325–347.

Loftus, E., 1984, Protocol analysis of responses to survey recall questions, in T. Jabine, M. Straf, J. Tanur, and R. Tourangeau (eds.), *Cognitive Aspects of Survey Methodology: Building a Bridge between Disciplines*. Washington, DC: National Academy Press.

Loftus, E., and Marburger, W., 1983, Since the eruption of Mt. St. Helens, has anyone beaten you up? Improving the accuracy of retrospective reports with landmark events, *Memory and Cognition* 114–120.

Lord, F., 1980, *Applications of Item Response Theory to Practical Testing Problems*. Mahwah, NJ: Lawrence Erlbaum.

Lord, F., and Novick, M., 1968, *Statistical Theories of Mental Test Scores*. Reading, MA: Addison-Wesley.

Louviere, J., 1991, Experimental choice analysis: introduction and overview, *Journal of Business Research* 23: 291–297.

Louviere, J., Hensher, D., and Swait, J., 2001, *Stated Choice Methods: Analysis and Applications*. Cambridge: Cambridge University Press.

Luce, R., 1986, *Response Times: Their Role in Inferring Elementary Mental Organization*. New York: Oxford University Press.

Luce, R., Smelser, N., and Gerstein, D., 1989, *Leading Edges in Social and Behavioral Science*. New York: Russell Sage.

Lupi, F., Kaplowitz, M., and Hoehn, J., 2002, The economic equivalency of drained and restored wetlands in Michigan, *American Journal of Agricultural Economics* 84: 1355–1361.

Lyberg, L., Biemer, P., Collins, M., de Leeuw, E., Dippo, C., Schwarz, N., and Trewin, D. (eds.), 1997, *Survey Measurement and Process Quality*. New York: Wiley.

Maccoby, E., and Maccoby, N., 1954, The interview: a tool of social science, in G. Lindzey (ed.), Handbook of Social Psychology, Vol. I, Theory and Method. Reading, MA: Addison-Wesley.

MacIntosh, R., 1998a, A confirmatory factor analysis of the affect balance scale in 38 nations: a research note, *Social Psychology Quarterly* 61: 83–91.

MacIntosh, R., 1998b, Global attitude measurement: an assessment of the World Values Survey Postmaterialism Scale, *American Sociological Review* 63: 452–464.

Mackenzie, J., 1993, A comparison of contingent preference models, *American Journal of Agricultural Economics* 75: 593–603.

MacKuen, M., and Turner, C., 1984, The popularity of presidents, 1963–1980, in C. Turner and E. Martin (eds.), *Surveying Subjective Phenomena*. New York: Russell Sage.

MacLachlan, J., Czepiel, J., and LaBarbera, P., 1979, Implementation of response latency measures, *Journal of Marketing Research* 16(4): 573–577.

Magliano, J., and Graesser, A., 1991, A three-pronged method for studying inference generation in literary text, *Poetics* 20: 193–232.

Manfreda, K., Vehovar, V., and Batagelj, Z., 2001, Web versus mail questionnaire for institutional surveys, presented at "The Challenge of the Internet," organized by the Association for Survey Computing, Latimer Conference Centre, Chesheem, England.

Mangione, T., 1995, *Mail Surveys: Improving the Quality*. Thousand Oaks, CA: Sage.

Mangione, T., Fowler, F., and Louis, T., 1992, A question characteristics and interviewer effects, *Journal of Official Statistics* 8: 293–307.

Market Research Society, 2003, Guidelines for research among children and young people. Retrieved January 17, 2003, at *http://www.mrs.org.uk/fr-code.htm*.

Markopoulos, P., and Bekker, M., 2002, How to compare usability testing methods with children participants, in M. Bekker, P. Markopoulos, and M. Kersten-Tsikalkina (eds.), *Interaction Design and Children: Proceedings of the International Workshop Interaction Design and Children*, August. Eindhoven/Maastricht, the Netherlands: Shaker Publishing, pp. 153–158.

Marquis, K., 1978, Inferring health interview response bias from imperfect record checks, *Proceedings of the ASA Section on Survey Research Methods*. Alexandria, VA: American Statistical Association, pp. 265–270.

Marquis, K., and Cannell, C., 1969, *A Study of Interviewer–Respondent Interaction in the Urban Employment Survey*. Ann Arbor, MI: Survey Research Center, Institute for Social Research, University of Michigan.

Marquis, K., Cannell, C., and Laurent, A., 1972, Reporting health events in household interviews: effects of reinforcement, question length, and reinterviews, *Vital and Health Statistics*, Series 2, No. 45, Data Evaluation and Methods Research. Rockville, MD: U.S. Department of Health and Human Services.

Marsh, H., and Bailey, M., 1991, Confirmatory factor analyses of multitrait–multimethod data: a comparison of alternative models, *Applied Psychological Measurement* 15: 47–70.

Martin, E., 1983, Surveys as social indicators: problems in monitoring trends, in P. Rossi, J. Wright, and A. Anderson (eds.), *Handbook of Survey Research*. San Diego, CA: Academic Press.

Martin, E., 1987, Some conceptual problems in the Current Population Survey, *Proceedings of the ASA Section on Survey Research Methods*. Alexandria, VA: American Statistical Association.

Martin, E., 2001, Privacy concerns and the census long form: some evidence from Census 2000, *Proceedings of the ASA Section on Survey Research Methods*. Alexandria, VA: American Statistical Association.

Martin, E., and Polivka, A., 1995, Diagnostics for redesigning questionnaires: measuring work in the Current Population Survey, *Public Opinion Quarterly* 59: 547–567.

Martin, E., Groves, R., Matlin, J., and Miller, C., 1986, *Report on the Development of Alternative Screening Procedures for the National Crime Survey*. Washington, DC: Bureau of Social Science Research.

Martin, E., Campanelli, P., and Fay, R., 1991, An application of Rasch analysis to questionnaire design: using vignettes to study the meaning of "Work" in the Current Population Survey, *The Statistician* 40: 265–276.

Martin, E., Hess, J., and Siegel, P., 1993, An empirical examination of the meaning of work, unpublished paper, U.S. Bureau of the Census, November 6.

Martin, E., Hess, J., and Siegel, P., 1995, Some effects of gender on the meaning of "Work": an empirical examination, in R. Simpson and I. Simpson (eds.), *Research in the Sociology of Work*, Vol. 5. Greenwich, CT: JAI Press.

Martin, E., Schechter, S., and Tucker, C., 1999, Interagency collaboration among the cognitive laboratories: past efforts and future opportunities, in *1998 Seminar on Interagency Coordination and Cooperation, Statistical Policy Working Paper 28*. Washington, DC: Federal Committee on Statistical Methodology, Office of Management and Budget, pp. 359–387.

Mathiowetz, N., 1998, Respondent expressions of uncertainty: data source for imputation, *Public Opinion Quarterly* 62: 47–56.

Mathiowetz, N., and Cannell, C., 1980, Coding interviewer behavior as a method of evaluating performance, *Proceedings of the ASA Section on Survey Research Methods*. Alexandria, VA: American Statistical Association.

Mathiowetz, N., and Dipko, S., 2000, A comparison of response error by adolescents and adults, *Medical Care* 38: 374–382.

Mathiowetz, N., and McGonagle, K., 2000, An assessment of the current state of dependent interviewing in household surveys, *Journal of Official Statistics* 16: 401–418.

Matlin, M., 1994, *Cognition*. Forth Worth, TX: Harcourt Brace.

Maxwell, J., 1996, *Qualitative Research Design: An Interactive Approach*. Thousand Oaks, CA: Sage.

McBeth, N., Pitts, S., and Johnston, S., 2001, Statistics New Zealand: recent developments in electronic data collection, *Proceedings of the Statistics Canada Symposium 2001: Achieving Data Quality in a Statistical Agency: A Methodological Perspective*.

McCabe, T., and Watson, A., 1994, Software complexity, *Crosstalk*. Retrieved at *http://www.stsc.hill.af.mil/crosstalk/1994/dec/complex.asp*, December.

McCarthy, J., 2001, Using respondent requests for help to develop quality data collection instruments: the 2000 Census of Agriculture Content Test, *Proceedings of the Annual Meeting of the American Statistical Association*, Alexandria, VA.

McCombs, M., and Reynolds, A., 1999, *The Poll with a Human Face: The National Issues Convention Experiment in Political Communication*. Mahwah, NJ: Lawrence Erlbaum.

McGorry, S., 2000, Measurement in cross-cultural environment: survey translation issues, *Qualitative Market Research* 3: 74–81.

McKay, R., Breslow, M., Sangster, R., Gabbard, S., Reynolds, R., Nakamoto, J., and Tarnai, J., 1996, Translating survey questionnaires: lessons learned, *New Directions for Evaluation* 70: 93–105.

McKinley, R., and Mills, C., 1985, A comparison of several goodness-of-fit statistics, *Applied Psychological Measurement* 9: 49–57.

McLachlan, G., and Peel, D., 2000, *Finite mixture models*. New York: Wiley

Medical Outcomes Trust, 1991, *Improving Medical Outcomes from the Patient's Point of View*. Boston: MOT.

Meeks, R., Lanier, A., Burrelli, J., and Fecso, R., 1998, Web-based data collection in NSF surveys, presented at the Joint Statistical Meetings of the American Statistical Association, Dallas, TX.

Meijer, R., and Sijtsma, K., 1995, Detection of aberrant item score patterns: a review of recent developments, *Applied Measurement in Education* 8: 261–272.

Mellenbergh, G., 1994, A unidimensional latent trait model for continuous item responses, *Multivariate Behavioral Research* 29: 223–236.

Memon, A., and Koehnken, G., 1992, Helping witnesses to remember more: the cognitive interview, *Expert Evidence: The International Digest of Human Behavior, Science, and Law* 1(2): 39–48.

Memon, A., Holley, A., Wark, L., Bull, R., and Koehnken, G., 1996, Reducing suggestibility in child witness interviews, *Applied Cognitive Psychology* 10: 503–518.

Menon, G., 1993, The effects of accessibility of information on judgments of behavioral frequencies, *Journal of Consumer Research* 20: 431–460.

Menon, G., and Yorkston, E., 2000, The use of memory and contextual cues in the formation of behavioral frequency judgments, in A. Stone, J. Turkkan, C. Bachrach, J. Jobe, H. Kurtzman, and V. Cain (eds.), *The Science of Self-Report: Implications for Research and Practice*. Mahwah, NJ: Lawrence Erlbaum.

Mieczkowski, T., 1991, The accuracy of self-reported drug use: an evaluation and analysis of new data, in R. Weisheit (ed.), *Drugs, Crime and the Criminal Justice System*. Cincinnati, OH: Anderson Publishing and the ACJS, pp. 275–302.

Miller, E., and Davis, W., 1994, Findings from the cognitive and field interview research on questions about "proof of paternity," unpublished report, U.S. Bureau of the Census, March 18.

Miller, J., Slomczynski, K., and Schoenberg, R., 1981, Assessing comparability of measurement in cross-national sociocultural settings, *Social Psychology Quarterly* 44: 178–191.

Mitchell, R., and Carson, R., 1989, *Using Surveys to Value Public Goods: The Contingent Valuation Method*. Washington, DC: Resources for the Future.

Mohler, P., Smith, T., and Harkness, J., 1998, Respondent's ratings of expressions from response scales: a two-country, two-language investigation on equivalence and translation, in J. Harkness (ed.), *Cross-Cultural Survey Equivalence*, Nachrichten Spezial 3. Mannheim, Germany: ZUMA.

Mohler, P. Ph., and Uher, R., 2003, Documenting comparative surveys for secondary analysis, in J. Harkness, F. Van de Vijver, and P. Ph. Mohler (eds.), *Cross-Cultural Survey Methods*. Hoboken, NJ: Wiley.

Molenaar, N., 1986, *Formleringseffecten in survey-interviews* [Formulation Effects in Survey Interviews]. Amsterdam: Vrije Universiteit Uitgeverij.

Monsour, N., 1985, Evaluation of the 1977 Economic Censuses of the United States, *Journal of Official Statistics* 1: 331–350.

Monsour, N., and Wolter, K., 1989, Evaluation of economic censuses at the United States Bureau of the Census, *Proceedings of the International Statistical Institute*, pp. 517–535.

Montalbán, P., 2002, Establishing a translation unit in a survey research organization, presented at the Annual Conference of the American Association for Public Opinion Research, St. Petersburg Beach, FL.

Moore, J., 1996, Person- vs. topic-based design for computer-assisted household survey instruments, presented at the International Conference on Computer-Assisted Survey Information Collection, San Antonio, TX.

Moore, J., 2001, Asset ownership screening procedures in the SIPP methods panel, *Proceedings of the ASA Section on Government Statistics*. Alexandria, VA: American Statistical Association.

Moore, J., and Marquis, K., 1989, Using administrative record data to evaluate the quality of survey estimates, *Survey Methodology* 15: 129–143.

Moore, J., and Moyer, L., 1998a, Questionnaire design effects on interview outcomes, *Proceedings of the ASA Section on Survey Research Methods*. Alexandria, VA: American Statistical Association, pp. 851–856.

Moore, J., and Moyer, L., 1998b, ACS/CATI person-based/topic-based field experiment, unpublished report, U.S. Bureau of the Census Bureau, July 29. (Also available as report #2002-04 in the Statistical Research Division's Research Report Series, *http://www.census.gov*).

Moore, J., Marquis, K., and Bogen, K., 1996, The SIPP Cognitive Research Evaluation Experiment: basic results and documentation, unpublished report, U.S. Bureau of the Census, January 11.

Moore, J., Stinson, L., and Welniak, E., Jr., 2000, Income measurement error in surveys: a review, *Journal of Official Statistics* 16: 331–361.

Morales, L., Reise, S., and Hays, R., 2000, Evaluating the equivalence of health care ratings by whites and Hispanics, *Medical Care* 38(5): 517–527.

Morgan, D., 1997, *Focus Groups as Qualitative Research*. Thousand Oaks, CA: Sage.

Morgan, M., Gibbs, S., Maxwell, K., and Britten, N., 2002, Hearing children's voices: methodological issues in conducting focus groups with children aged 7–11 years, *Qualitative Research* 2: 5–20.

Morrison, R., Stettler, K., and Anderson, A., 2002, Using vignettes in cognitive research on establishment surveys, presented at the International Conference on Questionnaire Development, Evaluation, and Testing Methods, Charleston, SC.

Morton-Williams, J., 1979, The use of "verbal interaction coding" for evaluating a questionnaire, *Quality and Quantity* 13: 59–75.

Morton-Williams, J., and Sykes, W., 1984, The use of interaction coding and follow-up interviews to investigate comprehension of survey questions, *Journal of Market Research Society* 2: 109–127.

Moser, C., and Kalton, G., 1971, *Survey Methods in Social Investigation*, 2nd ed. London: Heinemann.

Mosher, W., Pratt, W., and Duffer, A., 1994, CAPI, event histories, and incentives in the NSFG Cycle 5 pretest, *Proceedings of the ASA Section on Survey Methods Research*. Alexandria, VA: American Statistical Association.

Mosley, D., 1993, *Handbook of MIS Application Software Testing: Methods, Techniques and Tools for Assuring Quality through Testing*. Upper Saddle River, NJ: Prentice Hall/Yourdon Press.

Mote, V., and Anderson, R., 1965, An investigation of the effect of misclassification on the properties of chi-square tests in the analysis of categorical data, *Biometrika* 52(1–2): 95–109.

Moy, L., and Stinson, L., 1999, Two sides of a single coin? Dimensions of change suggested in different settings, *Proceedings of the ASA Section on Survey Research Methods*. Alexandria, VA: American Statistical Association, pp. 44–53.

Moyer, L., Fansler, N., Lee, M., and Von Thurn, D., 1997, How do people answer income questions? *Proceedings of the ASA Section on Survey Research Methods*. Alexandria, VA: American Statistical Association.

MS Interactive, 2001, Report on seniors pretest of the CMS Bounceback/SiteMight Survey content, report to Center for Medicare and Medicaid Services. Washington, DC: CMS.

Mueller, C., and Phillips, M., 2000, The genesis of an establishment survey: research and development for the Job Openings and Labor Turnover Survey at the BLS, *Proceedings of the ASA Survey Research Methods Section*. Alexandria, VA: American Statistical Association.

Mueller, C., and Wohlford, J., 2000, Developing a new business survey: Job Openings and Labor Turnover Survey at the Bureau of Labor Statistics, *Proceedings of the ASA Section on Survey Research Methods*, Alexandria, VA: American Statistical Association, pp. 360–365.

Murphy, E., Nichols, E., Anderson, A., Harley, M., and Pressley, K., 2001, Building usability into electronic data-collection forms for economic censuses and surveys, *Proceedings of the Federal Committee on Statistical Methodology Research Conference*, Arlington, VA, pp. 113–122.

Musciano, C., and Kennedy, B., 1998, *HTML: The Definitive Guide*. Sebastopol, CA: O'Reilly.

Muthén, L., and Muthén, B., 1999, *Mplus: The Comprehensive Modeling Program for Applied Researchers User's Guide*. Los Angeles: Muthén.

National Research Council, 1995, Committee on Characterization of Wetlands, *Wetlands: Characteristics and Boundaries*. Washington, DC: National Academy Press.

National Research Council, 2001, Committee on Mitigating Wetland Loss, *Compensating for Wetland Losses under the Clean Water Act*. Washington, DC: National Academy Press.

Neisser, U., and Winograd, E., 1988, *Remembering Reconsidered: Ecological and Traditional Approaches to the Study of Memory*. Cambridge: Cambridge University Press.

Nelson, N., 1976, Comprehension of spoken language by normal children as a function of speaking rate, sentence difficulty, and listeners' age and sex, *Child Development* 47: 299–303.

Neter, J., and Waksberg, J., 1964, A study of response errors in expenditures data from household interviews, *Journal of the American Statistical Association* 59: 17–55.

Newby, M., Amin, S., Diamond, I., and Naved, R., 1998, Survey experience among women in Bangladesh, *American Behavioral Scientist* 42: 252–275.

Newell, A., and Simon, H., 1972, *Human Problem Solving*. Englewood Cliffs, NJ: Prentice Hall.

Newman, S., and Stegehuis, P., 2002, Configuration management and advanced testing methods for large, complex Blaise instruments, presented at the Workshop on Survey Automation, Washington, DC.

Nicholls, W., II, and Groves, R., 1986, The status of computer-assisted telephone interviewing: I. Introduction and impact on cost and timeliness of survey data, *Journal of Official Statistics* 2: 93–115.

Nicholls, W., II, Baker, R., and Martin, J., 1997, The effect of new data collection technologies on survey data quality, in L. Lyberg, P. Biemer, M. Collins, E. de Leeuw, C. Dippo, N. Schwarz, and D. Trewin (eds.), *Survey Measurement and Process Quality*. New York: Wiley.

Nicholls, W., II, Mesenbourg, T., Jr., Andrews, S., and de Leeuw, E., 2000, Use of new data collection methods in establishment surveys, *Proceedings of the 2nd International Conference on Establishment Surveys*. Alexandria, VA: American Statistical Association, pp. 373–382.

Nichols, E., and Sedivi, B., 1998, Economic data collection via the Web: a Census Bureau case study, *Proceedings of the ASA Section on Survey Research Methods*. Alexandria, VA: American Statistical Association.

Nichols, E., Tedesco, H., and King, R., 1998, *Results from Usability Testing of Possible Electronic Questionnaires for the 1998 Library Media Center Public School Questionnaire Field Test*. Human–Computer Interaction Memorandum Series 20. Washington, DC: U.S. Bureau of the Census.

Nichols, E., Willimack, D., and Sudman, S., 1999, Who are the reporters: a study of government data providers in large, multi-unit companies, presented at the Joint Statistical Meetings of the American Statistical Association, Baltimore, MD.

Nichols, E., Murphy, E., and Anderson, A., 2001a, *Results from Cognitive and Usability Testing of Edit Messages for the 2002 Economic Census (First Round)*, Human–Computer Interaction Report Series 39. Washington, DC: Statistical Research Division, U.S. Bureau of the Census.

Nichols, E., Murphy, E., and Anderson, A., 2001b, *Usability Testing Results of the 2002 Economic Census Prototype RT-44401*, Human–Computer Interaction Report Series 49. Washington, DC: Statistical Research Division, U.S. Bureau of the Census.

Nida, E., 1964, *Toward a Science of Translating*. Leiden, The Netherlands: E.J. Brill.

Nielsen, J., 1999, *Designing Web Usability*. Indianapolis, IN: New Riders Publishing.

Nielsen, J., and Mack, R., 1994, *Usability Inspection Methods*. New York: Wiley.

Nisbett, R., and Wilson, T., 1977, Telling more than we know: verbal reports on mental processes, *Psychological Review* 84: 231–259.

Norman, D., 1986, Cognitive engineering, in D. Norman and S. Draper (eds.), *User Centered System Design: New Perspectives on Human–Computer Interaction*. Mahwah, NJ: Lawrence Erlbaum.

Nunnally, J., and Bernstein, I., 1994, *Psychometric Theory*. New York: McGraw-Hill.

Nuyts, K., Waege, H., Loosvelts, G., and Bulliet, J., 1997, The application of cognitive interviewing techniques in the development and testing of measurement instruments for survey research, *Tijdschrift voor Sociologie* 18: 477–500.

O'Brien, E., 2000a, Respondent role as a factor in establishment survey response, *Proceedings of the 2nd International Conference on Establishment Surveys*. Alexandria, VA: American Statistical Association, pp. 1462–1467.

O'Brien, E., 2000b, *A Cognitive Appraisal Methodology for Establishment Survey Questionnaires*, Statistical Policy Working Paper 30, 1999 Federal Committee on Statistical Methodology Research Conference: Complete Proceedings (Part 1 of 2). Washington, DC: Office of Management and Budget, pp. 307–316.

O'Brien, E., Fisher, S., Goldenberg, K., and Rosen, R., 2001, Application of cognitive methods to an establishment survey: a demonstration using the Current Employment Statistics Survey, *Proceedings of the Annual Meeting of the American Statistical Association*.

Oksenberg, L., Cannell, C., and Kalton, G., 1991, New strategies for pretesting survey questions, *Journal of Official Statistics* 7: 349–356.

O'Muircheartaigh, C., 1991, Simple response variance: estimation and determinants, in P. Biemer, R. Groves, L. Lyberg, N. Mathiowetz, and S. Sudman (eds.), *Measurement Errors in Surveys*. New York: Wiley.

O'Muircheartaigh, C., Krosnick, J., and Helic, A., 1998, Middle alternatives, acquiescence, and the quality of questionnaire data, unpublished NORC report.

Opaluch, J., Swallow, S., Weaver, T., Wessels, C., and Wichlens, D., 1993, Evaluating impacts from noxious waste facilities: including public preferences in current siting mechanisms, *Journal of Environmental Economics and Management* 24: 41–59.

Oppeneer, M., and Luppes, M., 1998, Improving communication with providers of statistical information in business surveys, *Proceedings of the GSS Methodology Conference*, June 29. London: Government Statistical Service.

O'Reilly, J., Hubbard, M., Lessler, J., Biemer, P., and Turner, C., 1994, Audio and video computer assisted self-interviewing: preliminary tests of new technology for data collection, *Journal of Official Statistics* 10: 197–214.

Orlando, M., and Marshall, G., 2002, Differential item functioning in a Spanish translation of the PTSD checklist: detection and evaluation of impact, *Psychological Assessment* 14: 50–59.

Orlando, M., and Thissen, D., 2000, Likelihood-based item-fit indices for dichotomous item response theory models, *Applied Psychological Measurement* 24: 50–64.

Orlando, M., Sherbourne, C., and Thissen, D., 2000, Summed-score linking using item response theory: application to depression measurement, *Psychological Assessment* 12: 354–359.

Orren, G., 1978, Presidential popularity ratings: another view, *Public Opinion* 1(May/June): 35.

Ostrom, T., and Gannan, K., 1996, Exemplar generation: assessing how respondents give meaning to rating scales, in N. Schwarz and S. Sudman (eds.), *Answering Questions: Methodology for Determining Cognitive and Communicative Processes in Survey Research.* San Francisco: Jossey-Bass.

Otter, M., 1993, Leesvaardigheid, leesonderwijs en buitenschools lezen: instrumentatie en effecten [Reading Ability, Education in Reading, and Reading outside the School Setting: Development of Measurement Instruments and a Study of Effect Size]. Amsterdam: SCO, University of Amsterdam.

Paben, S., 1998, The reinterview program for the BLS compensation surveys, presented at the Joint Statistical Meetings of the American Statistical Association, Dallas, TX.

Palmisano, M., 1988, The application of cognitive survey methodology to an establishment survey field test, *Proceedings of the ASA Section on Survey Research Methods.* Alexandria, VA: American Statistical Association, pp. 179–184.

Panter, A., and Reeve, B., 2002, Assessing tobacco beliefs among youth using item response theory models, *Drug and Alcohol Dependence* 68: 821–839.

Panter, A., Swygert, K., Dahlstrom, W., and Tanaka, J. 1997, Factor analytic approaches to personality item-level data, *Journal of Personality Assessment* 68: 561–589.

Parent, G., and Jamieson, R., 2000, The use of CAI for the collection of business surveys in Statistics Canada, *Proceedings of the 2nd International Conference on Establishment Surveys.* Alexandria, VA: American Statistical Association, pp. 383–392.

Pascale, J., 2001, Labor force participation in the Survey of Income and Program Participation (SIPP) Methods Panel, *Proceedings of the ASA Section on Government Statistics.* Alexandria, VA: American Statistical Association.

Pasick, R., Sabogal, F., Bird, J., D'Onofrio, C., Jenkins, C., Lee, M., Engelstad, L., and Hiatt, R., 1996, Problems and progress in translation of health survey questions: the pathways experience, *Health Education Quarterly* 23: 28–40.

Patterson, B., Dayton, C., and Grubard, B., 2002, *Latent Class Analysis of Complex Sample Survey Data: Application to Dietary Data. Journal of the American Statistical Association* 97: 721–741.

Payne, J., 1994, Thinking aloud: insights into information processing, *Psychological Science* 5: 241–248.

Pennell, B., Pennell, S., Holland, L., and Dinkelmann, K., in preparation, *Translation Protocol for the World Mental Health Initiative* (working title).

Perneger, T., Leplege, A., and Etter, J., 1999, Cross-cultural adaptation of a psychometric instrument: two methods compared, *Journal of Clinical Epidemiology* 52: 1037–1046.

Perry, W., 1995, *Effective Methods for Software Testing.* New York: Wiley.

Phipps, P., 1990, Applying cognitive theory to an establishment mail survey, *Proceedings of the ASA Section on Survey Research Methods.* Alexandria, VA: American Statistical Association, pp. 608–612.

Phipps, P., Butani, S., and Chun, Y., 1995, Research on establishment-survey questionnaire design, *Journal of Business and Economic Statistics* 13: 337–346.

Piaget, J., 1929, *The Child's Conception of the World*. London: Routledge. Also, *Introduction to the Child's Conception of the World*. New York: Harcourt. Original text: *La causalité physique chez l'enfant; la representation de la monde chez l'enfant*. Paris, 1927.

Piaget, J., 1932/1965, *The Moral Judgment of the Child*. New York: Free Press.

Pierzchala, M., 2002, Practitioner needs, reactions to model based testing approaches, presented at the Workshop on Survey Automation, Washington, DC.

Pierzchala, M., and Manners, T., 1998, Producing CAI instruments for a program of surveys, in M. Couper, R. Baker, J. Bethlehem, C. Clark, J. Martin, W. Nicholls, and J. O'Reilly (eds.), *Computer-Assisted Survey Information Collection*. New York: Wiley.

Polivka, A., and Rothgeb, J., 1993, Redesigning the questionnaire for the Current Population Survey, prepared for presentation at the annual meeting of the American Economics Association, Anaheim, CA, January.

Potaka, L., and Cochrane, S., 2002, Developing bilingual questionnaires: experiences from New Zealand in the development of the 2001 Maori language survey, unpublished report, New Zealand Statistics.

Poterba, J., and Summers, L., 1995, Unemployment benefits and labor market transitions: a multinomial logit model with errors in classification, *Review of Economics and Statistics* 77: 207–216.

Presser, S., and Blair, J., 1994, Survey pretesting: do different methods produce different results? in P. Marsden (ed.), *Sociological Methodology 1994*. San Francisco: Jossey-Bass, pages 73–104.

Presser, S., and Zhao, S., 1992, Attributes of questions and interviewers as determinants of interviewing performance, *Public Opinion Quarterly* 56: 236–240.

Presser, S., Blair, J., Mack, K., Ryan, C., and van Dyne, M., 1993, *Final Report on the University of Maryland–USDA Cooperative Agreement to Improve Reporting for Children in the Continuing Survey of Food Intakes by Individuals*. College Park, MD: Survey Research Center of the University of Maryland.

Prieto, A., 1992, A method for translation of instruments to other languages, *Adult Education Quarterly* 43: 1–14.

Prüfer, J., and Rexroth, M., 1996, *Verfahren zur Evaluation von Survey-Fragen: Ein Überblick*, ZUMA-Arbeitsbericht 95/5. Mannheim, Germany: ZUMA.

Prüfer, P., and Rexroth, M., 1985, Zur Anwendung der Interaction-Coding Technik, *ZUMA-Nachrichten* 17: 2–49.

Przeworski, A., and Teune, H., 1966, Equivalence in cross-national research, *Public Opinion Quarterly* 30: 551–568.

Pullman, C., 2000, Sense and census: Census 2000, the largest direct-mail program in U.S. history, teaches lessons about clarity and process, *Critique*, Summer: 54–61.

Ralph, A., Williams, C., and Campisi, A., 1997, Measuring peer interactions using the adolescent social interaction profile, *Journal of Applied Developmental Psychology* 18: 71–86.

Ramírez, C., 2002, Strategies for subject matter expert review in questionnaire design, presented at the International Conference on Questionnaire Development, Evaluation, and Testing Methods, Charleston, SC.

Ramos, M., and Sweet, E., 1995, *Results from 1993 Company Organization Survey (COS) Computerized Self-Administered Questionnaire (CSAQ) Pilot Test*, Working Papers in

Survey Methodology SM95/20. Washington, DC: Statistical Research Division, U.S. Bureau of the Census.

Ramos, M., Sedivi, B., and Sweet, E., 1998, Computerized self-administered questionnaires, in M. Couper, R. Baker, J. Bethlehem, C. Clark, J. Martin, W. Nicholls II, and J. O'Reilly (eds.), *Computer-Assisted Survey Information Collection*. New York: Wiley.

Ramsay, J., 1997, A functional approach to modeling test data, in W. J. van der Linder and R. Hambleton (eds.), *Handbook of Modern Item Response Theory*. New York: Springer, pp. 381–394.

Rasch, G., 1960, *Probabilistic Models for Some Intelligence and Attainment Tests*. Copenhagen: Denmark's Paedagogishe Institut. (Republished in 1980, Chicago: University of Chicago Press.)

Rayner, K., 1992, *Eye Movements and Visual Cognition: Scene Perception and Reading*. New York: Springer-Verlag.

Reckase, M., 1979, Unifactor latent trait models applied to multifactor tests: Results and implications, *Journal of Educational Statistics* 4: 207–230.

Redline, C., and Dillman, D., 2002, The influence of alternative visual designs on respondents' performance with branching instructions in self-administered questionnaires, in R. Groves, D. Dillman, J. Eltinge, and R. Little (eds.), *Survey Nonresponse*. New York: Wiley.

Redline, C., and Lankford, C., 2001, Eye-movement analysis: a new tool for evaluating the design of visually administered instruments (paper and Web), *Proceedings of the ASA Section on Survey Research Methods*. Alexandria, VA: American Statistical Association.

Redline, C., Smiley, R., Lee, M., DeMaio, T., and Dillman, D. A., 1999, Beyond concurrent interviews: an evaluation of cognitive interviewing techniques for self-administered questionnaires, *Proceedings of the ASA Section on Survey Research Methods*. Alexandria, VA: American Statistical Association.

Redline, C., Dillman, D., Smiley, R., Carley-Baxter, L., and Jackson, A., 2001, Making visible the invisible: an experiment with skip instructions on paper questionnaires, *Proceedings of the ASA Section on Survey Research Methods*. Alexandria, VA: American Statistical Association.

Redline, C., Dillman, D., Dajani, A., and Scaggs, M., 2003, Improving navigational performance in U.S. Census 2000 by altering the visually administered languages of branching instructions, *Journal of Official Statistics* 19: 403–420.

Reise, S., 1999, Personality measurement issues viewed through the eyes of IRT, in S. Embretson and S. Hershberger (eds.), *The New Rules of Measurement: What Every Psychologist and Educator Should Know*. Mahwah, NJ: Lawrence Erlbaum, pp. 219–242.

Reise, S., in press, Item response theory and its applications for cancer outcomes measurement, in J. Lipscomb, C. Gotay, and C. Snyder (eds.), *Outcomes Assessment in Cancer*. Cambridge: Cambridge University Press.

Reise, S., and Waller, N., 1993, Traitedness and the assessment of response pattern scalability, *Journal of Personality and Social Psychology* 65: 143–151.

Reise, S., Widaman, K., and Pugh, R., 1993, Confirmatory factor analysis and item response theory: two approaches for exploring measurement invariance, *Psychological Bulletin* 114: 552–566.

Reynes, E., and Lorant, J., 2001, Do comparative martial arts attract aggressive children? *Perceptual and Motor Skills* 94: 21–25.

Riley, A., Rebok, G., Forrest, C., Robertson, J., Green, B., and Starfield, B., 2001, Young children's reports of their health: a cognitive testing study, in M. Cynamon and R. Kulka (eds.), *Proceedings of the 7th Conference on Health Survey Research Methods*. Hyattsville, MD: U.S. Department of Health and Human Services.

Rips, L., Conrad, F., and Fricker, S., 2002, Straightening out the seam effect in panel surveys, unpublished manuscript.

Rivière, P., 2002, What makes business statistics special? *International Statistical Review* 70: 145–159.

Rodgers, W., Andrews, F., and Herzog, A., 1992, Quality of survey measures: a structural modeling approach, *Journal of Official Statistics* 8: 251–275.

Roediger, H., and Neely, J., 1982, Retrieval blocks in episodic and semantic memory, *Canadian Journal of Psychology* 36: 213–242.

Rogers, T., 1976, Interviews by telephone and in person: quality of response and field performance, *Public Opinion Quarterly* 40: 51–65.

Rogers, H., and Hattie, J., 1987, A Monte Carlo investigation of several person and item-fit statistics for item response models, *Applied Psychological Measurement* 11: 47–57.

Rosaria, S., and Robinson, H., 2000, Applying models in your testing process, *Information and Software Technology* 42: 815–824.

Rosen, R., and O'Connell, D., 1997, Developing an integrated system for mixed mode data collection in a large monthly establishment survey, *Proceedings of the ASA Section on Survey Research Methods*. Alexandria, VA: American Statistical Association, pp. 198–203.

Rosen, R., Manning, C., and Harrell, L., 1998, Web-based data collection in the Current Employment Statistics Survey, presented at the Joint Statistical Meetings of the American Statistical Association, Dallas, TX.

Rosenthal, M., and Hubble, D., 1993, Results from the National Crime Victimization Survey (NCVS) CATI experiment, *Proceedings of the ASA Section on Survey Research Methods*. Alexandria, VA: American Statistical Association, pp. 742–747.

Rossi, P., and Anderson, A., 1982, The factorial survey approach: an introduction, in P. Rossi and S. Nock (eds.), *Measuring Social Judgments: The Factorial Survey Approach*. Thousand Oaks, CA: Sage.

Rossi, P., Waite, E., Bose, C., and Berk, R., 1974, The seriousness of crimes: normative structure and individual differences, *American Sociological Review* 39: 224–237.

Rothgeb, J., Polivka, A., Creighton, K., and Cohany, S., 1991, Development of the proposed revised Current Population Survey, *Proceedings of the ASA Section on Survey Research Methods*. Alexandria, VA: American Statistical Association, pp. 56–65.

Rothgeb, J., Willis, G., and Forsyth, B., 2001, Questionnaire pretesting methods: Do different techniques and different organizations produce similar results? *Proceedings of the ASA Section on Survey Research Methods*. Alexandria, VA: American Statistical Association.

Rothwell, N., 1983. New Ways of Learning how to improve Self-Enumerative Questionnaires: A Demonstration Project, Unpublished MD., U.S. Bureau of the Census.

Rothwell, N., 1985. Laboratory and field response research studies for the 1980 Census of Population in the United States, *Journal of Official Statistics* 1: 137–157.

Rowlands, O., Eldridge, J., and Williams, S., 2002, Expert review followed by interviews with editing staff: effective first steps in the testing process for business surveys, presented at the International Conference on Questionnaire Development, Evaluation, and Testing Methods, Charleston, SC.

Royston, P., 1989, Using intensive interviews to evaluate questions, in F. Fowler, Jr. (ed.), *Health Survey Research Methods*, DHHS Publication PHS 89-3447. Washington, DC: U.S. Government Printing Office, pp. 3–7.

Royston, P., and Bercini, D., 1987. Questionnaire design research in a laboratory setting: results of testing cancer risk factor questions, *Proceedings of the ASA Section on Survey Research Methods*. Alexandria, VA: American Statistical Association, pp. 829–833.

Rubey, L., and Lupi, F., 1997, Predicting the effects of market reform in Zimbabwe: a stated preference approach, *American Journal of Agricultural Economics* 78: 89–99.

Rubin, J., 1994, *Handbook of Usability Testing*. New York: Wiley.

Russo, J., Johnson, E., and Stephens, D., 1989, The validity of verbal protocols, *Memory and Cognition* 17: 759–769.

Rutchik, R., and Freedman, S., 2002, Establishments as respondents: Is conventional cognitive interviewing enough?" presented at the International Conference on Questionnaire Development, Evaluation, and Testing Methods, Charleston, SC.

Ryan, A., Chan, D., Ployhart, R., and Slade, L., 1999, Employee attitude surveys in a multinational organization: considering language and culture in assessing measurement equivalence, *Personnel Psychology* 52: 37–58.

Sacks, H., Schegloff, E., and Jefferson, G., 1974, Simplest systematics for the organization of turn taking for conversation, *Language* 50: 696–735.

Samejima, F., 1969, Estimation of latent ability using a response pattern of graded scores, *Psychometrika Monographs* 34(4), Pt. 2, Whole No. 17.

Sanchez, M., 1992, Effects of questionnaire design on the quality of survey data, *Public Opinion Quarterly* 56: 206–217.

Sanders, D., 1994, Methodological considerations in comparative cross-national research, *International Social Science Journal* 46: 513–521.

Saner, L., and Pressley, K., 2000, Assessing the usability of an electronic establishment survey instrument in a natural use context, *Proceedings of the 2nd International Conference on Establishment Surveys*. Alexandria, VA: American Statistical Association, pp. 1634–1639.

Sardenberg, A., and Gloster, J., 2001, Testing a production Blaise computer-assisted telephone (CATI) instrument, presented at the 7th International Blaise Users Conference, Washington, DC, September.

Saris, W., 1990, The choice of a model for evaluation of measurement instruments, in W. Saris and A. van Meurs (eds.), *Evaluation of Measurement Instruments by Meta-analysis of Multitrait–Multimethod Studies*. Amsterdam: North-Holland.

Saris, W., 1998, The effects of measurement error in cross-cultural research, in J. Harkness (ed.), *Cross-Cultural Survey Equivalence*, Nachrichten Spezial 3. Mannheim, Germany: ZUMA.

Saris, W., 2003, Response function equality, in J. Harkness, F. van de Vijver, and P. Mohler (eds.), *Cross-Cultural Survey Methods*. New York: Wiley.

Saris, W., and Andrews, F., 1991, Evaluation of measurement instruments using a structural modeling approach, in P. Biemer, R. Groves, L. Lyberg, N. Mathiowetz and S. Sudman (eds.), *Measurement Errors in Surveys*. New York: Wiley.

Saris, W., and Gallhofer, I., 1998, Classificatie van surveyvragen, Tijdschrift voor communicatie wetenschap, 96–122.

Saris, W., and Gallhofer, I., 2003, Factors determining the reliability and validity of survey questions: a meta-analysis of MTMM studies, unpublished manuscript.

Sasaki, M., 1995, Research design of cross-national attitude surveys, *Behaviormetrika* 22: 99–114.

Sawyer, S., and Dillman, D., 2002, *How Graphical, Numerical, and Verbal Languages Affect the Completion of the Gallup Q-12 on Self-Administered Questionnaires: Results from 22 Cognitive Interviews and a Field Experiment*, Technical Report 02–26. Pullman, WA: Social and Economic Sciences Research Center, Washington State University.

Saywitz, K., 1987, Children's testimony: age related patterns of memory errors, in S. Ceci, M. Toglia, and D. Ross (eds.), *Children's Eyewitness Memory*. New York: Springer.

Schaeffer, N. C., 1991a, Conversation with a purpose—or conversation? Interaction in the standardized interview, in P. Biemer, R. Groves, L. Lyberg, N. Mathiowetz, and S. Sudman (eds.), *Measurement Errors in Surveys*. New York: Wiley.

Schaeffer, N. C., 1991b, Hardly Ever or Constantly? Group Comparisons Using Vague Quantifiers, *Public Opinion Quarterly*, 55: 395–423.

Schaeffer, N. C., and Maynard, D., 1996, From paradigm to prototype and back again: interactive aspects of cognitive processing in standardized survey interviews, in N. Schwarz and S. Sudman (eds.), *Answering Questions*. San Francisco: Jossey-Bass.

Schaeffer, N. C., and Maynard, D., 2002, Occasions for intervention: interactional resources for comprehension in standardized survey interviews, in D. Maynard, H. Houtkoop-Steenstra, N. C. Schaeffer, and J. van der Zouwen (eds.), *Standardization and Tacit Knowledge: Interaction and Practice in the Survey Interview*. New York: Wiley.

Schaeffer, N. C., and Presser, S., 2003, The science of asking questions, *Annual Review of Sociology* 29: 65–88.

Schaeffer, N. C., and Thomson, E., 1992, The discovery of grounded uncertainty: developing standardized questions about strength of fertility motivations, in P. Marsden (ed.), *Sociological Methodology* 22: 37–82.

Schechter, S., Stinson, L., and Moy, L., 1999, *Developing and Testing Aggregate Reporting Forms for Data on Race and Ethnicity*, Statistical Policy Working Paper 30, Federal Committee on Statistical Methodology Research Conference: Complete Proceedings (Part 1 of 2). Washington, DC: Office of Management and Budget, pp. 337–346.

Schegloff, E., 1996, Turn organization: one intersection of grammar and interaction, in E. Ochs, E. Schegloff, and S. Thompson (eds.), *Interaction and Grammar*, Cambridge: Cambridge University Press.

Schegloff, E., 2002, Survey interviews as talk-in-interaction, in D. Maynard, H. Houtkoop-Steenstra, N. Schaeffer, and J. van der Zouwen (eds.), *Standardization and Tacit Knowledge: Interaction and Practice in the Survey Interview*. New York: Wiley.

Scherpenzeel, A., 1995, *A Question of Quality: Evaluating Survey Questions by Multitrait–Multimethod Studies*. Amsterdam: University of Amsterdam Press.

Scherpenzeel, A., and Saris, W., 1997, The validity and reliability of survey questions: a meta-analysis of MTMM studies, *Sociological Methods and Research* 25: 344–383.

Scheuch, E., 1989, Theoretical implications of comparative survey research: why the wheel of cross-cultural methodology keeps on being reinvented, *International Sociology* 4: 147–167.

Schnell, R., and Kreuter, F., 2002, New software tools for questionnaire development documentation, presented at the International Conference on Questionnaire Development, Evaluation, and Testing, Charleston, SC.

Schober, M., and Conrad, F., 2002, A collaborative view of standardized survey interviews, in D. Maynard, H. Houtkoop-Steenstra, N. Schaeffer, and J. van der Zouwen (eds.), *Standardization and Tacit Knowledge: Interaction and Practice in the Survey Interview*. New York: Wiley.

Schooler, J., and Engstler-Schooler, T., 1990, Verbal overshadowing of visual memories: some things are better left unsaid, *Cognitive Psychology* 22: 36–71.

Schooler, J., Ohlsson, S., and Brooks, K., 1993, Thoughts beyond words: when language overshadows insight, *Journal of Experimental Psychology: General* 122: 166–183.

Schooler, C., Diakite, C., Vogel, J., Mounkoro, P., and Caplan, L., 1998, Conducting a complex sociological survey in rural Mali: three points of view, *American Behavioral Scientist* 42: 252–275.

Schoua-Glusberg, A., 1988, A focus-group approach to translating questionnaire items, presented at the Annual Conference of the American Association for Public Opinion Research, Toronto, Ontario, Canada.

Schoua-Glusberg, A., 1989, The Spanish version of the NHSB Questionnaire: Translation Issues. Paper commissioned under Order # 263 MD 835040 for presentation at the NIH AIDS and Sexual Behavior Change Conference.

Schoua-Glusberg, A., 1992, *Report on the Translation of the Questionnaire for the National Treatment Improvement Evaluation Study*. Chicago: National Opinion Research Center.

Schoua-Glusberg, A., 1997, Survey instrument translation: a test comparing the committee approach and back translation, presented at the Annual Meeting of the American Association for Public Opinion Research, Norfolk, VA.

Schuman, H., 1966, The random probe: a technique for evaluating the validity of closed questions, *American Sociological Review* 31: 218–222.

Schuman, H., and Presser, S., 1981, *Questions and Answers in Attitude Surveys: Experiments in Question Form, Wording, and Context*. New York: Academic Press.

Schuman, H., Steeh, C., Bobo, L., and Krysan, M., 1997, *Racial Attitudes in America: Trends and Interpretations*. Cambridge, MA: Harvard University Press.

Schwarz, N., 1997, Cognition, communication, and survey measurement, in R. Kopp, W. Pommerehne, and N. Schwarz (eds.), *Determining the Value of Non-marketed Goods*. Boston: Kluwer-Nijhoff.

Schwarz, N., 1999a, Self-reports: how the questions shape the answer, *American Psychologist* 54: 93–105.

Schwarz, N., 1999b, Cognitive research into survey measurement: its influence on survey methodology and cognitive theory, in M. Sirken, D. Herrmann, S. Schechter, N. Schwarz, J. Tanur, and R. Tourangeau (eds.), *Cognition and Survey Research*. New York: Wiley.

Schwarz, N., 2003, Culture-sensitive context effects: a challenge for cross-cultural surveys, in J. Harkness, F. van de Vijver, and P. Mohler (eds.), *Cross-Cultural Survey Methods*. New York: Wiley.

Schwarz, N., and Hippler, H., 1995, The numeric values of rating scales: a comparison of their impact in mail surveys and telephone interviews, *International Journal of Public Opinion Research* 7: 72–74.

Schwarz, N., Grayson, C., and Knaüper, B., 1998, Formal features of rating scales and the interpretation of question meaning, *International Journal of Public Opinion Research* 10: 177–183.

Schwarz, N., Park, D., Knäuper, B., and Sudman, S., 1999, *Cognition, Aging, and Self-Reports*. Philadelphia: Psychology Press.

Schwede, L., and Ellis, Y., 1997, *Children in Custody Questionnaire Redesign Project Final Report: Results of the Split-Panel and Feasibility Tests*, prepared by the U.S. Bureau of the Census for the Department of Justice.

Schwede, L., and Gallagher, C., 1996, *Children in Custody Questionnaire Redesign Project: Results from Phase 3 Cognitive Testing of the Roster Questionnaire*, prepared by the U.S. Bureau of the Census for the Department of Justice.

Schwede, L., and Moyer, L., 1996, *Children in Custody Questionnaire Redesign Project: Results from Phase 2 Questionnaire Development and Testing*, prepared by the U.S. Bureau of the Census for the Department of Justice.

Schwede, L., and Ott, K., 1995, *Children in Custody Questionnaire Redesign Project: Results from Phase 1*, prepared by the U.S. Bureau of the Census for the Department of Justice.

Scollon, R., and Scollon, S., 1995, *Intercultural Communication: A Discourse Approach*. Oxford: Blackwell.

Scott, J., 1997, Children as respondents: methods for improving data quality, in L. Lyberg, P. Biemer, M. Collins, E. de Leeuw, C. Dippo, N. Schwarz, and D. Trewin (eds.), *Survey Measurement and Process Quality*. New York: Wiley.

Scott, J., Brynin, M., and Smith, R., 1995, Interviewing children in the British household panel survey, in J. Hox, B. van der Meulen, J. Janssens, J. ter Laak, and L. Tavecchio (eds.), *Advances in Family Research*. Amsterdam: Thesis.

Sears, D., 1986, College sophomores in the laboratory: influences of a narrow data base on social psychology's view of human nature, *Journal of Personality and Social Psychology* 51: 515–530.

Sedivi, B., Nichols, E., and Kanarek, H., 2000, Web-based collection of economic data at the U.S. Census Bureau, *Proceedings of the 2nd International Conference on Establishment Surveys*. Alexandria, VA: American Statistical Association, pp. 459–468.

Selman, R., 1980, *The Growth of Interpersonal Understanding*. San Diego, CA: Academic Press.

Shaft, T., 1997, Responses to comprehension questions and verbal protocols as measures of computer program comprehension processes, *Behavior and Information Technology* 16: 320–336.

Sheatsley, P., 1983, Questionnaire construction and item writing, in P. Rossi, J. Wright, and A. Anderson (eds.), *Handbook of Survey Research*. San Diego, CA: Academic Press.

Shiffrin, R., and Schneider, W., 1977, Controlled and automatic human information processing: II. Perceptual learning, automatic attending and a general theory, *Psychological Review* 84: 127–190.

Sigelman, L., 1990, Answering the 1,000,000-person question: the measurement and meaning of presidential popularity, *Research in Micropolitics* 3: 209–226.

Sinaiko, H., and Brislin, R., 1973, Evaluating language translations: experiments on three assessment methods, *Journal of Applied Psychology* 57: 328–334.

Sinclair, M., and Gastwirth, J., 1996, On procedures for evaluating the effectiveness of reinterview survey methods: application to labor force data, *Journal of the American Statistical Association* 91: 961–969.

Sireci, S., 1999, Guidelines for adapting certification tests for use across multiple languages, *PES News* XIX: 2; *http://www.cesb.org/Guidelines%20for%20Adapting.htm.*

Sirken, M., and Schechter, S., 1999, Interdisciplinary survey methods research, in M. Sirken, D. Herrmann, S. Schechter, N. Schwarz, J. Tanur, and R. Tourangeau, (eds.), *Cognition and Survey Research*. New York: Wiley.

Sirken, M., Herrmann, D., Schechter, S., Schwarz, N., Tanur, J., and Tourangeau, R., 1999a, *Cognition and Survey Research*. New York: Wiley.

Sirken, M., Jabine, T., Willis, G., Martin, E., and Tucker, C. (eds.), 1999b. *A New Agenda for Interdisciplinary Research: Proceedings of the CASM II Seminar*. Hyattsville, MD: National Center for Health Statistics.

Sletto, R., 1950. Pretesting of questionnaires, *American Sociological Review* 5: 193–200.

Slovic, P., and Lichtenstein, S., 1971, Comparison of Bayesian and regression approaches to the study of information processing in judgment, *Organizational Behavior and Human Performance* 6: 649–744.

Smith, E., and Miller, F., 1978, Limits on perception of cognitive processes: a reply to Nisbett and Wilson, *Psychological Review* 85: 355–362.

Smith, S., 2000, How can we improve the training of cognitive interviewers? prepared for presentation at the Annual Meetings of the American Association for Public Opinion Research.

Smith, T., 1979, Happiness: time trends, seasonal variations, intersurvey differences, and other mysteries, *Social Psychology Quarterly* 42: 18–30.

Smith, T., 1982, Educated don't knows: an analysis of the relationship between education and item nonresponse, *Political Methodology* 8: 47–57.

Smith, T., 1984, Nonattitudes: a review and evaluation, in C. Turner and E. Martin (eds.), *Surveying Subjective Phenomenon*. New York: Russell Sage.

Smith, T., 1988a, *Rotation Design of the GSS*, GSS Methodological Report 52. Chicago: National Opinion Research Center.

Smith, T., 1988b, *The Ups and Downs of Cross-National Survey Research*, GSS Cross-National Report 8. Chicago: National Opinion Research Center.

Smith, T., 1989, Random probes of GSS questions, *International Journal of Public Opinion Research* 1: 305–325.

Smith, T., 1991a, *An Analysis of Missing Income Information on the General Social Survey*, GSS Methodological Report 71. Chicago: National Opinion Research Center.

Smith, T., 1991b, Thoughts on the nature of context effects, in N. Schwarz and S. Sudman (eds.), *Context Effects in Social and Psychological Research*. New York: Springer-Verlag.

Smith, T., 1994, An analysis of response patterns to the ten-point scalometer, *Proceedings of the ASA Section on Survey Research Methods*. Alexandria, VA: American Statistical Association.

Smith, T., 1995, Little things matter: a sampler of how differences in questionnaire format can affect survey responses, *Proceedings of the ASA Section on Survey Research Methods*. Alexandria, VA: American Statistical Association.

Smith, T., 1996, *Environmental and Scientific Knowledge around the World*, GSS Cross-National Report 16. Chicago: National Opinion Research Center.

Smith, T., 1997, *Improving Cross-National Survey Response by Measuring the Intensity of Response Categories*, GSS Cross-National Report 17. Chicago: National Opinion Research Center.

Snijkers, G., 1997, Computer-assisted qualitative interviewing: a method for cognitive pretesting of computerized questionnaires, *Bulletin de Methodologie Sociologique* 55: 93–107.

Snijkers, G., 2002, Cognitive laboratory experiences: on pretesting, computerised questionnaires and data quality, Ph.D. dissertation, University of Utrecht.

Snijkers, G., and Luppes, M., 2000, The best of two worlds: total design method and new Kontiv design of an operational model to improve respondent co-operation, *Proceedings of the 2nd International Conference on Establishment Surveys*. Alexandria, VA: American Statistical Association, pp. 361–372.

Solano-Flores, G., and Nelson-Barber, S., 2001, On the cultural validity of science assessments, *Journal of Research in Science Teaching* 38(5): 553–573.

Sparks, R., 1982, First shot at a "short-cue" screener, item 794 in Crime Survey Research Consortium teleconference, May 10.

Sparks, P., Triplett, T., Piazza, T., Mockovak, B., and Hill, L., 1998, Complex CAI instrument program, testing, and training, panel discussion, International Field Directors and Technologies Conference, St. Louis, MO, May 17–20.

Spector, P., 1976, Choosing response categories for summated rating scales, *Journal of Applied Psychology* 61: 374–375.

Sperber, D., and Wilson, D., 1986, *Relevance: Communication and Cognition* Oxford: Blackwell.

Sperry, S., Edwards, B., and Dulaney, R., 1998, Evaluating interviewer use of CAPI navigation features, in M. Couper, R. Baker, J. Bethlehem, C. Clark, J. Martin, W. Nicholls II, and J. O'Reilly (eds.), *Computer-Assisted Survey Information Collection*. New York: Wiley

Spruyt-Metz, D., 1999, *Adolescence, Affect, and Health*. London: Psychology Press.

Stettler, K., and Willimack, D., 2001, Designing a questionnaire on the confidentiality perceptions of business respondents, *Proceedings of the Statistics Canada Symposium 2001: Achieving Data Quality in a Statistical Agency: A Methodological Perspective*.

Stettler, K., Morrison, R., and Anderson, A., 2000, Results of cognitive interviews studying alternative formats for economic census forms, *Proceedings of the 2nd International Conference on Establishment Surveys*. Alexandria, VA: American Statistical Association, pp. 1646–1651.

Stettler, K., Willimack, D., and Anderson, A., 2001, Adapting cognitive interviewing methodologies to compensate for unique characteristics of establishments, *Proceedings of the Annual Meeting of the American Statistical Association*, Alexandria, VA.

Stewart, D., and Shamdasani, P., 1990, *Focus Groups: Theory and Practice*. London: Sage.

Strauss, A., and Corbin, J., 1990, *Basic of Qualitative Research: Grounded Theory Procedures and Techniques*. Thousand Oaks, CA: Sage.

Stussman, B., Willis, G., and Allen, K., 1993, Collecting information from teenagers: experiences from the cognitive lab, *Proceedings of the ASA Section on Survey Research Method's*, Alexandria, VA: American Statistical Association.

Suchman, L., 1987, *Plans and Situated Actions: The Problem of Human–Machine Communication*. New York: Cambridge University Press.

Suchman, L., and Jordan, B., 1990, Interactional troubles in face-to-face survey interviews, *Journal of the American Statistical Association* 85: 232–241.

Sudman, S., 1983, Applied sampling, in P. Rossi, J. Wright, and A. Anderson (eds.), *Handbook of Survey Research*. San Diego, CA: Academic Press.

Sudman, S., and Bradburn, N., 1973, Effects of time and memory factors on response in surveys, *Journal of the American Statistical Association* 68: 805–815.

Sudman, S., and Bradburn, N., 1974, *Response Effects in Surveys: A Review and Synthesis*. Chicago: Aldine.

Sudman, S., and Bradburn, N., 1982, *Asking Questions: A Practical Guide to Questionnaire Design*. San Francisco: Jossey-Bass.

Sudman, S., Bradburn, N., and Schwarz, N., 1996, *Thinking about Answers: The Application of Cognitive Processes to Survey Methodology*. San Francisco: Jossey-Bass.

Sudman, S., Willimack, D., Nichols, E., and Mesenbourg, T., Jr., 2000, Exploratory research at the U.S. Census Bureau on the survey response process in large companies, *Proceedings of the 2nd International Conference on Establishment Surveys*. Alexandria, VA: American Statistical Association, pp. 327–335.

Suen, H., 1990, *Principles of Test Theories*. Mahwah, NJ: Lawrence Erlbaum.

Surveillance, Epidemiology, and End Results (SEER) Program, 2001, SEER Web page, *http://seer.cancer.gov/*. Washington, DC: National Cancer Institute, National Institutes of Health.

Swallow, S., Spencer, M., Miller, C., Paton, P., Deegen, R., Whinstanley, L., and Shogren, J., 1998, Methods and applications for ecosystem valuation: a collage, in B. Kanninen (ed.), *Proceedings of the First Workshop in the Environmental Policy and Economics Workshop Series*. Washington, DC: Office of Research and Development and Office of Policy, U.S. Environmental Protection Agency.

Swanson, H., 1999, What develops in working memory? A life span perspective, *Developmental Psychology* 35: 986–1000.

Sweet, E., and Ramos, M., 1995, *Evaluation Results from a Pilot Test of a Computerized Self-Administered Questionnaire (CSAQ) for the 1994 Industrial Research and Development (R&D) Survey*, Working Papers in Survey Methodology SM95/22. Washington, DC: U.S. Bureau of the Census.

Sweet, E., Marquis, K., Sedevi, B., and Nash, F., 1997, *Results of Expert Review of Two Internet R&D Questionnaires*, Human–Computer Interaction Report Series 1. Washington, DC: Statistical Research Division, U.S. Bureau of the Census.

Sykes, W., 1997, *Quality Assurance Study on ONS Employment Questions: Qualitative Research*. London: Social Survey Division, Office for National Statistics.

Sykes, W., and Collins, M., 1992, Anatomy of the survey interview, *Journal of Official Statistics* 8: 277–291.

Sykes, W., and Morton-Williams, J., 1987, Evaluating survey questions, *Journal of Official Statistics* 3: 191–207.

Szalai, A., 1993, The organization and execution of cross-national survey research projects, *Historical Social Research* 18: 139–171.

Tanur, J., 1992, *Questions about Questions: Inquiries into the Cognitive Bases of Surveys*. New York: Russell Sage.

Tanzer, N., 2004, Developing tests for use in multiple languages and cultures: a plea for simultaneous development, in R. Hambleton, P. Merenda, and C. Spielberger (eds.), *Adapting Educational and Psychological Test for Cross-Cultural Assessment*. Mahwah, NJ: Lawrence Erlbaum.

Tanzer, N., Gittler, G., and Ellis, B., 1995, Cross-cultural validation of item complexity in a LLTM-calibrated spatial ability test, *European Journal of Psychology Assessment* 11: 170–183.

Tarnai, J., Kennedy, J., and Scudder, D., 1998, Organizational effects of CATI in small to medium survey centers, in M. Couper, R. Baker, J. Bethlehem, C. Clark, J. Martin, W. Nicholls, and J. O'Reilly (eds.), *Computer-Assisted Survey Information Collection*. New York: Wiley.

Tenenbein, A., 1979, A double sampling scheme for estimating from misclassified multinomial data with application to sampling inspection, *Technometrics* 14: 187–202.

Teresi, J., 2001, Statistical methods for examination of differential item functioning (DIF) with applications to cross-cultural measurement of functional, physical and mental health, *Journal of Mental Health and Aging* 7: 31–40.

Thissen, D., 1991, *MULTILOG User's Guide, Version 6.3*. Chicago: Scientific Software.

Thissen, D., and Steinberg, L., 1988, Data analysis using item response theory, *Psychological Bulletin* 104: 385–395.

Thissen, D., and Wainer, H., 2001, *Test Scoring*. Mahwah, NJ: Lawrence Erlbaum.

Thissen, D., Steinberg, L., and Wainer, H., 1993, Detection of differential item functioning using the parameters of item response models, in P. Holland and H. Wainer (eds.), *Differential Item Functioning*, Mahwah, NJ: Lawrence Erlbaum, pp. 67–113.

Thomas, P., 2001, Data collection development in the UK National Statistical Institute, *Proceedings of the Statistics Canada Symposium 2001: Achieving Data Quality in a Statistical Agency: A Methodological Perspective*.

Thompson, F., Subar A., Brown, C., Smith, A., Sharbaugh, C., Jobe, J., Mittl, B., Gibson, J., and Ziegler, R., 2002, Cognitive research enhances accuracy of food frequency questionnaire reports: results of an experimental validation study, *Journal of the American Dietetic Association* 102(2): 212–225.

Tien, F., 1999, The application of cognitive interview on survey research: an example of contingent valuation method, *Proceedings of the National Science Council, Republic of China* 9: 555–574.

Tomaskovic-Devey, D., Leiter, J., and Thompson, S., 1994, Organizational survey nonresponse, *Administrative Science Quarterly* 39: 439–457.

Tourangeau, R., 1984, Cognitive science and survey methods: a cognitive perspective, in T. Jabine, M. Straf, J. Tanur, and R. Tourangeau (eds.), *Cognitive Aspects of Survey Methodology: Building a Bridge between Disciplines*. Washington, DC: National Academy Press.

Tourangeau, R., 1999, Context effects to answers to attitude questions, in M. Sirken, D. Herrmann, S. Schechter, N. Schwarz, J. Tanur, and R. Tourangeau (eds.), *Cognition and Survey Research*. New York: Wiley.

Tourangeau, R., and Rasinski, K., 1988, Cognitive processes underlying context effects in attitude measurement, *Psychology Bulletin* 103: 299–314.

Tourangeau, R., and Smith, T., 1996, Asking sensitive questions: the impact of data collection mode, question format, and question context, *Public Opinion Quarterly* 60: 275–304.

Tourangeau, R., Rips, L., and Rasinski, K., 2000. *The Psychology of Survey Response*. Cambridge: Cambridge University Press.

Tourangeau, R., Couper, M., and Conrad, F., 2003, The design of Web surveys: visual and interactive features of Web questionnaires, presented at the Federal CASIC Workshops, Washington, DC.

Trabasso, T., and Suh, S., 1993, Understanding text: achieving explanatory coherence through on-line inferences and mental operations in working memory, *Discourse Processing* 16: 3–34.

Trussell, R., and Elinson, J., 1959, *Chronic Illness in a Rural Area*. Cambridge, MA: Harvard University Press.

Tucker, C., 1997, Measurement issues surrounding the use of cognitive methods in survey research, *Bulletin de Methodologie Sociologique* 55: 67–92.

Tucker, C., and Bennett, C., 1988, Procedural effects in the collection of consumer expenditure information, *Proceedings of the ASA Section on Survey Research Methods*. Alexandria, VA: American Statistical Association, pp. 256–261.

Tucker, C., Casady, R., and Lepkowski, J., 1991, An evaluation of the 1988 Current Point-of-Purchase CATI Feasibility Test, *Proceedings of the ASA Section on Survey Research Methods*. Alexandria, VA: American Statistical Association, pp. 508–513.

Tucker, C., Bloxham, J., Bowie, C., Esposito, J., Harris-Kojetin, B., Kostanich, D., Miller, S., Polivka, A., Robinson, E., and Stump, M., 1998, Improving field tests, unpublished report, U.S. Bureau of Labor Statistics and U.S. Bureau of the Census, January.

Tulving, E., 1984, Relations among components and processes of memory, *Behavioral and Brain Sciences* 7: 257–263.

Turner, C., Lessler, J., and Devore, J., 1992a, Effects of mode of administration and wording on reporting of drug use, in C. Turner, J. Lessler, and J. Gfroerer (eds.), *Survey Measurement of Drug Use: Methodological Studies*. Rockville MD: U.S. Department of Heath and Human Services.

Turner, C., Lessler, J., Geroge, G., Hubbard, M., and Witt, M., 1992b, Effects of mode of administration and wording on data quality, in C. Turner, J. Lessler, and J. Gfroerer (eds.), *Survey Measurement of Drug Abuse: Methodological Studies*. Rockville, MD: U.S. Department of Health and Human Services.

Turner, C., Ku, L., Rogers, S., Lindberg, L., Pleck, J., and Sonenstein, F., 1998, Adolescent sexual behavior, drug use, and violence: increased reporting with computer survey technology, *Science* 280: 867–873.

Underwood, B., 1972, Are we overloading memory? in A. Melton and E. Martin (eds.), *Coding Processes in Human Memory*. Washington, DC: Winston.

Underwood, C., Small, C., and Thomas, P., 2000, Improving the efficiency of data validation and editing activities for business surveys, *Proceedings of the 2nd International Conference on Establishment Surveys*. Alexandria, VA: American Statistical Association, pp. 1262–1267.

University of Essex, 2002, Legal and ethical issues in interviewing children. Retrieved January 17, 2003 at *http://www.qualidata.essex.ac.uk/creatingdata/guidelineschildren.asp*.

U.S. Bureau of the Census, 1985, *Evaluating Censuses of Population and Housing*, STD-ISP-TR-5. Washington, DC: U.S. Government Printing Office.

U.S. Bureau of the Census, 1987, *Economic Censuses and Census of Governments: Evaluation Studies*. Washington, DC: U.S. Government Printing Office.

U.S. Bureau of the Census, 1989, *Recordkeeping Practices Survey*. Washington, DC: U.S. Bureau of the Census.

U.S. Bureau of the Census, 1998, *Pretesting Policy and Options: Demographic Surveys at the Census Bureau*. Washington, DC: U.S. Government Printing Office.

U.S. Bureau of the Census, 2001a, Executive Steering Committee for Accuracy and Coverage Evaluation Policy, *Report of the Executive Steering Committee for Accuracy and Coverage Evaluation Policy on Adjustment for Non-redistricting Uses*. Washington, DC: U.S. Government Printing Office.

U.S. Bureau of the Census, 2001b, *Proceedings of the Questionnaire Evaluation Standards Workshop*. Washington, DC: U.S. Bureau of the Census.

U.S. Bureau of the Census, 2001c, *SIPP Users' Guide*. Washington, DC: U.S. Government Printing Office.

U.S. Bureau of the Census, 2003, Census 2000, Summary File 3, Tables P19, PCT13, and PCT14; Summary Tables on Language Use and English Ability: 2000 (PHC-T-20). Internet release data: February 25, 2003.

U.S. Department of Health and Human Services, 1997, *Development of Computer-Assisted Interviewing Procedures for the National Household Survey on Drug Abuse*, SAMHSA Methodology Series Report M-3. Washington, DC: U.S. Government Printing Office.

U.S. Federal Committee on Statistical Methodology, 1988, *Quality in Establishment Surveys*, Statistical Policy Working Paper 15. Washington, DC: Office of Management and Budget.

Usunier, J., 1999, *Marketing across Cultures*. Upper Saddle River, NJ: Prentice Hall.

Vaillancourt, P., 1973, Stability of children's survey responses, *Public Opinion Quarterly* 37: 373–387.

Van de Pol, F., and R., Langeheine, 1997, Separating change and measurement error in panel surveys with an application to labor market data, in L. Lyberg, P. Biemer, M. Collins, E. de Leeuw, C. Dippo, N. Schwarz, and D. Trewin (eds.), *Survey Measurement and Process Quality*. New York: Wiley.

Van de Vijver, F., 2003, Bias and equivalence: cross-cultural perspectives, in J. Harkness, F. van de Vijver, and P. Mohler (eds.), *Cross-Cultural Survey Methods*. New York: Wiley.

Van de Vijver, F., and Hambleton, R., 1996, Translating tests: some practical guidelines, *European Psychologist* 1: 89–99.

Van de Vijver, F., and Leung, K., 1997, *Methods and Data Analysis for Cross-Cultural Research*. Thousand Oaks, CA: Sage.

Van der Linden, W., and Hambleton, R. 1997, *Handbook of Modern Item Response Theory*. New York: Springer-Verlag.

Van der Veld, W., Saris, W., and Gallhofer, I., 2000, Survey quality prediction, presented at the ISA Methodology Conference, Cologne, Germany.

Van der Zouwen, J., 2000, An assessment of the difficulty of questions used in the ISSP-questionnaires, the clarity of their wording, and the comparability of the responses, *ZA-Information* 46: 96–114.

Van der Zouwen, J., and Dijkstra, W., 1995, Trivial and non-trivial question–answer sequences: types, determinants and effects on data quality, *Proceedings of the International Conference on Survey Measurement and Process Quality*. Alexandria, VA: American Statistical Association.

Van der Zouwen, J., and Dijkstra, W., 2002, Testing questionnaires using interaction coding, in D. Maynard, H. Houtkoop-Steenstra, N. Schaeffer, and J. van der Zouwen (eds.), *Standardization and Tacit Knowledge: Interaction and Practice in the Survey Interview*. New York: Wiley.

Van der Zouwen, J., Saris, W., Draisma, S., and Van der Veld, W., 2001, Assessing the quality of questionnaires: a comparison of three methods for the "ex ante" evaluation of survey questions, presented at the International Conference on Quality in Official Statistics, Stockholm, May 14–15.

Van Deth, J., 1999, Equivalence in comparative political research, in J. W. van Deth (ed.), *Comparative Politics: The Problem of Equivalence*. London: Routledge.

Van Hattum, M., and de Leeuw, E., 1999, A disk by mail survey of pupils in primary schools: data quality and logistics, *Journal of Official Statistics* 15: 413–430.

Van Herk, H., 2000, Equivalence in a cross-national context: methodological and empirical issues in marketing research, Ph.D. dissertation, Catholic University, Brabant.

Van Meurs, A., and Saris, W., 1990, Memory effects in MTMM studies, in W. Saris and A. van Meurs (eds.), *Evaluation of Measurement Instruments by Meta-analysis of Multitrait–Multimethod Studies*. Amsterdam: North-Holland, pp. 134–147.

Van Nest, J., 1987, Content evaluation pilot study, in *1982 Economic Censuses and Census of Governments: Evaluation Studies*. Washington, DC: U.S. Bureau of the Census.

Vehovar, V., and Manfreda, K., 2000, Costs and errors of Web surveys in establishment surveys, *Proceedings of the 2nd International Conference on Establishment Surveys*. Alexandria, VA: American Statistical Association, pp. 449–453.

Vermunt, J., 1996, *Log-Linear Event History Analysis: A General Approach with Missing Data, Latent Variables, and Unobserved Heterogeneity*. Tilburg, The Netherlands: Tilburg University Press.

Vermunt, J., 1997, *REM: A General Program for the Analysis of Categorical Data*. Tilburg, The Netherlands: Tilburg University.

Von Thurn, D., 1996, Report of the October 1995 CPS School Enrollment Debriefing Questionnaire, memorandum to J. Day, U.S. Census Bureau of the November 5.

Voss, K., Stem, D., Johnson, L., and Arce, C., 1996, An exploration of the comparability of semantic adjectives in three languages: a magnitude estimation approach, *International Marketing Review* 13: 44–58.

Wainer, H., 1996, How is reliability related to the quality of test scores? What is the effect of local dependence on reliability? *Educational Measurement: Issues and Practice* 15: 22–29.

Wainer, H., Thissen, D., and Mislevy, R., 2000, *Computerized Adaptive Testing: A Primer*. Mahwah, NJ: Lawrence Erlbaum.

Ward, K., and Duffield, J., 1992, *Natural Resource Damages: Law and Economics*. New York: Wiley Law Publications.

Ware, J., 2000, SF-36 health survey update, *Spine* 25: 3130–3139.

Ware, J., Bjorner, J., and Kosinski, M., 2000, Practical implications of item response theory and computerized adaptive testing: a brief summary of ongoing studies of widely used headache impact scales, *Medical Care* 38: 73–83.

Ware-Martin, A., 1999, *Introducing and Implementing Cognitive Interviewing Techniques at the Energy Information Administration: An Establishment Survey Environment*, Statistical Policy Working Paper 28, Seminar on Interagency Coordination and Cooperation. Washington, DC: Federal Committee on Statistical Methodology, Office of Management and Budget.

Ware-Martin, A., Adler, R., and Leach, N., 2000, Assessing the impact of the redesign of the Manufacturing Energy Consumption Survey, *Proceedings of the 2nd International Conference on Establishment Surveys*. Alexandria, VA: American Statistical Association, pp. 1488–1492.

Watkins, J., 2001, *Testing IT: An Off-the-Shelf Software Testing Process*. New York: Cambridge University Press.

Watson, P., Denny, S., Adair, V., Ameratunga, S., Clark, T., Crengle, S., Dixon, R., Fa'asisila, M., Merry, S., Robinson, E., and Sporle, A., 2001, Adolescents' perceptions of a health survey using multimedia computer-assisted self-administered interviews, *Australian and New Zealand Journal of Public Health* 25: 6.

Weeks, M., 1992, Computer-assisted survey information collection: a review of CASIC methods and their implications for survey operations, *Journal of Official Statistics* 8: 445–465.

Werner, O., and Campbell, D., 1970, Translating, working through interpreters, and the problem of decentering, in R. Naroll and R. Cohen (eds.), *A Handbook of Cultural Anthropology*. New York: American Museum of Natural History.

Whalen, J., 1995, A technology of order production: computer-aided dispatch in public safety communication, in P. ten Have and G. Psathas (eds.), *Situated Order: Studies in the Social Organization of Talk and Embodied Activities*. Washington, DC: International Institute for Ethnomethodology and Conversation Analysis and University Press of America.

Whitney, P., and Budd, D., 1996, Think-aloud protocols and the study of comprehension, *Discourse Processing* 21: 341–351.

Wiggins, L., 1973, *Panel Analysis: Latent Probability Models for Attitude and Behavior Processing*, Amsterdam: Elsevier.

Wilcox, C., Sigelman, L., and Cook, E., 1989, Some like it hot: individual differences in responses to group feeling thermometers, *Public Opinion Quarterly* 53: 246–257.

Willimack, D., and Nichols, E., 2001, Building an alternative response process model for business surveys, *Proceedings of the Annual Meeting of the American Statistical Association*, Alexandria, VA.

Willimack, D., Nichols, E., and Sudman, S., 1999, Understanding the questionnaire in business surveys, *Proceedings of the ASA Section on Survey Research Methods*. Alexandria, VA: American Statistical Association, pp. 889–894.

Willimack, D., Nichols, E., and Sudman, S., 2002, Understanding unit and item nonresponse in business surveys, in R. Groves, D. Dillman, J. Eltinge, and R. Little (eds.), *Survey Nonresponse*. New York: Wiley.

Willis, G., 1994, *Cognitive Interviewing and Questionnaire Design: A Training Manual*, Working Paper 7. Washington, DC: Cognitive Methods Staff, National Center for Health Statistics.

Willis, G., 1999, Cognitive interviewing: a "how to" guide, Research Triangle Institute, *http://appliedresearch.cancer.gov/areas/cognitive/interview.pdf*.

Willis, G., and Schechter, S., 1997, Evaluation of cognitive interviewing techniques: Do the results generalize in the field? *Bulletin de Methodologie Sociologique* 55: 40–66.

Willis, G., Royston, P., and Bercini, D., 1991, The use of verbal report methods in the development and testing of survey questionnaires, *Applied Cognitive Psychology* 5: 251–267.

Willis, G., DeMaio, T., and Harris-Kojetin, B., 1999a, Is the bandwagon headed to the methodological promised land? Evaluating the validity of cognitive interviewing techniques, in M. Sirken, D. Herrmann, S. Schechter, N. Schwarz, J. Tanur, and R. Tourangeau (eds.), *Cognition and Survey Research*. New York: Wiley.

Willis, G., Schechter, S., and Whitaker, K., 1999b, A comparison of cognitive interviewing, expert review, and behavior coding: what do they tell us? *Proceedings of the ASA Section on Survey Research Methods*. Alexandria, VA: American Statistical Association.

Willis, G., Reeve, B., and Barofsky, I., in press, The Use of Cognitive Interviewing Techniques in Quality of Life and Patient-Reported Outcomes Assessment, in J. Lipscomb, C. Gotay, C. Snyder (eds.), *Outcomes Assessment in Cancer*, Cambridge: Cambridge University Press.

Wilson, T., 1994, The proper protocol: validity and completeness of verbal reports, *Psychological Science* 5: 249–252.

Wilson, M., in press, Subscales and summary scales: issues in health-related outcomes, in J. Lipscomb, C. Gotay, and C. Snyder C (eds.), *Outcomes Assessment in Cancer*. Cambridge: Cambridge University Press.

Wilson, J., and Powell, M., 2001, *A Guide to Interviewing Children: Essential Skills for Counselors, Police Lawyers and Social Workers*. New York: Routledge.

Wilson, T., and Schooler, J., 1991, Thinking too much: introspection can reduce the quality of preferences and decisions, *Journal of Personality and Social Psychology* 60: 181–192.

Wilson, T., LaFleur, S., and Anderson, D., 1996, The validity and consequences of verbal reports about attitudes, in N. Schwarz and S. Sudman (eds.), *Answering Questions: Methodology for Determining Cognitive and Communicative Processes in Survey Research*. San Francisco: Jossey-Bass.

Wilss, W., 1996, *Knowledge and Skills in Translator Behavior*. Amsterdam: John Benjamins.

Winer, B., 1971, *Statistical Principles in Experimental Design*. New York: McGraw-Hill.

Wobus, P., and de la Puente, M., 1995, Results from telephone debriefing interviews: the Census Bureau's Spanish Forms Availability Test, *Proceedings of the ASA Section on Survey Research Methods*. Alexandria, VA: American Statistical Association, pp. 1040–1045.

Wolter, K., and Monsour, N., 1986, Conclusions from economic census evaluation studies, *Proceedings of the U.S. Census Bureau Annual Research Conference*, 2, 41–53.

Wothke, W., 1996, Models for multitrait–multimethod matrix analyses, in G. Marcoulides and R. Schumacker (eds.), *Advanced Structural Equation Modeling: Issues and Techniques*. Mahwah, NJ: Lawrence Erlbaum.

Young, C., 1999, What we know about "I don't know": an analysis of the relationship between "don't know" and education, presented to the American Association for Public Opinion Research, St. Petersburg Beach, FL.

Yu, C., and Muthén, B., 2001, Evaluation of model fit indices for latent variable models with categorical and continuous outcomes, technical report.

Zabel, J., 1994, *An Analysis of Attrition in the PSID and SIPP with an Application to a Model of Labor Market Behavior*, SIPP Working Paper Series 9403. Washington, DC: U.S. Bureau of the Census.

Zickar, M., and Drasgow, F., 1996, Detecting faking on a personality instrument using appropriateness measurement, *Applied Psychological Measurement* 20: 71–88.

Zill, N., 2001, Advantages and limitations of using children and adolescents as survey respondents, in M. Cynamon and R. Kulka (eds.), *Proceedings of the 7th Conference on Health Survey Research Methods*. Hyattsville, MD: U.S. Department of Health and Human Services.

Zukerberg, A., and Hess, J., 1996, *Uncovering Adolescent Perceptions: Experiences Conducting Cognitive Interviews with Adolescents*, SM 96/01. Washington, DC: U.S. Bureau of the Census. Also presented at the 1997 AAPOR Conference, Norfolk, VA.

Zukerberg, A., and Lee, M., 1997, Better Formatting for Lower Response Burden, Working Papers in Survey Methodology no. SM97/02. Washington, D.C.: U.S. Census Bureau, Statistical Research Division.

Zukerberg, A., Von Thurn, D., and Moore, J., 1995, Practical considerations in sample size selection for behavior coding pretest, in *1995 Proceedings of the Section on Survey Research Methods*. Alexandria, VA: American Statistical Association.

Zukerberg, A., Nichols, E., and Tedesco, H., 1999, Designing surveys for the next millennium: internet questionnaire design issues, presented at the Annual Conference of the American Association for Public Opinion Research, St. Petersburg Beach, FL.

Index

Methods for Testing and Evaluating Survey Questionnaires, Edited by Stanley Presser, Jennifer M. Rothgeb, Mick P. Couper, Judith T. Lessler, Elizabeth Martin, Jean Martin, and Eleanor Singer
ISBN 0-471-45841-4